HARCOURT BRACE COLLEGE OUTLINE SERIES

ARITHMETIC

W9-CHP-419

Alan Wise

Department of Mathematics, University of San Diego

Carol Wise

Harcourt Brace College Publishers

Fort Worth   Philadelphia   San Diego
New York   Orlando   Austin   San Antonio
Toronto   Montreal   London   Sydney   Tokyo

Printed in the United States of America

ISBN 0-15-601529-3

8 9 0 1 2 3 4 5 6 074 17 16 15 14 13 12 11 10

For our children—
Bob, Chris, Greg, Gary, and Denise

PREFACE

The purpose of this book is to present a complete course in arithmetic in the clear, concise form of an outline. This outline provides an in-depth review of the principles of arithmetic for independent study or pre-algebra review, and contains essential supplementary material for mastering arithmetic. Or, this outline can simply be used as a valuable, self-contained refresher course on the practical applications of arithmetic.

Although comprehensive enough to be used by itself for independent study, this outline is specifically designed to be used as a supplement to college courses and textbooks on the subject. Notice, however, that the topics in this outline are more narrowly defined than the topics in many textbooks. For instance, whereas percent applications are included in the discussion of percent in some books and excluded altogether in others, in this outline these applications have their own separate chapter. This isolation not only helps you to find the specific topics that you need to study, but also enables you to bypass those topics that you are already familiar with.

Regular features at the end of each chapter are also specially designed to supplement your textbook and course work in arithmetic.

RAISE YOUR GRADES This feature consists of a checkmarked list of open-ended thought questions to help you assimilate the material you have just studied. These thought questions invite you to compare concepts, interpret ideas, and examine the whys and wherefores of chapter material.

SUMMARY This feature consists of a brief restatement of the main ideas in each chapter, including definitions of key terms. Because it is presented in the efficient form of a numbered list, you can use it to refresh your memory quickly.

SOLVED PROBLEMS Each chapter of this outline contains a set of exercises and word problems and their step-by-step solutions. Undoubtedly the most valuable feature of this outline, these problems allow you to become proficient with the fundamental skills and applications of arithmetic. Along with the sample midterm and final examinations, they also give you ample exposure to the kinds of questions that you are likely to encounter on a typical college exam. To make the most of these solved problems, try writing your own solutions first. Then compare your answers to the detailed solutions provided in the book. If you have trouble understanding a particular problem, or set of problems, use the example reference numbers to locate and review the appropriate instruction and parallel step-by-step examples that were developed earlier in the chapter.

SUPPLEMENTARY EXERCISES Each chapter of this outline concludes with a set of drill exercises and practical applications and their answers. The supplementary exercises are designed to help you master and retain all the newly discussed skills and concepts presented in the given chapter, and also contain example references.

Of course there are other features of this outline that you will find very helpful, too. One is the format itself, which serves both as a clear guide to important ideas and as a convenient structure upon which to organize your knowledge. A second is the attention devoted to methodology and the practical applications of arithmetic. Yet a third is the careful listing of learning objectives in each section denoted by the uppercase letters A, B, C, D, or E.

We wish to thank the people who made this text possible, including Serena Hecker for painstakingly checking the solutions and answers to each exercise and word problem.

San Diego, California

ALAN WISE

CAROL WISE

CONTENTS

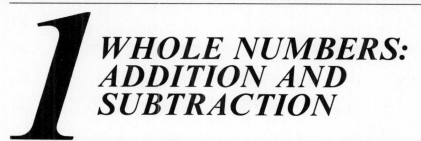

1 WHOLE NUMBERS: ADDITION AND SUBTRACTION

THIS CHAPTER IS ABOUT

☑ **Using Place Value**
☑ **Adding Whole Numbers**
☑ **Subtracting Whole Numbers**
☑ **Solving Word Problems (+, −)**

1-1. Using Place Value

A. Identify digits in whole numbers.

The numbers you use to count things are called the **counting numbers** or **natural numbers.**

EXAMPLE 1-1: List all the counting numbers.

Solution: The counting numbers are listed as: 1, 2, 3, 4, 5, 6, 7, 8, 9, 10, 11, 12, ···
or just: 1, 2, 3, ···

Note: The **ellipsis symbol** (···) following the counting numbers indicates that the counting number pattern shown continues on forever.

When the number **zero** (0) is combined with the counting numbers, they are called the **whole numbers.**

EXAMPLE 1-2: List all the whole numbers.

Solution: The whole numbers are listed as: 0, 1, 2, 3, 4, 5, 6, 7, 8, 9, 10, 11, 12, ···
or just: 0, 1, 2, 3, ···

Note: Every whole number is made up of one or more of the following ten **digits:** 0, 1, 2, 3, 4, 5, 6, 7, 8, and 9.

EXAMPLE 1-3: List the digits that make up the whole number 36,580.

Solution: The digits that make up 36,580 are 3, 6, 5, 8, and 0.

To help locate the correct digit given its **place-value name,** you can use a **place-value chart.**

EXAMPLE 1-4: In 750,408,391,062, draw a line under the digit that is in the ten-thousands place.

hundred billions	ten billions	billions	hundred millions	ten millions	millions	hundred thousands	ten thousands	thousands	hundreds	tens	ones
7	5	0	4	0	8	3	<u>9</u>	1	0	6	2

←———— place-value names

Solution: Draw a line under 9 *ten thousands.*

Note: The **commas** in 750,408,391,062 separate the number into **periods.**

EXAMPLE 1-5: Write 750,408,391,062 using period names instead of commas.

Billions	Millions	Thousands	Units ←
7 5 0 ,	4 0 8 ,	3 9 1 ,	0 6 2

period names

Solution: 750 billion 408 million 391 thousand 62

Note: All period names except for units must be written out. The whole number 62, for example, can be written as 62 units, or just 62.

To help identify the correct place-value name of a given digit, you can use a place-value chart.

EXAMPLE 1-6: Write the place-value name of the underlined digit in 246,137,958.

Billions			Millions			Thousands			Units ←		
hundred billions	ten billions	billions	hundred millions	ten millions	millions	hundred thousands	ten thousands	thousands	hundreds	tens	ones
			2	4	6	1	3	7	9	5	8

period names

place values

Solution: The place-value name of the underlined digit is *hundred millions.*

Once the place-value name of a given digit is identified, you can write the **value** of that digit.

EXAMPLE 1-7: What is the value of 5 in 2,035,819?

Solution: In 2,035,819, the value of 5 is 5 thousands, or 5000.

Note: In a four-digit number like 5000, you do not have to write the comma separating thousands and units. However, in all whole numbers with five digits or more, you must separate each period with a comma.

B. Write whole numbers.

To write a whole number when only **1-digit values** greater than zero are given, you must be careful to write a zero (0) each time a 1-digit value is missing.

EXAMPLE 1-8: Write the following as a whole number: 6 thousands 8 tens 5 ones

                                        missing value

Solution: 6 thousands 8 tens 5 ones = 6 thousands no hundreds 8 tens 5 ones

            = 6 thousands 0 hundreds 8 tens 5 ones ← 1-digit values

            = 6085 ← whole number

Note: "No hundreds" and "0 hundreds" mean the same thing.

The following **number-word names,** whole numbers, and one-digit values should be memorized.

Special Number Word Names

one	two	three	four	five	six
1	2	3	4	5	6
1 one	2 ones	3 ones	4 ones	5 ones	6 ones
seven	eight	nine	ten	eleven	twelve
7	8	9	10	11	12
7 ones	8 ones	9 ones	1 ten	1 ten 1 one	1 ten 2 ones
thirteen	fourteen	fifteen	sixteen	seventeen	eighteen
13	14	15	16	17	18
1 ten 3 ones	1 ten 4 ones	1 ten 5 ones	1 ten 6 ones	1 ten 7 ones	1 ten 8 ones
nineteen	twenty	thirty	forty	· · ·	ninety
19	20	30	40	· · ·	90
1 ten 9 ones	2 tens	3 tens	4 tens	· · ·	9 tens
one hundred	two hundred	three hundred	four hundred	· · ·	nine hundred
100	200	300	400	· · ·	900
1 hundred	2 hundreds	3 hundreds	4 hundreds	· · ·	9 hundreds
one thousand	two thousand	three thousand	four thousand	· · ·	nine thousand
1000	2000	3000	4000	· · ·	9000
1 thousand	2 thousands	3 thousands	4 thousands	· · ·	9 thousands

To write a whole number when the number-word name is given, you may find it helpful to first write one-digit values.

EXAMPLE 1-9: Write the following as a whole number: eight thousand thirty

$$
\overbrace{\qquad}^{\text{missing value}} \quad \overbrace{\qquad}^{\text{missing value}}
$$

Solution: eight thousand thirty = 8 thousands no hundreds 3 tens no ones

   = 8 thousands 0 hundreds 3 tens 0 ones ⟵— 1-digit values

   = 8030 ⟵— whole number

Similarly, to write the number-word name when the whole number is given, you may find it helpful to first write one-digit values.

EXAMPLE 1-10: Write 6207 as a number-word name.

Solution: 6207 = 6 thousands 2 hundreds 0 tens 7 ones ⟵— 1-digit values

   = six thousand two hundred seven ⟵— number-word name

Note: **A zero value** (like 0 tens) is never written as part of the number-word name.

When it takes two words to write a number-word name for whole numbers between 20 and 100, you must use a **hyphen** (-).

EXAMPLE 1-11: Write the number-word name for the following whole numbers: **(a)** 23   **(b)** 438 **(c)** 5089

Solution
(a) 23 = 2 tens 3 ones   **(b)** 438 = 4 hundreds 3 tens 8 ones

   = twenty-three   = four hundred thirty-eight

$$\overbrace{}^{\text{zero value}}$$

(c) $5089 = 5$ thousands $\overbrace{0 \text{ hundreds}}$ 8 tens 9 ones

= five thousand eighty-nine

Caution: Never use the word *and* when writing a whole-number word name.

EXAMPLE 1-12: Write the whole-number word name for 9007.

$$\overbrace{}^{\text{zero values}}$$

Solution: $9007 = 9$ thousands $\overbrace{0 \text{ hundreds } 0 \text{ tens}}$ 7 ones

= nine thousand seven (Do not write "nine thousand and seven.")

C. Compare whole numbers.

To compare two whole numbers, you identify which number is **greater** (larger) or which number is **less** (smaller). Any two whole numbers can be compared using one of the following three symbols:

$>$ means *is greater than* (3 is greater than 2, or $3 > 2$)

$<$ means *is less than* (2 is less than 3, or $2 < 3$)

$=$ means *is equal to* (3 is equal to 3, or $3 = 3$)

The symbol $=$ is called an **equality symbol.**

The symbols $>$, $<$, and $\neq$ (*greater than, less than,* and *not equal to*) are called **inequality symbols.**

To compare two whole numbers, you may find it helpful to write one number under the other while aligning like place values, like this:

407,845
428,231

EXAMPLE 1-13: Which is greater: 25,831,506, or 25,829,417?

Millions			Thousands			Units		
hundred millions	ten millions	millions	hundred thousands	ten thousands	thousands	hundreds	tens	ones
	2	5	8	3	1	5	0	6
	2	5	8	2	9	4	1	7

Write one number under the other while aligning like values.

same — different (3 ten thousands > 2 ten thousands)

Solution: 25,831,506 is greater than 25,829,417, or using the symbol, $25{,}831{,}506 > 25{,}829{,}417$ because 3 ten thousands $>$ 2 ten thousands.

D. Order whole numbers.

To **order two or more whole numbers,** you list them from largest to smallest or from smallest to largest.

To order whole numbers from largest to smallest, you write the largest number first, then the next largest number, and so on until you have listed all the given numbers.

EXAMPLE 1-14: Order the following whole numbers from largest to smallest: 8516, 263, 10,519, 8520, 900

Solution: 10,519 ⟵ largest number

8 520

8 516

900

263 ⟵ smallest number

To order whole numbers from smallest to largest, you just reverse the largest to smallest order.

EXAMPLE 1-15: Order the following whole numbers from smallest to largest: 7513, 2409, 2490, 5713, 7531, 2049

Solution: 2049 ⟵ smallest number

2409

2490

5713

7513

7531 ⟵ largest number

E. Round whole numbers.

When an exact answer is not needed for a problem, you may want to make the computation easier and **round the given whole numbers** before computing. The place to which you round the numbers will depend on how accurate your answer needs to be.

If the digit to the right of the digit to be rounded is 5 or more, then **round up** as follows:
1. Increase the digit to be rounded by 1.
2. Replace the digits to the right of the rounded digit with zeros.

EXAMPLE 1-16: Round 82,635 to the nearest thousand.

$$\overset{\text{thousands}}{\downarrow}$$

Solution: 82,635 = 8$\underline{2}$,635   Draw a line under the digit to be rounded.

= 8$\underline{2}$,635   Look at the next digit to the right. Is that digit 5 or more?
In this case, it is: 6 > 5

≈ 83,000   Then round up: Increase 2 by 1 to get 3.
                        Replace 6, 3, and 5 with zeros.

Note: The symbol ≈ means **is approximately equal to** and is used whenever you round a number.

If the digit to the right of the digit to be rounded is less than 5, then **round down** as follows:
1. Leave the digit to be rounded the same.
2. Replace the digits to the right of the rounded digit with zeros.

EXAMPLE 1-17: Round 82,635 to the nearest ten thousand.

$$\overset{\text{ten thousands}}{\downarrow}$$

Solution: 82,635 = $\underline{8}$2,635   Draw a line under the digit to be rounded.

= $\underline{8}$2,635   Look at the next digit to the right. Is that digit 5 or more?
In this case, it is not: 2 < 5

≈ 80,000   Then round down: Leave 8 the same.
                        Replace 2, 6, 3, and 5 with zeros.

1-2. Adding Whole Numbers

To join two or more amounts together, you **add.**

EXAMPLE 1-18: Write "☐☐ joined together with ☐☐☐" as addition.

Solution: ☐☐ joined together with ☐☐☐ equals: 2 + 3 ⟵— addition

The **addition problem** 2 + 3 is read as "two **plus** three." The symbol + is called an **addition sign** or **addition symbol.**

A. Add basic facts.

The following example shows one way to solve an addition problem.

EXAMPLE 1-19: Add 2 + 3.

Solution: 2 + 3 = 5 because ☐☐ joined together with ☐☐☐ equals ☐☐☐☐☐.

The **number sentence** 2 + 3 = 5 is called a **basic addition fact.** To add whole numbers quickly and accurately, you must know the 100 basic addition facts from memory [see Solved Problem 1-12]. The basic addition fact 2 + 3 = 5 is written in **horizontal form.** Basic addition facts can also be written in **vertical form.**

EXAMPLE 1-20: Write the following number sentence in vertical form: 2 + 3 = 5

Solution:
$$\begin{array}{c} 2 \\ +3 \\ \hline 5 \end{array} \text{ or } \begin{array}{c} 3 \\ +2 \\ \hline 5 \end{array}$$
⟵— 2 or 3 can be used as the top number.
⟵— The other number is used as the bottom number.
⟵— The answer (5) is written under the addition bar.

In 2 + 3 = 5 or $\begin{array}{c} 2 \\ +3 \\ \hline 5 \end{array}$, 2 and 3 are called **addends** and 5 is called the **sum.**

B. Add given horizontal form.

To add whole numbers with two or more digits given horizontal form, you first write the numbers in vertical form.

EXAMPLE 1-21: Add 215 + 63. ⟵— horizontal form

Solution

1. Write in vertical form. Line up like values

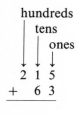

2. Add ones.
5 + 3 = 8 (ones)

3. Add tens.
1 + 6 = 7 (tens)

4. Add hundreds.
2 + 0 = 2 (hundreds)

$$\begin{array}{r} 2\ 1\ 5 \\ +\ 6\ 3 \\ \hline 8 \end{array} \qquad \begin{array}{r} 2\ 1\ 5 \\ +\ 6\ 3 \\ \hline 7\ 8 \end{array} \qquad \begin{array}{r} 2\ 1\ 5 \\ +\ 6\ 3 \\ \hline 2\ 7\ 8 \end{array} \text{⟵— sum}$$

C. Add with renaming.

Values like 12 ones and 13 tens are called **2-digit values.** A 2-digit value can always be renamed as two 1-digit values.

EXAMPLE 1-22: Rename as two 1-digit values: **(a)** 12 ones   **(b)** 13 tens

Solution

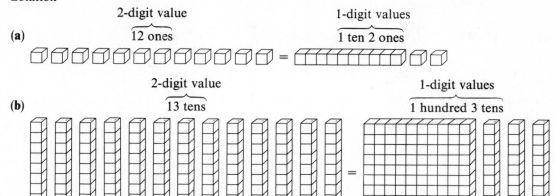

When adding, you must rename each 2-digit value, unless it is the last step.

EXAMPLE 1-23: Add 689 + 743.

Solution

1. Add ones.

$9 + 3 = 12$ (ones)

Rename ones:

12 ones = 1 ten 2 ones

2. Add tens.

$1 + 8 + 4 = 13$ (tens)

Rename tens:

13 tens = 1 hundred 3 tens

3. Add hundreds.

$1 + 6 + 7 = 14$ (hundreds)

Last step—Don't rename!

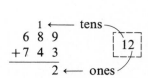

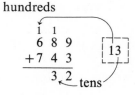

D. **Add more than two whole numbers.**

To add more than two whole numbers, add as you would with two whole numbers.

EXAMPLE 1-24: Add 98 + 7846 + 385 + 275.

Solution

1. Add ones.
$8 + 6 + 5 + 5 = 24$

```
    2
    9 8
  7 8 4 6
    3 8 5
+   2 7 5
─────────
        4
```

2. Add tens.
$2 + 9 + 4 + 8 + 7 = 30$

```
    3 2
    9 8
  7 8 4 6
    3 8 5
+   2 7 5
─────────
      0 4
```

3. Add hundreds.
$3 + 8 + 3 + 2 = 16$

```
  1 3 2
    9 8
  7 8 4 6
    3 8 5
+   2 7 5
─────────
    6 0 4
```

4. Add thousands.
$1 + 7 = 8$

```
  1 3 2
    9 8
  7 8 4 6
    3 8 5
+   2 7 5
─────────
  8 6 0 4
```

1-3. Subtracting Whole Numbers

You recall that to join two or more amounts together, you add. But what should be done when you want to do the opposite of add? That is, what should be done when you want to take one amount away from another amount? To take one amount away from another amount, you **subtract.**

EXAMPLE 1-25: Write "☐☐☐☐☐ take away ☐☐☐" as subtraction.

Solution: ☐☐☐☐☐ take away ☐☐☐ equals: $5 - 3$ ⟵ subtraction

The **subtraction problem** $5 - 3$ is read as "five **minus** three." The symbol $-$ is called a **subtraction sign** or **subtraction symbol.**

A. Subtract basic facts.

The following example shows one way to solve a subtraction problem.

EXAMPLE 1-26: Subtract $5 - 3$.

Solution: $5 - 3 = 2$ because

☐☐☐☐☐ take away ☐☐☐ equals ☐☐☒☒☒ or ☐☐

The number sentence $5 - 3 = 2$ is called a **basic subtraction fact.** To subtract whole numbers quickly and accurately, you must know the 100 basic subtraction facts from memory [see Solved Problem 1-16]. The basic subtraction fact $5 - 3 = 2$ is written in horizontal form. Basic subtraction facts can also be written in vertical form.

EXAMPLE 1-27: Write the following number sentence in vertical form: $5 - 3 = 2$

Solution: 5 ⟵ The first number in $5 - 3 = 2$ is the top number.

       -3 ⟵ The second number in $5 - 3 = 2$ is the bottom number.

        2 ⟵ The answer (2) is written under the subtraction bar.

In $5 - 3 = 2$ or $\begin{array}{r} 5 \\ -3 \\ \hline 2 \end{array}$, 5 is called the **minuend,** 3 is called the **subtrahend,** and 2 is called the **difference.**

Subtraction is related to addition in the following way:

$$5 - 3 = 2 \quad \text{because} \quad 2 + 3 = 5$$

The subtraction problem $5 - 3 = ?$ and the **missing addend problem** $? + 3 = 5$ are two different ways of stating the same problem. The answer to both problems is 2.

The subtraction problem $5 - 3 = ?$ asks: If you have 5 and take 3 away, how many are left?

The missing addend problem $? + 3 = 5$ asks: How many more are needed if you have 3 but want 5?

B. Subtract given horizontal form.

To subtract two whole numbers with two or more digits given horizontal form, you first write the numbers in vertical form.

EXAMPLE 1-28: Subtract $897 - 35$. ⟵ horizontal form

Solution

1. Write in vertical form.   **2.** Subtract ones.   **3.** Subtract tens.   **4.** Subtract hundreds.
    Line up like values. $7 - 5 = 2$ (ones) $9 - 3 = 6$ (tens) $8 - 0 = 8$ (hundreds)

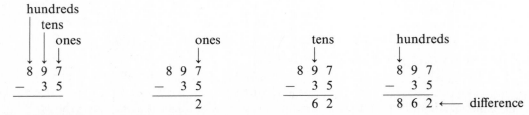

To **check subtraction,** you can use addition.

EXAMPLE 1-29: Check $897 - 35 = 862$. $\longleftarrow$ proposed difference

Think: $897 - 35 = 862$ if $862 + 35 = 897$

Solution:

$$\begin{array}{r} 8\ 6\ 2 \\ +\ \ 3\ 5 \\ \hline 8\ 9\ 7 \end{array}$$ $\longleftarrow$ 862 checks

Add the proposed difference (862) and the subtrahend (35) to see if you get the original minuend (897).

C. Subtract with one renaming.

The following example shows how to rename two 1-digit values (like 3 tens 5 ones) to get more ones.

EXAMPLE 1-30: Rename the following to get more ones: 3 tens 5 ones

Solution: 3 tens 5 ones = 2 tens 1 ten 5 ones   *Think:* 3 tens = 2 tens 1 ten

        = 2 tens 10 ones 5 ones   1 ten = 10 ones

        = 2 tens 15 ones $\longleftarrow$ more ones

When the digits in the ones column cannot be subtracted, you must rename to get more ones.

EXAMPLE 1-31: Subtract $935 - 216$.

Solution

1. Subtract ones? No. $5 - 6$ does not have a whole-number answer. You need more than 5 ones, so you must rename.

2. Rename to get more ones.
3 tens 5 ones = 2 tens 15 ones

3. Subtract as before.
$15 - 6 = 9$ (ones)
$2 - 1 = 1$ (tens)
$9 - 2 = 7$ (hundreds)

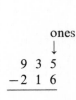

$$\begin{array}{r} 9\ 3\ 5 \\ -2\ 1\ 6 \\ \hline \end{array}$$

$$\begin{array}{r} 9\ \overset{2}{\cancel{3}}\ \overset{1}{5} \\ -2\ 1\ 6 \\ \hline \end{array}$$

$$\begin{array}{r} 9\ \overset{2}{\cancel{3}}\ \overset{1}{5} \\ -2\ 1\ 6 \\ \hline 7\ 1\ 9 \end{array}$$

When the digits in any given **place-value column** cannot be subtracted, you rename to get more of that place value in the minuend (top number).

EXAMPLE 1-32: Subtract $837 - 692$.

Solution

1. Subtract ones.
$7 - 2 = 5$ (ones)

2. Subtract tens? No. $3 - 9$ does not have a whole-number answer. You need more than 3 tens, so you must rename.

3. Rename to get more tens.
8 hundreds 3 tens = 7 hundreds 13 tens
Subtract as before.

$$\begin{array}{r} 8\ 3\ 7 \\ -6\ 9\ 2 \\ \hline 5 \end{array}$$

$$\begin{array}{r} 8\ 3\ 7 \\ -6\ 9\ 2 \\ \hline 5 \end{array}$$

$$\begin{array}{r} \overset{7}{8}\ \overset{1}{3}\ 7 \\ -6\ 9\ 2 \\ \hline 1\ 4\ 5 \end{array}$$

D. Subtract with two or more renamings.

To subtract with two or more renamings, you should write each renaming neatly to avoid errors.

EXAMPLE 1-33: Subtract 7516 − 3827.

Solution

1. Subtract ones.
16 − 7 = 9

2. Subtract tens.
10 − 2 = 8

3. Subtract hundreds.
14 − 8 = 6

4. Subtract thousands.
6 − 3 = 3

$$
\begin{array}{r}
7\ 5\ \overset{0}{\cancel{1}}\ \overset{1}{}6 \\
-3\ 8\ 2\ 7 \\
\hline
9
\end{array}
$$

$$
\begin{array}{r}
7\ \overset{4}{\cancel{5}}\ \overset{10}{}\overset{1}{}6 \\
-3\ 8\ 2\ 7 \\
\hline
8\ 9
\end{array}
$$

$$
\begin{array}{r}
\overset{6}{\cancel{7}}\ \overset{14}{\cancel{5}}\ \overset{10}{}\overset{1}{}6 \\
-3\ 8\ 2\ 7 \\
\hline
6\ 8\ 9
\end{array}
$$

$$
\begin{array}{r}
\overset{6}{\cancel{7}}\ \overset{14}{\cancel{5}}\ \overset{10}{}\overset{1}{}6 \\
-3\ 8\ 2\ 7 \\
\hline
3\ 6\ 8\ 9
\end{array}
$$

E. **Subtract across zeros.**

When the minuend (the top number) contains a zero, you will usually have to rename across zero in order to subtract.

EXAMPLE 1-34: Subtract 506 − 238.

Solution

1. Subtract ones? No.
6 − 8 does not have a
whole-number answer.
Rename 0 tens 6 ones
to get more ones? No.
You cannot take 1 away
from 0.

2. Rename across zero.
50 tens 6 ones = 49 tens 16 ones

3. Subtract as before.
16 − 8 = 8
9 − 3 = 6
4 − 2 = 2

$$
\begin{array}{c}
\text{No} \\
\downarrow \\
5\ 0\ 6 \\
-2\ 3\ 8
\end{array}
$$

$$
\begin{array}{c}
\ \ \text{tens} \\
\ \ \ \downarrow \quad \text{more ones} \\
\ \ \ \ \ \quad \downarrow \\
\overset{4}{}\overset{9}{}\overset{1}{} \\
\cancel{5}\ \cancel{0}\ 6 \\
-2\ 3\ 8
\end{array}
$$

$$
\begin{array}{r}
\overset{4}{}\overset{9}{}\overset{1}{} \\
\cancel{5}\ \cancel{0}\ 6 \\
-2\ 3\ 8 \\
\hline
2\ 6\ 8
\end{array}
$$

To rename across two or more zeros, rename in much the same way as you did across one zero.

EXAMPLE 1-35: Subtract 8000 − 654.

Solution

1. Rename across zeros.
800 tens 0 ones = 799 tens 10 ones

2. Subtract as before.

3. Check as before.

$$
\begin{array}{r}
\overset{7}{}\overset{9}{}\overset{9}{}\overset{1}{} \\
\cancel{8}\ \cancel{0}\ \cancel{0}\ 0 \\
-\ \ 6\ 5\ 4 \\
\hline
\end{array}
$$

$$
\begin{array}{r}
\overset{7}{}\overset{9}{}\overset{9}{}\overset{1}{} \\
\cancel{8}\ \cancel{0}\ \cancel{0}\ 0 \\
-\ \ 6\ 5\ 4 \\
\hline
7\ 3\ 4\ 6
\end{array}
$$

$$
\begin{array}{r}
\overset{1}{}\ \overset{1}{}\ \overset{1}{} \\
7\ 3\ 4\ 6 \\
+\ \ 6\ 5\ 4 \\
\hline
8\ 0\ 0\ 0 \longleftarrow 7346\ \text{checks}
\end{array}
$$

1-4. Solving Word Problems (+, −)

A. **Choose the correct operation: add or subtract (+, −).**

You should recall that to join two or more amounts together, you add (+) [see Section 1-2].
To take one amount away from another amount, you subtract (−) [see Section 1-3].

EXAMPLE 1-36: Choose the correct operation to solve these word problems.

1. *Read:* Emily ran 15 km in the first race. Then she ran 8 km in the second race. How far did Emily run all together?

1. *Read:* Peter worked 15 hours this week. Last week he worked 8 hours. How many more hours did Peter work this week than last week?

2. *Understand:* The question asks you to join amounts of 15 and 8 together.

3. *Decide:* To join amounts together, you **add.**

2. *Understand:* The question asks you to take the amount of 8 away from 15.

3. *Decide:* To take an amount away, you **subtract.**

B. Solve word problems using addition or subtraction (+, −).

To solve an addition word problem, you find how much two or more amounts equal all together.

To solve a subtraction word problem, you find (**a**) how much is left,

(**b**) how much more is needed,

(**c**) the difference between two amounts.

EXAMPLE 1-37: Solve the following word problem using addition or subtraction.

1. *Identify:*   Carol had $20. She spent $12.   Circle each fact.
How much money did Carol have left?   Underline the question.

2. *Understand:*   The question asks you to find how much is left when 12 is taken away from 20.

3. *Decide:*   To take 12 away from 20, you subtract.

4. *Compute:* $20 - 12 = 8$

5. *Interpret:*   The answer 8 means that after starting with $20 and then spending $12, Carol was left with **$8.**

6. *Check:*   Is $8 + $12 equal to the original $20? *Yes:* $8 + 12 = 20$

RAISE YOUR GRADES
Can you . . . ?

☑ list all the counting numbers (natural numbers)

☑ list all the whole numbers

☑ list the ten digits

☑ locate the correct digit in a whole number given its place-value name

☑ identify the correct place-value name of a given digit in a whole number

☑ write the value of a given digit in a whole number

☑ write the correct whole number when the values are given

☑ write the correct whole number when the number word name is given

☑ write the number word name for a given whole number

☑ compare two whole numbers using <, >, or =

☑ order two or more whole numbers from largest to smallest or smallest to largest

☑ round a whole number up or down

☑ answer all 100 basic addition facts correctly in less than 2 minutes

☑ add whole numbers given vertical form

☑ add whole numbers given horizontal form

☑ add two whole numbers when renaming is necessary

☑ add more than two whole numbers

☑ answer all 100 basic subtraction facts correctly in less than 3 minutes

☑ show how subtraction and addition are related

☑ subtract whole numbers given vertical form

☑ subtract whole numbers given horizontal form

☑ check subtraction using addition

☑ subtract whole numbers when renaming is necessary

☑ subtract across one or more zeros

☑ determine whether to add or subtract whole numbers in a word problem

☑ interpret the solution to a word problem in the words of the original problem

SUMMARY

1. The counting numbers (natural numbers) are: 1, 2, 3, $\cdots$
2. The whole numbers are: 0, 1, 2, 3, $\cdots$
3. The ten digits are: 0, 1, 2, 3, 4, 5, 6, 7, 8, and 9.
4. To locate digits and identify place-value names, you can use a place-value chart.

Whole-Number Place-Value Chart (to hundred billions)

Billions			Millions			Thousands			Units		
hundred billions	ten billions	billions	hundred millions	ten millions	millions	hundred thousands	ten thousands	thousands	hundreds	tens	ones

5. To write whole numbers, you can use the following special number word names.

Special Number Word Names

one	two	three	four	five	six
1	2	3	4	5	6
1 one	2 ones	3 ones	4 ones	5 ones	6 ones

seven	eight	nine	ten	eleven	twelve
7	8	9	10	11	12
7 ones	8 ones	9 ones	1 ten	1 ten 1 one	1 ten 2 ones

thirteen	fourteen	fifteen	sixteen	seventeen	eighteen
13	14	15	16	17	18
1 ten 3 ones	1 ten 4 ones	1 ten 5 ones	1 ten 6 ones	1 ten 7 ones	1 ten 8 ones

nineteen	twenty	thirty	forty	$\cdots$	ninety
19	20	30	40	$\cdots$	90
1 ten 9 ones	2 tens	3 tens	4 tens	$\cdots$	9 tens

one hundred	two hundred	three hundred	four hundred	$\cdots$	nine hundred
100	200	300	400	$\cdots$	900
1 hundred	2 hundreds	3 hundreds	4 hundreds	$\cdots$	9 hundreds

one thousand	two thousand	three thousand	four thousand	$\cdots$	nine thousand
1000	2000	3000	4000	$\cdots$	9000
1 thousand	2 thousands	3 thousands	4 thousands	$\cdots$	9 thousands

6. To compare two different whole numbers, you write < or > between them to show which number is larger or smaller.
7. To compare two whole numbers that are the same, you write = between them.
8. To order two or more whole numbers, you list them from largest to smallest or from smallest to largest.
9. To round a whole number when the digit to the right of the digit to be rounded is 5 or more, you round up as follows:
(**a**) Increase the digit to be rounded by 1.
(**b**) Replace digits to the right of the rounded digit with zeros.

10. To round a whole number when the digit to the right of the digit to be rounded is less than 5, you round down as follows:
    (a) Leave the digit to be rounded the same.
    (b) Replace digits to the right of the rounded digits with zeros.
11. Addition is used to join amounts together.

12. In $2 + 3 = 5$ or $\begin{array}{r} 2 \\ +3 \\ \hline 5 \end{array}$:

    (a) 2 and 3 are called addends.
    (b) 5 is called the sum.
13. To add given vertical form, you first add ones, then tens, then hundreds, and so on.
14. To add given horizontal form, you first write the numbers in vertical form.
15. When adding, you must rename 2-digit values (like 12 ones and 13 tens) unless it is the last step.
16. Subtraction is used to take one amount away from another amount.

17. In $5 - 3 = 2$ or $\begin{array}{r} 5 \\ -3 \\ \hline 2 \end{array}$:

    (a) 5 is called the minuend.
    (b) 3 is called the subtrahend.
    (c) 2 is called the difference.
18. To subtract given vertical form, you first subtract ones, then tens, then hundreds, and so on.
19. To subtract given horizontal form, you first write the numbers in vertical form.
20. To check a subtraction answer, you add the proposed difference to the subtrahend to see if you get the original minuend (top number).
21. When the digits in the ones column (or tens column, hundreds column, and so on) cannot be subtracted, you rename to get more ones (or tens, hundreds, and so on).
22. When the minuend (top number) has one or more zeros in it, you rename across those zeros in order to subtract.
23. To choose the correct operation in problem solving, you
    (a) add to join two or more amounts together,
    (b) subtract to take one amount away from another.
24. To solve addition word problems, you find how much two or more amounts equal all together.
25. To solve subtraction word problems, you find
    (a) how much is left,
    (b) how much more is needed,
    (c) the difference between two amounts.

SOLVED PROBLEMS

PROBLEM 1-1   Draw a line under the digit in each whole number that has the given place value:

(a) billions: 258,391,648,072
(b) tens: 169,418,720,000
(c) hundred thousands: 648,002,316,205
(d) ten millions: 715,800,321,296
(e) hundred billions: 403,489,267,185
(f) thousands: 350,813,765,048
(g) hundreds: 915,048,267,132
(h) millions: 512,346,992,817
(i) ten billions: 832,148,569,237
(j) ones: 209,136,004,218
(k) ten thousands: 108,655,431,789
(l) hundred millions: 621,856,413,729

Solution: Recall that to locate the correct digit given its place-value name, you can use a place-value chart [see Example 1-4]:

(a) 258,391,648,072
(b) 169,418,720,000
(c) 648,002,316,205
(d) 715,800,321,296
(e) 403,489,267,185
(f) 350,813,765,048
(g) 915,048,267,132
(h) 512,346,992,817
(i) 832,148,569,237
(j) 209,136,004,218
(k) 108,655,431,789
(l) 621,856,413,729

PROBLEM 1-2   Write each of the following whole numbers using period names instead of commas:

(a) 56,250   **(b)** 25,092   **(c)** 37,000   **(d)** 8,356,003   **(e)** 75,025,000   **(f)** 170,000,000
(g) 5,210,034,002   **(h)** 900,000,000,000

Solution: Recall that the given period names in order are billions, millions, thousands, and units [see Example 1-5]:

(a) 56 thousand 250   **(b)** 25 thousand 92   **(c)** 37 thousand   **(d)** 8 million 356 thousand 3
(e) 75 million 25 thousand   **(f)** 170 million   **(g)** 5 billion 210 million 34 thousand 2
(h) 900 billion

PROBLEM 1-3   Write the place-value name of each underlined digit:

(a) 352,167,859,203   **(b)** 658,214,039,715   **(c)** 102,314,876,925
(d) 971,058,030,219   **(e)** 764,302,125,893   **(f)** 561,458,215,613
(g) 287,050,923,000   **(h)** 476,451,873,029   **(i)** 876,915,254,003
(j) 301,210,582,907   **(k)** 961,785,859,619   **(l)** 234,569,213,458

Solution: Recall that to help identify the correct place-value name of a given digit, you can use a place-value chart [see Example 1-6]:

(a) billions   **(b)** ten thousands   **(c)** ones   **(d)** hundred millions   **(e)** hundreds
(f) millions   **(g)** hundred billions   **(h)** tens   **(i)** hundred thousands   **(j)** thousands
(k) ten billions   **(l)** ten millions

PROBLEM 1-4   Write the value of 5 in each whole number:

(a) 213,456,201,891   **(b)** 340,678,203,152   **(c)** 503,789,621,487
(d) 413,469,521,380   **(e)** 935,867,004,918   **(f)** 613,782,495,812
(g) 718,789,247,315   **(h)** 819,582,413,760   **(i)** 759,003,018,000
(j) 619,005,783,416   **(k)** 118,234,658,139   **(l)** 218,648,231,580

Solution: Recall that to write the value of a given digit, you first identify the place-value name of that digit [see Example 1-7]:

(a) 50,000,000   **(b)** 50   **(c)** 500,000,000,000   **(d)** 500,000   **(e)** 5,000,000,000   **(f)** 5000
(g) 5   **(h)** 500,000,000   **(i)** 50,000,000,000   **(j)** 5,000,000   **(k)** 50,000   **(l)** 500

PROBLEM 1-5   Write the correct whole number for each group of 1-digit values:

(a) 6 tens 5 ones   **(b)** 7 tens   **(c)** 8 ones   **(d)** 4 hundreds 3 tens 2 ones
(e) 2 hundreds 9 ones   **(f)** 1 hundred 6 tens   **(g)** 8 hundreds   **(h)** 7 thousands 9 hundreds

Solution: Recall that to write a whole number when 1-digit values are given, you must be careful to write a zero (0) each time a 1-digit value is missing [see Example 1-8]:

(a) 65   **(b)** 70   **(c)** 8   **(d)** 432   **(e)** 209   **(f)** 160   **(g)** 800   **(h)** 7900

PROBLEM 1-6   Write the correct whole number for each whole-number word name:

(a) eight   **(b)** eleven   **(c)** twenty   **(d)** one hundred   **(e)** fifty-four
(f) three hundred sixteen   **(g)** one thousand eighty   **(h)** seven hundred six

Solution: Recall that to write a whole number when the whole-number name is given, you may find it helpful to first write 1-digit values [see Example 1-9]: **(a)** 8   **(b)** 11   **(c)** 20   **(d)** 100
(e) 54   **(f)** 316   **(g)** 1080   **(h)** 706

PROBLEM 1-7   Write the correct whole-number word name for each whole number:
(a) 2   **(b)** 12   **(c)** 18   **(d)** 40   **(e)** 500   **(f)** 679   **(g)** 405   **(h)** 3090

Solution: Recall that to write the whole-number word name when the whole number is given, you may find it helpful to first write 1-digit values [see Example 1-10]: **(a)** two   **(b)** twelve
(c) eighteen   **(d)** forty   **(e)** five hundred   **(f)** six hundred seventy-nine
(g) four hundred five   **(h)** three thousand ninety

PROBLEM 1-8   Write the largest whole number of: **(a)** 24 and 25   **(b)** 200 and 199
(c) 1020 and 1100   **(d)** 306 and 360   **(e)** 5150 and 5151   **(f)** 25, 430, and 26,430
(g) 356,000,001 and 356,000,000   **(h)** 728,591,346,218 and 728,591,436,218

Solution: Recall that to compare two whole numbers, you may find it helpful to write one number
under the other while aligning like place values [see Example 1-13]: **(a)** 25   **(b)** 200
(c) 1100   **(d)** 360   **(e)** 5151   **(f)** 26,430   **(g)** 356,000,001   **(h)** 728,591,436,218

PROBLEM 1-9   List each group of whole numbers from largest to smallest:

(a) 580, 579, 597, 508   **(b)** 2031, 2000, 1999, 2301, 2130   **(c)** 28, 297, 1023, 300, 1022, 1000

Solution: Recall that to order whole numbers from largest to smallest, you write the largest number
first, then the next largest, and so on, until you have listed all the given numbers [see Example 1-14]:
(a) 597, 580, 579, 508   **(b)** 2301, 2130, 2031, 2000, 1999   **(c)** 1023, 1022, 1000, 300, 297, 28

PROBLEM 1-10   List each group of whole numbers from smallest to largest:

(a) 321, 123, 231, 132, 213, 312   **(b)** 5199, 5200, 5100, 4999, 5099
(c) 481, 2, 53, 1035, 480, 1036, 8   **(d)** 987, 988, 978, 789, 989

Solution: Recall that to order numbers from smallest to largest, you just reverse the largest to smallest
order [see Example 1-15]:

(a) 123, 132, 213, 231, 312, 321   **(b)** 4999, 5099, 5100, 5199, 5200
(c) 2, 8, 53, 480, 481, 1035, 1036   **(d)** 789, 978, 987, 988, 989

PROBLEM 1-11   Round each whole number to the given place:

(a) 785 to the nearest hundred   **(b)** 2043 to the nearest thousand
(c) 85,641 to the nearest ten thousand   **(d)** 39 to the nearest ten
(e) 731,258,413 to the nearest million   **(f)** 619,315,208 to the nearest hundred thousand
(g) 258,319,648,215 to the nearest billion   **(h)** 981,674,321,863 to the nearest hundred million
(i) 299 to the nearest ten   **(j)** 39,999 to the nearest hundred

Solution: Recall that if the digit to the right of the digit to be rounded is 5 or more, you round up.
If it's less than 5, you round down [see Examples 1-16 and 1-17]:

(a) 800   **(b)** 2000   **(c)** 90,000   **(d)** 40   **(e)** 731,000,000   **(f)** 619,300,000
(g) 258,000,000,000   **(h)** 981,700,000,000   **(i)** 300   **(j)** 40,000

PROBLEM 1-12   To practice the 100 basic addition facts:

1. Cover the solutions, one row at a time, with a blank sheet of paper.
2. Write the answer below each problem in that row on your blank sheet of paper.
3. Slide your answers down and compare them with the printed solutions.
4. Repeat any row in which you had an incorrect answer.
   [See Example 1-19.]

1	6	5	9	0	5	4	9	3	7
+3	+5	+0	+9	+9	+8	+1	+3	+4	+6
4	11	5	18	9	13	5	12	7	13
4	5	8	7	4	2	0	8	7	9
+7	+2	+5	+9	+3	+8	+5	+2	+2	+9
11	7	13	16	7	10	5	10	9	18
2	6	4	8	1	7	6	3	2	7
+1	+6	+5	+7	+7	+8	+3	+9	+5	+0
3	12	9	15	8	15	9	12	7	7

5	3	2	9	8	2	4	6	3	7
+6	+6	+2	+6	+1	+6	+9	+1	+2	+5
11	9	4	15	9	8	13	7	5	12
9	4	6	5	0	2	5	5	8	7
+0	+0	+9	+3	+6	+4	+7	+4	+4	+1
9	4	15	8	6	6	12	9	12	8
3	8	2	9	2	7	4	6	6	1
+0	+6	+3	+2	+0	+4	+4	+7	+2	+9
3	14	5	11	2	11	8	13	8	10
1	5	0	9	4	4	0	9	0	9
+5	+5	+3	+8	+2	+8	+8	+4	+2	+5
6	10	3	17	6	12	8	13	2	14
3	8	6	4	5	6	0	9	1	5
+3	+8	+0	+6	+1	+4	+7	+1	+6	+9
6	16	6	10	6	10	7	10	7	14
1	1	7	3	7	2	0	8	3	3
+0	+4	+3	+1	+7	+7	+0	+3	+5	+8
1	5	10	4	14	9	0	11	8	11
8	0	1	6	1	8	3	0	1	2
+9	+1	+2	+8	+8	+0	+7	+4	+1	+9
17	1	3	14	9	8	10	4	2	11

PROBLEM 1-13 Add two whole numbers with no renaming: **(a)** 32 + 56 **(b)** 92 + 83 **(c)** 201 + 25 **(d)** 325 + 413 **(e)** 729 + 830 **(f)** 2135 + 41 **(g)** 561 + 3218 **(h)** 9316 + 9270

Solution: Recall that to add whole numbers in horizontal form, you first write the numbers in vertical form [see Example 1-21]:

(a)
```
  3 2
+5 6
-----
  8 8
```

(b)
```
  9 2
+8 3
-----
1 7 5
```

(c)
```
  2 0 1
+   2 5
-------
  2 2 6
```

(d)
```
  3 2 5
+4 1 3
-------
  7 3 8
```

(e)
```
  7 2 9
+8 3 0
-------
1 5 5 9
```

(f)
```
  2 1 3 5
+     4 1
---------
  2 1 7 6
```

(g)
```
  3 2 1 8
+   5 6 1
---------
  3 7 7 9
```

(h)
```
  9 3 1 6
+9 2 7 0
---------
1 8,5 8 6
```

PROBLEM 1-14 Add two whole numbers with renaming: **(a)** 37 + 25 **(b)** 78 + 69 **(c)** 315 + 206 **(d)** 273 + 546 **(e)** 685 + 249 **(f)** 976 + 878 **(g)** 2859 + 729 **(h)** 6858 + 7578

Solution: Recall that when adding, you must rename each 2-digit value unless it is the last step [see Example 1-23]:

(a)
```
  1
  3 7
+2 5
-----
  6 2
```

(b)
```
  1
  7 8
+6 9
-----
1 4 7
```

(c)
```
    1
  3 1 5
+2 0 6
-------
  5 2 1
```

(d)
```
  1
  2 7 3
+5 4 6
-------
  8 1 9
```

(e)
```
  1 1
  6 8 5
+2 4 9
-------
  9 3 4
```

(f)
```
  1 1
  9 7 6
+8 7 8
-------
1 8 5 4
```

(g)
```
  1   1
  2 8 5 9
+   7 2 9
---------
  3 5 8 8
```

(h)
```
  1 1 1
  6 8 5 8
+7 5 7 8
---------
1 4,4 3 6
```

PROBLEM 1-15   Add more than two whole numbers: **(a)** 25 + 36 + 58   **(b)** 72 + 8 + 315
(c) 617 + 85 + 719   **(d)** 858 + 972 + 639   **(e)** 1035 + 483 + 5619   **(f)** 2514 + 34 + 8
(g) 12,502 + 2489 + 3488 + 25,869   **(h)** 8799 + 672 + 8431 + 52,789 + 482

Solution: Recall that to add more than two whole numbers, add as you would with two whole
numbers [see Example 1-24]:

(a)
```
    1
    2 5
    3 6
  + 5 8
  -------
  1 1 9
```

(b)
```
    1
    7 2
      8
  +3 1 5
  -------
  3 9 5
```

(c)
```
  1 2
  6 1 7
    8 5
  +7 1 9
  -------
  1 4 2 1
```

(d)
```
  1 1
  8 5 8
  9 7 2
  +6 3 9
  -------
  2 4 6 9
```

(e)
```
  1 1 1
  1 0 3 5
    4 8 3
  +5 6 1 9
  ---------
  7 1 3 7
```

(f)
```
      1
  2 5 1 4
      3 4
  +     8
  -------
  2 5 5 6
```

(g)
```
  1 2 2 2
  1 2,5 0 2
    2 4 8 9
    3 4 8 8
  +2 5,8 6 9
  -----------
  4 4,3 4 8
```

(h)
```
  2 3 3 2
    8 7 9 9
      6 7 2
    8 4 3 1
  5 2,7 8 9
  +     4 8 2
  -----------
  7 1,1 7 3
```

PROBLEM 1-16   To practice the 100 basic subtraction facts:

1. Cover the solutions, one row at a time, with a blank sheet of paper.
2. Write the answer below each problem in that row on your blank sheet of paper.
3. Slide your answers down and compare them with the printed solutions.
4. Repeat any row in which you had an incorrect answer.
   [See Example 1-26.]

$\begin{array}{r}9\\-0\\\hline 9\end{array}$	$\begin{array}{r}10\\-9\\\hline 1\end{array}$	$\begin{array}{r}5\\-0\\\hline 5\end{array}$	$\begin{array}{r}2\\-1\\\hline 1\end{array}$	$\begin{array}{r}17\\-8\\\hline 9\end{array}$	$\begin{array}{r}14\\-9\\\hline 5\end{array}$	$\begin{array}{r}8\\-6\\\hline 2\end{array}$	$\begin{array}{r}10\\-2\\\hline 8\end{array}$	$\begin{array}{r}6\\-3\\\hline 3\end{array}$	$\begin{array}{r}11\\-5\\\hline 6\end{array}$
$\begin{array}{r}11\\-9\\\hline 2\end{array}$	$\begin{array}{r}5\\-5\\\hline 0\end{array}$	$\begin{array}{r}10\\-1\\\hline 9\end{array}$	$\begin{array}{r}0\\-0\\\hline 0\end{array}$	$\begin{array}{r}5\\-4\\\hline 1\end{array}$	$\begin{array}{r}11\\-4\\\hline 7\end{array}$	$\begin{array}{r}9\\-3\\\hline 6\end{array}$	$\begin{array}{r}15\\-7\\\hline 8\end{array}$	$\begin{array}{r}16\\-9\\\hline 7\end{array}$	$\begin{array}{r}5\\-2\\\hline 3\end{array}$
$\begin{array}{r}9\\-8\\\hline 1\end{array}$	$\begin{array}{r}8\\-8\\\hline 0\end{array}$	$\begin{array}{r}14\\-5\\\hline 9\end{array}$	$\begin{array}{r}13\\-4\\\hline 9\end{array}$	$\begin{array}{r}8\\-2\\\hline 6\end{array}$	$\begin{array}{r}18\\-9\\\hline 9\end{array}$	$\begin{array}{r}14\\-7\\\hline 7\end{array}$	$\begin{array}{r}3\\-3\\\hline 0\end{array}$	$\begin{array}{r}6\\-1\\\hline 5\end{array}$	$\begin{array}{r}13\\-6\\\hline 7\end{array}$
$\begin{array}{r}10\\-8\\\hline 2\end{array}$	$\begin{array}{r}5\\-1\\\hline 4\end{array}$	$\begin{array}{r}7\\-0\\\hline 7\end{array}$	$\begin{array}{r}15\\-8\\\hline 7\end{array}$	$\begin{array}{r}7\\-7\\\hline 0\end{array}$	$\begin{array}{r}11\\-6\\\hline 5\end{array}$	$\begin{array}{r}11\\-3\\\hline 8\end{array}$	$\begin{array}{r}8\\-7\\\hline 1\end{array}$	$\begin{array}{r}3\\-0\\\hline 3\end{array}$	$\begin{array}{r}7\\-2\\\hline 5\end{array}$
$\begin{array}{r}11\\-8\\\hline 3\end{array}$	$\begin{array}{r}16\\-8\\\hline 8\end{array}$	$\begin{array}{r}6\\-4\\\hline 2\end{array}$	$\begin{array}{r}13\\-8\\\hline 5\end{array}$	$\begin{array}{r}5\\-3\\\hline 2\end{array}$	$\begin{array}{r}9\\-4\\\hline 5\end{array}$	$\begin{array}{r}13\\-5\\\hline 8\end{array}$	$\begin{array}{r}15\\-6\\\hline 9\end{array}$	$\begin{array}{r}9\\-5\\\hline 4\end{array}$	$\begin{array}{r}12\\-9\\\hline 3\end{array}$
$\begin{array}{r}12\\-8\\\hline 4\end{array}$	$\begin{array}{r}7\\-3\\\hline 4\end{array}$	$\begin{array}{r}7\\-1\\\hline 6\end{array}$	$\begin{array}{r}9\\-6\\\hline 3\end{array}$	$\begin{array}{r}9\\-7\\\hline 2\end{array}$	$\begin{array}{r}17\\-9\\\hline 8\end{array}$	$\begin{array}{r}16\\-7\\\hline 9\end{array}$	$\begin{array}{r}2\\-2\\\hline 0\end{array}$	$\begin{array}{r}9\\-2\\\hline 7\end{array}$	$\begin{array}{r}10\\-6\\\hline 4\end{array}$
$\begin{array}{r}15\\-9\\\hline 6\end{array}$	$\begin{array}{r}6\\-0\\\hline 6\end{array}$	$\begin{array}{r}1\\-0\\\hline 1\end{array}$	$\begin{array}{r}10\\-7\\\hline 3\end{array}$	$\begin{array}{r}7\\-5\\\hline 2\end{array}$	$\begin{array}{r}4\\-3\\\hline 1\end{array}$	$\begin{array}{r}11\\-2\\\hline 9\end{array}$	$\begin{array}{r}3\\-1\\\hline 2\end{array}$	$\begin{array}{r}10\\-5\\\hline 5\end{array}$	$\begin{array}{r}12\\-7\\\hline 5\end{array}$
$\begin{array}{r}6\\-5\\\hline 1\end{array}$	$\begin{array}{r}6\\-6\\\hline 0\end{array}$	$\begin{array}{r}12\\-4\\\hline 8\end{array}$	$\begin{array}{r}8\\-0\\\hline 8\end{array}$	$\begin{array}{r}9\\-1\\\hline 8\end{array}$	$\begin{array}{r}9\\-9\\\hline 0\end{array}$	$\begin{array}{r}12\\-3\\\hline 9\end{array}$	$\begin{array}{r}12\\-6\\\hline 6\end{array}$	$\begin{array}{r}3\\-2\\\hline 1\end{array}$	$\begin{array}{r}14\\-8\\\hline 6\end{array}$

7	1	8	8	13	4	12	7	14	2
−6	−1	−3	−5	− 9	−4	− 5	−4	− 6	−0
1	0	5	3	4	0	7	3	8	2

4	4	13	6	10	8	10	11	8	4
−0	−1	− 7	−2	− 3	−4	− 4	− 7	−1	−2
4	3	6	4	7	4	6	4	7	2

PROBLEM 1-17   Subtract whole numbers with no renaming and check each answer by adding:

(a) 86 − 63   (b) 725 − 10   (c) 599 − 203   (d) 855 − 45   (e) 7883 − 23
(f) 9873 − 8820   (g) 87,759 − 77,015   (h) 529,569 − 4342

Solution: Recall that to subtract whole numbers given horizontal form, you first rename in vertical form [see Example 1-28]:

(a)
```
   8 6
 −6 3
 ─────
   2 3
```
(b)
```
   7 2 5
 −   1 0
 ───────
   7 1 5
```
(c)
```
   5 9 9
 −2 0 3
 ───────
   3 9 6
```
(d)
```
   8 5 5
 −   4 5
 ───────
   8 1 0
```

(e)
```
   7 8 8 3
 −     2 3
 ─────────
   7 8 6 0
```
(f)
```
   9 8 7 3
 −8 8 2 0
 ─────────
   1 0 5 3
```
(g)
```
   8 7,7 5 9
 −7 7,0 1 5
 ───────────
   1 0,7 4 4
```
(h)
```
   5 2 9,5 6 9
 −       4 3 4 2
 ───────────────
   5 2 5,2 2 7
```

PROBLEM 1-18   Subtract whole numbers with one renaming and check each answer by adding:

(a) 91 − 75   (b) 390 − 62   (c) 374 − 119   (d) 407 − 395   (e) 661 − 90
(f) 1539 − 729   (g) 9166 − 1465   (h) 37,024 − 4104

Solution: Recall that when a digit in any given place-value column cannot be subtracted, you rename to get more of that place-value in the minuend (top number) [see Examples 1-31 and 1-32]:

(a)
```
   8 1
   9̸ 1
 −7 5
 ─────
   1 6
```
(b)
```
   8 1
 3 9̸ 0
 −  6 2
 ───────
 3 2 8
```
(c)
```
     6 1
 3 7̸ 4
 −1 1 9
 ───────
   2 5 5
```
(d)
```
   3 1
 4̸ 0 7
 −3 9 5
 ───────
     1 2
```

(e)
```
   5 1
   6̸ 6 1
 −  9 0
 ───────
   5 7 1
```
(f)
```
   0 1
 1̸ 5 3 9
 −  7 2 9
 ─────────
     8 1 0
```
(g)
```
   8 1
 9̸ 1 6 6
 −1 4 6 5
 ─────────
   7 7 0 1
```
(h)
```
     6 1
 3 7̸,0 2 4
 −  4 1 0 4
 ───────────
   3 2,9 2 0
```

PROBLEM 1-19   Subtract whole numbers with two or more renamings and check each answer by adding: (a) 941 − 569   (b) 3884 − 937   (c) 1853 − 1654   (d) 5621 − 2863 (e) 6930 − 6198   (f) 9743 − 4885   (g) 57,071 − 17,628   (h) 51,337 − 5789

Solution: Recall that to subtract with two or more renamings, you should write each renaming very neatly to avoid errors [see Example 1-33]:

(a)
```
 8 13 1
 9̸ 4̸ 1
 −5 6 9
 ───────
   3 7 2
```
(b)
```
 2 1  7 1
 3̸ 8̸ 8̸ 4
 −   9 3 7
 ─────────
   2 9 4 7
```
(c)
```
     7 14 1
 1 8 5̸ 3
 −1 6 5 4
 ─────────
       1 9 9
```
(d)
```
 4 15 11 1
 5̸ 6̸ 2̸ 1
 −2 8 6 3
 ───────────
   2 7 5 8
```

(e)
```
   8 12 1
 6 9̸ 3̸ 0
 −6 1 9 8
 ─────────
     7 3 2
```
(f)
```
 8 16 13 1
 9̸ 7̸ 4̸ 3
 −4 8 8 5
 ───────────
   4 8 5 8
```
(g)
```
 4 16 1   6 1
 5̸ 7̸,0 7̸ 1
 −1 7,6 2 8
 ───────────
   3 9,4 4 3
```
(h)
```
 4 10 12 12 1
 5̸ 1̸,3̸ 3̸ 7
 −   5 7 8 9
 ─────────────
   4 5,5 4 8
```

PROBLEM 1-20   Subtract whole numbers by renaming across one or more zeros and check each answer by adding: (a) 406 − 287   (b) 503 − 158   (c) 900 − 841   (d) 8005 − 287 (e) 6100 − 3619   (f) 70,002 − 2405   (g) 60,201 − 34,285   (h) 40,100 − 2639

Solution: Recall that when the minuend (top number) has one or more zeros, you will usually have to rename across those zeros [see Examples 1-34 and 1-35]:

(a)
$$\begin{array}{r} {}^{3}\ {}^{9}\ {}^{1} \\ 4\ 0\ 6 \\ -2\ 8\ 7 \\ \hline 1\ 1\ 9 \end{array}$$

(b)
$$\begin{array}{r} {}^{4}\ {}^{9}\ {}^{1} \\ 5\ 0\ 3 \\ -1\ 5\ 8 \\ \hline 3\ 4\ 5 \end{array}$$

(c)
$$\begin{array}{r} {}^{8}\ {}^{9}\ {}^{1} \\ 9\ 0\ 0 \\ -8\ 4\ 1 \\ \hline 5\ 9 \end{array}$$

(d)
$$\begin{array}{r} {}^{7}\ {}^{9}\ {}^{9}\ {}^{1} \\ 8\ 0\ 0\ 5 \\ -\ \ 2\ 8\ 7 \\ \hline 7\ 7\ 1\ 8 \end{array}$$

(e)
$$\begin{array}{r} {}^{5}\ {}^{10}\ {}^{9}\ {}^{1} \\ 6\ 1\ 0\ 0 \\ -3\ 6\ 1\ 9 \\ \hline 2\ 4\ 8\ 1 \end{array}$$

(f)
$$\begin{array}{r} {}^{6}\ {}^{9}\ {}^{9}\ {}^{9}\ {}^{1} \\ 7\ 0,0\ 0\ 2 \\ -\ \ 2\ 4\ 0\ 5 \\ \hline 6\ 7,5\ 9\ 7 \end{array}$$

(g)
$$\begin{array}{r} {}^{5}\ {}^{9}\ {}^{11}\ {}^{9}\ {}^{1} \\ 6\ 0,2\ 0\ 1 \\ -3\ 4,2\ 8\ 5 \\ \hline 2\ 5,9\ 1\ 6 \end{array}$$

(h)
$$\begin{array}{r} {}^{3}\ {}^{9}\ {}^{10}\ {}^{9}\ {}^{1} \\ 4\ 0,1\ 0\ 0 \\ -\ \ 2\ 6\ 3\ 9 \\ \hline 3\ 7,4\ 6\ 1 \end{array}$$

PROBLEM 1-21   Decide whether to add or subtract, and then solve the problem:

(a) Gary read 83 pages. Then he read 56 pages more. How many pages did Gary read in all?

(b) Denise drove 285 miles. Greg drove 190 miles. How much farther did Denise drive than Greg?

(c) Chris typed 105 pages. Then he typed another 26 pages. How many pages did Chris type all together?

(d) Nancy scored 405 points. Bob scored 135 points. What is the difference between Nancy and Bob's scores?

(e) Barbara needed to collect $345. She collected $186 of it. How much is left for Barbara to collect?

(f) Sue is 162 cm tall. Deirdre is 91 cm tall. How much taller is Sue than Deirdre?

(g) Richard worked 98 minutes, Peter worked 37 minutes, and Alan worked 25 minutes. How much longer did Richard work than Peter and Alan combined? (Careful! This problem takes more than one operation to solve.)

(h) Manuel had $582. He spent $135 on food, $87 for a car payment, and $56 for entertainment. How much did Manuel have then?

Solution: Recall that to join two amounts together, you add, and to take one amount away from another amount, you subtract [see Examples 1-36 and 1-37]:

(a) Add: $83 + 56 = 139$ (pages)

(b) Subtract: $285 - 190 = 95$ (miles)

(c) Add: $105 + 26 = 131$ (pages)

(d) Subtract: $405 - 135 = 270$ (points)

(e) Subtract: $345 - 186 = 159$ (dollars)

(f) Subtract: $162 - 91 = 71$ (cm)

(g) Add and then subtract: $37 + 25 = 62$ and $98 - 62 = 36$ (minutes),
or subtract and then subtract again: $98 - 37 = 61$ and $61 - 25 = 36$ (minutes)

(h) Add and then subtract: $135 + 87 + 56 = 278$ and $582 - 278 = 304$ (dollars),
or subtract, subtract, and subtract again: $582 - 135 = 447$, $447 - 87 = 360$ and $360 - 56 = 304$ (dollars)

Supplementary Exercises

PROBLEM 1-22   Draw a line under the digit in each whole number that has the given place value:

(a) thousands: 658,432,196,285

(b) ten millions: 269,015,347,250

(c) billions: 100,580,239,611

(d) tens: 591,036,549,872

(e) ten thousands: 300,000,000,000

(f) ten billions: 721,834,659,023

(g) ones: 487,291,635,205

(h) hundred billions: 721,834,659,023

(i) millions: 825,916,430,017   **(j)** hundreds: 200,310,508,200
(k) hundred millions: 534,816,729,885   **(l)** hundred thousands: 916,832,005,713

PROBLEM 1-23   Write each of the following whole numbers using period names instead of commas:

(a) 18,546   **(b)** 83,005   **(c)** 42,000   **(d)** 723,058   **(e)** 850,000   **(f)** 3,512,000
(g) 25,063,002   **(h)** 970,250,000,070

PROBLEM 1-24   Write the place-value name of each underlined digit:

(a) 285,316,487,915   **(b)** 817,639,854,203   **(c)** 105,213,698,540
(d) 903,142,586,200   **(e)** 305,002,913,617   **(f)** 785,413,826,918
(g) 463,218,915,834   **(h)** 692,413,876,518   **(i)** 588,765,432,017
(j) 261,823,415,286   **(k)** 199,381,418,275   **(l)** 960,108,235,872

PROBLEM 1-25   Write the value of 2 in each whole number:

(a) 862,153,089,451   **(b)** 615,843,569,205   **(c)** 318,792,013,459
(d) 213,085,619,318   **(e)** 186,753,048,629   **(f)** 734,206,518,349
(g) 963,825,619,004   **(h)** 425,871,643,019   **(i)** 563,871,942,018
(j) 618,594,231,075   **(k)** 185,619,348,752   **(l)** 791,843,621,587

PROBLEM 1-26   Write the correct whole number for each group of 1-digit values:

(a) 1 ten 5 ones   **(b)** 8 ones   **(c)** 3 tens
(d) 2 hundreds 9 tens 7 ones   **(e)** 4 hundreds 6 ones   **(f)** 5 hundreds 1 ten
(g) 9 hundreds   **(h)** 6 thousands 4 tens 2 ones   **(i)** 9 thousands

PROBLEM 1-27 Write the correct whole number for each whole-number word name:

(a) five   **(b)** twelve   **(c)** thirty   **(d)** two hundred   **(e)** forty-three   **(f)** four hundred ten
(g) two thousand seventy-two   **(h)** seven thousand six hundred thirteen
(i) five hundred eleven   **(j)** fifteen   **(k)** sixty-nine   **(l)** eighteen   **(m)** ninety-one
(n) nineteen   **(o)** eight hundred seventeen   **(p)** nine hundred eight

PROBLEM 1-28   Write the correct whole-number word name for each whole number:

(a) 8   **(b)** 11   **(c)** 20   **(d)** 100   **(e)** 54   **(f)** 316   **(g)** 1080   **(h)** 706   **(i)** 2005
(j) 3469   **(k)** 231   **(l)** 12   **(m)** 47   **(n)** 13   **(o)** 815   **(p)** 692   **(q)** 518   **(r)** 910
(s) 14   **(t)** 4017   **(u)** 19   **(v)** 73

PROBLEM 1-29   Write the smallest whole number of: **(a)** 36 and 35   **(b)** 101 and 110
(c) 4000 and 3999   **(d)** 7201 and 7210   **(e)** 4103 and 4113   **(f)** 256,001 and 257,000
(g) 100,000 and 99,999   **(h)** 999,999 and 100,000

PROBLEM 1-30   List each group of whole numbers from largest to smallest:

(a) 63, 27, 82, 59   **(b)** 285, 582, 825, 258, 528, 852   **(c)** 3105, 3200, 3104, 3199, 2999
(d) 786, 787, 878, 778, 877

PROBLEM 1-31   List each group of whole numbers from smallest to largest:

(a) 58, 59, 61, 53, 60   **(b)** 1001, 1100, 1010, 1000, 1011, 1110, 1101
(c) 657, 576, 765, 675, 567, 756   **(d)** 213, 315, 428, 99, 5, 2153

PROBLEM 1-32   Round each whole number to the nearest ten: **(a)** 53   **(b)** 95   **(c)** 69
(d) 870   **(e)** 2042   **(f)** 3927

PROBLEM 1-33   Round each whole number to the nearest hundred: **(a)** 649   **(b)** 863   **(c)** 985
(d) 418   **(e)** 2950   **(f)** 59,987

PROBLEM 1-34 Round each whole number to the nearest thousand: **(a)** 5369 **(b)** 4520 **(c)** 69,212 **(d)** 25,600 **(e)** 100,700 **(f)** 299,999 **(g)** 5,832,169 **(h)** 25,000,000

PROBLEM 1-35 Add basic facts in vertical form:

Row **(a)**
$$\begin{array}{ccccccccccc} 6 & 2 & 3 & 9 & 0 & 9 & 4 & 4 & 4 & 8 \\ +6 & +7 & +4 & +8 & +2 & +0 & +5 & +7 & +1 & +1 \end{array}$$

Row **(b)**
$$\begin{array}{ccccccccccc} 2 & 5 & 5 & 4 & 7 & 2 & 8 & 2 & 8 & 6 \\ +5 & +6 & +2 & +0 & +4 & +1 & +4 & +0 & +6 & +7 \end{array}$$

Row **(c)**
$$\begin{array}{ccccccccccc} 3 & 1 & 0 & 5 & 2 & 1 & 5 & 4 & 3 & 5 \\ +6 & +8 & +4 & +4 & +6 & +7 & +1 & +3 & +5 & +5 \end{array}$$

Row **(d)**
$$\begin{array}{ccccccccccc} 1 & 3 & 0 & 3 & 7 & 4 & 9 & 9 & 5 & 8 \\ +2 & +0 & +1 & +7 & +5 & +6 & +7 & +5 & +8 & +9 \end{array}$$

Row **(e)**
$$\begin{array}{ccccccccccc} 7 & 0 & 1 & 8 & 5 & 0 & 7 & 1 & 1 & 2 \\ +7 & +5 & +1 & +8 & +0 & +3 & +2 & +5 & +6 & +4 \end{array}$$

Row **(f)**
$$\begin{array}{ccccccccccc} 2 & 8 & 7 & 0 & 6 & 4 & 9 & 8 & 6 & 1 \\ +2 & +3 & +0 & +6 & +3 & +9 & +4 & +7 & +8 & +0 \end{array}$$

Row **(g)**
$$\begin{array}{ccccccccccc} 8 & 1 & 0 & 9 & 3 & 2 & 7 & 9 & 4 & 2 \\ +5 & +4 & +9 & +6 & +3 & +9 & +8 & +2 & +4 & +3 \end{array}$$

Row **(h)**
$$\begin{array}{ccccccccccc} 7 & 3 & 6 & 7 & 9 & 4 & 5 & 8 & 6 & 0 \\ +1 & +8 & +1 & +9 & +9 & +7 & +7 & +2 & +5 & +7 \end{array}$$

Row **(i)**
$$\begin{array}{ccccccccccc} 6 & 7 & 6 & 6 & 0 & 3 & 1 & 0 & 3 & 8 \\ +2 & +3 & +9 & +4 & +8 & +2 & +3 & +0 & +1 & +0 \end{array}$$

Row **(j)**
$$\begin{array}{ccccccccccc} 9 & 9 & 7 & 5 & 4 & 3 & 2 & 6 & 5 & 1 \\ +1 & +3 & +6 & +9 & +8 & +9 & +8 & +0 & +3 & +9 \end{array}$$

PROBLEM 1-36 Add basic facts in horizontal form:

Row **(a)** 4 + 2   0 + 5   9 + 8   5 + 7   5 + 5   6 + 2   7 + 6   4 + 7   1 + 8   2 + 0

Row **(b)** 9 + 1   8 + 4   0 + 8   9 + 9   8 + 0   3 + 2   7 + 0   3 + 3   0 + 1   3 + 6

Row **(c)** 6 + 1   8 + 8   8 + 5   7 + 3   5 + 3   4 + 5   6 + 8   0 + 4   1 + 4   8 + 9

Row **(d)** 3 + 5   9 + 5   1 + 3   4 + 6   5 + 6   9 + 7   4 + 4   6 + 9   8 + 6   6 + 0

Row **(e)** 5 + 1   1 + 1   5 + 9   1 + 0   3 + 4   4 + 0   0 + 2   2 + 5   7 + 4   4 + 2

Row **(f)** 0 + 6   6 + 7   0 + 0   9 + 6   2 + 8   1 + 6   2 + 9   9 + 0   7 + 8   0 + 7

Row (g)   2 + 7   5 + 0   7 + 9   0 + 3   1 + 5   8 + 1   4 + 8   6 + 5   5 + 4   7 + 2

Row (h)   8 + 2   3 + 9   3 + 1   3 + 7   7 + 5   8 + 3   8 + 7   4 + 1   4 + 9   3 + 8

Row (i)   3 + 0   7 + 7   6 + 4   2 + 1   7 + 1   6 + 3   1 + 2   5 + 2   0 + 9   6 + 6

Row (j)   9 + 3   1 + 9   2 + 3   2 + 6   2 + 4   1 + 7   9 + 2   5 + 8   9 + 4   2 + 2

PROBLEM 1-37   Add whole numbers: **(a)** 13 + 82   **(b)** 75 + 98   **(c)** 248 + 309
(d) 785 + 697   **(e)** 4028 + 3170   **(f)** 5816 + 4809   **(g)** 7019 + 8285   **(h)** 6487 + 9875
(i) 25 + 87 + 69 **(j)** 89 + 92 + 78 + 65   **(k)** 503 + 491 + 392   **(l)** 785 + 649 + 827 + 538
(m) 6240 + 5016 + 4395 + 7286 + 4917 + 8325

PROBLEM 1-38   Subtract basic facts in vertical form:

Row (a)	4 −2	8 −1	11 − 7	10 − 4	8 −4	10 − 3	6 −2	13 − 7	4 −1	4 −0
Row (b)	2 −0	14 − 6	7 −4	12 − 5	4 −4	13 − 9	8 −5	8 −3	1 −1	7 −6
Row (c)	14 − 8	3 −2	12 − 6	12 − 3	9 −9	9 −1	8 −0	12 − 4	6 −6	6 −5
Row (d)	12 − 7	10 − 5	3 −1	11 − 2	4 −3	7 −5	10 − 7	1 −0	6 −0	15 − 9
Row (e)	10 − 6	9 −2	2 −2	16 − 7	17 − 9	9 −7	9 −6	7 −1	7 −3	12 − 8
Row (f)	12 − 9	9 −5	15 − 6	13 − 5	9 −4	5 −3	13 − 8	6 −4	16 − 8	11 − 8
Row (g)	10 − 8	7 −2	3 −0	8 −7	11 − 3	11 − 6	7 −7	15 − 8	7 −0	5 −1
Row (h)	13 − 6	6 −1	3 −3	14 − 7	18 − 9	8 −2	13 − 4	14 − 5	8 −8	9 −8
Row (i)	5 −2	16 − 9	15 − 7	9 −3	11 − 4	5 −4	0 −0	10 − 1	5 −5	11 − 9
Row (j)	11 − 5	6 −3	10 − 2	8 −6	14 − 9	17 − 8	2 −1	5 −0	10 − 9	9 −0

PROBLEM 1-39 Subtract basic facts in horizontal form:

Row (**a**) 5 − 2 16 − 9 15 − 7 9 − 3 11 − 4 5 − 4 0 − 0 10 − 1 5 − 5 11 − 9

Row (**b**) 7 − 2 3 − 0 8 − 7 11 − 3 11 − 6 7 − 7 15 − 8 7 − 0 5 − 1 10 − 8

Row (**c**) 10 − 6 9 − 2 2 − 2 16 − 7 17 − 9 9 − 7 9 − 6 7 − 1 7 − 3 12 − 8

Row (**d**) 14 − 8 3 − 2 12 − 6 12 − 3 9 − 9 9 − 1 8 − 0 12 − 4 6 − 6 6 − 5

Row (**e**) 4 − 2 8 − 1 11 − 7 10 − 4 8 − 4 10 − 3 6 − 2 13 − 7 4 − 1 4 − 0

Row (**f**) 11 − 5 6 − 3 10 − 2 8 − 6 14 − 9 17 − 8 2 − 1 5 − 0 10 − 9 9 − 0

Row (**g**) 13 − 6 6 − 1 3 − 3 14 − 7 18 − 9 8 − 2 13 − 4 14 − 5 8 − 8 9 − 8

Row (**h**) 12 − 9 9 − 5 15 − 6 13 − 5 9 − 4 5 − 3 13 − 8 6 − 4 16 − 8 11 − 8

Row (**i**) 12 − 7 10 − 5 3 − 1 11 − 2 4 − 3 7 − 5 10 − 7 1 − 0 6 − 0 15 − 9

Row (**j**) 2 − 0 14 − 6 7 − 4 12 − 5 4 − 4 13 − 9 8 − 5 8 − 3 1 − 1 7 − 6

PROBLEM 1-40 Subtract whole numbers: (**a**) 78 − 23 (**b**) 82 − 59 (**c**) 987 − 256 (**d**) 863 − 129 (**e**) 528 − 130 (**f**) 825 − 739 (**g**) 4816 − 382 (**h**) 5291 − 1834 (**i**) 7214 − 685 (**j**) 98,514 − 23,802 (**k**) 302 − 89 (**l**) 4002 − 1235 (**m**) 80,000 − 24,567

PROBLEM 1-41 Solve the following word problems (+, −):

(**a**) Kristina had $105. She spent $57. How much did Kristina have left?

(**b**) Stephanie ran 200 m. Ryan ran 85 m. How much farther did Stephanie run than Ryan?

(**c**) Sean collected $249 one day. He collected $179 the next day. How much did Sean collect in all?

(**d**) Myrle collected $249. Evelyn collected $179. What is the difference between the amounts Myrle and Evelyn collected?

(**e**) Diane's car holds 86 liters of gasoline. Nelson's car holds 69 liters. How much do the two cars hold all together?

(**f**) Brett worked 85 minutes. Byron worked 129 minutes. How many combined minutes of work did both do?

(**g**) Carol opened a new checking account with $875. She then wrote checks for $25, $87, $129, and $17. How much was left in the account then?

(**h**) John earned $85 Monday, $39 Tuesday, $156 Wednesday, $83 Thursday, and $252 Friday. He spent $350 of his weekly earnings on rent. How much did John have then?

Answers to Supplementary Exercises

(**1-22**) (**a**) 658,432,196,285 (**b**) 269,015,347,250 (**c**) 100,580,239,611 (**d**) 591,036,549,872 (**e**) 300,000,000,000 (**f**) 721,834,659,023 (**g**) 487,291,635,205 (**h**) 721,834,659,023 (**i**) 825,916,430,017 (**j**) 200,310,508,200 (**k**) 534,816,729,885 (**l**) 916,832,005,713

(**1-23**) (**a**) 18 thousand 546 (**b**) 83 thousand 5 (**c**) 42 thousand (**d**) 723 thousand 58 (**e**) 850 thousand (**f**) 3 million 512 thousand (**g**) 25 million 63 thousand 2 (**h**) 970 billion 250 million 70

(**1-24**) (**a**) billions (**b**) hundreds (**c**) ten billions (**d**) ten millions (**e**) millions (**f**) hundred thousands (**g**) hundred millions (**h**) ten thousands (**i**) thousands (**j**) ones (**k**) hundred billions (**l**) tens

(1-25) **(a)** 2,000,000,000 **(b)** 200 **(c)** 2,000,000 **(d)** 200,000,000,000 **(e)** 20
(f) 200,000,000 **(g)** 20,000,000 **(h)** 20,000,000,000 **(i)** 2000 **(j)** 200,000 **(k)** 2
(l) 20,000

(1-26) **(a)** 15 **(b)** 8 **(c)** 30 **(d)** 297 **(e)** 406 **(f)** 510 **(g)** 900 **(h)** 6042 **(i)** 9000

(1-27) **(a)** 5 **(b)** 12 **(c)** 30 **(d)** 200 **(e)** 43 **(f)** 410 **(g)** 2072 **(h)** 7613
(i) 511 **(j)** 15 **(k)** 69 **(l)** 18 **(m)** 91 **(n)** 19 **(o)** 817 **(p)** 908

(1-28) **(a)** eight **(b)** eleven **(c)** twenty **(d)** one hundred **(e)** fifty-four
(f) three hundred sixteen **(g)** one thousand eighty **(h)** seven hundred six
(i) two thousand five **(j)** three thousand four hundred sixty-nine **(k)** two hundred thirty-one
(l) twelve **(m)** forty-seven **(n)** thirteen **(o)** eight hundred fifteen
(p) six hundred ninety-two **(q)** five hundred eighteen **(r)** nine hundred ten **(s)** fourteen
(t) four thousand seventeen **(u)** nineteen **(v)** seventy-three

(1-29) **(a)** 35 **(b)** 101 **(c)** 3999 **(d)** 7201 **(e)** 4103 **(f)** 256,001 **(g)** 99,999
(h) 100,000

(1-30) **(a)** 82, 63, 59, 27 **(b)** 852, 825, 582, 528, 285, 258 **(c)** 3200, 3199, 3105, 3104, 2999
(d) 878, 877, 787, 786, 778

(1-31) **(a)** 53, 58, 59, 60, 61 **(b)** 1000, 1001, 1010, 1011, 1100, 1101, 1110
(c) 567, 576, 657, 675, 756, 765 **(d)** 5, 99, 213, 315, 428, 2153

(1-32) **(a)** 50 **(b)** 100 **(c)** 70 **(d)** 870 **(e)** 2040 **(f)** 3930

(1-33) **(a)** 600 **(b)** 900 **(c)** 1000 **(d)** 400 **(e)** 3000 **(f)** 60,000

(1-34) **(a)** 5000 **(b)** 5000 **(c)** 69,000 **(d)** 26,000 **(e)** 101,000 **(f)** 300,000
(g) 5,832,000 **(h)** 25,000,000

(1-35)

Row **(a)**	12	9	7	17	2	9	9	11	5	9
Row **(b)**	7	11	7	4	11	3	12	2	14	13
Row **(c)**	9	9	4	9	8	8	6	7	8	10
Row **(d)**	3	3	1	10	12	10	16	14	13	17
Row **(e)**	14	5	2	16	5	3	9	6	7	6
Row **(f)**	4	11	7	6	9	13	13	15	14	1
Row **(g)**	13	5	9	15	6	11	15	11	8	5
Row **(h)**	8	11	7	16	18	11	12	10	11	7
Row **(i)**	8	10	15	10	8	5	4	0	4	8
Row **(j)**	10	12	13	14	12	12	10	6	8	10

(1-36)

Row **(a)**	6	5	17	12	10	8	13	11	9	2
Row **(b)**	10	12	8	18	8	5	7	6	1	9
Row **(c)**	7	16	13	10	8	9	14	4	5	17
Row **(d)**	8	14	4	10	11	16	8	15	14	6
Row **(e)**	6	2	14	1	7	4	2	7	11	6
Row **(f)**	6	13	0	15	10	7	11	9	15	7
Row **(g)**	9	5	16	3	6	9	12	11	9	9
Row **(h)**	10	12	4	10	12	11	15	5	13	11
Row **(i)**	3	14	10	3	8	9	3	7	9	12
Row **(j)**	12	10	5	8	6	8	11	13	13	4

(1-37) **(a)** 95 **(b)** 173 **(c)** 557 **(d)** 1482 **(e)** 7198 **(f)** 10,625 **(g)** 15,304
(h) 16,362 **(i)** 181 **(j)** 324 **(k)** 1386 **(l)** 2799 **(m)** 36,179

(1-38)

Row (a)	2	7	4	6	4	7	4	6	3	4
Row (b)	2	8	3	7	0	4	3	5	0	1
Row (c)	6	1	6	9	0	8	8	8	0	1
Row (d)	5	5	2	9	1	2	3	1	6	6
Row (e)	4	7	0	9	8	2	3	6	4	4
Row (f)	3	4	9	8	5	2	5	2	8	3
Row (g)	2	5	3	1	8	5	0	7	7	4
Row (h)	7	5	0	7	9	6	9	9	0	1
Row (i)	3	7	8	6	7	1	0	9	0	2
Row (j)	6	3	8	2	5	9	1	5	1	9

(1-39)

Row (a)	3	7	8	6	7	1	0	9	0	2
Row (b)	5	3	1	8	5	0	7	7	4	2
Row (c)	4	7	0	9	8	2	3	6	4	4
Row (d)	6	1	6	9	0	8	8	8	0	1
Row (e)	2	7	4	6	4	7	4	6	3	4
Row (f)	6	3	8	2	5	9	1	5	1	9
Row (g)	7	5	0	7	9	6	9	9	0	1
Row (h)	3	4	9	8	5	2	5	2	8	3
Row (i)	5	5	2	9	1	2	3	1	6	6
Row (j)	2	8	3	7	0	4	3	5	0	1

(1-40) **(a)** 55   **(b)** 23   **(c)** 731   **(d)** 734   **(e)** 398   **(f)** 86   **(g)** 4434   **(h)** 3457
(i) 6529   **(j)** 74,712   **(k)** 213   **(l)** 2767   **(m)** 55,433

(1-41) **(a)** $48   **(b)** 115 m   **(c)** $428   **(d)** $70   **(e)** 155 liters   **(f)** 214 minutes
(g) $617   **(h)** $265

2 WHOLE NUMBERS: MULTIPLICATION

THIS CHAPTER IS ABOUT

☑ **Multiplying Basic Facts**
☑ **Multiplying by One Digit**
☑ **Multiplying by Two or More Digits**
☑ **Shortcuts in Multiplying**
☑ **Solving Word Problems (+, −, ×)**

2-1. Multiplying Basic Facts

A short way to write repeated addends is **multiplication.**

EXAMPLE 2-1: Write the repeated addends 4 + 4 + 4 as a multiplication problem.

Solution: You can think of 4 + 4 + 4 as: 3 equal amounts of 4, or 3 × 4.

The multiplication problem 3 × 4 can be represented by **sets** of objects.

EXAMPLE 2-2: Represent the following multiplication problem using sets: 3 × 4

Solution

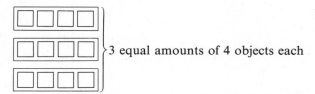

3 equal amounts of 4 objects each

The multiplication problem 3 × 4 can also be represented by a **rectangular array** of objects.

EXAMPLE 2-3: Represent the following multiplication problem using a rectangular array: 3 × 4

Solution

3 rows of 4 objects each

Note: In any case, 3 × 4 = 12 because: 4 + 4 + 4 = 8 + 4 = 12

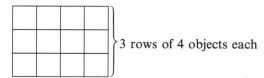

The number sentence $3 \times 4 = 12$ is called a **basic multiplication fact.** Basic multiplication can be written in horizontal form: $3 \times 4 = 12$. Or it can be written in vertical form: $\begin{array}{r} 4 \\ \times 3 \\ \hline 12 \end{array}$

In $\begin{array}{r} 4 \\ \times 3 \\ \hline 12 \end{array}$, 4 is called the **multiplicand,** 3 is called the **multiplier,** and both 3 and 4 can also be called **factors.**

12 is called the **product.**

To multiply whole numbers quickly and accurately, you must know the 100 basic multiplication facts from memory [see Solved Problem 2-2].

2-2. Multiplying by One Digit

To multiply a whole number with two or more digits by a one-digit whole number, you multiply in vertical form.

A. Multiply given vertical form.

To multiply by one digit given vertical form, you first multiply ones, then tens, then hundreds, and so on.

EXAMPLE 2-4: Multiply:  $\begin{array}{r} 512 \\ \times 3 \end{array}$ ⎬ $\longleftarrow$ vertical form

Solution

1. Multiply ones.

$3 \times 2 = 6$ (ones)

$$\begin{array}{r} \text{ones} \\ \downarrow \\ 5\ 1\ 2 \\ \times \qquad 3 \\ \hline 6 \\ \uparrow \\ \text{ones} \end{array}$$

2. Multiply tens.

$3 \times 1 = 3$ (tens)

$$\begin{array}{r} \text{tens} \\ \downarrow \\ 5\ 1\ 2 \\ \times \qquad 3 \\ \hline 3\ 6 \\ \uparrow \\ \text{tens} \end{array}$$

3. Multiply hundreds.

$3 \times 5 = 15$ (hundreds)

$$\begin{array}{r} \text{hundreds} \\ \downarrow \\ 5\ 1\ 2 \\ \times \qquad 3 \\ \hline 1\ 5\ 3\ 6 \\ \uparrow \\ \text{hundreds} \end{array} \longleftarrow \text{product}$$

B. Multiply in horizontal form.

To multiply given horizontal form, you must first change to vertical form.

EXAMPLE 2-5: Multiply 6043×2.   horizontal form

Solution

1. Change to vertical form.

$$\begin{array}{r} 6\ 0\ 4\ 3 \\ \times \qquad 2 \\ \hline \end{array}$$

2. Multiply as before.

$$\begin{array}{r} 6\ 0\ 4\ 3 \\ \times \qquad 2 \\ \hline 1\ 2{,}0\ 8\ 6 \end{array}$$

C. **Multiply with renaming.**

When multiplying, you must rename 2-digit values (like 35 ones, 18 tens, or 21 hundreds) unless it is the last step in the problem.

EXAMPLE 2-6: Rename the 2-digit values for the following multiplication problem: 437×5

Solution

1. Multiply ones.
$5 \times 7 = 35$ (ones)
Rename:
35 ones = 3 tens 5 ones

2. Multiply tens.
$5 \times 3 = 15$ (tens)
Add renaming:
$15 + 3 = 18$ (tens)
Rename:
18 tens = 1 hundred 8 tens

3. Multiply hundreds.
$5 \times 4 = 20$ (hundreds)
Add renaming:
$20 + 1 = 21$ (hundreds)
Last step—Don't rename!

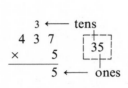

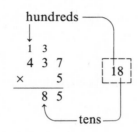

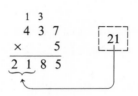

2-3. **Multiplying by Two or More Digits**

To multiply by two or more digits, you first multiply to get **partial products** and then add the partial products.

A. **Multiply by two digits.**

To multiply by two digits, you first multiply by ones and then by tens.

EXAMPLE 2-7: Multiply 625×43.

Solution

1. Multiply by ones.
$625 \times 3 = 1875$ (ones)

2. Multiply by tens.
$625 \times 4 = 2500$ (tens)

3. Add partial products.
$1875 + 25,000 = 26,875$

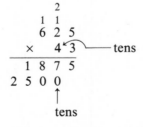

Note: It is very important to always write the first digit on the right of each partial product directly below the digit that you are multiplying by in the multiplier. That is, to multiply by the digit in the tens place, you write the first digit of that partial product in the tens place.

B. **Multiply by more than two digits.**

To multiply by more than two digits, you first find each partial product and then add the partial products.

EXAMPLE 2-8: Multiply 316 × 258.

Solution

1. Find each partial product.
   258 × 6 (ones) = 1548 (ones)
   258 × 1 (ten) = 258 (tens)
   258 × 3 (hundreds) = 774 (hundreds)

2. Add partial products.
   1548 + 2580 + 77,400 = 81,528

```
      1 2
      3 4
      2 5 8
    ×3 1 6  ──── hundreds
    ───────
    1 5 4 8
      2 5 8
    7 7 4
      ↑
    hundreds
```

```
      1 2
      3 4
      2 5 8
    ×3 1 6
    ───────
    1 5 4 8
      2 5 8
  + 7 7 4
  ─────────
    8 1,5 2 8
```

2-4. Shortcuts in Multiplying

To save time and effort while multiplying, you can use the following shortcuts.

A. Multiply whole numbers that end in one or more zeros.

When a whole-number factor ends in one or more zeros, you can use a shortcut by first grouping and bringing down those zeros before multiplying.

EXAMPLE 2-9: Multiply the following using a shortcut: 500 × 29,000

Solution

1. Group zeros together.

2. Shortcut! Bring down zeros.

3. Multiply as before.
   5 × 29 = 145

```
         zeros
         ⌒
     2 9|0 0 0
   ×   5|0 0
```

```
     2 9|0 0 0
   ×   5|0 0
   ───────────
         |0 0 0 0 0
```

```
          4
     2 9,0 0 0
   ×     5 0 0
   ─────────────
   1 4,5 0 0,0 0 0
```

B. Multiply by a whole number with one or more zeros in the middle.

When a whole-number factor has one or more zeros in the middle, you can use a shortcut by first writing that factor on the bottom as the multiplier.

EXAMPLE 2-10: Multiply the following using a shortcut: 658 × 7003 ⟵ factor with zeros in the middle

Solution

1. Multiply by ones.
   658 × 3 = 1974 (ones)

2. Shortcut! Multiply by zeros.
   658 × 0 = 0 (tens)
   658 × 0 = 0 (hundreds)

3. Multiply by thousands and add.
   658 × 7 = 4606 (thousands)
   1974 + 4,606,000 = 4,607,974

```
      1 2
      6 5 8
    ×7 0 0 3  ⟵ multiplier
    ───────
    1 9 7 4
        ↑
      ones
```

```
      1 2
      6 5 8
    ×7 0 0 3
    ───────
    1 9 7 4
      0 0
      ↑ ↑
      | hundreds
    tens
```

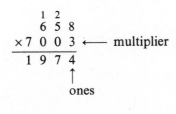

```
        4 5
        1 2
        6 5 8
      ×7 0 0 3  ──── thousands
      ───────
      1 9 7 4
    4 6 0 6 0 0
    ─────────────
    4,6 0 7,9 7 4
              ↑
          thousands
```

Writing the factor with zeros in the middle on the bottom as the multiplier is a shortcut because you have to write and add fewer partial products. Compare the shortcut method to the long method in the following examples.

$$
\begin{array}{r}
6\ 5\ 8 \\
\times 7\ 0\ 0\ 3 \quad\longleftarrow \text{ shortcut} \\
\hline
1\ 9\ 7\ 4 \\
4\ 6\ 0\ 6\ 0\ 0 \\
\hline
4{,}6\ 0\ 7{,}9\ 7\ 4
\end{array}
\Big\}\text{ two partial products}
$$

$$
\begin{array}{r}
7\ 0\ 0\ 3 \quad\longleftarrow \text{ long method} \\
\times \quad 6\ 5\ 8 \\
\hline
5\ 6\ 0\ 2\ 4 \\
3\ 5\ 0\ 1\ 5 \\
4\ 2\ 0\ 1\ 8 \\
\hline
4{,}6\ 0\ 7{,}9\ 7\ 4
\end{array}
\Big\}\text{ three partial products}
$$

2-5. Solving Word Problems $(+, -, \times)$

A. Choose the correct operation $(+, -, \times)$.

To join two or more amounts together, you add $(+)$ [see Section 1-2].
To take one amount away from another amount, you subtract $(-)$ [see Section 1-3].
To join two or more equal amounts together, you multiply $(\times)$ [see Section 2-1].

EXAMPLE 2-11: Choose the correct operation to solve these word problems.

(a) There are 6 balls in one can. There are 3 balls in another can. How many balls are there in all?

Understand: The question asks you to join amounts of 6 and 3 together.

Decide: To join amounts together, you **add.**

(b) There are 6 cans. There are 3 balls in each can. How many balls are there in all?

Understand: The question asks you to join 6 equal amounts of 3 together.

Decide: To join equal amounts together, you **multiply.**

B. Solve word problems $(+, -, \times)$.

To solve addition word problems, you find how much two or more amounts are all together.

To solve subtraction word problems, you find **(a)** how much is left;
(b) how much more is needed;
(c) the difference between two amounts.

To solve multiplication word problems, you find **(a)** how much two or more equal amounts are all together;
(b) the product of one amount times another amount.

EXAMPLE 2-12: Solve the following word problem using addition, subtraction, or multiplication.

1. *Identify:* The high school track is (1320 feet (ft)) around.   Circle each fact.
A one mile race is (4 times) around the track.
How many feet are in one mile?   Underline the question.

2. *Understand:* The question asks you to find 4 times 1320 ft.

3. *Decide:* To find the product of one amount times another amount, you multiply.

4. *Compute:* $4 \times 1320 = 5280$

5. *Interpret:* 5280 means that 4 times around the 1320-ft track is **5280 ft.** $\longleftarrow$ solution

RAISE YOUR GRADES

Can you . . . ?

☑ answer all 100 basic multiplication facts correctly in less than 4 minutes
☑ interpret multiplication facts in at least three different ways
☑ identify the three parts of a multiplication problem
☑ write a multiplication problem in vertical form given horizontal form
☑ determine when to rename when multiplying
☑ determine where to write each renaming
☑ determine where the partial products should be written when multiplying by two or more digits
☑ determine when a multiplication shortcut should be used
☑ multiply whole numbers that end in one or more zeros
☑ multiply by a whole number with one or more zeros in the middle
☑ determine whether to add, subtract, or multiply in a word problem
☑ interpret the number answer to a word problem in the words of the original problem

SUMMARY

1. Multiplication can be used to represent (a) repeated addition;
   (b) equal amounts of like objects;
   (c) rectangular arrays.

2. In $\frac{\begin{array}{r}4\\\times 3\end{array}}{12}$: (a) 4 is the multiplicand or factor;
   (b) 3 is the multiplier or factor;
   (c) 12 is the product.

3. To multiply by one digit given vertical form, you first multiply ones, then tens, then hundreds, and so on.

4. To multiply given horizontal form, you first change to vertical form.

5. When multiplying, you must rename 2-digit values (like 35 ones, 18 tens, or 21 hundreds) unless it is the last step.

6. To multiply by two digits, you first multiply by ones and then by tens.

7. To multiply by more than two digits, you find each partial product and then add all the partial products.

8. To multiply using a shortcut, one of the whole-number factors has to
   (a) end in one or more zeros;
   (b) have one or more zeros in the middle.

9. To choose the correct operation in problem-solving, you
   (a) add to join two or more amounts together;
   (b) subtract to take one amount away from another;
   (c) multiply to join equal amounts together.

10. To solve addition word problems, you find how much two or more amounts are all together.

11. To solve subtraction word problems, you find:
    (a) how much is left;
    (b) how much more is needed;
    (c) the difference between two amounts.

12. To solve multiplication word problems, you find:
    (a) how much two or more equal amounts are all together;
    (b) the product of one amount times another amount.

SOLVED PROBLEMS

PROBLEM 2-1   Write each of the following repeated addends as a multiplication problem:

(a) $3 + 3$ **(b)** $4 + 4$ **(c)** $5 + 5 + 5$ **(d)** $9 + 9 + 9$ **(e)** $2 + 2 + 2 + 2$
(f) $7 + 7 + 7 + 7$ **(g)** $8 + 8 + 8 + 8 + 8$ **(h)** $1 + 1 + 1 + 1 + 1$
(i) $0 + 0 + 0 + 0 + 0 + 0$ **(j)** $6 + 6 + 6 + 6 + 6 + 6 + 6$

Solution: Recall that you can think of $4 + 4 + 4$ as 3 equal amounts of 4 or, 3×4 [see Example 2-1]: **(a)** 2×3 **(b)** 2×4 **(c)** 3×5 **(d)** 3×9 **(e)** 4×2 **(f)** 4×7 **(g)** 5×8
(h) 5×1 **(i)** 6×0 **(j)** 7×6

PROBLEM 2-2   To practice the 100 basic multiplication facts:

1. Cover the solutions, one row at a time, with a blank sheet of paper.
2. Write your answer below each problem in that row on your blank sheet of paper.
3. Slide your answers down and compare them with the printed solutions.
4. Repeat any row in which you had an incorrect answer.
   [See Examples 2-2 and 2-3.]

2×4	1×9	1×6	5×3	1×5	6×0	7×2	2×8	0×3	3×9
8	9	6	15	5	0	14	16	0	27
5×0	4×8	8×8	5×9	1×1	7×6	0×5	9×3	7×7	9×1
0	32	64	45	1	42	0	27	49	9
8×9	8×0	5×8	3×1	9×5	0×0	9×7	1×3	4×6	3×2
72	0	40	3	45	0	63	3	24	6
7×5	0×8	3×7	6×4	0×1	6×9	3×0	7×3	1×2	6×2
35	0	21	24	0	54	0	21	2	12
5×5	0×7	3×5	6×5	4×3	8×2	5×1	5×7	1×7	4×2
25	0	15	30	12	16	5	35	7	8
2×6	9×9	5×4	7×9	0×4	6×1	1×8	3×8	3×6	7×1
12	81	20	63	0	6	8	24	18	7
6×7	2×3	8×6	4×4	2×0	9×2	8×4	7×8	2×1	2×9
42	6	48	16	0	18	32	56	2	18
7×4	3×3	4×0	9×6	5×2	0×9	5×6	1×4	2×5	8×5
28	9	0	54	10	0	30	4	10	40
8×1	1×0	4×1	6×8	4×7	8×7	4×5	9×4	9×0	4×9
8	0	4	48	28	56	20	36	0	36

0	6	9	0	3	7	2	8	6	2
$\times 2$	$\times 3$	$\times 8$	$\times 6$	$\times 4$	$\times 0$	$\times 7$	$\times 3$	$\times 6$	$\times 2$
0	18	72	0	12	0	14	24	36	4

PROBLEM 2-3   Multiply given vertical form:

(a)	34	(b)	21	(c)	70	(d)	61	(e)	132	(f)	412	(g)	1687
	$\times 2$		$\times 6$		$\times 5$		$\times 9$		$\times 3$		$\times 4$		$\times 1$

(h)	3010	(i)	40,132	(j)	96,584
	$\times 8$		$\times 2$		$\times 0$

Solution: Recall that to multiply by one digit, you first multiply ones, then tens, then hundreds, and so on [see Example 2-5]:

(a)	34	(b)	21	(c)	70	(d)	61	(e)	132	(f)	412	(g)	1687
	$\times 2$		$\times 6$		$\times 5$		$\times 9$		$\times 3$		$\times 4$		$\times 1$
	68		126		350		549		396		1648		1687

(h)	3010	(i)	40,132	(j)	96,584
	$\times 8$		$\times 2$		$\times 0$
	24,080		80,264		0

PROBLEM 2-4   Multiply given horizontal form: **(a)** 3×12 **(b)** 2×34 **(c)** 9×81
(d) 7×60 **(e)** 4×102 **(f)** 2×824 **(g)** 1×8519 **(h)** 0×7258 **(i)** $2 \times 24,013$
(j) $3 \times 91,203$

Solution: Recall that to multiply given horizontal form, you first change to vertical form [see Example 2-6]:

(a)	12	(b)	34	(c)	81	(d)	60	(e)	102	(f)	824	(g)	8519
	$\times 3$		$\times 2$		$\times 9$		$\times 7$		$\times 4$		$\times 2$		$\times 1$
	36		68		729		420		408		1648		8519

(h)	7258	(i)	24,013	(j)	91,203
	$\times 0$		$\times 2$		$\times 3$
	0		48,026		273,609

PROBLEM 2-5   Multiply when renaming is necessary: **(a)** 3×25 **(b)** 2×319 **(c)** 8×591
(d) 6×487 **(e)** 4×2018 **(f)** 7×6051 **(g)** 9×2900 **(h)** 2×5186 **(i)** 3×4815
(j) 5×6480 **(k)** 4×3859 **(l)** $5 \times 20,419$ **(m)** $8 \times 32,584$

Solution: Recall that you must rename each 2-digit value when multiplying unless it is the last step [see Example 2-7]:

(a)	$\overset{1}{25}$	(b)	$\overset{1}{319}$	(c)	$\overset{7}{591}$	(d)	$\overset{54}{487}$	(e)	$\overset{3}{2018}$	(f)	$\overset{3}{6051}$	(g)	$\overset{8}{2900}$
	$\times 3$		$\times 2$		$\times 8$		$\times 6$		$\times 4$		$\times 7$		$\times 9$
	75		638		4728		2922		8072		42,357		26,100

(h)	$\overset{11}{5186}$	(i)	$\overset{2\,1}{4815}$	(j)	$\overset{24}{6480}$	(k)	$\overset{323}{3859}$	(l)	$\overset{2\,4}{20,419}$	(m)	$\overset{24\,63}{32,584}$
	$\times 2$		$\times 3$		$\times 5$		$\times 4$		$\times 5$		$\times 8$
	10,372		14,445		32,400		15,436		102,095		260,672

PROBLEM 2-6 Multiply by two digits: **(a)** 12×34 **(b)** 15×23 **(c)** 41×58
(d) 53×26 **(e)** 23×132 **(f)** 14×513 **(g)** 72×645 **(h)** 87×905 **(i)** 96×513
(j) 75×486

Solution: Recall that you first multiply by ones and then by tens [see Example 2-8]:

(a)
```
     34
   ×12
   ─────
     68
   34
   ─────
    408
```

(b)
```
    1
     23
   ×15
   ─────
    115
    23
   ─────
    345
```

(c)
```
     3
     58
   ×41
   ─────
     58
    232
   ─────
   2378
```

(d)
```
    3
    1
     26
   ×53
   ─────
     78
    130
   ─────
   1378
```

(e)
```
     132
   ×  23
   ─────
     396
     264
   ─────
    3036
```

(f)
```
     1
     513
   ×  14
   ─────
    2052
     513
   ─────
    7182
```

(g)
```
      33
       1
     645
   ×  72
   ─────
    1 290
    45 15
   ─────
   46,440
```

(h)
```
      4
      3
     905
   ×  87
   ─────
    6 335
    72 40
   ─────
   78,735
```

(i)
```
      12
       1
     513
   ×  96
   ─────
    3 078
    46 17
   ─────
   49,248
```

(j)
```
      64
      43
     486
   ×  75
   ─────
    2 430
    34 02
   ─────
   36,450
```

PROBLEM 2-7 Multiply by more than two digits: **(a)** 213×123 **(b)** 312×234
(c) 412×287 **(d)** 351×386 **(e)** 485×285 **(f)** 649×782 **(g)** 321×3021
(h) 328×4719 **(i)** 4213×4232 **(j)** 3849×5764

Solution: Recall that to multiply by more than two digits, you first find each partial product and then add the partial products [see Example 2-9]:

(a)
```
     1 2 3
   ×2 1 3
   ───────
     3 6 9
   1 2 3
  2 4 6
  ───────
  2 6,1 9 9
```

(b)
```
   1 1
   2 3 4
  ×3 1 2
  ───────
   4 6 8
   2 3 4
  7 0 2
  ───────
  7 3,0 0 8
```

(c)
```
  3 2
  1 1
  2 8 7
 ×4 1 2
 ───────
  5 7 4
  2 8 7
1 1 4 8
 ───────
1 1 8,2 4 4
```

(d)
```
  2 1
  4 3
  3 8 6
 ×3 5 1
 ───────
  3 8 6
 1 9 3 0
1 1 5 8
 ───────
1 3 5,4 8 6
```

(e)
```
  3 2
  6 4
  4 2
  2 8 5
 ×4 8 5
 ───────
 1 4 2 5
 2 2 8 0
1 1 4 0
 ───────
1 3 8,2 2 5
```

(f)
```
  4 1
  3
  7 1
  7 8 2
 ×6 4 9
 ───────
 7 0 3 8
 3 1 2 8
4 6 9 2
 ───────
5 0 7,5 1 8
```

(g)
```
  3 0 2 1
 ×  3 2 1
 ─────────
  3 0 2 1
  6 0 4 2
 9 0 6 3
 ─────────
 9 6 9,7 4 1
```

(h)
```
  2   2
  1   1
  5 1 7
  4 7 1 9
 ×  3 2 8
 ─────────
  3 7 7 5 2
  9 4 3 8
1 4 1 5 7
 ─────────
1,5 4 7,8 3 2
```

(i)
```
     1
  4 2 3 2
 ×4 2 1 3
 ─────────
 1 2 6 9 6
 4 2 3 2
 8 4 6 4
1 6 9 2 8
 ─────────
1 7,8 2 9,4 1 6
```

(j)
```
  2 1 1
  6 5 3
  3 2 1
  6 5 3
  5 7 6 4
 ×3 8 4 9
 ─────────
 5 1 8 7 6
 2 3 0 5 6
 4 6 1 1 2
1 7 2 9 2
 ─────────
2 2,1 8 5,6 3 6
```

PROBLEM 2-8 Multiply using a shortcut: **(a)** 20×36 **(b)** 45×80 **(c)** 60×90
(d) 300×58 **(e)** 700×320 **(f)** 90×400 **(g)** 500×800 **(h)** 7000×258
(i) 900×3500 **(j)** 3000×6000

Solution: Recall that when a whole-number factor ends in one or more zeros, you can use a shortcut by first grouping and bringing down the zeros before multiplying [see Example 2-10]:

(a)
$$\begin{array}{r} \overset{1}{36} \\ \times\ 20 \\ \hline 720 \end{array}$$
(b)
$$\begin{array}{r} \overset{4}{45} \\ \times\ 80 \\ \hline 3600 \end{array}$$
(c)
$$\begin{array}{r} 90 \\ \times 60 \\ \hline 5400 \end{array}$$
(d)
$$\begin{array}{r} \overset{2}{5}8 \\ \times\ 300 \\ \hline 17{,}400 \end{array}$$
(e)
$$\begin{array}{r} \overset{1}{32}0 \\ \times\ 700 \\ \hline 224{,}000 \end{array}$$
(f)
$$\begin{array}{r} 400 \\ \times 90 \\ \hline 36{,}000 \end{array}$$

(g)
$$\begin{array}{r} 800 \\ \times 500 \\ \hline 400{,}000 \end{array}$$
(h)
$$\begin{array}{r} \overset{45}{258} \\ \times\ 7000 \\ \hline 1{,}806{,}000 \end{array}$$
(i)
$$\begin{array}{r} \overset{4}{3500} \\ \times\ 900 \\ \hline 3{,}150{,}000 \end{array}$$
(j)
$$\begin{array}{r} 6000 \\ \times 3000 \\ \hline 18{,}000{,}000 \end{array}$$

PROBLEM 2-9 Multiply using a shortcut: **(a)** 203×345 **(b)** 502×420 **(c)** 601×500
(d) 2043×3271 **(e)** 3208×4526 **(f)** 4005×8216 **(g)** 5007×9250 **(h)** 6002×4500
(i) 8003×2000 **(j)** $87{,}256 \times 100{,}001$

Solution: Recall that when a whole-number factor has one or more zeros in the middle, you can use a shortcut by first writing that factor on the bottom as the multiplier [see Example 2-11]:

(a)
$$\begin{array}{r} \overset{1}{\overset{11}{345}} \\ \times 203 \\ \hline 1\,035 \\ 69\,00 \\ \hline 70{,}035 \end{array}$$
(b)
$$\begin{array}{r} \overset{1}{420} \\ \times 502 \\ \hline 840 \\ 210\,0 \\ \hline 210{,}840 \end{array}$$
(c) (shortest method)
$$\begin{array}{r} 601 \\ \times\ 500 \\ \hline 300{,}500 \end{array}$$
(d)
$$\begin{array}{r} \overset{1}{\overset{12}{\overset{2}{3271}}} \\ \times 2043 \\ \hline 9\,813 \\ 130\,84 \\ 6\,542\,0 \\ \hline 6{,}682{,}653 \end{array}$$
(e)
$$\begin{array}{r} \overset{1\ 1}{\overset{1\ 1}{\overset{424}{4526}}} \\ \times 3208 \quad \text{thousands} \\ \hline 36\,208 \\ 905\,20 \\ \hline 13{,}578 \quad \text{Careful!} \\ \hline 14{,}519{,}408 \end{array}$$

$\uparrow$
thousands

(f)
$$\begin{array}{r} \overset{2}{\overset{1\ 3}{8216}} \\ \times 4005 \\ \hline 41\,080 \\ 32\,864\,00 \\ \hline 32{,}905{,}080 \end{array}$$
(g)
$$\begin{array}{r} \overset{12}{\overset{13}{9250}} \\ \times 5007 \\ \hline 64\,750 \\ 46\,250\,0 \\ \hline 46{,}314{,}750 \end{array}$$
(h)
$$\begin{array}{r} \overset{3}{\overset{1}{4500}} \\ \times 6002 \\ \hline 9\,000 \\ 27\,000 \\ \hline 27{,}009{,}000 \end{array}$$
(i)
$$\begin{array}{r} 8003 \\ \times\ 2000 \\ \hline 16{,}006{,}000 \end{array}$$
(j)
$$\begin{array}{r} 87{,}256 \\ \times 100{,}001 \\ \hline 87\,256 \\ 8\,725\,600\,00 \\ \hline 8{,}725{,}687{,}256 \end{array}$$

PROBLEM 2-10 Decide what to do (add, subtract, or multiply) and then solve the problem:

(a) There are 42 people in one line. There are 6 people in another line. How many people are in line all together?

(b) There are 28 people in the school bus. There are 9 people in the school van. How many more people are in the bus than in the van?

(c) There are 35 people at a party. By midnight 16 people had gone home. How many people were left at the party then?

(d) One tank holds 72 gallons. Another tank holds 56 gallons. What is the difference in the capacity of the two tanks?

(e) There are 12 eggs in one dozen. There are 12 dozen eggs in one flat. How many single eggs are in one flat?

(f) It takes 5 minutes for Chuck to walk once around a certain track. How long will it take Chuck to walk around the track 20 times?

(g) A certain drive-in theatre admitted 520 cars on Saturday night. If each car contained 3 people, how many people attended the drive-in on Saturday night?

(h) Each minute is 60 seconds long. Each hour is 60 minutes long. How many seconds are in one hour? How many seconds does a person work in an 8-hour work day?

(i) There are 8 boxes that contain 12 cans each. If 18 cans are sold, how many cans are left in the 8 boxes? (Careful! This problem takes more than one operation to solve.)

(j) Last week Emily ran 6 miles each week day and 10 miles on each weekend day. How many total miles did Emily run last week?

Solution

(a) add: $42 + 6 = 48$ (people)

(b) subtract: $28 - 9 = 19$ (people)

(c) subtract: $35 - 16 = 19$ (people)

(d) subtract: $72 - 56 = 16$ (gallons)

(e) multiply: $12 \times 12 = 144$ (eggs)

(f) multiply: $5 \times 20 = 100$ (minutes)

(g) multiply: $3 \times 520 = 1560$ (people)

(h) multiply: $60 \times 60 = 3600$ (seconds); multiply again: $8 \times 3600 = 28{,}800$ (seconds)

(i) multiply and then subtract: $8 \times 12 = 96$ and $96 - 18 = 78$ (cans)

(j) multiply, multiply again, and then add: $5 \times 6 = 30$, $2 \times 10 = 20$, and $30 + 20 = 50$ (miles) [see Section 2-5].

Supplementary Exercises

PROBLEM 2-11   Multiply basic facts in vertical form:

Row (a)

6	2	3	9	0	9	4	4	4	8
×6	×7	×4	×8	×2	×0	×5	×7	×1	×1

Row (b)

2	5	5	4	7	2	8	2	8	6
×5	×6	×2	×0	×4	×1	×4	×0	×6	×7

Row (c)

3	1	0	5	2	1	5	4	3	5
×6	×8	×4	×4	×6	×7	×1	×3	×5	×5

Row (d)

1	3	0	3	7	4	9	9	5	8
×2	×0	×1	×7	×5	×6	×7	×5	×8	×9

Row (e)

7	0	1	8	5	0	7	1	1	2
×7	×5	×1	×8	×0	×3	×2	×5	×6	×4

Row (f)

2	8	7	0	6	4	9	8	6	1
×2	×3	×0	×6	×3	×9	×4	×7	×8	×0

Row (g)

8	1	0	9	3	2	7	9	4	2
×5	×4	×9	×6	×3	×9	×8	×2	×4	×3

Row (h)

7	3	6	7	9	4	5	8	6	0
×1	×8	×1	×9	×9	×2	×7	×2	×5	×7

Row (i)

6	7	6	6	0	3	1	0	3	8
×2	×3	×9	×4	×8	×2	×3	×0	×1	×0

Row (j)	9	9	7	5	4	3	2	6	5	1
	×1	×3	×6	×9	×8	×9	×8	×0	×3	×9

PROBLEM 2-12   Multiply basic facts in horizontal form:

Row (a)	4 × 2	0 × 5	9 × 8	5 × 7	5 × 5	6 × 2	7 × 6	4 × 7	1 × 8	2 × 0
Row (b)	9 × 1	8 × 4	0 × 8	9 × 9	8 × 0	3 × 2	7 × 0	3 × 3	0 × 1	3 × 6
Row (c)	6 × 1	8 × 8	8 × 5	7 × 3	5 × 3	4 × 5	6 × 8	0 × 4	1 × 4	8 × 9
Row (d)	3 × 5	9 × 5	1 × 3	4 × 6	5 × 6	9 × 7	4 × 4	6 × 9	8 × 6	6 × 0
Row (e)	5 × 1	1 × 1	5 × 9	1 × 0	3 × 4	4 × 0	0 × 2	2 × 5	7 × 4	4 × 3
Row (f)	0 × 6	6 × 7	0 × 0	9 × 6	2 × 8	1 × 6	2 × 9	9 × 0	7 × 8	0 × 7
Row (g)	2 × 7	5 × 0	7 × 9	0 × 3	1 × 5	8 × 1	4 × 8	6 × 5	5 × 4	7 × 2
Row (h)	8 × 2	3 × 9	3 × 1	3 × 7	7 × 5	8 × 3	8 × 7	4 × 1	4 × 9	3 × 8
Row (i)	3 × 0	7 × 7	6 × 4	2 × 1	7 × 1	6 × 3	1 × 2	5 × 2	0 × 9	6 × 6
Row (j)	9 × 3	1 × 9	2 × 3	2 × 6	2 × 4	1 × 7	9 × 2	5 × 8	9 × 4	2 × 2

PROBLEM 2-13   Multiply by one digit: **(a)** 2 × 43   **(b)** 9 × 38   **(c)** 2 × 80,143
(d) 5 × 64,980   **(e)** 4 × 21   **(f)** 3 × 408   **(g)** 7 × 10,231   **(h)** 46,105 × 8   **(i)** 6 × 51
(j) 671 × 5   **(k)** 5847 × 6   **(l)** 7 × 80   **(m)** 5 × 211   **(n)** 9 × 352   **(o)** 501 × 3
(p) 7891 × 4   **(q)** 8 × 3102   **(r)** 1 × 7685   **(s)** 5810 × 6   **(t)** 7 × 7059   **(u)** 0 × 6528
(v) 4 × 9182   **(w)** 8739 × 8

PROBLEM 2-14   Multiply by two or more digits: **(a)** 32 × 41   **(b)** 312 × 132   **(c)** 63 × 75
(d) 314 × 483   **(e)** 29 × 580   **(f)** 485 × 639   **(g)** 53 × 978   **(h)** 725 × 8169
(i) 18 × 39   **(j)** 238 × 432   **(k)** 14 × 512   **(l)** 693 × 528   **(m)** 45 × 803
(n) 213 × 3102   **(o)** 581 × 657   **(p)** 16 × 715   **(q)** 512 × 3018   **(r)** 85 × 608
(s) 295 × 347

PROBLEM 2-15   Multiply using shortcuts: **(a)** 30 × 45   **(b)** 304 × 581   **(c)** 62 × 70
(d) 340 × 602   **(e)** 40 × 50   **(f)** 705 × 400   **(g)** 600 × 72   **(h)** 3058 × 6143
(i) 900 × 450   **(j)** 5106 × 2432   **(k)** 70 × 800   **(l)** 6003 × 5193   **(m)** 500 × 300
(n) 5820 × 4002   **(o)** 357 × 2000   **(p)** 3600 × 5002   **(q)** 2700 × 600   **(r)** 6000 × 7009
(s) 8000 × 4000   **(t)** 200,005 × 35,600

PROBLEM 2-16   Solve word problems (+, −, ×):

(a) There are 83 people in one plane. There are 112 people in another plane. How many people are in both planes all together?

(b) Mount McKinley is 6194 meters high. Mount Everest is 8848 meters high. How much higher is Mount Everest than Mount McKinley?

(c) It is 450 miles from San Diego to San Jose. How far is it to make the round trip?

(d) Juan earns $8 per hour. How much does he earn during a 40-hour work week?

(e) Sam is 150 cm tall. Sue is 162 cm tall. What is the difference in their heights?

(f) Carol has $359. She bought food for $178. How much did Carol have left then?

(g) There are 50 pennies in each roll. There are 25 rolls all together. How many pennies in all?

(h) Ben works 8 hours for 5 days in a row. He then works 6 hours for 2 days in a row. How many hours did Ben work in the 7 days?

(i) Jane is 16 years old. Fred is 18 years old. Alice is twice as old as Jane and Fred are together. How old is Alice?

(j) At Joe's Store, carrots are 49¢ a bunch and lettuce is 69¢ a head. How much are 8 bunches of carrots and 7 heads of lettuce?

Answers to Supplementary Exercises

(2-11)

Row (a)	36	14	12	72	0	0	20	28	4	8
Row (b)	10	30	10	0	28	2	32	0	48	42
Row (c)	18	8	0	20	12	7	5	12	15	25
Row (d)	2	0	0	21	35	24	63	45	40	72
Row (e)	49	0	1	64	0	0	14	5	6	8
Row (f)	4	24	0	0	18	36	36	56	48	0
Row (g)	40	4	0	54	9	18	56	18	16	6
Row (h)	7	24	6	63	81	8	35	16	30	0
Row (i)	12	21	54	24	0	6	3	0	3	0
Row (j)	9	27	42	45	32	27	16	0	15	9

(2-12)

Row (a)	8	0	72	35	25	12	42	28	8	0
Row (b)	9	32	0	81	0	6	0	9	0	18
Row (c)	6	64	40	21	15	20	48	0	4	72
Row (d)	15	45	3	24	30	63	16	54	48	0
Row (e)	5	1	45	0	12	0	0	10	28	12
Row (f)	0	42	0	54	16	6	18	0	56	0
Row (g)	14	0	63	0	5	8	32	30	20	14
Row (h)	16	27	3	21	35	24	56	4	36	24
Row (i)	0	49	24	2	7	18	2	10	0	36
Row (j)	27	9	6	12	8	7	18	40	36	4

(2-13) **(a)** 86   **(b)** 342   **(c)** 160,286   **(d)** 324,900   **(e)** 84   **(f)** 1224   **(g)** 71,617 **(h)** 368,840   **(i)** 306   **(j)** 3355   **(k)** 35,082   **(l)** 560   **(m)** 1055   **(n)** 3168   **(o)** 1503 **(p)** 31,564   **(q)** 24,816   **(r)** 7685   **(s)** 34,860   **(t)** 49,413   **(u)** 0   **(v)** 36,728 **(w)** 69,912

(2-14) **(a)** 1312   **(b)** 41,184   **(c)** 4725   **(d)** 151,662   **(e)** 16,820   **(f)** 309,915 **(g)** 51,834   **(h)** 5,922,525   **(i)** 702   **(j)** 102,816   **(k)** 7168   **(l)** 365,904   **(m)** 36,135 **(n)** 660,726   **(o)** 381,717   **(p)** 11,440   **(q)** 1,545,216   **(r)** 51,680   **(s)** 102,365

(2-15) **(a)** 1350   **(b)** 176,624   **(c)** 4340   **(d)** 204,680   **(e)** 2000   **(f)** 282,000 **(g)** 43,200   **(h)** 18,785,294   **(i)** 405,000   **(j)** 12,417,792   **(k)** 56,000   **(l)** 31,173,579 **(m)** 150,000   **(n)** 23,291,640   **(o)** 714,000   **(p)** 18,007,200   **(q)** 1,620,000 **(r)** 42,054,000   **(s)** 32,000,000   **(t)** 7,120,178,000

(2-16) **(a)** 195 people   **(b)** 2654 meters   **(c)** 900 miles   **(d)** $320   **(e)** 12 cm   **(f)** $181 **(g)** 1250 cents or $12.50   **(h)** 52 hours   **(i)** 68 years old   **(j)** $8.75

3 WHOLE NUMBERS: DIVISION

THIS CHAPTER IS ABOUT

☑ **Dividing Basic Facts**
☑ **Dividing by One Digit**
☑ **Dividing by Two Digits**
☑ **Dividing by Three or More Digits**
☑ **Factoring Whole Numbers**
☑ **Solving Word Problems (+, −, ×, ÷)**

3-1. Dividing Basic Facts

You should recall that to join two or more equal amounts together, you multiply. But what should be done when you want to do the opposite of multiply? That is, what should be done when you want to separate an amount into two or more equal amounts?

To separate an amount into two or more equal amounts, you **divide.**

EXAMPLE 3-1: Write "□□□□□□ separated into equal amounts of □□□" as division.

Solution: □□□□□□ separated into equal amounts of □□□ equals: 6 ÷ 3 ⟵ division

The **division problem** 6 ÷ 3 is read as "six **divided by** three." The symbol ÷ is called a **division sign** or **division symbol.** The following example shows one way to solve a division problem.

EXAMPLE 3-2: 6 ÷ 3 = 2 because

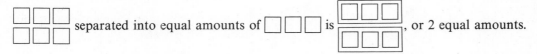 separated into equal amounts of □□□ is [ ], or 2 equal amounts.

The number sentence 6 ÷ 3 = 2 is called a **basic division fact.** To divide quickly and accurately, you must know the 90 basic division facts from memory [see Solved Problem 3-2]. The basic division fact 6 ÷ 3 = 2 is written in horizontal form. Basic division facts can also be written in **division box form.**

EXAMPLE 3-3: Write 6 ÷ 3 = 2 in division box form.

Solution:
$$3\overline{)6}^{\,2} \longleftarrow \text{the first number in } 6 \div 3 = 2 \text{ goes in the division box}$$
↑
the second number in 6 ÷ 3 = 2 goes in front of the division box

In 6 ÷ 3 = 2, or $3\overline{)6}^{\,2}$, 3 is the **divisor**, 6 is the **dividend**, and 2 is the **quotient.**

Every basic division fact can be related to a basic multiplication fact, and vice versa.

EXAMPLE 3-4: Write the related basic multiplication fact for $6 \div 3 = 2$.

Solution: $6 \div 3 = 2$ because $2 \times 3 = 6$ ⟵ related basic multiplication fact

The division problem $6 \div 3 = ?$ and the related **missing factor problem** $? \times 3 = 6$ are two different ways of stating the same problem. The answer to both problems is 2.

EXAMPLE 3-5: Write $6 \div 3 = ?$ and $? \times 3 = 6$ in words.

Solution: $6 \div 3 = ?$ asks: If you separate 6 into equal amounts of 3 each, how many equal amounts are there in all?

$? \times 3 = 6$ asks: How many equal amounts of 3 each are needed to make 6 all together?

If a division problem is given, you can always write the related missing factor problem.

EXAMPLE 3-6: Write the related missing factor problem for $8 \div 2 = ?$.

Solution: $8 \div 2 = ?$ means $? \times 2 = 8$ ⟵ related missing factor problem

To help find a basic division fact, you can use the related missing factor problem and your multiplication facts.

EXAMPLE 3-7: Use a missing factor problem to solve the following division problem: $42 \div 7$

Solution: $42 \div 7 = ?$ means $? \times 7 = 42$ ⟵ missing factor problem

and $\mathbf{6} \times 7 = 42$ ⟵ basic multiplication fact

so $42 \div 7 = \mathbf{6}$ ⟵ related basic division fact

3-2. Dividing by One Digit

To divide by a one-digit whole number when there are two or more digits in the dividend, you always start the division process by trying to divide into the first digit in the dividend.

A. Find a one-digit quotient with remainder.

When a divisor does not divide a dividend exactly, there will always be a nonzero **remainder**.

EXAMPLE 3-8: Divide $35 \div 4$. ⟵ horizontal form

Solution

1. Write division box form.

2. Divide tens? No. There are no fours in 3.

3. Divide ones? Yes. How many fours in 35 ones? Are there 7? Yes: $4 \times 7 = 28$ Are there 8? Yes: $4 \times 8 = 32$ Are there 9? No: $4 \times 9 = 36$ ⟵ too big There are **8** fours in 35.

$$4 \overline{)35}$$

No!
$$4 \overline{)35}$$
↑
tens

$$\overset{8}{4 \overline{)35}}$$
↑
ones

4. Multiply.   **5.** Subtract.   **6.** Compare.   **7.** Write the remainder.

$4 \times 8 = 32$ $35 - 32 = 3$ Is the remainder 3 less than the divisor 4? Yes: $3 < 4$, so 8 is correct.

$$
4\overline{)3\ 5} \quad\quad 4\overline{)\ 3\ 5} \quad\quad \boxed{3<}\ 4\overline{)\ 3\ 5} \xleftarrow{} \text{correct} \quad\quad 4\overline{)\ 3\ 5}\ \uparrow
$$

(columns showing: 8 over $4)35$, 32; 8 over $4)35$, -32, 3; $8 \leftarrow$ correct over $4)35$, -32, 3; $8\ R3$ over $4)35$, -32, 3 remainder)

Note: In $35 \div 4 = 8\ R3$, 8 is the quotient, 3 is the remainder, 8 R3 is the **division answer,** and R is the symbol for "remainder."

To check a division answer when the remainder is not zero, you multiply the proposed quotient by the divisor and then add the remainder.

EXAMPLE 3-9: Check $35 \div 4 = 8\ R3$ $\longleftarrow$ proposed division answer

Solution:

$$
\begin{array}{r}
8 \leftarrow \text{quotient} \\
\times 4 \leftarrow \text{divisor} \\
\hline
3\ 2 \\
+\ \ 3 \leftarrow \text{remainder} \\
\hline
3\ 5 \leftarrow 8\ R3\ \text{checks}
\end{array}
$$

Multiply the proposed quotient (8) by the divisor (4) and then add the remainder (3) to see if you get the original dividend (35).

Caution: For a division answer to be correct, the remainder must always be less than the divisor.

EXAMPLE 3-10: Find the error in the following step-by-step solution.

1. Divide.   **2.** Multiply.   **3.** Subtract.   **4.** Write remainder.   **5.** Check.

$$
8\overline{)5\ 6} \quad 8\overline{)5\ 6}\ (48) \quad 8\overline{)\ 5\ 6}\ (-48)\ 8 \quad 8\overline{)\ 5\ 6}\ (-48)\ 8 \uparrow
$$

(columns: 6 over 8)56; 6 over 8)56, 48; 6 over 8)56, −48, 8; 6 R8 over 8)56, −48, 8; Check: $6 \leftarrow$ quotient, $\times 8 \leftarrow$ divisor, 48, $+\ 8 \leftarrow$ remainder, $5\ 6 \leftarrow 6\ R8$ checks?)

Solution: The error is in step 1, but becomes obvious in step 4:

$$
6\ R8 \leftarrow \text{wrong (the remainder is not less than the divisor)} \qquad 7 \leftarrow \text{correct}
$$
$$
8\overline{)5\ 6} \qquad\qquad\qquad 8\overline{)5\ 6} \quad \text{quotient}
$$

Note: In Example 3-10, the error would have been caught following the "Subtract" step if the "Compare" step had not been omitted. **NEVER omit the "Compare" step!**

EXAMPLE 3-11: Show how the "Compare" Step will catch the error in Example 3-10.

Solution

1. Divide.   **2.** Multiply.   **3.** Subtract.   **4.** Compare.

$$
8\overline{)5\ 6} \quad 8\overline{)5\ 6} \quad 8\overline{)\ 5\ 6}\ (-48)\ 8 \quad \boxed{8=}\ 8\overline{)\ 5\ 6}
$$

(columns: 6 over 8)56; 6 over 8)56; 6 over 8)56, −48, 8; $6 \leftarrow$ wrong over 8)56)

Think: 8 (the remainder) is not less than 8 (the divisor) which means 6 is not correct—it has to be larger. So you increase the quotient 6 by one to 7 and start the division process over again.

The correct solution to 56 ÷ 8 is found as follows:

1. Divide, multiply, and subtract.

$$
\begin{array}{r}
7 \\
8\overline{)\,5\ 6} \\
-5\ 6 \\
\hline
0
\end{array}
$$

2. Compare.

$$
\boxed{0<} \quad
\begin{array}{r}
7 \;\longleftarrow\; \text{correct} \\
8\overline{)\,5\ 6} \\
-5\ 6 \\
\hline
0
\end{array}
$$

3. Write the remainder.

$$
\begin{array}{r}
7 \text{ R0 or just } 7 \\
8\overline{)\,5\ 6} \\
-5\ 6 \\
\hline
0
\end{array}
$$

4. Check.

$$
\begin{array}{r}
7 \;\longleftarrow\; \text{quotient} \\
\times\,8 \;\longleftarrow\; \text{divisor} \\
\hline
5\ 6 \;\longleftarrow\; 7 \text{ checks}
\end{array}
$$

Note 1: When the remainder is zero, it is not necessary to write the remainder as part of the division answer: 7 R0 = 7

Note 2: To check a division answer when the remainder is zero, you multiply the proposed quotient by the divisor to see if you get the original dividend [see step 4 in the previous example].

Note 3: When there is one digit in the divisor, the six basic division steps are
1. Divide **2.** Multiply **3.** Subtract **4.** Compare **5.** Write the remainder (if not zero) **6.** Check.

B. Find two or more digits in the quotient.

To divide when there are two or more digits in the quotient, you must **bring down** one or more digits from the dividend.

EXAMPLE 3-12: Divide the following: 184 ÷ 5

Solution

1. Divide hundreds? No. There are no fives in 1.

No!
$$5\overline{)\,1\ 8\ 4}$$
↑
hundreds

2. Divide tens? Yes.

$$
\begin{array}{r}
3 \\
5\overline{)\,15}
\end{array}
\text{ means }
\begin{array}{r}
\text{about 3} \\
5\overline{)\,18}
\end{array}
$$

tens
↓
$$
\begin{array}{r}
3 \\
5\overline{)\,1\ 8\ 4}
\end{array}
$$
↑
tens

3. Multiply, subtract and compare.

correct
↓
$$
\boxed{3<} \quad
\begin{array}{r}
3 \\
5\overline{)\,1\ 8\ 4} \\
-1\ 5 \\
\hline
3
\end{array}
$$
↑
tens

4. Bring down:
3 tens 4 ones = 34 ones

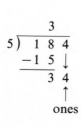

$$
\begin{array}{r}
3 \\
5\overline{)\,1\ 8\ 4} \\
-1\ 5\ \downarrow \\
\hline
3\ 4
\end{array}
$$
↑
ones

5. Divide ones? Yes.

$$
\begin{array}{r}
6 \\
5\overline{)\,30}
\end{array}
\text{ means }
\begin{array}{r}
\text{about 6} \\
5\overline{)\,34}
\end{array}
$$

ones
↓
$$
\begin{array}{r}
3\ 6 \\
5\overline{)\,1\ 8\ 4} \\
-1\ 5 \\
\hline
3\ 4
\end{array}
$$
Think: $5\overline{)\,34}$

6. Multiply, subtract, compare, and write the remainder.

correct
↓
$$
\boxed{4<} \quad
\begin{array}{r}
3\ 6 \text{ R4} \\
5\overline{)\,1\ 8\ 4} \\
-1\ 5 \\
\hline
3\ 4 \\
-3\ 0 \\
\hline
4
\end{array}
$$

7. Check.

```
       3
     3 6  ⟵ quotient
   ×   5  ⟵ divisor
   ───────
   1·8 0
   +     4  ⟵ remainder
   ───────
   1 8 4  ⟵ 36 R4 checks
```

Note: When there are no more digits to bring down from the dividend and the remainder is less than the divisor, you write the remainder (if not zero) as part of the division answer and then check.

C. Divide with zeros in the quotient.

When you cannot divide after bringing down a digit from the dividend, write a zero above that digit in the quotient.

EXAMPLE 3-13: Divide the following: $617 \div 3$

Solution

1. Divide hundreds. **2.** Bring down tens. **3.** Divide tens? No.
There are no threes in 1.

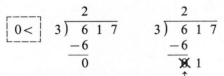

```
            2                    2
  ┌ ─ ─ ┐                            2 0  ⟵ write a zero in the
  │ 0<  │  3) 6 1 7      3) 6 1 7     3) 6 1 7    quotient to show you
  └ ─ ─ ┘     −6            −6           −6       cannot divide
            ─────         ─────        ─────
              0            ✗ 1         ✗ 1     Think: 3) 1
```

When zero is the first digit of a whole number, you can cross it out.

4. Bring down ones. **5.** Divide ones.

```
            2 0                           2 0 5 R2
  3) 6 1 7              ┌ ─ ─ ┐  3) 6 1 7
     −6                 │ 2<  │     −6
  ─────                 └ ─ ─ ┘  ─────
    ✗ 1 7                        ✗ 1 7   Think: 3) 17
                                − 1 5
                                ─────
                                    2
```

6. Check.

```
     2 0 5  ⟵ quotient
   ×     3  ⟵ divisor
   ─────────
     6 1 5
   +     2  ⟵ remainder
   ─────────
     6 1 7  ⟵ 205 R2 checks
```

3-3. Dividing by Two Digits

To divide by a two-digit divisor, you can use the first digit of the divisor to **find an estimate** for each digit of the quotient.

A. Find a one-digit quotient.

To find an estimate for a one-digit quotient, you can use basic division facts.

EXAMPLE 3-14: Divide the following: 260 ÷ 31

Solution

1. Find an estimate.

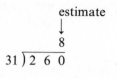

$$\frac{8}{3\overline{)24}} \qquad \text{about 8} \atop 3\overline{)26}$$

means about 8
31 $\overline{)260}$

estimate
↓
8
31 $\overline{)260}$

2. Multiply, subtract compare, and write the remainder.

correct
↓
$$\begin{array}{r} 8\,\text{R}12 \\ 31\overline{)\ 2\ 6\ 0} \\ -2\ 4\ 8 \leftarrow 8 \times 31 \\ \hline 1\ 2 \end{array}$$

$\boxed{12 <}$

3. Check.

$$\begin{array}{r} 3\ 1 \leftarrow \text{divisor} \\ \times\ \ 8 \leftarrow \text{quotient} \\ \hline 2\ 4\ 8 \\ +\ \ 1\ 2 \leftarrow \text{remainder} \\ \hline 2\ 6\ 0 \leftarrow 8\ \text{R}12\ \text{checks} \end{array}$$

Note 1: Another way to state "$\text{about 8} \atop 3\overline{)26}$ means $\text{about 8} \atop 31\overline{)260}$" is "There are about 8 threes in 26, which means there should be about 8 thirty-ones in 260."

Note 2: When there are two or more digits in the divisor, the six basic division steps are
1. Find an estimate   **2.** Multiply   **3.** Subtract   **4.** Compare
5. Write the remainder (if not zero)   **6.** Check.

Caution: When the first digit of the divisor is used to find an estimate, you may get an estimate that is too large. If an estimate is too large, you will not be able to subtract. When this happens, you must decrease the estimate until you can subtract.

EXAMPLE 3-15: Divide the following: 365 ÷ 46

Solution

1. Find an estimate.

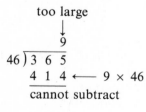

$$\frac{9}{4\overline{)36}} \text{ means } \text{about 9} \atop 46\overline{)365}$$

too large
↓
$$\begin{array}{r} 9 \\ 46\overline{)3\ 6\ 5} \\ 4\ 1\ 4 \leftarrow 9 \times 46 \\ \hline \text{cannot subtract} \end{array}$$

2. Decrease the estimate.
9 − 1 = 8 (new estimate)

too large also
↓
$$\begin{array}{r} 8 \\ 46\overline{)3\ 6\ 5} \\ 3\ 6\ 8 \leftarrow 8 \times 46 \\ \hline \text{cannot subtract} \end{array}$$

3. Decrease the estimate again.
8 − 1 = 7 (new estimate)

correct
↓
$$\begin{array}{r} 7\,\text{R}43 \\ 46\overline{)\ 3\ 6\ 5} \\ -3\ 2\ 2 \leftarrow 7 \times 46 \\ \hline 4\ 3 \end{array}$$

$\boxed{43 <}$

4. Check.

$$\begin{array}{r} 4 \\ 4\ 6 \leftarrow \text{divisor} \\ \times\ \ 7 \leftarrow \text{quotient} \\ \hline 3\ 2\ 2 \\ +\ \ 4\ 3 \leftarrow \text{remainder} \\ \hline 3\ 6\ 5 \leftarrow 7\ \text{R}43\ \text{checks} \end{array}$$

B. **Find two or more digits in the quotient.**

To divide by two digits when there are two or more digits in the quotient, you can use the first digit of the divisor to find an estimate for each digit in the quotient.

EXAMPLE 3-16: Divide the following: 8932 ÷ 29

Solution

1. Find an estimate.

$$\frac{4}{2\overline{)8}} \text{ means } \frac{\text{about 4}}{29\overline{)89}}$$

```
              too large
                 ↓
                 4
    29 ) 8 9 3 2
         1 1 6  ←—— 4 × 29
    ────────────
    cannot subtract
```

2. Decrease the estimate.

$4 - 1 = 3$ (new estimate)

```
              correct
                 ↓
                 3
    29 ) 8 9 3 2
        −8 7  ←—— 3 × 29
    ────────────
             2
```

3. Bring down and find an estimate.
There are no twenty-nines in 23.

```
         write zero
              ↓
              3 0
    29 ) 8 9 3 2
        −8 7 ↓
    ────────────
             2 3      Think: 29 ) 23
```

4. Bring down and find an estimate.

$$\frac{1}{2\overline{)2}} \text{ means } \frac{\text{about 11}}{2\overline{)23}} \text{ means } \frac{\text{about 9}}{29\overline{)232}}$$

```
              too large
                 ↓
              3 0 9
    29 ) 8 9 3 2
        −8 7  ↓
    ────────────
             2 3 2
             2 6 1  ←—— 9 × 29
    ────────────────
    cannot subtract
```

5. Decrease the estimate.
$9 - 1 = 8$ (new estimate)

```
               correct
                  ↓
               3 0 8
    29 ) 8 9 3 2
        −8 7
    ────────────
             2 3 2
            −2 3 2  ←—— 8 × 29
    ────────────────
                 0
```

6. Check.

```
       3 0 8  ←—— quotient
    ×    2 9  ←—— divisor
    ──────────
    2 7 7 2
    6 1 6
    ──────────
    8 9 3 2  ←—— 308 checks
```

Note: In Step 4 of Example 3-16, the estimate for $2\overline{)23}$ was the two-digit number 11. When an estimate has two digits (greater than 9), you first decrease the estimate to the one-digit number 9 and then see if you can subtract.

3-4. Dividing by Three or More Digits

To divide by three or more digits, you can use the first digit of the divisor to find an estimate for each digit in the quotient.

EXAMPLE 3-17: Divide the following: $9500 \div 253$

1. Find an estimate.

$$\frac{4}{2\overline{)8}} \text{ means } \frac{\text{about 4}}{2\overline{)9}} \text{ means } \frac{\text{about 4}}{253\overline{)950}}$$

```
              too large
                 ↓
                 4
    253 ) 9 5 0 0
          1 0 1 2  ←—— 4 × 253
    ─────────────────
    cannot subtract
```

2. Decrease the estimate.

$4 - 1 = 3$ (new estimate)

```
                       correct
                          ↓
                          3
    ┌─────┐
    │ 191<│   253 ) 9 5 0 0
    └─────┘        −7 5 9  ←—— 3 × 253
               ──────────────
                    1 9 1
```

3. Bring down and find an estimate.

$$\overset{9}{2\,\overline{)\,18}} \text{ means } \overset{\text{about 9}}{2\,\overline{)\,19}} \text{ means } \overset{\text{about 9}}{253\,\overline{)\,1910}}$$

$$
\begin{array}{r}
\text{too large} \\
\downarrow \\
3\ 9 \\
253\,\overline{)\ 9\ 5\ 0\ 0} \\
-7\ 5\ 9 \\
\hline
1\ 9\ 1\ 0 \\
2\ 2\ 7\ 7 \longleftarrow 9 \times 253 \\
\hline
\text{cannot subtract}
\end{array}
$$

4. Decrease the estimate until you can subtract.

$$
\begin{array}{r}
\text{correct} \\
\downarrow \\
3\ 7 \text{ R139} \\
\boxed{139<}\quad 253\,\overline{)\ 9\ 5\ 0\ 0} \\
-7\ 5\ 9 \\
\hline
1\ 9\ 1\ 0 \\
-1\ 7\ 7\ 1 \\
\hline
1\ 3\ 9
\end{array}
$$

5. Check.

$$
\begin{array}{r}
2\ 5\ 3 \longleftarrow \text{ divisor} \\
\times\ \ \ 3\ 7 \longleftarrow \text{ quotient} \\
\hline
1\ 7\ 7\ 1 \\
+7\ 5\ 9 \\
\hline
9\ 3\ 6\ 1 \\
+\ \ 1\ 3\ 9 \longleftarrow \text{ remainder} \\
\hline
9\ 5\ 0\ 0 \longleftarrow 37 \text{ R139 checks}
\end{array}
$$

3-5. Factoring Whole Numbers

Recall that in $3 \times 4 = 12$, 3 and 4 are called factors of the product 12.

Note: A given whole number is a factor of a second whole number if the given whole number divides the second whole number evenly.

EXAMPLE 3-18: Which one of the following are factors of 12? **(a)** 2 **(b)** 5

Solution
(a) 2 is a factor of 12 because 2 divides 12 evenly: $12 \div 2 = 6$ or $6 \times 2 = 12$
(b) 5 is not a factor of 12 because 5 does not divide 12 evenly: $12 \div 5 = 2$ R2

To find all the whole-number factors of another whole number, you can use division.

EXAMPLE 3-19: Find all the whole-number factors of 12.

Solution:
$12 \div 1 = 12$ means 1 is a factor of 12.
$12 \div 2 = 6$ means 2 is a factor of 12.
$12 \div 3 = 4$ means 3 is a factor of 12.
$12 \div 4 = 3$ means 4 is a factor of 12.
$12 \div 5 = 2$ R2 means 5 is not a factor of 12.
$12 \div 6 = 2$ means 6 is a factor of 12.
$12 \div 7 = 1$ R5 means 7 is not a factor of 12.
$12 \div 8 = 1$ R4 means 8 is not a factor of 12.
$12 \div 9 = 1$ R3 means 9 is not a factor of 12.
$12 \div 10 = 1$ R2 means 10 is not a factor of 12.
$12 \div 11 = 1$ R1 means 11 is not a factor of 12.
$12 \div 12 = 1$ means 12 is a factor of 12.

Note: The factors of 12 are 1, 2, 3, 4, 6 and 12.

A whole number greater than zero that has exactly two different factors is called a **prime number.**

EXAMPLE 3-20: List all the prime numbers from 0 to 10.

Solution

0 is not a prime number because 0 has more than two different factors.
($0 \div 1 = 0, 0 \div 2 = 0, 0 \div 3 = 0, \cdots$ means every nonzero whole number is a factor of 0).

1 is not a prime number because 1 has only one different factor.
($1 \div 1 = 1$ only means the only factor of 1 is 1).

2 is a prime number because 2 has exactly two different factors.
($2 \div 1 = 2$ and $2 \div 2 = 1$ means the factors of 2 are 1 and 2).

3 is a prime number because 3 has exactly two different factors.
($3 \div 1 = 3$ and $3 \div 3 = 1$ means the factors of 3 are 1 and 3).

4 is not a prime number because 4 has more than two different factors.
($4 \div 1 = 4, 4 \div 2 = 2$, and $4 \div 4 = 1$ means the factors of 4 are 1, 2, and 4).

5 is a prime number because 5 has exactly two different factors.
($5 \div 1 = 5$ and $5 \div 5 = 1$ means the factors of 5 are 1 and 5).

6 is not a prime number because 6 has more than two different factors.
($6 \div 1 = 6, 6 \div 2 = 3, 6 \div 3 = 2$, and $6 \div 6 = 1$ means the factors of 6 are 1, 2, 3, and 6).

7 is a prime number because 7 has exactly two different factors.
($7 \div 1 = 7$, and $7 \div 7 = 1$ means the factors of 7 are 1 and 7).

8 is not a prime number because 8 has more than two different factors.
($8 \div 1 = 8, 8 \div 2 = 4, 8 \div 4 = 2$, and $8 \div 8 = 1$ means the factors of 8 are 1, 2, 4, and 8).

9 is not a prime number because 9 has more than two different factors.
($9 \div 1 = 9, 9 \div 3 = 3$, and $9 \div 9 = 1$ means the factors of 9 are 1, 3, and 9).

10 is not a prime number because 10 has more than two different factors.
($10 \div 1 = 10, 10 \div 2 = 5, 10 \div 5 = 2$, and $10 \div 10 = 1$ means the factors of 10 are 1, 2, 5, and 10).

Note: The prime numbers from 0 to 10 are 2, 3, 5, and 7.

A whole number is **even** if it ends in 0, 2, 4, 6, or 8.
A whole number is **odd** if it ends in 1, 3, 5, 7, or 9.

EXAMPLE 3-21: Identify 0, 5, 12, 29, 136, and 407 as odd or even.

Solution: 0 is even because it ends in 0.
5 is odd because it ends in 5.
12 is even because it ends in 2.
29 is odd because it ends in 9.
136 is even because it ends in 6.
407 is odd because it ends in 7.

Note: The only even prime number is 2. All of the prime numbers greater than 2 are odd because every even number greater than 2 has at least three different factors of 1, 2, and the even number itself.

Caution: Not all odd numbers greater than 1 are prime numbers.

EXAMPLE 3-22: Find an odd number greater than 1 that is not prime.

Solution: 9 is an odd number because it ends in 9. However, 9 is not a prime number because 9 has more than two different factors (1, 3, 9).

Note 1: Other odd numbers that are not prime are 1, 9, 15, 21, 25, 27, 33, 35, $\cdots$.

Note 2: Every whole number is either even or odd.

A factor of a given whole number that is also a prime number is called a **prime factor.**

EXAMPLE 3-23: Find all the prime factors of 12.

Solution: The factors of 12 are 1, 2, 3, 4, and 6 [see Example 3-19]. The only prime factors of 12 are 2 and 3.

A whole number greater than zero with more than two different factors is called a **composite number.**

EXAMPLE 3-24: List all the composite numbers between 0 and 10.

Solution: 1 is not a composite number because 1 has only 1 different factor.
2 is not a composite number because 2 has exactly two different factors.
3 is not a composite number because 3 has exactly two different factors.
4 is a composite number because 4 has more than two different factors.
5 is not a composite number because 5 has exactly two different factors.
6 is a composite number because 6 has more than two different factors.
7 is not a composite number because 7 has exactly two different factors.
8 is a composite number because 8 has more than two different factors.
9 is a composite number because 9 has more than two different factors.

Note 1: The composite numbers between 0 and 10 are 4, 6, 8, and 9.

Note 2: The whole numbers 0 and 1 are neither prime nor composite numbers.

Note 3: No prime number is a composite number because prime numbers have exactly two different factors and composite numbers always have more than two different factors.

Note 4: Every whole number is either 0, 1, prime, or composite.

To **factor** a whole number, you write it as a product of primes and/or composite numbers.

EXAMPLE 3-25: Factor 12 in as many different ways as possible.

Solution: $12 = 1 \times 12$
$12 = 2 \times 6$
$12 = 3 \times 4$
$12 = 2 \times 2 \times 3$
} 12 can be factored in exactly four different ways.

Note: The **factorizations** 1×12 and 12×1 are not considered to be different factorizations, but **equivalent factorizations.**

A product of factors that are all prime numbers is called a **product of primes.**

EXAMPLE 3-26: Identify 12 as a product of primes choosing from the factorizations from Example 3-25.

Solution: 12 is not factored as a product of primes as 1×12 because 1 and 12 are not prime.
12 is not factored as a product of primes as 2×6 because 6 is not prime.
12 is not factored as a product of primes as 3×4 because 4 is not prime.
12 is factored as a product of primes as $2 \times 2 \times 3$ because 2 and 3 are both prime numbers.

Note: The factorizations $2 \times 2 \times 3$, $2 \times 3 \times 2$, and $3 \times 2 \times 2$ are all considered equivalent factorizations because each contains two factors of 2 and one factor of 3.

The following statement is one of the most important facts about the arithmetic of whole numbers.

The Fundamental Rule of Arithmetic

Every composite number can be factored as a product of primes in exactly one way (except for equivalent factorizations).

To factor a composite number as a product of primes, you can use division.

EXAMPLE 3-27: Factor 12 as a product of primes using division.

prime divisor

Solution: $12 \div 2 = 6$ ⟵— composite quotient (Continue the division process.)

$6 \div 2 = 3$ ⟵— prime quotient (*Stop!*)

$12 = 2 \times 2 \times 3$ ⟵— product of primes

Note 1: When the quotient is a composite number, you continue the division process by dividing that quotient by another prime number.

Note 2: When the quotient is a prime number, you stop the division process and write each prime divisor and the prime quotient as the product of primes.

To factor a whole number as a product of primes using division, it is usually easier to use a shortcut division method.

EXAMPLE 3-28: Factor the following numbers as products of primes using a shortcut division method: **(a)** 15 **(b)** 18 **(c)** 56

Solution

(a)

$$3 \underline{|\, 15}$$
$$5$$

$$15 = 3 \times 5$$

(b)

$$2 \underline{|\, 18}$$
$$3 \underline{|\, 9}$$
$$3$$

$$18 = 2 \times 3 \times 3$$

(c)

$$2 \underline{|\, 56}$$
$$2 \underline{|\, 28}$$
$$2 \underline{|\, 14}$$
$$7$$

$$56 = 2 \times 2 \times 2 \times 7$$

Note: To use the shortcut division method, you turn the division box upside down, then divide by a prime number, and then write the quotient below the division box:

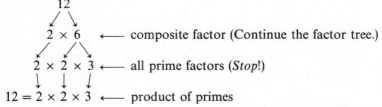

3 ⟵— prime quotient (*Stop!*)

$12 = 2 \times 2 \times 3$ ⟵— product of primes

To factor a composite number as a product of primes, you can also use a **factor tree.**

EXAMPLE 3-29: Factor 12 as a product of primes using a factor tree.

Solution:

12

2×6 ⟵— composite factor (Continue the factor tree.)

$2 \times 2 \times 3$ ⟵— all prime factors (*Stop!*)

$12 = 2 \times 2 \times 3$ ⟵— product of primes

Note 1: When a factor in the bottom row of a factor tree is a composite number, you continue the factor tree by factoring the composite number.

Note 2: When all the factors in the bottom row of a factor tree are prime numbers, you stop the factor tree and write each prime factor in the bottom row as the product of primes.

Note 3: When a factor tree is continued with one or more prime numbers in the bottom row, each prime number is brought down to the next row.

Factor trees for composite numbers always have two or more rows.

EXAMPLE 3-30: Factor the following numbers using a factor tree: **(a)** 15   **(b)** 18   **(c)** 56

Solution

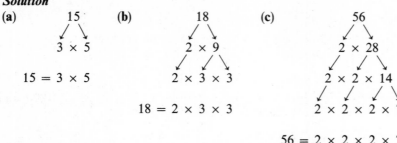

(a)   15
          3 × 5
      15 = 3 × 5

(b)   18
          2 × 9
          2 × 3 × 3
      18 = 2 × 3 × 3

(c)   56
          2 × 28
          2 × 2 × 14
          2 × 2 × 2 × 7
      56 = 2 × 2 × 2 × 7

3-6. Solving Word Problems (+, −, ×, ÷)

To solve the word problems in this section, you must add, subtract, multiply, and/or divide.

A. Choose the correct operation (+, −, ×, ÷).

To join two or more amounts together, you add [see Section 1-2].
To take one amount away from another amount, you subtract [see Section 1-3].
To join two or more equal amounts together, you multiply [see Section 2-1].
To separate an amount into two or more equal amounts, you divide [see Section 3-1].

EXAMPLE 3-31: Choose the correct operation to solve the following word problems.

1. *Read:*   Evelyn needs 40 woodscrews.
                    Each package contains
                    5 woodscrews. How many
                    packages should Evelyn buy?

2. *Understand:* The question asks you to
                     separate 40 into equal
                     amounts of 5.

3. *Decide:*   To separate into equal
                    amounts, you **divide.**

1. *Read:*   Myrle needs 40 woodscrews.
                    Each woodscrew costs 5 cents.
                    How much will the woodscrews
                    cost Myrle?

2. *Understand:* The question asks you to join
                     40 equal amounts of 5 together.

3. *Decide:*   To join equal amounts together,
                    you **multiply.**

B. Solve word problems (+, −, ×, ÷).

To solve addition word problems, you find
(a) how much two or more amounts are all together;
(b) the sum of one amount plus another amount.

To solve subtraction word problems, you find
(a) how much is left;
(b) how much more is needed;
(c) the difference between two amounts.

To solve multiplication word problems, you find
(a) how much two or more equal amounts are all together;
(b) the product of one amount times another amount.

To solve a division word problem, you find
(a) how many equal amounts there are in all;
(b) how many are in each equal amount;
(c) how many times larger one amount is than another amount.

EXAMPLE 3-32: Solving the following word problem using addition, subtraction, multiplication, and/or division.

1. *Identify:*   A restaurant needs at least (1000 eggs)   *Circle* each fact.
   for the week. <u>How many whole dozen</u>   *Underline* the question.
   <u>eggs should be ordered if there are</u>
   <u>(12 eggs)</u> per dozen?

2. *Understand:* The question asks you to find how many equal amounts of 12 there are in 1000, and then add 1 equal amount of 12 if there are any remaining.

3. *Decide:*   To find how many equal amounts there are in all, you divide.
   To find if there is anything left over, you look for a nonzero remainder.

4. *Compute:*

$$
\begin{array}{r}
8\ \ 3\ \text{R4} \longleftarrow \text{nonzero remainder} \\
12\overline{)\ 1\ 0\ 0\ 0} \\
-\ 9\ 6 \\
\hline
4\ 0 \\
-3\ 6 \\
\hline
4
\end{array}
$$

5. *Interpret:* 83 R4 means that to receive at least 1000 eggs while ordering only whole dozens, the restaurant will need to order 84 (83 + 1) dozen eggs. $\longleftarrow$ solution

6. *Check:*   Is 84 dozen eggs at least 1000 eggs? Yes: 84 dozen = 84 × 12 = 1008 (eggs)

RAISE YOUR GRADES

Can you . . . ?

☑ answer all 90 basic division facts correctly in less than 5 minutes
☑ identify each part of a division problem
☑ write a division problem in division box form given horizontal form
☑ write the related multiplication problem for a division problem
☑ determine when a one-digit quotient is too large
☑ determine when a one-digit quotient is too small
☑ write a nonzero remainder in the proper place in a division answer
☑ check a division answer when the remainder is zero
☑ check a division answer when the remainder is not zero
☑ state the six basic division steps for a one-digit divisor
☑ determine when to bring down digits from the dividend
☑ determine when to write a zero in the quotient
☑ use the first digit of a divisor to find an estimate for each digit in the quotient
☑ determine when to decrease an estimate
☑ state the six basic division steps for a divisor that contains two or more digits
☑ identify prime and composite numbers
☑ identify even and odd numbers
☑ factor a whole number in as many different ways as possible
☑ factor a composite number as a product of primes using division or a factor tree
☑ determine whether to add, subtract, multiply, or divide in a word problem
☑ interpret the answer to a word problem in the words of the original problem
☑ check to see if the solution to a word problem makes sense in the original problem

SUMMARY

1. Division is used to separate an amount into two or more equal amounts.

2. In $35 \div 4 = 8\,R3$ or $4\,\overline{)\,35\,}^{\,8\,R3}$:
   (a) 4 is the divisor
   (b) 35 is the dividend
   (c) 8 is the quotient
   (d) 3 is the remainder
   (e) 8 R3 is the division answer
   (f) R is the symbol for "remainder"
   (g) $\div$ is the division symbol
   (h) $\overline{)}$ is the division box

3. To help find a basic division fact, you can use the related missing factor problem and basic multiplication facts.

4. To divide by a one-digit whole number when there are two or more digits in the dividend, you always start the division process by trying to divide into the first digit in the dividend.

5. To check a division answer when the remainder is zero, you multiply the proposed quotient by the divisor to see if you get the original dividend.

6. To check a division answer when the remainder is not zero, you multiply the proposed quotient by the divisor and then add the remainder to see if you get the original dividend.

7. To divide when there are two or more digits in the quotient, you must bring down one or more digits from the dividend.

8. When there are no more digits to bring down from the dividend and the remainder is less than the divisor, you write the remainder (if not zero) as part of the division answer.

9. When you cannot divide after bringing down a digit from the dividend, write a zero in the quotient above that digit.

10. The six basic division steps for a one-digit divisor are 1. Divide   2. Multiply   3. Subtract   4. Compare   5. Write the remainder (if not zero)   6. Check.

11. To divide by a divisor that contains two or more digits, you can use the first digit of the divisor to find an estimate for each digit in the quotient.

12. When the first digit of a divisor is used to find an estimate, you may get an estimate that is too large. If an estimate is too large, you will not be able to subtract. When this happens, you must decrease the estimate until you can subtract.

13. When an estimate is greater than 9, you first decrease the estimate to 9 and then see if you can subtract.

14. A given whole number is a factor of a second whole number if the given whole number divides the second whole number evenly.

15. A whole number greater than zero that has exactly two different factors is called a prime number.

16. A whole number greater than zero with more than two different factors is called a composite number.

17. A whole number that ends in 0, 2, 4, 6, or 8 is called an even number.

18. A whole number that ends in 1, 3, 5, 7, or 9 is called odd number.

19. A factor of a given whole number that is also a prime number is called a prime factor.

20. To factor a whole number, you write it as a product of primes and/or composite numbers.

21. A product of factors that are all prime numbers is called a product of primes.

22. **The Fundamental Rule of Arithmetic:** Every composite number can be factored as a product of primes in exactly one way, not counting equivalent factorizations.

23. To factor a composite number as a product of primes, you can use division or a factor tree.

24. To solve a word problem using the correct operation, you
    (a) add to join two or more amounts together;
    (b) subtract to take one amount away from another amount;
    (c) multiply to join equal amounts together;
    (d) divide to separate an amount into two or more equal amounts.

25. To solve addition word problems, you find
    (a) how much two or more amounts are all together;
    (b) the sum of one amount plus another amount.

26. To solve subtraction word problems, you find
    (a) how much is left;
    (b) how much more is needed;
    (c) the difference between two amounts.
27. To solve multiplication word problems, you find
    (a) how much two or more equal amounts are all together;
    (b) the product of one amount times another amount.
28. To solve division word problems, you find
    (a) how many equal amounts there are in all;
    (b) how many are in each equal amount;
    (c) how many times larger one amount is than another amount.

SOLVED PROBLEMS

PROBLEM 3-1   Write the related missing factor problem for each division problem.

(a) $21 \div 7 = ?$ (d) $54 \div 6 = ?$ (g) $6 \div 6 = ?$ (j) $14 \div 2 = ?$ (m) $64 \div 8 = ?$
(b) $48 \div 6 = ?$ (e) $72 \div 9 = ?$ (h) $9 \div 1 = ?$ (k) $56 \div 7 = ?$ (n) $63 \div 7 = ?$
(c) $0 \div 4 = ?$ (f) $24 \div 3 = ?$ (i) $10 \div 5 = ?$ (l) $81 \div 9 = ?$ (o) $45 \div 5 = ?$

Solution: Recall that the division problem $6 \div 3 = ?$ and the related missing factor problem $? \times 3 = 6$ are two different ways of stating the same problem [see Example 3-6]:

(a) $21 \div 7 = ?$ means $? \times 7 = 21$ (b) $48 \div 6 = ?$ means $? \times 6 = 48$
(c) $0 \div 4 = ?$ means $? \times 4 = 0$ (d) $54 \div 6 = ?$ means $? \times 6 = 54$
(e) $72 \div 9 = ?$ means $? \times 9 = 72$ (f) $24 \div 3 = ?$ means $? \times 3 = 24$
(g) $6 \div 6 = ?$ means $? \times 6 = 6$ (h) $9 \div 1 = ?$ means $? \times 1 = 9$
(i) $10 \div 5 = ?$ means $? \times 5 = 10$ (j) $14 \div 2 = ?$ means $? \times 2 = 14$
(k) $56 \div 7 = ?$ means $? \times 7 = 56$ (l) $81 \div 9 = ?$ means $? \times 9 = 81$
(m) $64 \div 8 = ?$ means $? \times 8 = 64$ (n) $63 \div 7 = ?$ means $? \times 7 = 63$
(o) $45 \div 5 = ?$ means $? \times 5 = 45$

PROBLEM 3-2   To practice the 90 basic division facts:

1. Cover the solutions one row at a time with a blank sheet of paper.
2. Write the answer above each problem in that row on your blank sheet of paper.
3. Slide your answers up and compare them with the printed solutions.
4. Repeat any row in which you had an incorrect answer.
   [see Example 3-7.]

$\frac{2}{6\overline{)12}}$	$\frac{4}{9\overline{)36}}$	$\frac{2}{1\overline{)2}}$	$\frac{0}{2\overline{)0}}$	$\frac{3}{7\overline{)21}}$	$\frac{6}{9\overline{)54}}$	$\frac{1}{2\overline{)2}}$	$\frac{4}{4\overline{)16}}$	$\frac{4}{1\overline{)4}}$	$\frac{4}{6\overline{)24}}$
$\frac{1}{3\overline{)3}}$	$\frac{6}{8\overline{)48}}$	$\frac{0}{8\overline{)0}}$	$\frac{2}{9\overline{)18}}$	$\frac{0}{4\overline{)0}}$	$\frac{8}{3\overline{)24}}$	$\frac{1}{5\overline{)5}}$	$\frac{5}{3\overline{)15}}$	$\frac{1}{1\overline{)1}}$	$\frac{2}{5\overline{)10}}$
$\frac{7}{5\overline{)35}}$	$\frac{9}{2\overline{)18}}$	$\frac{6}{6\overline{)36}}$	$\frac{0}{6\overline{)0}}$	$\frac{8}{6\overline{)48}}$	$\frac{2}{4\overline{)8}}$	$\frac{8}{1\overline{)8}}$	$\frac{4}{8\overline{)32}}$	$\frac{3}{2\overline{)6}}$	$\frac{6}{2\overline{)12}}$
$\frac{9}{7\overline{)63}}$	$\frac{0}{7\overline{)0}}$	$\frac{2}{3\overline{)6}}$	$\frac{5}{5\overline{)25}}$	$\frac{8}{8\overline{)64}}$	$\frac{1}{9\overline{)9}}$	$\frac{6}{4\overline{)24}}$	$\frac{4}{2\overline{)8}}$	$\frac{4}{3\overline{)12}}$	$\frac{4}{5\overline{)20}}$
$\frac{3}{6\overline{)18}}$	$\frac{8}{4\overline{)32}}$	$\frac{7}{7\overline{)49}}$	$\frac{8}{9\overline{)72}}$	$\frac{0}{3\overline{)0}}$	$\frac{1}{7\overline{)7}}$	$\frac{5}{7\overline{)35}}$	$\frac{6}{1\overline{)6}}$	$\frac{7}{2\overline{)14}}$	$\frac{9}{4\overline{)36}}$
$\frac{3}{9\overline{)27}}$	$\frac{6}{3\overline{)18}}$	$\frac{7}{4\overline{)28}}$	$\frac{9}{6\overline{)54}}$	$\frac{3}{4\overline{)12}}$	$\frac{0}{1\overline{)0}}$	$\frac{5}{1\overline{)5}}$	$\frac{3}{3\overline{)9}}$	$\frac{6}{7\overline{)42}}$	$\frac{5}{8\overline{)40}}$

$$\begin{array}{r} 2 \\ 7\overline{)14} \end{array} \quad \begin{array}{r} 7 \\ 9\overline{)63} \end{array} \quad \begin{array}{r} 7 \\ 6\overline{)42} \end{array} \quad \begin{array}{r} 1 \\ 6\overline{)6} \end{array} \quad \begin{array}{r} 9 \\ 8\overline{)72} \end{array} \quad \begin{array}{r} 9 \\ 1\overline{)9} \end{array} \quad \begin{array}{r} 7 \\ 3\overline{)21} \end{array} \quad \begin{array}{r} 1 \\ 4\overline{)4} \end{array} \quad \begin{array}{r} 1 \\ 8\overline{)8} \end{array} \quad \begin{array}{r} 0 \\ 9\overline{)0} \end{array}$$

$$\begin{array}{r} 3 \\ 1\overline{)3} \end{array} \quad \begin{array}{r} 5 \\ 4\overline{)20} \end{array} \quad \begin{array}{r} 5 \\ 9\overline{)45} \end{array} \quad \begin{array}{r} 3 \\ 8\overline{)24} \end{array} \quad \begin{array}{r} 7 \\ 1\overline{)7} \end{array} \quad \begin{array}{r} 9 \\ 5\overline{)45} \end{array} \quad \begin{array}{r} 5 \\ 6\overline{)30} \end{array} \quad \begin{array}{r} 7 \\ 8\overline{)56} \end{array} \quad \begin{array}{r} 3 \\ 5\overline{)15} \end{array} \quad \begin{array}{r} 8 \\ 5\overline{)40} \end{array}$$

$$\begin{array}{r} 4 \\ 7\overline{)28} \end{array} \quad \begin{array}{r} 9 \\ 9\overline{)81} \end{array} \quad \begin{array}{r} 8 \\ 2\overline{)16} \end{array} \quad \begin{array}{r} 5 \\ 2\overline{)10} \end{array} \quad \begin{array}{r} 9 \\ 3\overline{)27} \end{array} \quad \begin{array}{r} 2 \\ 8\overline{)16} \end{array} \quad \begin{array}{r} 8 \\ 7\overline{)56} \end{array} \quad \begin{array}{r} 0 \\ 5\overline{)0} \end{array} \quad \begin{array}{r} 2 \\ 2\overline{)4} \end{array} \quad \begin{array}{r} 6 \\ 5\overline{)30} \end{array}$$

PROBLEM 3-3   Divide by a one-digit divisor to find a one-digit quotient with remainder.

(a) $19 \div 2$	(b) $61 \div 8$	(c) $28 \div 5$	(d) $5 \div 2$	(e) $10 \div 6$
(f) $14 \div 5$	(g) $18 \div 4$	(h) $13 \div 2$	(i) $62 \div 7$	(j) $29 \div 3$
(k) $31 \div 4$	(l) $39 \div 7$	(m) $41 \div 6$	(n) $25 \div 3$	(o) $38 \div 9$

Solution: Recall that when a divisor does not divide a dividend exactly, then there will always be a nonzero remainder [see Example 3-8]:

(a)
$$\begin{array}{r} 9 \text{ R}1 \\ 2\overline{)19} \\ -18 \\ \hline 1 \end{array}$$
(b)
$$\begin{array}{r} 7 \text{ R}5 \\ 8\overline{)61} \\ -56 \\ \hline 5 \end{array}$$
(c)
$$\begin{array}{r} 5 \text{ R}3 \\ 5\overline{)28} \\ -25 \\ \hline 3 \end{array}$$
(d)
$$\begin{array}{r} 2 \text{ R}1 \\ 2\overline{)5} \\ -4 \\ \hline 1 \end{array}$$
(e)
$$\begin{array}{r} 1 \text{ R}4 \\ 6\overline{)10} \\ -6 \\ \hline 4 \end{array}$$

(f)
$$\begin{array}{r} 2 \text{ R}4 \\ 5\overline{)14} \\ -10 \\ \hline 4 \end{array}$$
(g)
$$\begin{array}{r} 4 \text{ R}2 \\ 4\overline{)18} \\ -16 \\ \hline 2 \end{array}$$
(h)
$$\begin{array}{r} 6 \text{ R}1 \\ 2\overline{)13} \\ -12 \\ \hline 1 \end{array}$$
(i)
$$\begin{array}{r} 8 \text{ R}6 \\ 7\overline{)62} \\ -56 \\ \hline 6 \end{array}$$
(j)
$$\begin{array}{r} 9 \text{ R}2 \\ 3\overline{)29} \\ -27 \\ \hline 2 \end{array}$$

(k)
$$\begin{array}{r} 7 \text{ R}3 \\ 4\overline{)31} \\ -28 \\ \hline 3 \end{array}$$
(l)
$$\begin{array}{r} 5 \text{ R}4 \\ 7\overline{)39} \\ -35 \\ \hline 4 \end{array}$$
(m)
$$\begin{array}{r} 6 \text{ R}5 \\ 6\overline{)41} \\ -36 \\ \hline 5 \end{array}$$
(n)
$$\begin{array}{r} 8 \text{ R}1 \\ 3\overline{)25} \\ -24 \\ \hline 1 \end{array}$$
(o)
$$\begin{array}{r} 4 \text{ R}2 \\ 9\overline{)38} \\ -36 \\ \hline 2 \end{array}$$

PROBLEM 3-4   Divide by a one-digit divisor to find two or more digits in the quotient.

(a) $96 \div 8$	(b) $28 \div 2$	(c) $375 \div 7$	(d) $325 \div 9$	(e) $359 \div 4$
(f) $3367 \div 7$	(g) $3138 \div 6$	(h) $2000 \div 3$	(i) $7070 \div 9$	(j) $197 \div 2$
(k) $7239 \div 3$	(l) $21,144 \div 6$	(m) $17,410 \div 4$	(n) $39,349 \div 5$	(o) $193,080 \div 8$

Solution: Recall that to divide when there are two or more digits in the quotient, you must bring down one or more digits from the dividend [see Example 3-12]:

(a)
$$\begin{array}{r} 12 \\ 8\overline{)96} \\ -8 \\ \hline 16 \\ -16 \\ \hline 0 \end{array}$$
(b)
$$\begin{array}{r} 14 \\ 2\overline{)28} \\ -2 \\ \hline 8 \\ -8 \\ \hline 0 \end{array}$$
(c)
$$\begin{array}{r} 53 \text{ R}4 \\ 7\overline{)375} \\ -35 \\ \hline 25 \\ -21 \\ \hline 4 \end{array}$$
(d)
$$\begin{array}{r} 36 \text{ R}1 \\ 9\overline{)325} \\ -27 \\ \hline 55 \\ -54 \\ \hline 1 \end{array}$$
(e)
$$\begin{array}{r} 89 \text{ R}3 \\ 4\overline{)359} \\ -32 \\ \hline 39 \\ -36 \\ \hline 3 \end{array}$$

(f)
$$\begin{array}{r} 481 \\ 7\overline{)3367} \\ -28 \\ \hline 56 \\ -56 \\ \hline 7 \\ -7 \\ \hline 0 \end{array}$$
(g)
$$\begin{array}{r} 523 \\ 6\overline{)3138} \\ -30 \\ \hline 13 \\ -12 \\ \hline 18 \\ -18 \\ \hline 0 \end{array}$$
(h)
$$\begin{array}{r} 666 \text{ R}2 \\ 3\overline{)2000} \\ -18 \\ \hline 20 \\ -18 \\ \hline 20 \\ -18 \\ \hline 2 \end{array}$$
(i)
$$\begin{array}{r} 785 \text{ R}5 \\ 9\overline{)7070} \\ -63 \\ \hline 77 \\ -72 \\ \hline 50 \\ -45 \\ \hline 5 \end{array}$$
(j)
$$\begin{array}{r} 98 \text{ R}1 \\ 2\overline{)197} \\ -18 \\ \hline 17 \\ -16 \\ \hline 1 \end{array}$$

```
       2413              3 524              4 352 R2            7 869 R4             24,135
(k) 3) 7239    (l) 6) 21,144    (m) 4) 17,410      (n) 5) 39,349      (o) 8) 193,080
     -6                -18                -16                  -35                  -16
     ──                ──                 ──                   ──                   ──
      12                31                 14                   43                   33
     -12               -30                -12                  -40                  -32
     ──                ──                 ──                   ──                   ──
      03                14                 21                   34                   10
      -3               -12                -20                  -30                   -8
     ──                ──                 ──                   ──                   ──
      09                24                 10                   49                   28
      -9               -24                 -8                  -45                  -24
     ──                ──                 ──                   ──                   ──
       0                 0                  2                    4                   40
                                                                                   -40
                                                                                   ──
                                                                                    0
```

PROBLEM 3-5 Divide by a one-digit divisor to find zeros in the quotient.

(a) $40 \div 4$ (b) $187 \div 9$ (c) $309 \div 3$ (d) $1240 \div 6$ (e) $2100 \div 7$

(f) $4004 \div 5$ (g) $42,324 \div 6$ (h) $44,140 \div 7$ (i) $18,420 \div 2$ (j) $24,400 \div 8$

(k) $72,070 \div 9$ (l) $18,602 \div 3$ (m) $56,000 \div 8$ (n) $300,125 \div 5$ (o) $81,615 \div 4$

Solution: Recall that when you cannot divide after bringing down a digit from the dividend, write a zero above that digit in the quotient [see Example 3-13]:

```
      10              20 R7              103            206 R4             300
(a) 4) 40    (b) 9) 187     (c) 3) 309     (d) 6) 1240     (e) 7) 2100
    -4             -18            -3             -12               -21
    ──             ──            ──             ──                ──
    00             07            009            040               000
                                  -9            -36
                                 ──             ──
                                  0              4
```

```
     800 R4           7 054            6 305 R5          9 210            3 050
(f) 5) 4004   (g) 6) 42,324   (h) 7) 44,140   (i) 2) 18,420   (j) 8) 24,400
    -40            -42              -42             -18             -24
    ──             ──              ──              ──              ──
    004            032              21              04              040
                   -30             -21              -4              -40
                   ──             ──               ──              ──
                    24             040              02              00
                   -24             -35              -2
                   ──             ──               ──
                    0               5              00
```

```
    8 007 R7         6 200 R2           7 000          60,025          20,403 R3
(k) 9) 72,070  (l) 3) 18,602   (m) 8) 56,000  (n) 5) 300,125   (o) 4) 81,615
    -72            -18             -56             -30              -8
    ──             ──             ──              ──               ──
    0070           006            0000            0012             016
    -63            -6                              -10              -16
    ──             ──                              ──               ──
     7             002                             25               015
                                                   -25              -12
                                                   ──               ──
                                                    0                3
```

PROBLEM 3-6 Divide by a two-digit divisor to find a one-digit quotient.

(a) $480 \div 60$ (b) $600 \div 80$ (c) $155 \div 31$ (d) $199 \div 22$ (e) $82 \div 41$

(f) $225 \div 52$ (g) $36 \div 12$ (h) $69 \div 11$ (i) $80 \div 45$ (j) $450 \div 56$

(k) $190 \div 27$ (l) $210 \div 38$ (m) $112 \div 18$ (n) $100 \div 19$ (o) $80 \div 28$

Solution: Recall that to divide by a two-digit divisor to find a one-digit quotient, you can use the first digit of the divisor to find an estimate for the one-digit quotient.

Caution: When the estimate is too large, you must decrease the estimate until you can subtract [see Examples 3-14 and 3-16]:

(a)
$$\begin{array}{r} 8 \\ 6\overline{)48} \end{array}$$

$$\begin{array}{r} 8 \\ 60\overline{)480} \\ -480 \quad\longleftarrow\; 8 \times 60 \\ \hline 0 \end{array}$$

(b) about 7
$$\begin{array}{r} 7 \\ 8\overline{)60} \end{array}$$

$$\begin{array}{r} 7\ \text{R}40 \\ 80\overline{)600} \\ -560 \quad\longleftarrow\; 7 \times 80 \\ \hline 40 \end{array}$$

(c)
$$\begin{array}{r} 5 \\ 3\overline{)15} \end{array}$$

$$\begin{array}{r} 5 \\ 31\overline{)155} \\ -155 \quad\longleftarrow\; 5 \times 31 \\ \hline 0 \end{array}$$

(d) about 9
$$\begin{array}{r} 9 \\ 2\overline{)19} \end{array}$$

$$\begin{array}{r} 9\ \text{R}1 \\ 22\overline{)199} \\ -198 \quad\longleftarrow\; 9 \times 22 \\ \hline 1 \end{array}$$

(e)
$$\begin{array}{r} 2 \\ 4\overline{)8} \end{array}$$

$$\begin{array}{r} 2 \\ 41\overline{)82} \\ -82 \quad\longleftarrow\; 2 \times 41 \\ \hline 0 \end{array}$$

(f) about 4
$$\begin{array}{r} 4 \\ 5\overline{)22} \end{array}$$

$$\begin{array}{r} 4\ \text{R}17 \\ 52\overline{)225} \\ -208 \quad\longleftarrow\; 4 \times 52 \\ \hline 17 \end{array}$$

(g)
$$\begin{array}{r} 3 \\ 1\overline{)3} \end{array}$$

$$\begin{array}{r} 3 \\ 12\overline{)36} \\ -36 \quad\longleftarrow\; 3 \times 12 \\ \hline 0 \end{array}$$

(h)
$$\begin{array}{r} 6 \\ 1\overline{)6} \end{array}$$

$$\begin{array}{r} 6\ \text{R}3 \\ 11\overline{)69} \\ -66 \quad\longleftarrow\; 6 \times 11 \\ \hline 3 \end{array}$$

(i)
$$\begin{array}{r} 2 \\ 4\overline{)8} \end{array}$$

$$\begin{array}{r} \cancel{2} \\ 45\overline{)80} \\ -90 \quad\longleftarrow\; 2 \times 45 \\ \hline \text{cannot} \\ \text{subtract} \end{array}$$

$$\begin{array}{r} 1\ \text{R}35 \\ 45\overline{)80} \\ -45 \quad\longleftarrow\; 1 \times 45 \\ \hline 35 \end{array}$$

(j)
$$\begin{array}{r} 9 \\ 5\overline{)45} \end{array}$$

$$\begin{array}{r} \cancel{9} \\ 56\overline{)450} \\ -504 \quad\longleftarrow\; 9 \times 56 \\ \hline \text{cannot} \\ \text{subtract} \end{array}$$

$$\begin{array}{r} 8\ \text{R}2 \\ 56\overline{)450} \\ -448 \quad\longleftarrow\; 8 \times 56 \\ \hline 2 \end{array}$$

(k) about 9
$$\begin{array}{r} 9 \\ 2\overline{)19} \end{array}$$

$$\begin{array}{r} \cancel{9} \\ 27\overline{)190} \\ -243 \quad\longleftarrow\; 9 \times 27 \\ \hline \text{cannot} \\ \text{subtract} \end{array}$$

$$\begin{array}{r} \cancel{8} \\ 27\overline{)190} \\ -216 \quad\longleftarrow\; 8 \times 27 \\ \hline \text{cannot} \\ \text{subtract} \end{array}$$

$$\begin{array}{r} 7\ \text{R}1 \\ 27\overline{)190} \\ -189 \quad\longleftarrow\; 7 \times 27 \\ \hline 1 \end{array}$$

(l)
$$\begin{array}{r} 7 \\ 3\overline{)21} \end{array}$$

$$\begin{array}{r} \cancel{7} \\ 38\overline{)210} \\ -266 \quad\longleftarrow\; 7 \times 38 \\ \hline \text{cannot} \\ \text{subtract} \end{array}$$

$$\begin{array}{r} \cancel{6} \\ 38\overline{)210} \\ -228 \quad\longleftarrow\; 6 \times 38 \\ \hline \text{cannot} \\ \text{subtract} \end{array}$$

$$\begin{array}{r} 5\ \text{R}20 \\ 38\overline{)210} \\ -190 \quad\longleftarrow\; 5 \times 38 \\ \hline 20 \end{array}$$

(m)
$$\begin{array}{r} 11 \longrightarrow 9 \\ 1\overline{)11} \end{array}$$

$$\begin{array}{r} \cancel{9} \\ 18\overline{)112} \\ -162 \quad\longleftarrow\; 9 \times 18 \\ \hline \text{cannot} \\ \text{subtract} \end{array}$$

$$\begin{array}{r} \cancel{8} \\ 18\overline{)112} \\ -144 \quad\longleftarrow\; 8 \times 18 \\ \hline \text{cannot} \\ \text{subtract} \end{array}$$

$$\begin{array}{r} \cancel{7} \\ 18\overline{)112} \\ -126 \quad\longleftarrow\; 7 \times 18 \\ \hline \text{cannot} \\ \text{subtract} \end{array}$$

$$\begin{array}{r} 6\ \text{R}4 \\ 18\overline{)112} \\ -108 \quad\longleftarrow\; 6 \times 18 \\ \hline 4 \end{array}$$

(n)
$$\begin{array}{r} 10 \longrightarrow 9 \\ 1\overline{)\,10} \end{array}$$

$$\begin{array}{r} \cancel{9} \\ 19\overline{)\,100} \\ -171 \longleftarrow 9 \times 19 \\ \hline \text{cannot} \\ \text{subtract} \end{array}$$

$$\begin{array}{r} \cancel{8} \\ 19\overline{)\,100} \\ -152 \longleftarrow 8 \times 19 \\ \hline \text{cannot} \\ \text{subtract} \end{array}$$

$$\begin{array}{r} \cancel{7} \\ 19\overline{)\,100} \\ -133 \longleftarrow 7 \times 19 \\ \hline \text{cannot} \\ \text{subtract} \end{array}$$

$$\begin{array}{r} \cancel{6} \\ 19\overline{)\,100} \\ -114 \longleftarrow 6 \times 19 \\ \hline \text{cannot} \\ \text{subtract} \end{array}$$

$$\begin{array}{r} 5\,R5 \\ 19\overline{)\,100} \\ -\ 95 \longleftarrow 5 \times 19 \\ \hline 5 \end{array}$$

(o)
$$\begin{array}{r} 4 \\ 2\overline{)\,8} \end{array}$$

$$\begin{array}{r} \cancel{4} \\ 28\overline{)\,80} \\ -112 \longleftarrow 4 \times 28 \\ \hline \text{cannot} \\ \text{subtract} \end{array}$$

$$\begin{array}{r} \cancel{3} \\ 28\overline{)\,80} \\ -84 \longleftarrow 3 \times 28 \\ \hline \text{cannot} \\ \text{subtract} \end{array}$$

$$\begin{array}{r} 2\,R24 \\ 28\overline{)\,80} \\ -56 \longleftarrow 2 \times 28 \\ \hline 24 \end{array}$$

PROBLEM 3-7 Divide by a two-digit divisor to find two or more digits in the quotient.

(a) $1000 \div 40$	(b) $1730 \div 50$	(c) $1281 \div 21$	(d) $2200 \div 31$	(e) $2350 \div 11$
(f) $5472 \div 12$	(g) $22{,}720 \div 25$	(h) $55{,}480 \div 76$	(i) $348{,}900 \div 84$	(j) $379{,}260 \div 63$
(k) $599{,}742 \div 99$	(l) $69{,}862 \div 58$	(m) $110{,}780 \div 29$	(n) $134{,}000 \div 19$	(o) $90{,}000 \div 18$

Solution: Recall that to divide by a two-digit divisor, you can use the first digit of the divisor to find an estimate for each digit in the quotient.

Caution: When the estimate is too large, you must decrease the estimate until you can subtract [see Example 3-16]:

(a)
$$\begin{array}{r} 25 \\ 40\overline{)\,1000} \\ -\ 80 \longleftarrow 2 \times 40 \\ \hline 200 \\ -200 \longleftarrow 5 \times 40 \\ \hline 0 \end{array}$$

(b)
$$\begin{array}{r} 34\,R30 \\ 50\overline{)\,1730} \\ -150 \longleftarrow 3 \times 50 \\ \hline 230 \\ -200 \longleftarrow 4 \times 50 \\ \hline 30 \end{array}$$

(c)
$$\begin{array}{r} 61 \\ 21\overline{)\,1281} \\ -126 \longleftarrow 6 \times 21 \\ \hline 21 \\ -21 \longleftarrow 1 \times 21 \\ \hline 0 \end{array}$$

(d)
$$\begin{array}{r} 70\,R30 \\ 31\overline{)\,2200} \\ -217 \longleftarrow 7 \times 31 \\ \hline 30 \end{array}$$

(e)
$$\begin{array}{r} 213\,R7 \\ 11\overline{)\,2350} \\ -22 \longleftarrow 2 \times 11 \\ \hline 15 \\ -11 \longleftarrow 1 \times 11 \\ \hline 40 \\ -33 \longleftarrow 3 \times 11 \\ \hline 7 \end{array}$$

(f)
$$\begin{array}{r} 456 \\ 12\overline{)\,5472} \\ -48 \longleftarrow 4 \times 12 \\ \hline 67 \\ -60 \longleftarrow 5 \times 12 \\ \hline 72 \\ -72 \longleftarrow 6 \times 12 \\ \hline 0 \end{array}$$

(g)
$$\begin{array}{r} 908\,R20 \\ 25\overline{)\,22{,}720} \\ -22\,5 \longleftarrow 9 \times 25 \\ \hline 220 \\ -200 \longleftarrow 8 \times 25 \\ \hline 20 \end{array}$$

(h)
$$\begin{array}{r} 730 \\ 76\overline{)\,55{,}480} \\ -53\,2 \longleftarrow 7 \times 76 \\ \hline 2\,28 \\ -2\,28 \longleftarrow 3 \times 76 \\ \hline 0 \end{array}$$

(i)
$$\begin{array}{r} 4\,153\,R48 \\ 84\overline{)\,348{,}900} \\ -336 \longleftarrow 4 \times 84 \\ \hline 12\,9 \\ -\ 84 \longleftarrow 1 \times 84 \\ \hline 4\,50 \\ -4\,20 \longleftarrow 5 \times 84 \\ \hline 300 \\ -252 \longleftarrow 3 \times 84 \\ \hline 48 \end{array}$$

$$\begin{array}{r}6\,020\\63\overline{)379{,}260}\end{array}$$

(j) 63) 379,260
 −378 ←— 6 × 63
 1 26
 −1 26 ←— 2 × 63
 0

(k) 99) 599,742
 −594 ←— 6 × 99
 5 74
 −4 95 ←— 5 × 99
 792
 −792 ←— 8 × 99
 0
(quotient 6 058)

(l) 58) 69,862
 −58 ←— 1 × 58
 11 8
 −11 6 ←— 2 × 58
 262
 −232 ←— 4 × 58
 30
(quotient 1 204 R30)

(m) 29) 110,780
 − 87 ←— 3 × 29
 23 7
 −23 2 ←— 8 × 29
 58
 −58 ←— 2 × 29
 0
(quotient 3 820)

(n) 19) 134,000
 −133 ←— 7 × 19
 1 00
 − 95 ←— 5 × 19
 50
 −38 ←— 2 × 19
 12
(quotient 7 052 R12)

(o) 18) 90,000
 −90 ←— 5 × 18
 0000
(quotient 5 000)

PROBLEM 3-8 Divide by divisors containing three or more digits.

(a) 1200 ÷ 300   **(b)** 7300 ÷ 200   **(c)** 3360 ÷ 420   **(d)** 48,000 ÷ 510
(e) 1645 ÷ 235   **(f)** 10,565 ÷ 123   **(g)** 199,424 ÷ 608   **(h)** 357,000 ÷ 709
(i) 5915 ÷ 845   **(j)** 15,354 ÷ 756   **(k)** 40,000 ÷ 5000   **(l)** 169,825 ÷ 4000
(m) 1,755,000 ÷ 2500   **(n)** 2,484,000 ÷ 3900   **(o)** 1,439,570 ÷ 2845

Solution: Recall that to divide by three or more digits, you can use the first digit of the divisor to find an estimate for each digit in the quotient [see Example 3-17]:

(a) 3)12 → 4; 300) 1200, −1200 ←— 4 × 300, 0

(b) about 3: 2)7; about 6: 2)13; 200) 7300 = 36 R100, −600 ←— 3 × 200, 1300, −1200 ←— 6 × 200, 100

(c) about 8: 4)33; 420) 3360 = 8, −3360 ←— 8 × 420, 0

(d) about 9: 5)48; about 4: 5)21; 510) 48,000 = 94 R60, −45 90 ←— 9 × 510, 2 100, −2 040 ←— 4 × 510, 60

(e) 2)16 → 8 → 7; 235) 1645 = 7, −1645 ←— 7 × 235, 0

(f) 1)10 → 10 → 9 → 8; 1)7 → 7 → 6 → 5; 123) 10,565 = 85 R110, − 9 84 ←— 8 × 123, 725, −615 ←— 5 × 123, 110

(g) about 3 | 6)19   328 | 608) 199,424

about 2 | 6)17   −182 4 ←—— 3 × 608

          17 02

          −12 16 ←—— 2 × 608

           4 864

8 | 6)48   −4 864 ←—— 8 × 608

             0

(h) 5 | 7)35   503 R373 | 709) 357,000

          −354 5 ←—— 5 × 709

about 3 | 7)25   2 500

          −2 127 ←—— 3 × 709

            373

(i) about 7 | 8)59   7 | 845) 5915

         −5915 ←—— 7 × 845

           0

(j) about 2 | 7)15   20 R234 | 756) 15,354

         −15 12 ←—— 2 × 756

          234

(k) 8 | 5)40   8 | 5000) 40,000

        −40,000 ←—— 8 × 5000

          0

(l) 4 | 4)16   42 R1825 | 4000) 169,825

        −160 00 ←—— 4 × 4000

2 | 4)9   9 825

        −8 000 ←—— 2 × 4000

         1 825

(m) 8 → 7 | 2)17   702 | 2500) 1,755,000

        −1 750 0 ←—— 7 × 2500

2 | 2)5   5 000

        −5 000 ←—— 2 × 2500

         0

(n) 8 → 7 → 6 | 3)24   636 R3600 | 3900) 2,484,000

        −2 340 0 ←—— 6 × 3900

4 → 3 | 3)14   144 00

        −117 00 ←—— 3 × 3900

9 → 8 → 7 → 6 | 3)27   27 000

        −23 400 ←—— 6 × 3900

         3 600

(o) 7 → 6 → 5 | 2)14   506 | 2845) 1,439,570

        −1 422 5 ←—— 5 × 2845

8   7   6 | 2)17   17 070

        −17 070 ←—— 6 × 2845

         0

PROBLEM 3-9 Find all the whole-number factors of each given number.

(a) 1   **(b)** 2   **(c)** 3   **(d)** 4   **(e)** 5   **(f)** 6   **(g)** 7   **(h)** 8   **(i)** 9   **(j)** 10
(k) 11   **(l)** 12   **(m)** 13   **(n)** 14   **(o)** 15   **(p)** 16   **(q)** 18   **(r)** 20   **(s)** 24
(t) 30   **(u)** 54   **(v)** 72   **(w)** 100   **(x)** 108   **(y)** 112   **(z)** 184

Solution: Recall a given whole number is a factor of a second whole number if the given whole number divides the given whole number evenly [see Example 3-19]:

(a) 1   **(b)** 1, 2   **(c)** 1, 3   **(d)** 1, 2, 4   **(e)** 1, 5   **(f)** 1, 2, 3, 6   **(g)** 1, 7   **(h)** 1, 2, 4, 8

(i) 1, 3, 9 **(j)** 1, 2, 5, 10 **(k)** 1, 11 **(l)** 1, 2, 3, 4, 6, 12 **(m)** 1, 13 **(n)** 1, 2, 7, 14
(o) 1, 3, 5, 15 **(p)** 1, 2, 4, 8, 16 **(q)** 1, 2, 3, 6, 9, 18 **(r)** 1, 2, 4, 5, 10, 20
(s) 1, 2, 3, 4, 6, 8, 12, 24 **(t)** 1, 2, 3, 5, 6, 10, 15, 30 **(u)** 1, 2, 3, 6, 9, 18, 27, 54
(v) 1, 2, 3, 4, 6, 8, 9, 12, 18, 24, 36, 72 **(w)** 1, 2, 4, 5, 10, 20, 25, 50, 100
(x) 1, 2, 3, 4, 6, 9, 12, 18, 27, 36, 54, 108 **(y)** 1, 2, 4, 7, 8, 14, 16, 28, 56, 112
(z) 1, 2, 4, 8, 23, 46, 92, 184

PROBLEM 3-10 Circle each prime number. Cross out each composite number.

0	1	2	3	4	5	6	7	8	9
10	11	12	13	14	15	16	17	18	19
20	21	22	23	24	25	26	27	28	29
30	31	32	33	34	35	36	37	38	39
40	41	42	43	44	45	46	47	48	49
50	51	52	53	54	55	56	57	58	59
60	61	62	63	64	65	66	67	68	69
70	71	72	73	74	75	76	77	78	79
80	81	82	83	84	85	86	87	88	89
90	91	92	93	94	95	96	97	98	99

Solution: Recall that a prime number has exactly two different factors and a composite number has more than two different factors [see Examples 3-20 and 3-24]:

0	1	(2)	(3)	~~4~~	(5)	~~6~~	(7)	~~8~~	~~9~~
~~10~~	(11)	~~12~~	(13)	~~14~~	~~15~~	~~16~~	(17)	~~18~~	(19)
~~20~~	~~21~~	~~22~~	(23)	~~24~~	~~25~~	~~26~~	~~27~~	~~28~~	(29)
~~30~~	(31)	~~32~~	~~33~~	~~34~~	~~35~~	~~36~~	(37)	~~38~~	~~39~~
~~40~~	(41)	~~42~~	(43)	~~44~~	~~45~~	~~46~~	(47)	~~48~~	~~49~~
~~50~~	~~51~~	~~52~~	(53)	~~54~~	~~55~~	~~56~~	~~57~~	~~58~~	(59)
~~60~~	(61)	~~62~~	~~63~~	~~64~~	~~65~~	~~66~~	(67)	~~68~~	~~69~~
~~70~~	(71)	~~72~~	(73)	~~74~~	~~75~~	~~76~~	~~77~~	~~78~~	(79)
~~80~~	~~81~~	~~82~~	(83)	~~84~~	~~85~~	~~86~~	~~87~~	~~88~~	(89)
~~90~~	~~91~~	~~92~~	~~93~~	~~94~~	~~95~~	~~96~~	(97)	~~98~~	~~99~~

PROBLEM 3-11 Identify each whole number as odd or even: **(a)** 0 **(b)** 1 **(c)** 5 **(d)** 8 **(e)** 107 **(f)** 2469 **(g)** 3792

Solution: Recall that even numbers end in 0, 2, 4, 6, or 8 and odd numbers end in 1, 3, 5, 7, or 9 [see Example 3-21]: **(a)** even **(b)** odd **(c)** odd **(d)** even **(e)** odd **(f)** odd **(g)** even

PROBLEM 3-12 Factor each whole number in as many different ways as possible: **(a)** 0 **(b)** 1 **(c)** 2 **(d)** 4 **(e)** 18 **(f)** 100

Solution: Recall that to factor a whole number, you write it as a product of primes and/or composite numbers [see Example 3-25]:

(a) $0 \times 1, 0 \times 2, 0 \times 3, \cdots$ (and infinitely many more different ways)
(b) 1×1 **(c)** 1×2 **(d)** $1 \times 4, 2 \times 2$ **(e)** $1 \times 18, 2 \times 9, 3 \times 6, 2 \times 3 \times 3$
(f) $1 \times 100, 2 \times 50, 4 \times 25, 5 \times 20, 10 \times 10, 2 \times 2 \times 25, 2 \times 5 \times 10, 4 \times 5 \times 5, 2 \times 2 \times 5 \times 5$

PROBLEM 3-13 Factor each composite number as a product of primes using division: **(a)** 4 **(b)** 8 **(c)** 10 **(d)** 16 **(e)** 21 **(f)** 24 **(g)** 26 **(h)** 28 **(i)** 32 **(j)** 34 **(k)** 36 **(l)** 39 **(m)** 42 **(n)** 45 **(o)** 48 **(p)** 50 **(q)** 52 **(r)** 55 **(s)** 58 **(t)** 62 **(u)** 64 **(v)** 66 **(w)** 68 **(x)** 69 **(y)** 72 **(z)** 75

Solution: [see Examples 3-27 and 3-28]:

(a) 2 ⌊ 4,   2 × 2
       ‾2‾

(b) 2 ⌊ 8,   2 × 2 × 2
       2 ⌊ 4
         ‾2‾

(c) 2 ⌊ 10,   2 × 5
        ‾5‾

(d) 2 ⌊ 16,   2 × 2 × 2 × 2
       2 ⌊ 8
       2 ⌊ 4
         ‾2‾

(e) 3 ⌊ 21,   3 × 7
        ‾7‾

(f) 2 ⌊ 24,   2 × 2 × 2 × 3
       2 ⌊ 12
       2 ⌊ 6
         ‾3‾

(g) 2 ⌊ 26,   2 × 13
        ‾13‾

(h) 2 ⌊ 28,   2 × 2 × 7
       2 ⌊ 14
         ‾7‾

(i) 2 ⌊ 32,   2 × 2 × 2 × 2 × 2
       2 ⌊ 16
       2 ⌊ 8
       2 ⌊ 4
         ‾2‾

(j) 2 ⌊ 34,   2 × 17
        ‾17‾

(k) 2 ⌊ 36,   2 × 2 × 3 × 3
       2 ⌊ 18
       3 ⌊ 9
         ‾3‾

(l) 3 ⌊ 39,   3 × 13
        ‾13‾

(m) 2 ⌊ 42,   2 × 3 × 7
       3 ⌊ 21
         ‾7‾

(n) 3 ⌊ 45,   3 × 3 × 5
       3 ⌊ 15
         ‾5‾

(o) 2 ⌊ 48,   2 × 2 × 2 × 2 × 3
       2 ⌊ 24
       2 ⌊ 12
       2 ⌊ 6
         ‾3‾

(p) 2 ⌊ 50,   2 × 5 × 5
       5 ⌊ 25
         ‾5‾

(q) 2 ⌊ 52,   2 × 2 × 13
       2 ⌊ 26
         ‾13‾

(r) 5 ⌊ 55,   5 × 11
        ‾11‾

(s) 2 ⌊ 58,   2 × 29
        ‾29‾

(t) 2 ⌊ 62,   2 × 31
        ‾31‾

(u) 2 ⌊ 64,   2 × 2 × 2 × 2 × 2 × 2
       2 ⌊ 32
       2 ⌊ 16
       2 ⌊ 8
       2 ⌊ 4
         ‾2‾

(v) 2 ⌊ 66,   2 × 3 × 11
       3 ⌊ 33
         ‾11‾

(w) 2 ⌊ 68,   2 × 2 × 17
       2 ⌊ 34
         ‾17‾

(x) 3 ⌊ 69,   3 × 23
        ‾23‾

(y) 2 ⌊ 72,   2 × 2 × 2 × 3 × 3
       2 ⌊ 36
       2 ⌊ 18
       3 ⌊ 9
         ‾3‾

(z) 3 ⌊ 75,   3 × 5 × 5
       5 ⌊ 25
         ‾5‾

PROBLEM 3-14   Factor each composite number as a product of primes using a factor tree: **(a)** 6
(b) 9   **(c)** 14   **(d)** 20   **(e)** 22   **(f)** 25   **(g)** 27   **(h)** 33   **(i)** 35   **(j)** 38   **(k)** 40
(l) 44   **(m)** 46   **(n)** 49   **(o)** 51   **(p)** 54   **(q)** 57   **(r)** 60   **(s)** 63   **(t)** 65   **(u)** 70
(v) 74   **(w)** 76   **(x)** 77   **(y)** 78   **(z)** 80

Solution: [see Examples 3-29 and 3-30]:

(a) 6
2 × 3

(b) 9
3 × 3

(c) 14
2 × 7

(d) 20
2 × 10
2 × 2 × 5

(e) 22
2 × 11

(f) 25
5 × 5

(g) 27
3 × 9
3 × 3 × 3

(h) 33
3 × 11

(i) 35
5 × 7

(j) 38
2 × 19

(k) 40
2 × 20
2 × 2 × 10
2 × 2 × 2 × 5

(l) 44
2 × 22
2 × 2 × 11

(m) 46
2 × 23

(n) 49
7 × 7

(o) 51
3 × 17

(p) 54
2 × 27
2 × 3 × 9
2 × 3 × 3 × 3

(q) 57
3 × 19

(r) 60
2 × 30
2 × 2 × 15
2 × 2 × 3 × 5

(s) 63
3 × 21
3 × 3 × 7

(t) 65
5 × 13

(u) 70
2 × 35
2 × 5 × 7

(v) 74
2 × 37

(w) 76
2 × 38
2 × 2 × 19

(x) 77
7 × 11

(y) 78
2 × 39
2 × 3 × 13

(z) 80
2 × 40
2 × 2 × 20
2 × 2 × 2 × 10
2 × 2 × 2 × 2 × 5

PROBLEM 3-15   Decide what to do (add, subtract, multiply, or divide) and then solve the problem:

(a) Donald weighed 200 pounds. He lost 27 pounds. How much did Donald weigh then?

(b) Karen weighed 108 pounds. She gained 17 pounds. How much did Karen weigh then?

(c) Warren sold 125 books last week. Each book cost $25. How many dollars worth of books did Warren sell last week?

(d) Marlene sold $125 worth of books last week. Each book cost $25. How many books did Marlene sell last week?

(e) Charles sold 125 books last week. His sales amounted to $2500. How much did each book cost?

(f) Susan sold 125 books last week and Paul 25 books. What was the difference between the number of books they sold?

(g) Leslie sold 125 books last week and Bill sold 25 books. How many books did they sell together?

(h) There are 144 eggs in one flat. There are 12 eggs in one dozen. How many dozen eggs are in one flat?

(i) There are 144 eggs in one flat. A restaurant orders 12 flats. How many eggs were ordered?

(j) A board is 255 cm long. It is cut so that each piece is 15 cm in length. How many pieces are there?

(k) A board is 364 cm in length. It is cut in 28 pieces of equal length. How long is each piece?

(l) Ann purchased 18 gallons of gasoline for $2 per gallon. How much did she pay for the gasoline?

(m) Bob purchased 18 liters of gasoline for $9 (900 cents). How much did Bob pay for each liter?

(n) There are 216 rolls in 27 bags. If there are the same number of rolls in each bag, how many rolls are in each bag?

(o) Aaron worked 7 hours each of 5 days last week. He earned $245 for the week. How much does Aaron earn per hour?

(p) Natalie worked 40 hours at $8 per hour. Then she worked 26 hours at double her hourly rate. How much did Natalie earn in all?

Solution: [see Examples 3-31 and 3-32]:

(a) subtract: $200 - 27 = 173$ (pounds)

(b) add: $108 + 17 = 125$ (pounds)

(c) multiply: $125 \times 25 = 3125$ (dollars)

(d) divide: $125 \div 25 = 5$ (books)

(e) divide: $2500 \div 125 = 20$ (dollars)

(f) subtract: $125 - 25 = 100$ (books)

(g) add: $125 + 25 = 150$ (books)

(h) divide: $144 \div 12 = 12$ (dozen)

(i) multiply: $144 \times 12 = 1728$ (eggs)

(j) divide: $255 \div 15 = 17$ (pieces)

(k) divide: $364 \div 28 = 13$ (cm)

(l) multiply: $18 \times 2 = 36$ (dollars)

(m) divide: $900 \div 18 = 50$ (cents)

(n) divide: $216 \div 27 = 8$ (rolls)

(o) multiply and then divide: $7 \times 5 = 35$, and $245 \div 35 = 7$ (dollars)

(p) multiply three times and then add: $8 \times 40 = 320$, $8 \times 2 = 16$, $26 \times 16 = 416$, and $320 + 416 = 736$ (dollars)

Supplementary Exercises

PROBLEM 3-16   Divide basic facts given division box form:

Row **(a)**	3)27	8)16	7)56	5)0	2)4	5)30	5)15	5)40	8)8	9)0
Row **(b)**	7)28	9)81	2)16	2)10	9)45	8)24	1)7	5)45	1)3	4)20
Row **(c)**	6)30	8)56	3)21	4)4	7)14	9)63	6)42	6)6	8)72	1)9
Row **(d)**	8)40	7)42	3)9	1)5	1)0	4)12	6)54	4)28	3)18	9)27
Row **(e)**	4)36	2)14	1)6	7)35	7)7	3)0	9)72	7)49	4)32	6)18
Row **(f)**	5)20	3)12	2)8	4)24	9)9	8)64	5)25	3)6	7)0	7)63
Row **(g)**	2)12	2)6	8)32	1)8	4)8	6)48	6)0	6)36	2)18	5)35
Row **(h)**	5)10	1)1	3)15	5)5	3)24	4)0	9)18	8)0	8)48	3)3
Row **(i)**	6)24	1)4	4)16	2)2	9)54	7)21	2)0	1)2	9)36	6)12

PROBLEM 3-17   Divide basic facts given horizontal form:

Row (a)	24 ÷ 6	10 ÷ 5	12 ÷ 2	20 ÷ 5	36 ÷ 4	40 ÷ 8	30 ÷ 6	28 ÷ 7	27 ÷ 3	4 ÷ 1
Row (b)	1 ÷ 1	6 ÷ 2	12 ÷ 3	14 ÷ 2	42 ÷ 7	56 ÷ 8	81 ÷ 9	16 ÷ 8	16 ÷ 4	15 ÷ 3
Row (c)	32 ÷ 8	8 ÷ 2	6 ÷ 1	9 ÷ 3	21 ÷ 3	16 ÷ 2	56 ÷ 7	2 ÷ 2	5 ÷ 5	8 ÷ 1
Row (d)	24 ÷ 4	35 ÷ 7	5 ÷ 1	4 ÷ 4	10 ÷ 2	0 ÷ 5	21 ÷ 7	0 ÷ 4	48 ÷ 6	64 ÷ 8
Row (e)	0 ÷ 3	12 ÷ 4	63 ÷ 9	24 ÷ 8	30 ÷ 5	2 ÷ 1	0 ÷ 8	36 ÷ 6	6 ÷ 3	49 ÷ 7
Row (f)	28 ÷ 4	6 ÷ 6	45 ÷ 5	40 ÷ 5	36 ÷ 9	48 ÷ 8	18 ÷ 2	0 ÷ 7	32 ÷ 4	18 ÷ 3
Row (g)	72 ÷ 8	3 ÷ 1	8 ÷ 8	12 ÷ 6	3 ÷ 3	35 ÷ 5	63 ÷ 7	18 ÷ 6	27 ÷ 9	9 ÷ 1
Row (h)	20 ÷ 4	0 ÷ 9	54 ÷ 9	24 ÷ 3	8 ÷ 4	9 ÷ 9	7 ÷ 7	0 ÷ 1	14 ÷ 7	45 ÷ 9
Row (i)	4 ÷ 2	0 ÷ 2	18 ÷ 9	0 ÷ 6	25 ÷ 5	72 ÷ 9	54 ÷ 6	42 ÷ 6	7 ÷ 1	15 ÷ 5

PROBLEM 3-18   Divide by a one-digit divisor:

(a) 15 ÷ 2   (b) 35 ÷ 4   (c) 23 ÷ 6   (d) 50 ÷ 8
(e) 45 ÷ 3   (f) 78 ÷ 5   (g) 609 ÷ 7   (h) 990 ÷ 9
(i) 246 ÷ 2   (j) 1069 ÷ 5   (k) 1740 ÷ 4   (l) 3070 ÷ 7
(m) 4248 ÷ 6   (n) 7475 ÷ 9   (o) 19,896 ÷ 8   (p) 24,230 ÷ 3
(q) 37,128 ÷ 7   (r) 12,059 ÷ 6

PROBLEM 3-19   Divide by a two-digit divisor:

(a) 100 ÷ 20   (b) 250 ÷ 30   (c) 33 ÷ 11   (d) 200 ÷ 21
(e) 912 ÷ 38   (f) 1050 ÷ 29   (g) 2250 ÷ 45   (h) 1089 ÷ 18
(i) 15,872 ÷ 64   (j) 27,900 ÷ 72   (k) 6885 ÷ 17   (l) 15,139 ÷ 28
(m) 251,906 ÷ 31   (n) 127,090 ÷ 42   (o) 78,033 ÷ 19   (p) 195,350 ÷ 39
(q) 138,040 ÷ 68   (r) 632,075 ÷ 79

PROBLEM 3-20   Divide by three or more digits:

(a) 2400 ÷ 300   (b) 1900 ÷ 200   (c) 2460 ÷ 410   (d) 4000 ÷ 520
(e) 13,041 ÷ 621   (f) 10,852 ÷ 312   (g) 30,972 ÷ 534   (h) 43,000 ÷ 653
(i) 266,875 ÷ 875   (j) 62,500 ÷ 297   (k) 151,200 ÷ 189   (l) 143,500 ÷ 685
(m) 6000 ÷ 2000   (n) 23,000 ÷ 4000   (o) 54,180 ÷ 3010   (p) 128,000 ÷ 5008
(q) 207,230 ÷ 2438   (r) 260,000 ÷ 2875

PROBLEM 3-21   Factor each whole number in as many different ways as possible:

(a) 3   (b) 5   (c) 6   (d) 8   (e) 10   (f) 20   (g) 36

PROBLEM 3-22   Factor each composite number as a product of primes:

(a) 12   (b) 82   (c) 15   (d) 85   (e) 18   (f) 87   (g) 88   (h) 90   (i) 91   (j) 92
(k) 93   (l) 94   (m) 95   (n) 96   (o) 98   (p) 99   (q) 100   (r) 102   (s) 86
(t) 56   (u) 84   (v) 120   (w) 81   (x) 144   (y) 150   (z) 180

PROBLEM 3-23   Solve the following word problems (+, −, ×, ÷):

(a) A certain car traveled 440 miles in 8 hours at a constant speed. What was the constant speed?

(b) A certain car traveled 375 miles on 15 gallons of gasoline. How many miles per gallon (mpg) did the car get?

(c) A certain train traveled 5 hours at 65 miles per hour (mph). How many miles did the train travel?

(d) A certain train traveled 315 miles at a constant rate of 45 mph. How long did the trip take?

(e) A certain bus traveled 720 miles while averaging 15 mpg. How many gallons of fuel were used?

(f) A certain bus used 50 gallons of fuel while averaging 16 mpg. How far did the bus travel?

(g) A certain plane traveled 825 miles on the first part of a trip and 975 miles on the second part. How far did the plane travel in all?

(h) The air distance between two cities is 1285 km. The driving distance between the same two cities is 1487 km. How much farther is the driving distance than the air distance?

(i) Linda earns $9 per hour. She worked 35 hours this week. How much did Linda earn this week?

(j) Paul earned $392 this week. He earns $8 per hour. How many hours did Paul work this week?

(k) Pat earned $504 this week while working 42 hours. How much does she earn per hour?

(l) A certain TV set costs $600. If there are 12 equal payments, how much is each payment for the TV set?

(m) A certain clothes dryer costs $500. Each payment is $25. How many payments must be made in all?

(n) A certain car is purchased for no money down and $285 per month for 36 months. How much will the car cost?

(o) There are 365 days in a normal year. How many complete 7-day weeks are in a normal year?

(p) There are 30 days in a business month. There are 12 business months in a business year. How many days in a business year?

(q) Beth worked 8 hours on Monday, 6 hours on Tuesday, 8 hours on Wednesday, 7 hours on Thursday, and 5 hours on Friday. She earns $15 per hour. How much did Beth earn Monday through Friday?

(r) Marcia earns $25 per hour. Frank earns $18 per hour. How much more does Marcia earn during a forty-hour work week than Frank?

Answers to Supplementary Exercises

(3-16)

Row (a)	9	2	8	0	2	6	3	8	1	0
Row (b)	4	9	8	5	5	3	7	9	3	5
Row (c)	5	7	7	1	2	7	7	1	9	9
Row (d)	5	6	3	5	0	3	9	7	6	3
Row (e)	9	7	6	5	1	0	8	7	8	3
Row (f)	4	4	4	6	1	8	5	2	0	9
Row (g)	6	3	4	8	2	8	0	6	9	7
Row (h)	2	1	5	1	8	0	2	0	6	1
Row (i)	4	4	4	1	6	3	0	2	4	2

(3-17)

Row (a)	4	2	6	4	9	5	5	4	9	4
Row (b)	1	3	4	7	6	7	9	2	4	5
Row (c)	4	4	6	3	7	8	8	1	1	8
Row (d)	6	5	5	1	5	0	3	0	8	8
Row (e)	0	3	7	3	6	2	0	6	2	7
Row (f)	7	1	9	8	4	6	9	0	8	6
Row (g)	9	3	1	2	1	7	9	3	3	9
Row (h)	5	0	6	8	2	1	1	0	2	5
Row (i)	2	0	2	0	5	8	9	7	7	3

(3-18) (a) 7 R1   (b) 8 R3   (c) 3 R5   (d) 6 R2   (e) 15   (f) 15 R3   (g) 87
(h) 110   (i) 123   (j) 213 R4   (k) 435   (l) 438 R4   (m) 708   (n) 830 R5
(o) 2487   (p) 8076 R2   (q) 5304   (r) 2009 R5

(3-19) **(a)** 5 **(b)** 8 R10 **(c)** 3 **(d)** 9 R11 **(e)** 24 **(f)** 36 R6 **(g)** 50 **(h)** 60 R9 **(i)** 248 **(j)** 387 R36 **(k)** 405 **(l)** 540 R19 **(m)** 8126 **(n)** 3025 R40 **(o)** 4107 **(p)** 5008 R38 **(q)** 2030 **(r)** 8000 R75

(3-20) **(a)** 8 **(b)** 9 R100 **(c)** 6 **(d)** 7 R360 **(e)** 21 **(f)** 34 R244 **(g)** 58 **(h)** 65 R555 **(i)** 305 **(j)** 210 R130 **(k)** 800 **(l)** 209 R335 **(m)** 3 **(n)** 5 R3000 **(o)** 18 **(p)** 25 R2800 **(q)** 85 **(r)** 90 R1250

(3-21) **(a)** 1×3 **(b)** 1×5 **(c)** $1 \times 6, 2 \times 3$ **(d)** $1 \times 8, 2 \times 4$ **(e)** $1 \times 10, 2 \times 5$ **(f)** $1 \times 20, 2 \times 10, 4 \times 5, 2 \times 2 \times 5$ **(g)** $1 \times 36, 2 \times 18, 3 \times 12, 4 \times 9, 6 \times 6, 2 \times 2 \times 9, 2 \times 3 \times 6, 3 \times 3 \times 4, 2 \times 2 \times 3 \times 3$

(3-22) **(a)** $2 \times 2 \times 3$ **(b)** 2×41 **(c)** 3×5 **(d)** 5×17 **(e)** $2 \times 3 \times 3$ **(f)** 3×29 **(g)** $2 \times 2 \times 2 \times 11$ **(h)** $2 \times 3 \times 3 \times 5$ **(i)** 7×13 **(j)** $2 \times 2 \times 23$ **(k)** 3×31 **(l)** 2×47 **(m)** 5×19 **(n)** $2 \times 2 \times 2 \times 2 \times 2 \times 3$ **(o)** $2 \times 7 \times 7$ **(p)** $3 \times 3 \times 11$ **(q)** $2 \times 2 \times 5 \times 5$ **(r)** $2 \times 3 \times 17$ **(s)** 2×43 **(t)** $2 \times 2 \times 2 \times 7$ **(u)** $2 \times 2 \times 3 \times 7$ **(v)** $2 \times 2 \times 2 \times 3 \times 5$ **(w)** $3 \times 3 \times 3 \times 3$ **(x)** $2 \times 2 \times 2 \times 2 \times 3 \times 3$ **(y)** $2 \times 3 \times 5 \times 5$ **(z)** $2 \times 2 \times 3 \times 3 \times 5$

(3-23) **(a)** 55 mph **(b)** 25 mpg **(c)** 325 miles **(d)** 7 hours **(e)** 48 gallons **(f)** 800 miles **(g)** 1800 miles **(h)** 202 km **(i)** $315 **(j)** 49 hours **(k)** $12 per hour **(l)** $50 **(m)** 20 payments **(n)** $10,260 **(o)** 52 complete weeks **(p)** 360 days **(q)** $510 **(r)** $280

4 FRACTIONS AND MIXED NUMBERS: RENAME AND SIMPLIFY

THIS CHAPTER IS ABOUT

☑ **Understanding Fractions and Mixed Numbers**
☑ **Renaming Fractions and Mixed Numbers**
☑ **Simplifying Fractions and Mixed Numbers**

4-1. Understanding Fractions and Mixed Numbers

A. Understand fractions.

If a and b are whole numbers ($b \neq 0$), then $\frac{a}{b}$ is called a **common fraction,** or just a **fraction.**

EXAMPLE 4-1: Use the whole numbers 2 and 3 to make two different fractions.

Solution: $\frac{2}{3}$ and $\frac{3}{2}$ ⟵── fractions

In the fraction $\frac{2}{3}$, 2 is called the **numerator,** 3 is called the **denominator,** and the horizontal bar that separates 2 and 3 is called the **fraction bar.**

EXAMPLE 4-2: Name the three different parts of the fraction $\frac{0}{5}$.

Solution: $\dfrac{0}{5}$ ⟵── numerator
⟵── fraction bar
⟵── denominator

Note: If $\frac{a}{b}$ is a fraction, then by definition the denominator b cannot be zero ($b \neq 0$).

EXAMPLE 4-3: Which one of the following is not defined as a fraction: (a) $\frac{0}{5}$ (b) $\frac{5}{0}$

Solution
(a) $\frac{0}{5}$ is defined as a fraction because 0 and 5 are whole numbers and $5 \neq 0$.
(b) $\frac{5}{0}$ is not defined because the denominator is zero.

When the numerator is less than the denominator, the fraction is called a **proper fraction.**

EXAMPLE 4-4: Which of the following are proper fractions: (a) $\frac{3}{4}$ (b) $\frac{9}{9}$ (c) $\frac{5}{2}$
(a) $\frac{3}{4}$ is a proper fraction because 3 is less than 4 ($3 < 4$).
(b) $\frac{9}{9}$ is not a proper fraction because 9 is not less than 9 ($9 = 9$).
(c) $\frac{5}{2}$ is not a proper fraction because 5 is not less than 2 ($5 > 2$).

To name part of a whole, you can use a proper fraction.

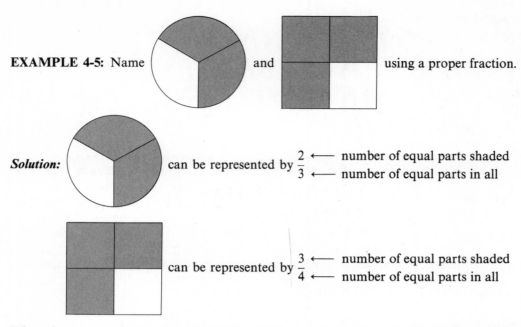

EXAMPLE 4-5: Name ⬭ and ▣ using a proper fraction.

Solution: ⬭ can be represented by $\dfrac{2}{3}$ ⟵ number of equal parts shaded / ⟵ number of equal parts in all

▣ can be represented by $\dfrac{3}{4}$ ⟵ number of equal parts shaded / ⟵ number of equal parts in all

When the numerator is greater than or equal to the denominator, the fraction is called an **improper fraction.**

EXAMPLE 4-6: Which of the following are improper fractions: **(a)** $\frac{2}{3}$ **(b)** $\frac{5}{5}$ **(c)** $\frac{4}{1}$

Solution
(a) $\frac{2}{3}$ is not an improper fraction because 2 is not greater than or equal to 3 ($2 < 3$).
(b) $\frac{5}{5}$ is an improper fraction because 5 is greater than or equal to 5 ($5 = 5$).
(c) $\frac{4}{1}$ is an improper fraction because 4 is greater than or equal to 1 ($4 > 1$).

To read or write the word name of a fraction, you use the whole-number name of the numerator, plus the name of the denominator, as follows:
(a) wholes if the denominator is 1;
(b) halves if the denominator is 2;
(c) thirds if the denominator is 3;
(d) fourths if the denominator is 4;
(e) fifths if the denominator is 5;
and so on.

EXAMPLE 4-7: How would you read or write the following fractions? $\frac{9}{1}, \frac{5}{2}, \frac{2}{3}, \frac{3}{4}, \frac{7}{8}, \frac{11}{10}$

Solution: Read or write $\frac{9}{1}$ as "nine-wholes."   Read or write $\frac{5}{2}$ as "five-halves."
Read or write $\frac{2}{3}$ as "two-thirds."   Read or write $\frac{3}{4}$ as "three-fourths."
Read or write $\frac{7}{8}$ as "seven-eighths."   Read or write $\frac{11}{10}$ as "eleven-tenths."

Caution: When writing the word name for a fraction, always place a hyphen between the word name for the numerator and the word name for the denominator.

Note: When the numerator of the fraction is 1, the denominator is singular, and should be read or written that way.

EXAMPLE 4-8: How would you read or write the following fractions? $\frac{1}{2}, \frac{1}{3}, \frac{1}{4}, \frac{1}{5}, \frac{1}{6}$

Solution: Read or write $\frac{1}{2}$ as "one-half."   ⟵ *half* is the singular version of *halves*
Read or write $\frac{1}{3}$ as "one-third."   Read or write $\frac{1}{4}$ as "one-fourth."
Read or write $\frac{1}{5}$ as "one-fifth."   Read or write $\frac{1}{6}$ as "one-sixth."

B. Understand mixed numbers.

A whole number joined together with a fraction is called a **mixed number.**

EXAMPLE 4-9: Use 2 and $\frac{3}{4}$ to write a mixed number.

Solution: 2 and $\frac{3}{4}$ can be written as the mixed number $2\frac{3}{4}$.

In the mixed number $2\frac{3}{4}$, 2 is called the **whole-number part** and $\frac{3}{4}$ is called the **fraction part.** To name one or more wholes and part of a whole together, you can use a mixed number.

EXAMPLE 4-10:

Name  using a mixed number.

Solution:

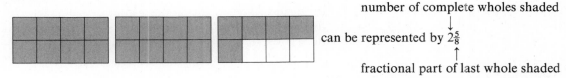

number of complete wholes shaded

↓

can be represented by $2\frac{5}{8}$

↑

fractional part of last whole shaded

Caution: Always include the word *and* between the whole-number part and the fraction part when reading or writing a mixed number.

EXAMPLE 4-11: How would you read or write the word name for $2\frac{3}{4}$?

Solution: Read or write $2\frac{3}{4}$ as "two and three-fourths."

Note: Mixed numbers are a short way to write the sum of a whole number and a fraction.

EXAMPLE 4-12: Rename $2\frac{3}{4}$ as a sum of a whole number and a fraction.

Solution: $2\frac{3}{4} = 2 + \frac{3}{4}$ ◄——— sum

EXAMPLE 4-13: Rename $5 + \frac{2}{3}$ as a mixed number.

Solution: $5 + \frac{2}{3} = 5\frac{2}{3}$ ◄——— mixed number

4-2. Renaming Fractions and Mixed Numbers

A. Identify equal fractions.

Two fractions that represent the same part of a whole are called **equal fractions.**

EXAMPLE 4-14: Are $\frac{3}{4}$ and $\frac{6}{8}$ equal fractions?

Solution: $\frac{3}{4}$ and $\frac{6}{8}$ are equal fractions because $\frac{3}{4}$ and $\frac{6}{8}$ represents the same part of a whole:

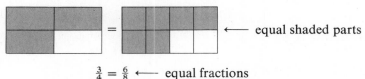 ◄——— equal shaded parts

$\frac{3}{4} = \frac{6}{8}$ ◄——— equal fractions

Two given fractions are equal if their **cross products** are equal.

EXAMPLE 4-15: Are $\frac{12}{18}$ and $\frac{10}{15}$ equal fractions?

Solution

$15 \times 12 = 180$ $\qquad\qquad$ $18 \times 10 = 180$ ⟵ cross product

$$\frac{12}{18} \times \frac{10}{15}$$ $\qquad$ Cross multiply.

$\qquad\qquad 180 = 180 \qquad$ Compare cross products.

Therefore, $\frac{12}{18}$ and $\frac{10}{15}$ are equal fractions because their cross products are equal.

Note: $\frac{6}{8} \neq \frac{8}{10}$ because their cross products are not equal: $10 \times 6 = 60$

$\qquad\qquad\qquad\qquad\qquad\qquad\qquad\qquad\qquad\qquad\qquad 8 \times 8 = 64$

B. Rename division problems as fractions.

Every division problem can be renamed as a fraction.

EXAMPLE 4-16: Rename $8 \div 3$ as a fraction.

Solution: $8 \div 3$ can be renamed as $\dfrac{8 \;\longleftarrow\; \text{dividend}}{3} \begin{array}{l} \longleftarrow \text{divided by} \\ \longleftarrow \text{divisor} \end{array}$

Note: To rename a division problem as a fraction, you
1. Write the dividend as the numerator.
2. Write the division symbol ($\div$) as the fraction bar (——).
3. Write the divisor as the denominator.

To rename a fraction as a division problem, you just reverse the previous steps as follows:
1. Write the numerator as the dividend.
2. Write the fraction bar as the division symbol.
3. Write the denominator as the divisor.

EXAMPLE 4-17: Rename $\frac{5}{2}$ as a division problem.

Solution: $\frac{5}{2}$ can be renamed as $5 \div 2 \qquad$ or $\qquad 2\overline{)5}$

denominator

numerator

C. Rename whole numbers as equal fractions.

To rename a whole number as an equal fraction, you can write the whole number over 1.

EXAMPLE 4-18: Rename 5 as an equal fraction.

Solution: $5 = 5 \div 1 = \dfrac{5}{1}$ ⟵ a fraction that is equal to 5

Note: There are many ways to rename a whole number as a fraction.

EXAMPLE 4-19: Rename 5 as an equal fraction in several different ways.

Solution: $5 = 5 \div 1 = \dfrac{5}{1}$

$\qquad\quad 5 = 10 \div 2 = \dfrac{10}{2}$

$\qquad\quad 5 = 15 \div 3 = \dfrac{15}{3} \Big\}$ fractions that are equal to 5

$\qquad\quad 5 = 20 \div 4 = \dfrac{20}{4}$

$\qquad\qquad\qquad$ and so on

Agreement: Unless otherwise stated, "rename a whole number as an equal fraction" means you rename the whole number as an equal fraction with a denominator of 1.

To rename a fraction with a denominator of 1 as an equal whole number, you write just the numerator.

EXAMPLE 4-20: Rename $\frac{3}{1}$ as an equal whole number.

Solution: $\frac{3}{1} = 3 \div 1 = 3$ ⟵ whole number that is equal to $\frac{3}{1}$

D. Rename mixed numbers as fractions.

Every mixed number can be renamed as an equal improper fraction.

EXAMPLE 4-21: Rename the following mixed numbers as equal improper fractions:

(a) $2\frac{3}{4}$ (b) $5\frac{7}{8}$

Solution

(a) $2\frac{3}{4} = \frac{11}{4}$ ⟵ $4 \times 2 + 3 = 8 + 3 = 11$
$\phantom{(a) 2\frac{3}{4} = \frac{11}{4}}$ ⟵ same denominator

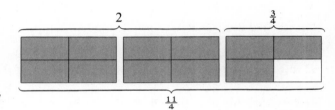

(b) $5\frac{7}{8} = \frac{47}{8}$ ⟵ $8 \times 5 + 7 = 40 + 7 = 47$
$\phantom{(b) 5\frac{7}{8} = \frac{47}{8}}$ ⟵ same denominator

Note: The numerator is the only number that changes when renaming a mixed number; the denominator always remains the same. To find the numerator of an equal improper fraction when renaming a mixed number:
1. Multiply the denominator of the fraction part by the whole-number part.
2. Add that product to the numerator of the fraction part.

To rename an improper fraction as an equal mixed number using division:
1. Divide the numerator by the denominator.
2. The quotient then becomes the whole-number part of the mixed number.
3. The remainder is the numerator of the fraction part.
4. The divisor is the denominator of the fraction part, which always remains the same when renaming.

EXAMPLE 4-22: Rename the following improper fractions as equal mixed numbers using division:
(a) $\frac{23}{4}$ (b) $\frac{14}{3}$

Solution

(a) $\frac{23}{4} = 5\frac{3}{4}$ because: $\frac{23}{4} = 23 \div 4$ or

$$4\overline{)23} \quad 5\tfrac{3}{4} \begin{array}{l} \leftarrow \text{remainder} \\ \leftarrow \text{divisor} \end{array}$$
$$\underline{-20}$$
$$3$$

(b) $\frac{14}{3} = 4\frac{2}{3}$ because: $\frac{14}{3} = 14 \div 3$ or

$$3\overline{)14} \quad 4\tfrac{2}{3} \begin{array}{l} \leftarrow \text{remainder} \\ \leftarrow \text{divisor} \end{array}$$
$$\underline{-12}$$
$$2$$

Note: The answer to the division problem $23 \div 4$ can be written in two forms:
1. remainder form: 5 R3   2. mixed number form: $5\frac{3}{4}$

4-3. Simplifying Fractions and Mixed Numbers

A. Simplify proper fractions.

A fraction for which the numerator and denominator do not share a common factor other than 1 is called a **fraction in lowest terms.**

EXAMPLE 4-23: Which of the following fractions is in lowest terms? $\frac{3}{4}$, $\frac{4}{6}$

Solution: $\frac{3}{4}$ is in lowest terms because 3 and 4 do not share a common factor other than 1.

$$\frac{3}{4} = \frac{3}{2 \times 2} \quad \longrightarrow \text{ no common factors other than 1}$$

$\frac{4}{6}$ is not in lowest terms because 4 and 6 do share a common factor of 2.

$$\frac{4}{6} = \frac{2 \times 2}{3 \times 2} \quad \longrightarrow \text{ common factor of 2}$$

When you add, subtract, multiply, or divide and get a fraction or a mixed number for an answer, you should always **simplify** that answer by writing it in **simplest form.** The simplest form of a fraction or mixed number can be
1. a proper fraction in lowest terms
2. a whole number
3. a mixed number when the fraction part is a proper fraction in lowest terms

To **simplify a proper fraction** that is not in lowest terms, you **reduce the fraction** to lowest terms using the following rule:

Fundamental Rule for Fractions

If a, b, and c are whole numbers ($b \neq 0$ and $c \neq 0$), then $\dfrac{a \times c}{b \times c} = \dfrac{a}{b}$.

Note: According to the Fundamental Rule for Fractions, the value of a fraction will not change when you divide both the numerator and denominator by the same nonzero number.

To simplify a fraction that is not in lowest terms, you first factor the numerator and denominator, and then eliminate all the common factors.

EXAMPLE 4-24: Simplify the following fractions: **(a)** $\frac{6}{8}$ **(b)** $\frac{6}{9}$ **(c)** $\frac{4}{8}$

Solution

(a) $\dfrac{6}{8} = \dfrac{3 \times 2}{4 \times 2} = \dfrac{3}{4}$

(b) $\dfrac{6}{9} = \dfrac{2 \times 3}{3 \times 3} = \dfrac{2}{3} \longleftrightarrow$ simplest form

(c) $\dfrac{4}{8} = \dfrac{1 \times 4}{2 \times 4} = \dfrac{1}{2}$

Caution: A fraction is not in simplest form until all common whole-number factors other than 1 have been eliminated.

EXAMPLE 4-25: Simplify the following fraction: $\frac{12}{18}$

Solution
Wrong Method

$$\frac{12}{18} = \frac{6 \times 2}{9 \times 2} = \frac{6}{9} \longleftarrow \text{ not in lowest terms (6 and 9 have a common factor of 3)}$$

Correct Method

$$\frac{12}{18} = \frac{2 \times \cancel{6}}{3 \times \cancel{6}} = \frac{2}{3} \longleftarrow \text{lowest terms}$$

Caution: If all factors are eliminated in either the numerator or denominator, put the number 1 in place of the eliminated factors.

EXAMPLE 4-26: Simplify the following fraction: $\frac{2}{6}$

Solution

Wrong Method

$$\frac{2}{6} = \frac{\cancel{2}}{\cancel{2} \times 3} = \frac{0}{3} \longleftarrow \text{Wrong! Always replace the eliminated factors with 1.}$$

Correct Method

$$\frac{2}{6} = \frac{\cancel{2}}{\cancel{2} \times 3} = \frac{1}{3} \text{ or } \frac{2}{6} = \frac{\cancel{2} \times 1}{\cancel{2} \times 3} = \frac{1}{3} \longleftarrow \text{correct}$$

Caution: To eliminate a factor in the numerator of a fraction, you must eliminate a like factor in the denominator. Similarly, to eliminate a factor in the denominator, you must eliminate a like factor in the numerator.

EXAMPLE 4-27: Simplify the following fraction: $\frac{16}{36}$

Solution

Wrong Method

$$\frac{16}{36} = \frac{\cancel{4} \times \cancel{4}}{3 \times 3 \times 4} \longleftarrow \text{Wrong! One factor must be from the denominator.}$$

Wrong Method

$$\frac{16}{36} = \frac{4 \times 4}{\cancel{3} \times \cancel{3} \times 4} \longleftarrow \text{Wrong! One factor must be from the numerator.}$$

Correct Method

$$\frac{16}{36} = \frac{4 \times \cancel{4}}{3 \times 3 \times \cancel{4}} = \frac{4}{3 \times 3} \text{ or } \frac{4}{9} \longleftarrow \text{correct}$$

B. Simplify improper fractions.

To **simplify an improper fraction,** you first reduce to lowest terms (when possible) and then rename the improper fraction as a mixed number in lowest terms, or as a whole number.

EXAMPLE 4-28: Simplify the following fractions: **(a)** $\frac{10}{6}$ **(b)** $\frac{45}{15}$

Solution

(a) $\dfrac{10}{6} = \dfrac{5 \times \cancel{2}}{3 \times \cancel{2}} = \dfrac{5}{3} = 1\dfrac{2}{3}$ ⎫

⎬ simplest form

(b) $\dfrac{45}{15} = \dfrac{3 \cdot \cancel{15}}{1 \times \cancel{15}} = \dfrac{3}{1} = 3$ ⎭

C. Simplify mixed numbers.

To **simplify a mixed number** that is not in simplest form, you simplify the fraction part.

EXAMPLE 4-29: Simplify the following mixed numbers: **(a)** $5\frac{8}{10}$ **(b)** $4\frac{11}{3}$ **(c)** $3\frac{18}{2}$

Solution

(a) $5\frac{8}{10} = 5 + \frac{8}{10} = 5 + \frac{4 \times \cancel{2}}{5 \times \cancel{2}} = 5 + \frac{4}{5} \qquad = 5\frac{4}{5}$

(b) $4\frac{11}{3} = 4 + \frac{11}{3} = 4 + 3\frac{2}{3} \qquad = 4 + 3 + \frac{2}{3} = 7 + \frac{2}{3} = 7\frac{2}{3}$ → simplest form

(c) $3\frac{18}{2} = 3 + \frac{18}{2} = 3 + \frac{9 \times \cancel{2}}{1 \times \cancel{2}} = 3 + \frac{9}{1} \qquad = 3 + 9 = 12$

RAISE YOUR GRADES

Can you . . . ?

☑ identify the three parts of a fraction
☑ identify fractions that are not defined
☑ identify proper fractions
☑ name part of a whole using a proper fraction
☑ identify improper fractions
☑ read a fraction aloud correctly
☑ write a fraction in words correctly
☑ identify the two parts of a mixed number
☑ name one or more wholes and part of a whole together using a mixed number
☑ read a mixed number aloud correctly
☑ write a mixed number in words correctly
☑ rename a mixed number as the sum of a whole number and a fraction
☑ rename the sum of a whole number and a fraction as a mixed number
☑ identify equal fractions
☑ rename a division problem as a fraction
☑ rename a fraction as a division problem
☑ rename a whole number as an equal fraction
☑ rename a fraction with a denominator of 1 as an equal whole number
☑ rename a mixed number as an equal improper fraction
☑ rename an improper fraction as an equal mixed number
☑ write the answer to a division problem in remainder form or mixed-number form
☑ identify a fraction in lowest terms
☑ simplify a proper fraction that is not in lowest terms
☑ simplify an improper fraction
☑ simplify a mixed number that is not in simplest form

SUMMARY

1. If a and b are whole numbers ($b \neq 0$), then $\frac{a}{b}$ is called a fraction.

2. In the fraction $\frac{2}{3}$, 2 is called the numerator, 3 is called the denominator, and the horizontal bar that separates 2 and 3 is called the fraction bar.

3. Fractions that have a denominator of 0 are not defined.

4. When the numerator is less than the denominator, the fraction is called a proper fraction.

5. To name part of a whole, you can use a proper fraction.

6. When the numerator is greater than or equal to the denominator, the fraction is called an improper fraction.

7. To read a fraction, you read the whole-number name of the numerator plus the name of the denominator, as follows:
   (a) whole(s) if the denominator is 1
   (b) halves or half if the denominator is 2
   (c) third(s) if the denominator is 3
   (d) fourth(s) if the denominator is 4
   (e) fifth(s) if the denominator is 5
   and so on.
8. A whole number joined together with a fraction is called a mixed number.
9. In the mixed number $2\frac{3}{4}$, 2 is called the whole-number part and $\frac{3}{4}$ is called the fraction part.
10. To name one or more wholes and part of a whole together, you use a mixed number.
11. Mixed numbers are a short way to write the sum of a whole number and a fraction.
12. Two fractions that represent the same part of a whole are called equal fractions.
13. Two given fractions are equal if their cross products are equal.
14. Every division problem can be renamed as a fraction: $2 \div 3 = \frac{2}{3}$.
15. Every fraction can be renamed as a division problem: $\frac{2}{3} = 2 \div 3$.
16. To rename a whole number as an equal fraction, you can write the whole number over 1.
17. To rename a fraction with a denominator of 1 as an equal whole number, you write just the numerator.
18. Every mixed number can be renamed as an equal improper fraction: $2\frac{3}{4} = \dfrac{4 \times 2 + 3}{4} = \dfrac{11}{4}$.
19. Every improper fraction can be renamed as an equal mixed number: $\dfrac{11}{4} = 11 \div 4 = 2\frac{3}{4}$.
20. The answer to a division problem like $11 \div 4$ can be written in remainder form or in mixed number form: 2 R3 or $2\frac{3}{4}$.
21. A fraction for which the numerator and denominator do not share a common factor other than 1 is called a fraction in lowest terms.
22. When you add, subtract, multiply, or divide and get a fraction or mixed number for an answer, you should always simplify that answer by writing it in simplest form. The simplest form of a fraction or mixed number can be:
   (a) a proper fraction in lowest terms
   (b) a whole number
   (c) a mixed number when the fraction part is a proper fraction in lowest terms.
23. To simplify a proper fraction that is not in lowest terms, you reduce the fraction to lowest terms using the following rule:
   Fundamental Rule for Fractions
   If a, b, and c are whole numbers ($b \neq 0$ and $c \neq 0$), then: $\dfrac{a \times c}{b \times c} = \dfrac{a}{b}$.
24. To eliminate a factor in the numerator of a fraction, you must eliminate a like factor in the denominator. Similarly, to eliminate a factor in the denominator, you must eliminate a like factor in the numerator.
25. To simplify an improper fraction, you first reduce to lowest terms (when possible) and then rename the improper fraction as a mixed number in lowest terms, or as a whole number.
26. To simplify a mixed number that is not in simplest form, you simplify the fraction part.

SOLVED PROBLEMS

PROBLEM 4-1   Identify each fraction as proper, improper, or not defined:

(a) $\frac{2}{3}$ (b) $\frac{3}{2}$ (c) $\frac{4}{4}$ (d) $\frac{0}{4}$ (e) $\frac{4}{0}$ (f) $\frac{8}{1}$ (g) $\frac{6}{6}$ (h) $\frac{1}{0}$ (i) $\frac{0}{1}$ (j) $\frac{3}{100}$

Solution: Recall that a proper fraction has a numerator that is less than the denominator, an improper fraction has a numerator that is greater than or equal to the denominator, and a fraction is not defined when the denominator is zero [see Examples 4-3, 4-4, and 4-6]:

(**a**) proper   (**b**) improper   (**c**) improper   (**d**) proper   (**e**) not defined   (**f**) improper
(**g**) improper   (**h**) not defined   (**i**) proper   (**j**) proper

PROBLEM 4-2   Name the shaded part of each figure using a proper fraction:

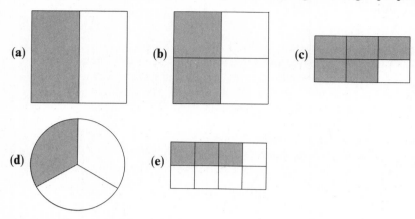

Solution: Recall that the numerator names the number of equal parts that are shaded and the denominator names the number of equal parts in all [see Example 4-5]: (**a**) $\frac{1}{2}$ (**b**) $\frac{2}{4}$ (**c**) $\frac{5}{6}$ (**d**) $\frac{1}{3}$
(**e**) $\frac{3}{8}$

PROBLEM 4-3   Write the correct word name for each fraction: (**a**) $\frac{1}{2}$ (**b**) $\frac{4}{3}$ (**c**) $\frac{3}{4}$ (**d**) $\frac{6}{5}$
(**e**) $\frac{5}{6}$ (**f**) $\frac{1}{8}$ (**g**) $\frac{8}{9}$ (**h**) $\frac{7}{10}$ (**i**) $\frac{11}{12}$ (**j**) $\frac{29}{100}$

Solution: Recall that to read or write the word name of a fraction, you use the whole-number word name of the numerator followed by: *whole(s)* if the denominator is 1; *half* or *halves* if the denominator is 2; *third(s)* if the denominator is 3; and so on [see Examples 4-7 and 4-8]: (**a**) one-half
(**b**) four-thirds   (**c**) three-fourths   (**d**) six-fifths   (**e**) five-sixths   (**f**) one-eighth
(**g**) eight-ninths   (**h**) seven-tenths   (**i**) eleven-twelfths   (**j**) twenty-nine-hundredths

PROBLEM 4-4   Name each group of shaded figures using a mixed number:

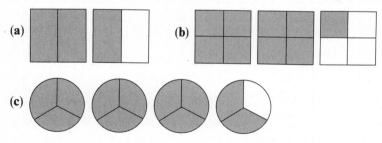

Solution: Recall that each completely shaded whole represents one [see Example 4-10]:
(**a**) $1\frac{1}{2}$ (**b**) $2\frac{1}{4}$ (**c**) $3\frac{2}{3}$

PROBLEM 4-5   Write the correct word name for each mixed number: (**a**) $1\frac{1}{3}$ (**b**) $2\frac{1}{4}$ (**c**) $8\frac{3}{2}$
(**d**) $5\frac{7}{6}$ (**e**) $3\frac{5}{8}$ (**f**) $9\frac{3}{10}$ (**g**) $4\frac{13}{12}$ (**h**) $10\frac{2}{9}$ (**i**) $25\frac{4}{15}$ (**j**) $99\frac{99}{100}$

Solution: Recall that you always write the word *and* between the whole-number part and the fraction part of the mixed number [see Example 4-11]: (**a**) one and one-third   (**b**) two and one-fourth
(**c**) eight and three-halves   (**d**) five and seven-sixths   (**e**) three and five-eighths
(**f**) nine and three-tenths   (**g**) four and thirteen-twelfths   (**h**) ten and two-ninths
(**i**) twenty-five and four-fifteenths   (**j**) ninety-nine and ninety-nine-hundredths

PROBLEM 4-6   Rename each mixed number as a sum: (**a**) $5\frac{1}{2}$ (**b**) $3\frac{1}{4}$ (**c**) $2\frac{5}{4}$ (**d**) $4\frac{9}{5}$
(**e**) $9\frac{3}{5}$ (**f**) $11\frac{7}{8}$ (**g**) $15\frac{5}{9}$ (**h**) $20\frac{3}{10}$ (**i**) $5\frac{1}{25}$ (**j**) $16\frac{39}{100}$

Solution: Recall that mixed numbers are just a short way to write the sum of a whole number and a fraction [see Example 4-12]: (a) $5 + \frac{1}{2}$ (b) $3 + \frac{1}{4}$ (c) $2 + \frac{5}{4}$ (d) $4 + \frac{9}{5}$ (e) $9 + \frac{3}{5}$ (f) $11 + \frac{7}{8}$ (g) $15 + \frac{5}{9}$ (h) $20 + \frac{3}{10}$ (i) $5 + \frac{1}{25}$ (j) $16 + \frac{39}{100}$

PROBLEM 4-7 Rename each sum as a mixed number: (a) $4 + \frac{3}{4}$ (b) $2 + \frac{5}{2}$ (c) $8 + \frac{1}{3}$ (d) $6 + \frac{4}{5}$ (e) $8 + \frac{3}{7}$ (f) $10 + \frac{11}{6}$ (g) $25 + \frac{9}{10}$ (h) $100 + \frac{97}{100}$ (i) $1 + \frac{1}{4}$ (j) $99 + \frac{999}{1000}$

Solution: Recall that the sum of a whole number and a fraction is just another way to write a mixed number [see Example 4-13]: (a) $4\frac{3}{4}$ (b) $2\frac{5}{2}$ (c) $8\frac{1}{3}$ (d) $6\frac{4}{5}$ (e) $8\frac{3}{7}$ (f) $10\frac{11}{6}$ (g) $25\frac{9}{10}$ (h) $100\frac{97}{100}$ (i) $1\frac{1}{4}$ (j) $99\frac{999}{1000}$

PROBLEM 4-8 Are the following pairs of fractions equal or not equal?

(a) $\frac{1}{2}$ and $\frac{2}{4}$ (b) $\frac{1}{2}$ and $\frac{2}{3}$ (c) $\frac{0}{3}$ and $\frac{0}{5}$ (d) $\frac{1}{4}$ and $\frac{4}{8}$ (e) $\frac{1}{4}$ and $\frac{2}{8}$ (f) $\frac{3}{5}$ and $\frac{6}{10}$ (g) $\frac{3}{4}$ and $\frac{6}{8}$ (h) $\frac{5}{8}$ and $\frac{15}{16}$ (i) $\frac{9}{12}$ and $\frac{3}{4}$ (j) $\frac{12}{18}$ and $\frac{4}{5}$ (k) $\frac{8}{8}$ and $\frac{120}{120}$

Solution: Recall that two given fractions are equal only when their cross products are equal [see Examples 4-14 and 4-15]: (a) $4 = 4$; equal (b) $3 \neq 4$; not equal (c) $0 = 0$; equal (d) $8 \neq 16$; not equal (e) $8 = 8$; equal (f) $30 = 30$; equal (g) $24 = 24$; equal (h) $80 \neq 120$; not equal (i) $36 = 36$; equal (j) $60 \neq 72$; not equal (k) $960 = 960$; equal

PROBLEM 4-9 Rename each division problem as a fraction: (a) $2 \div 3$ (b) $3 \div 2$ (c) $8 \div 5$ (d) $1 \div 2$ (e) $0 \div 6$ (f) $15 \div 4$ (g) $99 \div 100$ (h) $7 \div 10$ (i) $1 \div 8$ (j) $243 \div 17$

Solution: Recall that to rename a division problem as a fraction, you write the dividend as the numerator, write the division symbol ($\div$) as the fraction bar (——), and write the divisor as the denominator [see Example 4-16]: (a) $\frac{2}{3}$ (b) $\frac{3}{2}$ (c) $\frac{8}{5}$ (d) $\frac{1}{2}$ (e) $\frac{0}{6}$ (f) $\frac{15}{4}$ (g) $\frac{99}{100}$ (h) $\frac{7}{10}$ (i) $\frac{1}{8}$ (j) $\frac{243}{17}$

PROBLEM 4-10 Rename each fraction as a division problem: (a) $\frac{7}{10}$ (b) $\frac{1}{2}$ (c) $\frac{3}{5}$ (d) $\frac{8}{3}$ (e) $\frac{5}{4}$ (f) $\frac{17}{8}$ (g) $\frac{1}{10}$ (h) $\frac{100}{11}$ (i) $\frac{49}{100}$ (j) $\frac{357}{25}$

Solution: Recall that to rename a fraction as a division problem, you write the numerator as the dividend, write the fraction bar (——) as the division symbol ($\div$), and write the denominator as the divisor [see Example 4-17]: (a) $7 \div 10$ (b) $1 \div 2$ (c) $3 \div 5$ (d) $8 \div 3$ (e) $5 \div 4$ (f) $17 \div 8$ (g) $1 \div 10$ (h) $100 \div 11$ (i) $49 \div 100$ (j) $357 \div 25$

PROBLEM 4-11 Rename each whole number as an equal fraction: (a) 2 (b) 8 (c) 1 (d) 0 (e) 25

Solution: Recall that to rename a whole number as an equal fraction, you write the whole number over 1 [see Example 4-18]: (a) $\frac{2}{1}$ (b) $\frac{8}{1}$ (c) $\frac{1}{1}$ (d) $\frac{0}{1}$ (e) $\frac{25}{1}$

PROBLEM 4-12 Rename each mixed number as an equal fraction: (a) $1\frac{1}{2}$ (b) $4\frac{2}{3}$ (c) $5\frac{3}{4}$ (d) $3\frac{1}{4}$ (e) $6\frac{3}{8}$ (f) $2\frac{6}{5}$ (g) $7\frac{5}{2}$ (h) $3\frac{9}{10}$ (i) $2\frac{49}{50}$ (j) $4\frac{7}{100}$

Solution: Recall that to find the numerator of an equal fraction when renaming a mixed number, you multiply the whole-number part by the denominator of the fraction part and then add that product to the numerator of the fraction part [see Example 4-21]:

(a) $2 \times 1 + 1 = 3; \frac{3}{2}$ (b) $3 \times 4 + 2 = 14; \frac{14}{3}$ (c) $4 \times 5 + 3 = 23; \frac{23}{4}$

(d) $4 \times 3 + 1 = 13; \frac{13}{4}$ (e) $8 \times 6 + 3 = 51; \frac{51}{8}$ (f) $5 \times 2 + 6 = 16; \frac{16}{5}$

(g) $2 \times 7 + 5 = 19; \frac{19}{2}$ (h) $10 \times 3 + 9 = 39; \frac{39}{10}$ (i) $50 \times 2 + 49 = 149; \frac{149}{50}$

(j) $100 \times 4 + 7 = 407; \frac{407}{100}$

PROBLEM 4-13 Rename each improper fraction as an equal mixed number: (a) $\frac{9}{2}$ (b) $\frac{16}{3}$ (c) $\frac{7}{4}$ (d) $\frac{14}{5}$ (e) $\frac{70}{6}$ (f) $\frac{11}{8}$ (g) $\frac{47}{9}$ (h) $\frac{87}{10}$ (i) $\frac{25}{12}$ (j) $\frac{347}{100}$

Solution: Recall that to rename an improper fraction as an equal mixed number using division, divide the numerator by the denominator. The quotient then becomes the whole-number part of the mixed number, the remainder is the numerator of the fraction part, and the divisor is the denominator of the fraction part, which never changes [see Example 4-22]:

(a) $2\overline{)\,9}\,;4\frac{1}{2}$
$\quad\underline{-8}$
$\quad\;\;1$

(b) $3\overline{)\,16}\,;5\frac{1}{3}$
$\quad\underline{-15}$
$\quad\;\;1$

(c) $4\overline{)\,7}\,;1\frac{3}{4}$
$\quad\underline{-4}$
$\quad\;\;3$

(d) $5\overline{)\,14}\,;2\frac{4}{5}$
$\quad\underline{-10}$
$\quad\;\;4$

(e) $6\overline{)\,70}\,;11\frac{4}{6}=11\frac{2}{3}$
$\quad\underline{-6}$
$\quad\;\;10$
$\quad\underline{-\;6}$
$\quad\;\;\;4$

(f) $8\overline{)\,11}\,;1\frac{3}{8}$
$\quad\underline{-8}$
$\quad\;\;3$

(g) $9\overline{)\,47}\,;5\frac{2}{9}$
$\quad\underline{-45}$
$\quad\;\;2$

(h) $10\overline{)\,87}\,;8\frac{7}{10}$
$\quad\underline{-80}$
$\quad\;\;7$

(i) $12\overline{)\,25}\,;2\frac{1}{12}$
$\quad\underline{-24}$
$\quad\;\;1$

(j) $100\overline{)\,347}\,;3\frac{47}{100}$
$\quad\underline{-300}$
$\quad\;\;47$

PROBLEM 4-14   Simplify each proper fraction: (a) $\frac{2}{4}$ (b) $\frac{4}{6}$ (c) $\frac{3}{12}$ (d) $\frac{5}{15}$ (e) $\frac{6}{8}$
(f) $\frac{4}{20}$ (g) $\frac{3}{24}$ (h) $\frac{4}{10}$ (i) $\frac{18}{48}$ (j) $\frac{15}{25}$

Solution: Recall that to simplify a proper fraction that is not in lowest terms, you reduce the fraction to lowest terms using the Fundamental Rule for Fractions [see Examples 4-24, 4-25, 4-26, and 4-27]:

(a) $\frac{2}{4}=\frac{1\times\not2}{2\times\not2}=\frac{1}{2}$

(b) $\frac{4}{6}=\frac{\not2\times2}{\not2\times3}=\frac{2}{3}$

(c) $\frac{3}{12}=\frac{1\times\not3}{\not3\times4}=\frac{1}{4}$

(d) $\frac{5}{15}=\frac{1\times\not5}{3\times\not5}=\frac{1}{3}$

(e) $\frac{6}{8}=\frac{\not2\times3}{\not2\times4}=\frac{3}{4}$

(f) $\frac{4}{20}=\frac{1\times\not4}{\not4\times5}=\frac{1}{5}$

(g) $\frac{3}{24}=\frac{1\times\not3}{\not3\times8}=\frac{1}{8}$

(h) $\frac{4}{10}=\frac{\not2\times2}{\not2\times5}=\frac{2}{5}$

(i) $\frac{18}{48}=\frac{3\times\not6}{\not6\times8}=\frac{3}{8}$

(j) $\frac{15}{25}=\frac{3\times\not5}{5\times\not5}=\frac{3}{5}$

PROBLEM 4-15   Simplify each improper fraction: (a) $\frac{7}{2}$ (b) $\frac{7}{3}$ (c) $\frac{14}{8}$ (d) $\frac{42}{9}$ (e) $\frac{100}{16}$
(f) $\frac{32}{10}$ (g) $\frac{63}{24}$ (h) $\frac{47}{10}$ (i) $\frac{35}{25}$

Solution: Recall that to simplify an improper fraction, you first reduce to lowest terms when possible and then rename the improper fraction in lowest terms as a mixed number [see Example 4-28]:

(a) $\frac{7}{2}=2\overline{)\,7}\,;3\frac{1}{2}$
$\qquad\;\;\underline{-6}$
$\qquad\;\;\;\;1$

(b) $\frac{7}{3}=3\overline{)\,7}\,;2\frac{1}{3}$
$\qquad\;\;\underline{-6}$
$\qquad\;\;\;\;1$

(c) $\frac{14}{8}=\frac{\not2\times7}{\not2\times4}=\frac{7}{4};4\overline{)\,7}\,;1\frac{3}{4}$
$\qquad\qquad\qquad\qquad\;\underline{-4}$
$\qquad\qquad\qquad\qquad\;\;3$

(d) $\frac{42}{9}=\frac{\not3\times14}{\not3\times3}=\frac{14}{3};3\overline{)\,14}\,;4\frac{2}{3}$
$\qquad\qquad\qquad\qquad\;\;\underline{-12}$
$\qquad\qquad\qquad\qquad\;\;\;2$

(e) $\frac{100}{16}=\frac{\not4\times25}{\not4\times4}=\frac{25}{4};4\overline{)\,25}\,;6\frac{1}{4}$
$\qquad\qquad\qquad\qquad\;\;\underline{-24}$
$\qquad\qquad\qquad\qquad\;\;\;1$

(f) $\frac{32}{10}=\frac{\not2\times16}{\not2\times5}=\frac{16}{5};5\overline{)\,16}\,;3\frac{1}{5}$
$\qquad\qquad\qquad\qquad\;\;\underline{-15}$
$\qquad\qquad\qquad\qquad\;\;\;1$

(g) $\frac{63}{24}=\frac{\not3\times21}{\not3\times8}=\frac{21}{8};8\overline{)\,21}\,;2\frac{5}{8}$
$\qquad\qquad\qquad\qquad\;\;\underline{-16}$
$\qquad\qquad\qquad\qquad\;\;\;5$

(h) $\frac{47}{10}=10\overline{)\,47}\,;4\frac{7}{10}$
$\qquad\qquad\;\;\underline{-40}$
$\qquad\qquad\;\;\;7$

(i) $\frac{35}{25} = \frac{\cancel{5} \times 7}{\cancel{5} \times 5} = \frac{7}{5}; \ 5\overline{)7}\ ^{1}; \ 1\frac{2}{5}$
$\underline{-5}$
2

PROBLEM 4-16 Simplify each mixed number: **(a)** $3\frac{4}{8}$ **(b)** $5\frac{6}{9}$ **(c)** $4\frac{3}{2}$ **(d)** $8\frac{5}{3}$ **(e)** $1\frac{8}{4}$
(f) $2\frac{15}{5}$ **(g)** $6\frac{24}{30}$ **(h)** $7\frac{20}{12}$ **(i)** $5\frac{80}{5}$ **(j)** $99\frac{101}{100}$

Solution: Recall that to simplify a mixed number that is not in simplest form, you simplify the fraction part [see Example 4-29]:

(a) $3\frac{4}{8} = 3 + \frac{4}{8} = 3 + \frac{1}{2} = 3\frac{1}{2}$ **(b)** $5\frac{6}{9} = 5 + \frac{6}{9} = 5 + \frac{2}{3} = 5\frac{2}{3}$
(c) $4\frac{3}{2} = 4 + \frac{3}{2} = 4 + 1\frac{1}{2} = 4 + 1 + \frac{1}{2} = 5 + \frac{1}{2} = 5\frac{1}{2}$
(d) $8\frac{5}{3} = 8 + \frac{5}{3} = 8 + 1\frac{2}{3} = 8 + 1 + \frac{2}{3} = 9 + \frac{2}{3} = 9\frac{2}{3}$ **(e)** $1\frac{8}{4} = 1 + \frac{8}{4} = 1 + 2 = 3$
(f) $2\frac{15}{5} = 2 + \frac{15}{5} = 2 + 3 = 5$ **(g)** $6\frac{24}{30} = 6 + \frac{24}{30} = 6 + \frac{4}{5} = 6\frac{4}{5}$
(h) $7\frac{20}{12} = 7 + \frac{20}{12} = 7 + \frac{5}{3} = 7 + 1\frac{2}{3} = 7 + 1 + \frac{2}{3} = 8 + \frac{2}{3} = 8\frac{2}{3}$ **(i)** $5\frac{80}{5} = 5 + \frac{80}{5} = 5 + 16 = 21$
(j) $99\frac{101}{100} = 99 + \frac{101}{100} = 99 + 1\frac{1}{100} = 99 + 1 + \frac{1}{100} = 100 + \frac{1}{100} = 100\frac{1}{100}$

Supplementary Exercises
PROBLEM 4-17 Are the following pairs of fractions equal or not equal?

(a) $\frac{3}{4}$ and $\frac{6}{8}$ **(b)** $\frac{2}{3}$ and $\frac{8}{12}$ **(c)** $\frac{1}{2}$ and $\frac{4}{6}$ **(d)** $\frac{4}{5}$ and $\frac{12}{15}$ **(e)** $\frac{1}{3}$ and $\frac{6}{21}$ **(f)** $\frac{1}{4}$ and $\frac{5}{20}$
(g) $\frac{1}{5}$ and $\frac{2}{10}$ **(h)** $\frac{3}{8}$ and $\frac{8}{24}$ **(i)** $\frac{7}{10}$ and $\frac{20}{30}$ **(j)** $\frac{5}{12}$ and $\frac{15}{36}$ **(k)** $\frac{6}{8}$ and $\frac{30}{40}$ **(l)** $\frac{4}{6}$ and $\frac{18}{24}$
(m) $\frac{8}{12}$ and $\frac{24}{36}$ **(n)** $\frac{4}{12}$ and $\frac{16}{60}$ **(o)** $\frac{6}{10}$ and $\frac{18}{30}$

PROBLEM 4-18 Rename each division problem as a fraction:

(a) $1 \div 2$ **(b)** $7 \div 8$ **(c)** $5 \div 3$ **(d)** $8 \div 5$ **(e)** $2 \div 9$ **(f)** $5 \div 6$ **(g)** $12 \div 1$
(h) $0 \div 4$ **(i)** $3 \div 7$ **(j)** $7 \div 10$ **(k)** $11 \div 12$ **(l)** $3 \div 20$ **(m)** $27 \div 50$
(n) $49 \div 100$ **(o)** $253 \div 1000$

PROBLEM 4-19 Rename each fraction as a division problem:

(a) $\frac{5}{8}$ **(b)** $\frac{2}{3}$ **(c)** $\frac{6}{5}$ **(d)** $\frac{7}{1}$ **(e)** $\frac{0}{9}$ **(f)** $\frac{3}{2}$ **(g)** $\frac{1}{6}$ **(h)** $\frac{3}{4}$ **(i)** $\frac{8}{7}$ **(j)** $\frac{3}{10}$ **(k)** $\frac{5}{12}$
(l) $\frac{7}{18}$ **(m)** $\frac{21}{25}$ **(n)** $\frac{199}{100}$ **(o)** $\frac{249}{500}$

PROBLEM 4-20 Rename each whole or mixed number as an equal fraction:

(a) 2 **(b)** 5 **(c)** $3\frac{1}{2}$ **(d)** $1\frac{2}{3}$ **(e)** $2\frac{3}{4}$ **(f)** $4\frac{3}{5}$ **(g)** 6 **(h)** 0 **(i)** $2\frac{1}{3}$ **(j)** $5\frac{1}{4}$
(k) $3\frac{1}{5}$ **(l)** $2\frac{7}{10}$ **(m)** 1 **(n)** 20 **(o)** $25\frac{49}{100}$

PROBLEM 4-21 Rename each improper fraction as an equal mixed number:

(a) $\frac{3}{2}$ **(b)** $\frac{5}{4}$ **(c)** $\frac{8}{5}$ **(d)** $\frac{5}{3}$ **(e)** $\frac{11}{6}$ **(f)** $\frac{13}{8}$ **(g)** $\frac{17}{9}$ **(h)** $\frac{13}{10}$ **(i)** $\frac{7}{2}$ **(j)** $\frac{11}{4}$
(k) $\frac{10}{3}$ **(l)** $\frac{25}{8}$ **(m)** $\frac{24}{5}$ **(n)** $\frac{37}{6}$ **(o)** $\frac{51}{8}$

PROBLEM 4-22 Simplify each fraction or mixed number:

(a) $\frac{2}{6}$ **(b)** $\frac{4}{8}$ **(c)** $\frac{8}{4}$ **(d)** $\frac{12}{9}$ **(e)** $3\frac{4}{6}$ **(f)** $2\frac{10}{8}$ **(g)** $\frac{9}{12}$ **(h)** $\frac{8}{12}$ **(i)** $\frac{10}{6}$ **(j)** $\frac{6}{1}$
(k) $1\frac{8}{10}$ **(l)** $3\frac{9}{18}$ **(m)** $\frac{12}{18}$ **(n)** $\frac{35}{20}$ **(o)** $5\frac{13}{3}$

Answers to Supplementary Exercises

(4-17) **(a)** equal **(b)** equal **(c)** not equal **(d)** equal **(e)** not equal **(f)** equal **(g)** equal **(h)** not equal **(i)** not equal **(j)** equal **(k)** equal **(l)** not equal **(m)** equal **(n)** not equal **(o)** equal

(4-18) **(a)** $\frac{1}{2}$ **(b)** $\frac{7}{8}$ **(c)** $\frac{5}{3}$ **(d)** $\frac{8}{5}$ **(e)** $\frac{2}{9}$ **(f)** $\frac{5}{6}$ **(g)** $\frac{12}{1}$ **(h)** $\frac{0}{4}$ **(i)** $\frac{3}{7}$ **(j)** $\frac{7}{10}$ **(k)** $\frac{11}{12}$ **(l)** $\frac{3}{20}$ **(m)** $\frac{27}{50}$ **(n)** $\frac{49}{100}$ **(o)** $\frac{253}{1000}$

(4-19) **(a)** $5 \div 8$ **(b)** $2 \div 3$ **(c)** $6 \div 5$ **(d)** $7 \div 1$ **(e)** $0 \div 9$ **(f)** $3 \div 2$ **(g)** $1 \div 6$ **(h)** $3 \div 4$ **(i)** $8 \div 7$ **(j)** $3 \div 10$ **(k)** $5 \div 12$ **(l)** $7 \div 18$ **(m)** $21 \div 25$ **(n)** $199 \div 100$ **(o)** $249 \div 500$

(4-20) **(a)** $\frac{2}{1}$ **(b)** $\frac{5}{1}$ **(c)** $\frac{7}{2}$ **(d)** $\frac{5}{3}$ **(e)** $\frac{11}{4}$ **(f)** $\frac{23}{5}$ **(g)** $\frac{6}{1}$ **(h)** $\frac{0}{1}$ **(i)** $\frac{7}{3}$ **(j)** $\frac{21}{4}$ **(k)** $\frac{16}{5}$ **(l)** $\frac{27}{10}$ **(m)** $\frac{1}{1}$ **(n)** $\frac{20}{1}$ **(o)** $\frac{2549}{100}$

(4-21) **(a)** $1\frac{1}{2}$ **(b)** $1\frac{1}{4}$ **(c)** $1\frac{3}{5}$ **(d)** $1\frac{2}{3}$ **(e)** $1\frac{5}{6}$ **(f)** $1\frac{5}{8}$ **(g)** $1\frac{8}{9}$ **(h)** $1\frac{3}{10}$ **(i)** $3\frac{1}{2}$ **(j)** $2\frac{3}{4}$ **(k)** $3\frac{1}{3}$ **(l)** $3\frac{1}{8}$ **(m)** $4\frac{4}{5}$ **(n)** $6\frac{1}{6}$ **(o)** $6\frac{3}{8}$

(4-22) **(a)** $\frac{1}{3}$ **(b)** $\frac{1}{2}$ **(c)** 2 **(d)** $1\frac{1}{3}$ **(e)** $3\frac{2}{3}$ **(f)** $3\frac{1}{4}$ **(g)** $\frac{3}{4}$ **(h)** $\frac{2}{3}$ **(i)** $1\frac{2}{3}$ **(j)** 6 **(k)** $1\frac{4}{5}$ **(l)** $3\frac{1}{2}$ **(m)** $\frac{2}{3}$ **(n)** $1\frac{3}{4}$ **(o)** $9\frac{1}{3}$

5 FRACTIONS AND MIXED NUMBERS: MULTIPLICATION AND DIVISION

THIS CHAPTER IS ABOUT

☑ **Multiplying with Fractions and Mixed Numbers**
☑ **Finding Reciprocals**
☑ **Dividing with Fractions and Mixed Numbers**
☑ **Solving Word Problems Containing Fractions and Mixed Numbers (×, ÷)**

5-1. Multiplying with Fractions and Mixed Numbers

To find a fractional part of an amount, you **multiply by a fraction.**

EXAMPLE 5-1: Rename the following as multiplication: $\frac{5}{6}$ of $\frac{3}{4}$

Solution: $\frac{5}{6}$ of $\frac{3}{4} = \frac{5}{6} \times \frac{3}{4}$ ⟵ multiplication

Note: When the word *of* appears between two numbers, you can usually replace the word with a multiplication symbol (×).

To join equal fractional parts together, you also multiply with fractions.

EXAMPLE 5-2: Write $\frac{2}{3} + \frac{2}{3} + \frac{2}{3} + \frac{2}{3}$ as multiplication.

Solution: Think of $\frac{2}{3} + \frac{2}{3} + \frac{2}{3} + \frac{2}{3}$ as 4 equals amounts of $\frac{2}{3}$, or $4 \times \frac{2}{3}$ ⟵ multiplication

A. Multiply with fractions.

To multiply with fractions, you can use the following rule:

Multiplication Rule for Fractions

If $\dfrac{a}{b}$ and $\dfrac{c}{d}$ are fractions, then $\dfrac{a}{b} \times \dfrac{c}{d} = \dfrac{a \times c}{b \times d}$.

Note: To multiply fractions, you multiply the numerators first and then the denominators.

EXAMPLE 5-3: Multiply the following fractions: $\frac{5}{6} \times \frac{3}{4}$

Solution: $\dfrac{5}{6} \times \dfrac{3}{4} = \dfrac{5 \times 3}{6 \times 4}$ Use the Multiplication Rule for Fractions.

$= \dfrac{15}{24}$ Multiply the numerators.
Multiply the denominators.

$= \dfrac{5 \times \cancel{3}}{8 \times \cancel{3}}$ Eliminate common factors.

$= \dfrac{5}{8}$ ⟵ simplest form

Note: If you factor first and eliminate all common factors before multiplying, the product will always be in lowest terms after multiplying.

EXAMPLE 5-4: Eliminate common factors before multiplying the following fractions: $\frac{5}{6} \times \frac{3}{4}$

Solution: $\dfrac{5}{6} \times \dfrac{3}{4} = \dfrac{5}{2 \times 3} \times \dfrac{3}{2 \times 2}$ Factor.

$\qquad\qquad = \dfrac{5}{2 \times \cancel{3}} \times \dfrac{\cancel{3}}{4}$ Eliminate all common factors.

$\qquad\qquad = \dfrac{5 \times 1}{2 \times 4}$ Multiply fractions.

$\qquad\qquad = \dfrac{5}{8}$ ⟵ simplest form

To multiply two or more fractions, you eliminate common factors, then multiply the numerators and multiply the denominators, simplifying when possible.

EXAMPLE 5-5: Multiply the following fractions: $\frac{2}{3} \times \frac{3}{5} \times \frac{15}{4}$

Solution: $\dfrac{2}{3} \times \dfrac{3}{5} \times \dfrac{15}{4} = \dfrac{\cancel{2}}{3} \times \dfrac{3}{5} \times \dfrac{3 \times 5}{\cancel{2} \times 2}$ Eliminate all common factors.

$\qquad\qquad = \dfrac{1}{\cancel{3}} \times \dfrac{\cancel{3}}{5} \times \dfrac{3 \times 5}{2}$

$\qquad\qquad = \dfrac{1}{1} \times \dfrac{1}{\cancel{5}} \times \dfrac{3 \times \cancel{5}}{2}$

$\qquad\qquad = \dfrac{1}{1} \times \dfrac{1}{1} \times \dfrac{3}{2}$

$\qquad\qquad = \dfrac{1 \times 1 \times 3}{1 \times 1 \times 2}$ Multiply the numerators. Multiply the denominators.

$\qquad\qquad = \dfrac{3}{2}$ Simplify.

$\qquad\qquad = 1\frac{1}{2}$ ⟵ simplest form

To multiply with fractions and whole numbers, you rename each whole number as an equal fraction.

EXAMPLE 5-6: Multiply the following whole number and fraction: $4 \times \frac{2}{3}$

Solution: $4 \times \dfrac{2}{3} = \dfrac{4}{1} \times \dfrac{2}{3}$ Rename: $4 = \dfrac{4}{1}$

$\qquad\qquad = \dfrac{4 \times 2}{1 \times 3}$ Multiply fractions as before.

$\qquad\qquad = \dfrac{8}{3}$

$\qquad\qquad = 2\frac{2}{3}$ ⟵ simplest form

B. Multiply with mixed numbers.
To multiply with mixed numbers, you rename each mixed number as an equal fraction.

EXAMPLE 5-7: Multiply the folllowing mixed number, whole number, and fraction: $2\frac{1}{2} \times 4 \times \frac{3}{5}$

Solution: $2\frac{1}{2} \times 4 \times \frac{3}{5} = \frac{5}{2} \times \frac{4}{1} \times \frac{3}{5}$ Rename: $2\frac{1}{2} = \frac{5}{2}$ and $4 = \frac{4}{1}$

$$= \frac{\cancel{5}}{\cancel{2}} \times \frac{\cancel{2} \times 2}{1} \times \frac{3}{\cancel{5}}$$ Eliminate all common factors.

$$= \frac{1 \times 2 \times 3}{1 \times 1 \times 1}$$ Multiply fractions.

$$= \frac{6}{1}$$ Simplify.

$$= 6 \longleftarrow \text{ simplest form}$$

5-2. Finding Reciprocals

When the product of two fractions is 1, each fraction is called a **reciprocal** of the other.

EXAMPLE 5-8: Interpret the following equation in terms of reciprocals: $\frac{3}{4} \times \frac{4}{3} = 1$

Solution: $\frac{3}{4} \times \frac{4}{3} = 1$ means that $\frac{3}{4}$ is the reciprocal of $\frac{4}{3}$ and that $\frac{4}{3}$ is the reciprocal of $\frac{3}{4}$.

When you divide with fractions, you will need to find the reciprocal of the divisor.

A. Find the reciprocal of a fraction.

To find the reciprocal of a fraction, you interchange the whole numbers in the numerator and denominator.

EXAMPLE 5-9: Find the reciprocals of the following fractions: **(a)** $\frac{2}{3}$ **(b)** $\frac{8}{5}$

Solution
(a) The reciprocal of $\frac{2}{3}$ is $\frac{3}{2}$ (or $1\frac{1}{2}$) because $\frac{2}{3} \times \frac{3}{2} = 1$.
(b) The reciprocal of $\frac{8}{5}$ is $\frac{5}{8}$ because $\frac{8}{5} \times \frac{5}{8} = 1$.

B. Find the reciprocal of a whole number.

To find the reciprocal of a whole number, you rename the whole number as an equal fraction.

EXAMPLE 5-10: Find the reciprocals of the following whole numbers: **(a)** 3   **(b)** 0

Solution
(a) The reciprocal of 3 (or $\frac{3}{1}$) is $\frac{1}{3}$ because $3 \times \frac{1}{3} = \frac{3}{1} \times \frac{1}{3} = 1$.
(b) The reciprocal of 0 (or $\frac{0}{1}$) is not defined because $\frac{1}{0}$ (division by zero) is not defined.

Note: Zero is the only whole number that does not have a reciprocal.

C. Find the reciprocal of a mixed number.

To find the reciprocal of a mixed number, you rename the mixed number as an equal fraction.

EXAMPLE 5-11: Find the reciprocals of the following mixed numbers: **(a)** $3\frac{1}{4}$ **(b)** $2\frac{5}{8}$

Solution
(a) The reciprocal of $3\frac{1}{4}$ (or $\frac{13}{4}$) is $\frac{4}{13}$ because $3\frac{1}{4} \times \frac{4}{13} = \frac{13}{4} \times \frac{4}{13} = 1$.
(b) The reciprocal of $2\frac{5}{8}$ (or $\frac{21}{8}$) is $\frac{8}{21}$ because $2\frac{5}{8} \times \frac{8}{21} = \frac{21}{8} \times \frac{8}{21} = 1$.

5-3. Dividing with Fractions and Mixed Numbers

To find an amount given a fractional part of that amount you **divide by a fraction**.

EXAMPLE 5-12: Write the following as a division problem with fractions: $\frac{1}{8}$ is $\frac{1}{4}$ of an amount

Solution: $\frac{1}{8}$ is $\frac{1}{4}$ of an amount means $\frac{1}{8} \div \frac{1}{4}$ = the amount, or the amount = $\frac{1}{8} \div \frac{1}{4}$ ⟵— division

Note: When the word *is* appears between two numbers, you can usually replace the word with a division symbol ($\div$).

To separate a fractional amount into equal parts, you also divide by a fraction.

EXAMPLE 5-13: Write the following as a division problem with fractions: $\frac{3}{4}$ separated into 8 equal parts

Solution: $\frac{3}{4}$ separated into 8 equal parts means $\frac{3}{4} \div 8$, or $\frac{3}{4} \div \frac{8}{1}$ ⟵— division

A. Divide with fractions.

To divide with fractions, you can use the following rule:

Division Rule for Fractions

If $\frac{a}{b}$ and $\frac{c}{d}$ are fractions ($c \neq 0$), then $\frac{a}{b} \div \frac{c}{d} = \frac{a}{b} \times \frac{d}{c}$.

Note: To divide fractions, you change the divisor to its reciprocal and then multiply.

EXAMPLE 5-14: Divide the following fractions: $\frac{1}{8} \div \frac{1}{4}$

change to multiplication

Solution: $\frac{1}{8} \div \frac{1}{4} = \frac{1}{8} \times \frac{4}{1}$ Use the Division Rule for Fractions.

write the reciprocal of the divisor

$$= \frac{1}{2 \times 4} \times \frac{4 \times 1}{1}$$ Multiply as before [see Example 4-29].

$$= \frac{1}{2} \times \frac{1}{1}$$

$$= \frac{1}{2} \longleftarrow \text{simplest form}$$

Note: $\frac{1}{8} \div \frac{1}{4} = \frac{1}{2}$ means $\frac{1}{8}$ is $\frac{1}{4}$ of $\frac{1}{2}$.

To check division, you multiply the proposed quotient by the original divisor to see if you get the original dividend.

EXAMPLE 5-15: Check the following division problem: $\frac{1}{8} \div \frac{1}{4} = \frac{1}{2}$

original divisor		proposed quotient				original dividend		

Solution: $\frac{1}{4} \times \frac{1}{2} = \frac{1 \times 1}{4 \times 2} = \frac{1}{8} \longleftarrow \frac{1}{2}$ checks

Caution: Do not try to eliminate common factors before changing a division problem to a multiplication problem.

EXAMPLE 5-16: Divide the following fractions: $\frac{3}{5} \div \frac{2}{3}$

Solution

Wrong Method

$$\frac{3}{5} \div \frac{2}{3} = \frac{\cancel{3}}{5} \div \frac{2}{\cancel{3}}$$ ⟵ *Wrong!* Never eliminate common factors when there is a division symbol in the problem.

$$= \frac{1}{5} \div \frac{2}{1}$$

$$= \frac{1}{5} \times \frac{1}{2}$$

$$= \frac{1}{10}$$ ⟵ wrong answer

Check: $\dfrac{2}{3} \times \dfrac{1}{10} = \dfrac{\cancel{2}}{3} \times \dfrac{1}{\cancel{2} \times 5}$

$$= \frac{1}{15}$$ ⟵ not the original dividend $\dfrac{3}{5}$ $\left(\text{so } \dfrac{1}{10} \text{ does not check}\right)$

Correct Method

$$\frac{3}{5} \div \frac{2}{3} = \frac{3}{5} \times \frac{3}{2}$$ Change to multiplication and invert the divisor.

$$= \frac{3 \times 3}{5 \times 2}$$

$$= \frac{9}{10}$$ ⟵ correct answer

Check: $\dfrac{2}{3} \times \dfrac{9}{10} = \dfrac{\cancel{2}}{\cancel{3}} \times \dfrac{\cancel{3} \times 3}{\cancel{2} \times 5}$

$$= \frac{3}{5}$$ ⟵ original dividend $\left(\dfrac{9}{10} \text{ checks}\right)$

To divide a fraction by a whole number, you rename the whole number as an equal fraction.

EXAMPLE 5-17: Divide the following fraction and whole number: $\dfrac{3}{4} \div 8$

Solution: $\dfrac{3}{4} \div 8 = \dfrac{3}{4} \div \dfrac{8}{1}$ Rename: $8 = \dfrac{8}{1}$

$$= \frac{3}{4} \times \frac{1}{8}$$ Multiply by the reciprocal of the divisor.

$$= \frac{3 \times 1}{4 \times 8}$$ Multiply as before.

$$= \frac{3}{32}$$ ⟵ proposed answer

Check: $8 \times \dfrac{3}{32} = \dfrac{8}{1} \times \dfrac{3}{32}$ Rename.

$$= \frac{\cancel{8}}{1} \times \frac{3}{4 \times \cancel{8}}$$ Multiply as before.

$$= \frac{3}{4}$$ ⟵ original dividend $\left(\dfrac{3}{32} \text{ checks}\right)$

To divide a whole number by a fraction, you rename the whole number as an equal fraction.

EXAMPLE 5-18: Divide the following whole number and fraction: $6 \div \frac{2}{3}$

Solution: $6 \div \frac{2}{3} = \frac{6}{1} \div \frac{2}{3}$ Rename.

$$= \frac{6}{1} \times \frac{3}{2}$$ Multiply by the reciprocal of the divisor.

$$= \frac{\cancel{2} \times 3}{1} \times \frac{3}{\cancel{2}}$$ Multiply as before.

$$= \frac{9}{1}$$

$$= 9 \longleftarrow \text{ proposed answer}$$

Check: $\frac{2}{3} \times 9 = \frac{2}{3} \times \frac{9}{1}$ Rename.

$$= \frac{2}{\cancel{3}} \times \frac{\cancel{3} \times 3}{1}$$ Multiply as before.

$$= \frac{6}{1}$$

$$= 6 \longleftarrow \text{ original dividend (9 checks)}$$

B. Divide with mixed numbers.

To divide with mixed numbers, you rename each mixed number as an equal fraction.

EXAMPLE 5-19: Divide the following mixed numbers: $2\frac{2}{5} \div 1\frac{1}{5}$

Solution: $2\frac{2}{5} \div 1\frac{1}{5} = \frac{12}{5} \div \frac{6}{5}$ Rename: $2\frac{2}{5} = \frac{12}{5}$ and $1\frac{1}{5} = \frac{6}{5}$

$$= \frac{12}{5} \times \frac{5}{6}$$ Multiply by the reciprocal of the divisor.

$$= \frac{2 \times \cancel{6}}{\cancel{5}} \times \frac{\cancel{5}}{\cancel{6}}$$ Multiply as before.

$$= \frac{2}{1}$$

$$= 2 \longleftarrow \text{ proposed answer}$$

Check: $1\frac{1}{5} \times 2 = \frac{6}{5} \times \frac{2}{1}$ Rename and multiply as before.

$$= \frac{12}{5}$$

$$= 2\frac{2}{5} \longleftarrow \text{ original dividend (2 checks)}$$

5-4. Solving Word Problems Containing Fractions and Mixed Numbers ($\times$, $\div$)

A. Solve word problems containing fractions ($\times$, $\div$).

To solve multiplication word problems containing fractions, you find
(a) a fractional part of a given amount;

(b) how much two or more equal fractional parts are all together;

(c) the product of one fractional part times another fractional part.

To solve division word problems containing fractions, you find

(a) an amount given a fractional part of it;

(b) how many equal fractional parts there are in all;

(c) how many are in each equal fractional part;

(d) how many times larger one fractional part is than another fractional part.

To help identify certain multiplication word problems containing fractions, you can look for two numbers with the word *of* between them. To help identify certain division word problems containing fractions, you can look for two numbers with the word *is* between them.

EXAMPLE 5-20: Solve the following word problems containing fractions:

1. *Identify:* Ralph was awake $\left(\frac{2}{3}\right)$ of the day. He worked $\left(\frac{1}{2}\right)$ of the time he was awake. <u>What part of a day did Ralph work?</u>

2. *Understand:* The question asks you to find $\frac{1}{2}$ **of** $\frac{2}{3}$.

3. *Decide:* To find a fractional part of an amount, you **multiply.**

4. *Compute:* $\frac{1}{2}$ of $\frac{2}{3}$

$$\frac{1}{2} \times \frac{2}{3} = \frac{1}{2} \times \frac{\cancel{2}}{3}$$
$$= \frac{1}{3}$$

5. *Picture:*

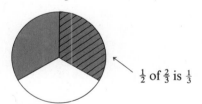

$\frac{1}{2}$ of $\frac{2}{3}$ is $\frac{1}{3}$

6. *Interpret:* $\frac{1}{3}$ means that Ralph worked $\frac{1}{3}$ **of the day.**

7. *Check:* Is the time that Ralph was awake ($\frac{2}{3}$ of the day) twice as long as the time he worked ($\frac{1}{3}$ of the day)?
Yes: $2 \times \frac{1}{3} = \frac{2}{3}$

1. *Identify:* Norma worked $\left(\frac{1}{2}\right)$ of the day. The time she worked was $\left(\frac{2}{3}\right)$ of the time she was awake. <u>What part of the day was Norma awake?</u>

2. *Understand:* The question asks you to find what part of $\frac{2}{3}$ **is** $\frac{1}{2}$.

3. *Decide:* To find an amount given a fractional part of it, you **divide.**

4. *Compute:* $\frac{1}{2}$ is $\frac{2}{3}$ of what part?

$$\frac{1}{2} \div \frac{2}{3} = \frac{1}{2} \times \frac{3}{2}$$
$$= \frac{3}{4}$$

5. *Picture:*

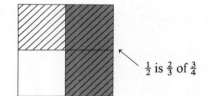

$\frac{1}{2}$ is $\frac{2}{3}$ of $\frac{3}{4}$

6. *Interpret:* $\frac{3}{4}$ means that Norma was awake $\frac{3}{4}$ **of the day.**

7. *Check:* Is the time Norma worked ($\frac{2}{3}$ of $\frac{3}{4}$ of the day) equal to $\frac{1}{2}$ of the day?
Yes: $\frac{2}{3} \times \frac{3}{4} = \frac{1}{2}$

B. Solve word problems containing mixed numbers ($\times$, $\div$).

To solve a multiplication or a division word problem containing mixed numbers, it is usually helpful to first solve a similar problem containing easier whole numbers, and then use the same method to solve the original problem.

EXAMPLE 5-21: Solve the following word problem containing mixed numbers:

1. *Identify:* Connie used $\left(22\frac{1}{4} \text{ gallons}\right)$ of gas to drive $\left(311\frac{1}{2} \text{ miles}\right)$. <u>How many miles per gallon (mpg) did Connie get on this trip?</u>

Circle each fact.

Underline the question.

2. *Think:*   Connie used **20** gallons of gas to drive **300** miles. How many miles per gallon (mpg) did Connie get on this trip?   Use a similar problem containing easier whole numbers.

3. *Understand:*   The question asks you to find how many equal amounts of 20 there are in 300.

4. *Decide:*   To find how many equal amounts of 20 there are in 300, you **divide.**

5. *Compute using whole numbers:* $300 \div 20 = 15$ ⟵ approximate answer

6. *Compute using mixed numbers:* $311\frac{1}{2} \div 22\frac{1}{4} = \dfrac{623}{2} \div \dfrac{89}{4}$ Replace the easier whole numbers with the original mixed numbers and compute as before.

$$= \frac{623}{2} \times \frac{4}{89}$$

$$= \frac{7 \times \cancel{89}}{\cancel{2}} \times \frac{\cancel{2} \times 2}{\cancel{89}}$$

$$= \frac{7}{1} \times \frac{2}{1}$$

$$= 14 \text{ ⟵ exact answer}$$

7. *Interpret:*   14 means that Connie got 14 mpg on this trip: $311\frac{1}{2}$ miles on $22\frac{1}{4}$ gallons of gas.

RAISE YOUR GRADES

Can you . . . ?

☑ multiply fractions using the Multiplication Rule for Fractions
☑ eliminate all common factors before multiplying fractions
☑ multiply with fractions and whole numbers
☑ multiply with mixed numbers
☑ find the reciprocal of a fraction
☑ find the reciprocal of a whole number
☑ find the reciprocal of a mixed number
☑ check a proposed reciprocal
☑ divide fractions using the Division Rule for Fractions
☑ check division using multiplication
☑ divide a fraction by a whole number
☑ divide a whole number by a fraction
☑ divide with mixed numbers
☑ determine whether to multiply or divide in a word problem containing fractions
☑ determine whether to multiply or divide in a word problem containing mixed numbers

SUMMARY

1. To find a fractional part of an amount, you multiply by a fraction.
2. When the word *of* appears between two numbers, you can usually replace the word with a multiplication symbol (×).
3. To join equal fractional parts together, you also multiply with fractions.
4. To multiply fractions, you can use the following rule:

Multiplication Rule for Fractions

If $\dfrac{a}{b}$ and $\dfrac{c}{d}$ are fractions, then $\dfrac{a}{b} \times \dfrac{c}{d} = \dfrac{a \times c}{b \times d}$.

5. The Multiplication Rule for Fractions states that to multiply fractions, you multiply the numerators first and then the denominators.

6. If you factor first and eliminate all common factors before multiplying, the product will always be in lowest terms after multiplying.

7. To multiply with fractions and whole numbers, you rename each whole number as an equal fraction.

8. To multiply with mixed numbers, you rename each mixed number as an equal fraction.

9. When the product of two fractions is 1, each fraction is called a reciprocal of the other.

10. To find the reciprocal of a fraction, you interchange the whole numbers in the numerator and denominator.

11. To find the reciprocal of a whole number or a mixed number, you rename the number as an equal fraction and then invert the whole numbers in the numerator and denominator.

12. To find an amount given a fractional part of that amount you divide by a fraction.

13. When the word *is* appears between two numbers, you can usually replace the word with a division symbol ($\div$).

14. To separate a fractional amount into equal parts, you also divide with fractions.

15. To divide with fractions, you can use the following rule:

Division Rule for Fractions

If $\dfrac{a}{b}$ and $\dfrac{c}{d}$ are fractions ($c \neq 0$), then $\dfrac{a}{b} \div \dfrac{c}{d} = \dfrac{a}{b} \times \dfrac{d}{c}$.

16. The Division Rule for Fractions states that to divide fractions, you change the divisor to its reciprocal and then multiply.

17. To check division, you multiply the proposed quotient by the original divisor to see if you get the original dividend.

18. Never try to eliminate common factors before changing a division problem to a multiplication problem.

19. To divide a fraction by a whole number (or a whole number by a fraction), you rename the whole number as an equal fraction.

20. To divide with mixed numbers, you rename each mixed number as an equal fraction.

21. To solve multiplication word problems containing fractions, you find
    (a) a fractional part of a given amount;
    (b) how much two or more equal fractional parts are all together;
    (c) the product of one fractional part times another fractional part.

22. To solve division word problems containing fractions, you find
    (a) an amount given a fractional part of it;
    (b) how many equal fractional parts there are in all;
    (c) how many are in each equal fractional part;
    (d) how many times larger one fractional part is than another fractional part.

23. To help identify certain multiplication word problems containing fractions, you can look for two numbers separated by the word *of*.

24. To help identify certain division word problems containing fractions, you can look for two numbers separated by the word *is*.

25. To solve a multiplication or division word problem containing mixed numbers, it is usually helpful to first solve a similar problem containing easier whole numbers, and then use the same method to solve the original problem.

SOLVED PROBLEMS

PROBLEM 5-1   Rename each fractional statement as multiplication:

(a) $\frac{1}{2}$ of $\frac{3}{4}$ (b) $\frac{2}{3}$ of $\frac{5}{8}$ (c) $\frac{1}{3}$ of 5   (d) $\frac{1}{4}$ of 10   (e) $\frac{7}{10}$ of $2\frac{1}{2}$ (f) $\frac{2}{5}$ of $2\frac{3}{4}$

Solution: Recall that to find a fractional part of an amount you multiply. Also, you can usually replace the word *of* with a multiplication symbol ($\times$) [see Example 5-1]:

(a) $\frac{1}{2} \times \frac{3}{4}$ **(b)** $\frac{2}{3} \times \frac{5}{8}$ **(c)** $\frac{1}{3} \times 5$ **(d)** $\frac{1}{4} \times 10$ **(e)** $\frac{7}{10} \times 2\frac{1}{2}$ **(f)** $\frac{2}{5} \times 2\frac{3}{4}$

PROBLEM 5-2 Rename each repeated addition problem as multiplication:

(a) $\frac{1}{3} + \frac{1}{3}$ **(b)** $\frac{1}{2} + \frac{1}{2} + \frac{1}{2}$ **(c)** $\frac{3}{4} + \frac{3}{4} + \frac{3}{4} + \frac{3}{4} + \frac{3}{4}$

Solution: Recall that to join equal fractional parts together, you multiply [see Example 5-2]:

(a) 2 equal amounts of $\frac{1}{3}$, or $2 \times \frac{1}{3}$ **(b)** 3 equal amounts of $\frac{1}{2}$, or $3 \times \frac{1}{2}$
(c) 5 equal amounts of $\frac{3}{4}$, or $5 \times \frac{3}{4}$

PROBLEM 5-3 Multiply with fractions: **(a)** $\frac{2}{3} \times \frac{1}{5}$ **(b)** $\frac{1}{2} \times \frac{3}{4}$ **(c)** $\frac{2}{5} \times \frac{5}{9}$ **(d)** $\frac{1}{3} \times \frac{6}{5}$
(e) $\frac{2}{3} \times \frac{9}{2}$ **(f)** $\frac{8}{3} \times \frac{3}{4}$ **(g)** $\frac{5}{3} \times 2$ **(h)** $\frac{5}{8} \times 8$ **(i)** $5 \times \frac{1}{2}$ **(j)** $6 \times \frac{2}{3}$ **(k)** $\frac{2}{3} \times \frac{3}{4} \times \frac{5}{6}$
(l) $\frac{16}{5} \times 4 \times \frac{1}{10} \times \frac{5}{4}$

Solution: Recall that if you eliminate all common factors before you multiply fractions, then the product will always be in lowest terms after multiplying [see Examples 5-3 through 5-6]:

(a) $\dfrac{2}{3} \times \dfrac{1}{5} = \dfrac{2 \times 1}{3 \times 5} = \dfrac{2}{15}$ **(b)** $\dfrac{1}{2} \times \dfrac{3}{4} = \dfrac{1 \times 3}{2 \times 4} = \dfrac{3}{8}$ **(c)** $\dfrac{2}{\cancel{5}} \times \dfrac{\cancel{5}}{9} = \dfrac{2}{1} \times \dfrac{1}{9} = \dfrac{2}{9}$

(d) $\dfrac{1}{3} \times \dfrac{6}{5} = \dfrac{1}{\cancel{3}} \times \dfrac{2 \times \cancel{3}}{5} = \dfrac{1}{1} \times \dfrac{2}{5} = \dfrac{2}{5}$ **(e)** $\dfrac{\cancel{2}}{3} \times \dfrac{9}{\cancel{2}} = \dfrac{1}{\cancel{3}} \times \dfrac{\cancel{3} \times 3}{1} = \dfrac{1}{1} \times \dfrac{3}{1} = \dfrac{3}{1} = 3$

(f) $\dfrac{8}{\cancel{3}} \times \dfrac{\cancel{3}}{4} = \dfrac{2 \times \cancel{4}}{1} \times \dfrac{1}{\cancel{4}} = \dfrac{2}{1} \times \dfrac{1}{1} = \dfrac{2}{1} = 2$ **(g)** $\dfrac{5}{3} \times 2 = \dfrac{5}{3} \times \dfrac{2}{1} = \dfrac{10}{3} = 3\dfrac{1}{3}$

(h) $\dfrac{5}{8} \times 8 = \dfrac{5}{\cancel{8}} \times \dfrac{\cancel{8}}{1} = \dfrac{5}{1} \times \dfrac{1}{1} = \dfrac{5}{1} = 5$ **(i)** $5 \times \dfrac{1}{2} = \dfrac{5}{1} \times \dfrac{1}{2} = \dfrac{5}{2} = 2\dfrac{1}{2}$

(j) $6 \times \dfrac{2}{3} = \dfrac{6}{1} \times \dfrac{2}{3} = \dfrac{2 \times \cancel{3}}{1} \times \dfrac{2}{\cancel{3}} = \dfrac{2}{1} \times \dfrac{2}{1} = \dfrac{4}{1} = 4$

(k) $\dfrac{2}{\cancel{3}} \times \dfrac{\cancel{3}}{4} \times \dfrac{5}{6} = \dfrac{\cancel{2}}{1} \times \dfrac{1}{\cancel{2} \times 2} \times \dfrac{5}{6} = \dfrac{1}{1} \times \dfrac{1}{2} \times \dfrac{5}{6} = \dfrac{1 \times 1 \times 5}{1 \times 2 \times 6} = \dfrac{5}{12}$

(l) $\dfrac{16}{\cancel{5}} \times 4 \times \dfrac{1}{10} \times \dfrac{\cancel{5}}{4} = \dfrac{16}{1} \times \dfrac{\cancel{4}}{1} \times \dfrac{1}{10} \times \dfrac{1}{\cancel{4}} = \dfrac{\cancel{2} \times 8}{1} \times \dfrac{1}{1} \times \dfrac{1}{\cancel{2} \times 5} \times \dfrac{1}{1} = \dfrac{8}{5} = 1\dfrac{3}{5}$

PROBLEM 5-4 Multiply with mixed numbers: **(a)** $2\frac{1}{4} \times \frac{5}{6}$ **(b)** $1\frac{1}{2} \times \frac{1}{3}$ **(c)** $\frac{2}{5} \times 1\frac{3}{4}$
(d) $1\frac{1}{2} \times \frac{1}{2}$ **(e)** $\frac{3}{8} \times 2\frac{1}{3}$ **(f)** $\frac{3}{5} \times 7\frac{1}{2}$ **(g)** $4\frac{1}{2} \times \frac{2}{3}$ **(h)** $7\frac{1}{2} \times 1\frac{3}{5}$ **(i)** $2\frac{1}{3} \times 4\frac{1}{2}$
(j) $6 \times 2\frac{1}{2}$ **(k)** $1\frac{1}{3} \times 12 \times \frac{1}{2}$ **(l)** $\frac{3}{10} \times 3\frac{3}{4} \times 2$

Solution: Recall that to multiply with mixed numbers, you first rename each mixed number and whole number as an equal fraction and then multiply [see Example 5-7]:

(a) $2\dfrac{1}{4} \times \dfrac{5}{6} = \dfrac{9}{4} \times \dfrac{5}{6} = \dfrac{3 \times \cancel{3}}{4} \times \dfrac{5}{2 \times \cancel{3}} = \dfrac{15}{8} = 1\dfrac{7}{8}$ **(b)** $1\dfrac{1}{2} \times \dfrac{1}{3} = \dfrac{\cancel{3}}{2} \times \dfrac{1}{\cancel{3}} = \dfrac{1}{2}$

(c) $\dfrac{2}{5} \times 1\dfrac{3}{4} = \dfrac{2}{5} \times \dfrac{7}{4} = \dfrac{\cancel{2}}{5} \times \dfrac{7}{\cancel{2} \times 2} = \dfrac{7}{10}$ **(d)** $1\dfrac{1}{2} \times \dfrac{1}{2} = \dfrac{3}{2} \times \dfrac{1}{2} = \dfrac{3}{4}$

(e) $\dfrac{3}{8} \times 2\dfrac{1}{3} = \dfrac{\cancel{3}}{8} \times \dfrac{7}{\cancel{3}} = \dfrac{7}{8}$ **(f)** $\dfrac{3}{5} \times 7\dfrac{1}{2} = \dfrac{3}{5} \times \dfrac{15}{2} = \dfrac{3}{\cancel{5}} \times \dfrac{3 \times \cancel{5}}{2} = \dfrac{9}{2} = 4\dfrac{1}{2}$

(g) $4\dfrac{1}{2} \times \dfrac{2}{3} = \dfrac{9}{\cancel{2}} \times \dfrac{\cancel{2}}{3} = \dfrac{\cancel{3} \times 3}{1} \times \dfrac{1}{\cancel{3}} = \dfrac{3}{1} = 3$

(h) $7\dfrac{1}{2} \times 1\dfrac{3}{5} = \dfrac{15}{2} \times \dfrac{8}{5} = \dfrac{3 \times \cancel{5}}{\cancel{2}} \times \dfrac{\cancel{2} \times 4}{\cancel{5}} = \dfrac{12}{1} = 12$

(i) $2\frac{1}{3} \times 4\frac{1}{2} = \frac{7}{3} \times \frac{9}{2} = \frac{7}{\cancel{3}} \times \frac{\cancel{3} \times 3}{2} = \frac{21}{2} = 10\frac{1}{2}$

(j) $6 \times 2\frac{1}{2} = \frac{6}{1} \times \frac{5}{2} = \frac{\cancel{2} \times 3}{1} \times \frac{5}{\cancel{2}} = \frac{15}{1} = 15$

(k) $1\frac{1}{3} \times 12 \times \frac{1}{2} = \frac{4}{3} \times \frac{12}{1} \times \frac{1}{2} = \frac{\cancel{2} \times 2}{\cancel{3}} \times \frac{\cancel{3} \times 4}{1} \times \frac{1}{\cancel{2}} = \frac{8}{1} = 8$

(l) $\frac{3}{10} \times 3\frac{3}{4} \times 2 = \frac{3}{10} \times \frac{15}{4} \times \frac{2}{1} = \frac{3}{\cancel{2} \times \cancel{5}} \times \frac{3 \times \cancel{5}}{4} \times \frac{\cancel{2}}{1} = \frac{9}{4} = 2\frac{1}{4}$

PROBLEM 5-5 Find and check the reciprocal of each fraction: **(a)** $\frac{1}{2}$ **(b)** $\frac{2}{3}$ **(c)** $\frac{5}{4}$ **(d)** $\frac{0}{1}$

Solution: Recall that to find the reciprocal of a fraction, you interchange the whole numbers in the numerator and denominator. To check the reciprocal of a fraction, you multiply the two fractions to see if the product is 1 [see Example 5-9]:

(a) The reciprocal of $\frac{1}{2}$ is $\frac{2}{1}$ (or 2) because $\frac{1}{2} \times \frac{2}{1} = 1$.
(b) The reciprocal of $\frac{2}{3}$ is $\frac{3}{2}$ because $\frac{2}{3} \times \frac{3}{2} = 1$.
(c) The reciprocal of $\frac{5}{4}$ is $\frac{4}{5}$ because $\frac{5}{4} \times \frac{4}{5} = 1$.
(d) The reciprocal of $\frac{0}{1}$ is not defined because the denominator of a fraction can never be zero.

PROBLEM 5-6 Find and check the reciprocal of each whole number: **(a)** 2 **(b)** 6 **(c)** 0 **(d)** 15

Solution: Recall that to find the reciprocal of a whole number, you first rename the whole number as an equal fraction and then find the reciprocal of that fraction [see Example 5-10]:

(a) The reciprocal of 2 (or $\frac{2}{1}$) is $\frac{1}{2}$ because $2 \times \frac{1}{2} = \frac{2}{1} \times \frac{1}{2} = 1$.
(b) The reciprocal of 6 (or $\frac{6}{1}$) is $\frac{1}{6}$ because $6 \times \frac{1}{6} = \frac{6}{1} \times \frac{1}{6} = 1$.
(c) The reciprocal of 0 is not defined because $0 = \frac{0}{1}$ and $\frac{1}{0}$ (its reciprocal) is not defined.
(d) The reciprocal of 15 (or $\frac{15}{1}$) is $\frac{1}{15}$ because $15 \times \frac{1}{15} = \frac{15}{1} \times \frac{1}{15} = 1$.

PROBLEM 5-7 Find and check the reciprocal of each mixed number: **(a)** $2\frac{1}{3}$ **(b)** $3\frac{1}{2}$ **(c)** $2\frac{3}{4}$

Solution: Recall that to find the reciprocal of a mixed number, you first rename the mixed number as an equal fraction and then find the reciprocal of that fraction [see Example 5-11]:

(a) The reciprocal of $2\frac{1}{3}$ (or $\frac{7}{3}$) is $\frac{3}{7}$ because $2\frac{1}{3} \times \frac{3}{7} = \frac{7}{3} \times \frac{3}{7} = 1$.
(b) The reciprocal of $3\frac{1}{2}$ (or $\frac{7}{2}$) is $\frac{2}{7}$ because $3\frac{1}{2} \times \frac{2}{7} = \frac{7}{2} \times \frac{2}{7} = 1$.
(c) The reciprocal of $2\frac{3}{4}$ (or $\frac{11}{4}$) is $\frac{4}{11}$ because $2\frac{3}{4} \times \frac{4}{11} = \frac{11}{4} \times \frac{4}{11} = 1$.

PROBLEM 5-8 Rename each fractional statement as division:

(a) $\frac{2}{3}$ is $\frac{1}{4}$ of an amount. **(b)** $\frac{2}{3}$ of an amount is $\frac{1}{2}$. **(c)** 2 is $\frac{5}{8}$ of an amount.
(d) $\frac{1}{2}$ of an amount is 5. **(e)** $3\frac{1}{4}$ is $\frac{1}{3}$ of an amount. **(f)** $\frac{1}{10}$ of an amount is $1\frac{1}{2}$.

Solution: Recall that to find an amount given a fractional part of it, you divide. Also, when the word *is* appears between two numbers, you can usually replace the word with a division symbol ($\div$) [see Example 5-12]:

(a) $\frac{2}{3} \div \frac{1}{4}$ **(b)** $\frac{1}{2}$ is $\frac{2}{3}$ of amount, or $\frac{1}{2} \div \frac{2}{3}$ **(c)** $2 \div \frac{5}{8}$
(d) 5 is $\frac{1}{2}$ of an amount, or $5 \div \frac{1}{2}$ **(e)** $3\frac{1}{4} \div \frac{1}{3}$ **(f)** $1\frac{1}{2}$ is $\frac{1}{10}$ of an amount, or $1\frac{1}{2} \div \frac{1}{10}$

PROBLEM 5-9 Divide the following fractions. Check each proposed quotient using multiplication:

(a) $\frac{3}{4} \div \frac{2}{3}$ **(b)** $\frac{3}{8} \div \frac{2}{3}$ **(c)** $\frac{2}{3} \div \frac{5}{6}$ **(d)** $\frac{1}{6} \div \frac{2}{5}$ **(e)** $\frac{1}{2} \div \frac{1}{3}$ **(f)** $\frac{1}{3} \div \frac{1}{2}$
(g) $\frac{3}{5} \div \frac{3}{5}$ **(h)** $\frac{0}{3} \div \frac{7}{9}$ **(i)** $\frac{5}{6} \div 3$ **(j)** $\frac{2}{3} \div 4$ **(k)** $2 \div \frac{1}{2}$ **(l)** $9 \div \frac{3}{4}$

Solution: Recall that to divide with fractions, you change the divisor to its reciprocal and then multiply. To check division, you multiply the proposed quotient by the original divisor to see if you get the original dividend [See Examples 5-14 through 5-18]:

(a) $\dfrac{3}{4} \div \dfrac{2}{3} = \dfrac{3}{4} \times \dfrac{3}{2} = \dfrac{9}{8} = 1\dfrac{1}{8}$ *Check:* $\dfrac{2}{3} \times 1\dfrac{1}{8} = \dfrac{2}{3} \times \dfrac{9}{8} = \dfrac{\cancel{2}}{\cancel{3}} \times \dfrac{\cancel{3} \times 3}{\cancel{2} \times 4} = \dfrac{3}{4}$

(b) $\dfrac{3}{8} \div \dfrac{2}{3} = \dfrac{3}{8} \times \dfrac{3}{2} = \dfrac{9}{16}$ *Check:* $\dfrac{2}{3} \times \dfrac{9}{16} = \dfrac{\cancel{2}}{\cancel{3}} \times \dfrac{\cancel{3} \times 3}{\cancel{2} \times 8} = \dfrac{3}{8}$

(c) $\dfrac{2}{3} \div \dfrac{5}{6} = \dfrac{2}{3} \times \dfrac{6}{5} = \dfrac{2}{\cancel{3}} \times \dfrac{2 \times \cancel{3}}{5} = \dfrac{4}{5}$ *Check:* $\dfrac{\cancel{5}}{6} \times \dfrac{4}{\cancel{5}} = \dfrac{1}{\cancel{2} \times 3} \times \dfrac{\cancel{2} \times 2}{1} = \dfrac{2}{3}$

(d) $\dfrac{1}{6} \div \dfrac{2}{5} = \dfrac{1}{6} \times \dfrac{5}{2} = \dfrac{5}{12}$ *Check:* $\dfrac{2}{\cancel{5}} \times \dfrac{\cancel{5}}{12} = \dfrac{\cancel{2}}{1} \times \dfrac{1}{\cancel{2} \times 6} = \dfrac{1}{6}$

(e) $\dfrac{1}{2} \div \dfrac{1}{3} = \dfrac{1}{2} \times \dfrac{3}{1} = \dfrac{3}{2} = 1\dfrac{1}{2}$ *Check:* $\dfrac{1}{3} \times 1\dfrac{1}{2} = \dfrac{1}{\cancel{3}} \times \dfrac{\cancel{3}}{2} = \dfrac{1}{2}$

(f) $\dfrac{1}{3} \div \dfrac{1}{2} = \dfrac{1}{3} \times \dfrac{2}{1} = \dfrac{2}{3}$ *Check:* $\dfrac{1}{\cancel{2}} \times \dfrac{\cancel{2}}{3} = \dfrac{1}{3}$

(g) $\dfrac{3}{5} \div \dfrac{3}{5} = \dfrac{3}{5} \times \dfrac{5}{3} = 1$ *Check:* $\dfrac{3}{5} \times 1 = \dfrac{3}{5}$

(h) $\dfrac{0}{3} \div \dfrac{7}{9} = 0 \div \dfrac{7}{9} = 0$ *Check:* $\dfrac{7}{9} \times 0 = 0 = \dfrac{0}{3}$

(i) $\dfrac{5}{6} \div 3 = \dfrac{5}{6} \div \dfrac{3}{1} = \dfrac{5}{6} \times \dfrac{1}{3} = \dfrac{5}{18}$ *Check:* $3 \times \dfrac{5}{18} = \dfrac{\cancel{3}}{1} \times \dfrac{5}{\cancel{3} \times 6} = \dfrac{5}{6}$

(j) $\dfrac{2}{3} \div 4 = \dfrac{2}{3} \div \dfrac{4}{1} = \dfrac{2}{3} \times \dfrac{1}{4} = \dfrac{\cancel{2}}{3} \times \dfrac{1}{\cancel{2} \times 2} = \dfrac{1}{6}$ *Check:* $4 \times \dfrac{1}{6} = \dfrac{4}{1} \times \dfrac{1}{6} = \dfrac{\cancel{2} \times 2}{1} \times \dfrac{1}{\cancel{2} \times 3} = \dfrac{2}{3}$

(k) $2 \div \dfrac{1}{2} = \dfrac{2}{1} \div \dfrac{1}{2} = \dfrac{2}{1} \times \dfrac{2}{1} = \dfrac{4}{1} = 4$ *Check:* $\dfrac{1}{2} \times 4 = \dfrac{1}{2} \times \dfrac{4}{1} = \dfrac{1}{\cancel{2}} \times \dfrac{\cancel{2} \times 2}{1} = \dfrac{2}{1} = 2$

(l) $9 \div \dfrac{3}{4} = \dfrac{9}{1} \div \dfrac{3}{4} = \dfrac{9}{1} \times \dfrac{4}{3} = \dfrac{\cancel{3} \times 3}{1} \times \dfrac{4}{\cancel{3}} = \dfrac{12}{1} = 12$ *Check:* $\dfrac{3}{4} \times 12 = \dfrac{3}{4} \times \dfrac{12}{1} = \dfrac{3}{\cancel{4}} \times \dfrac{3 \times \cancel{4}}{1} = \dfrac{9}{1} = 9$

PROBLEM 5-10 Divide with mixed numbers. Check each proposed quotient using multiplication:

(a) $2\frac{1}{2} \div \frac{1}{4}$ **(b)** $4\frac{1}{2} \div 2$ **(c)** $\frac{3}{4} \div 1\frac{1}{2}$ **(d)** $1\frac{1}{2} \div \frac{2}{3}$ **(e)** $2\frac{1}{2} \div 1\frac{1}{2}$ **(f)** $3\frac{3}{4} \div 4\frac{1}{2}$

(g) $9 \div 2\frac{1}{4}$ **(h)** $4\frac{1}{4} \div 6$ **(i)** $2\frac{5}{8} \div 3\frac{1}{2}$ **(j)** $4\frac{1}{2} \div \frac{3}{4}$ **(k)** $0 \div 5\frac{3}{4}$ **(l)** $6\frac{7}{8} \div 1$

Solution: Recall that to divide with mixed numbers, you first rename each mixed number (and each whole number) as an equal fraction and then divide. To check division, you multiply the proposed quotient by the original divisor to see if you get the original dividend [see Example 5-19]:

(a) $2\dfrac{1}{2} \div \dfrac{1}{4} = \dfrac{5}{2} \div \dfrac{1}{4} = \dfrac{5}{2} \times \dfrac{4}{1} = \dfrac{5}{\cancel{2}} \times \dfrac{\cancel{2} \times 2}{1} = \dfrac{10}{1} = 10$ *Check:* $\dfrac{1}{4} \times 10 = \dfrac{1}{4} \times \dfrac{10}{1} = \dfrac{5}{2} = 2\dfrac{1}{2}$

(b) $4\dfrac{1}{2} \div 2 = \dfrac{9}{2} \div \dfrac{2}{1} = \dfrac{9}{2} \times \dfrac{1}{2} = \dfrac{9}{4} = 2\dfrac{1}{4}$ *Check:* $2 \times 2\dfrac{1}{4} = \dfrac{2}{1} \times \dfrac{9}{4} = \dfrac{\cancel{2}}{1} \times \dfrac{9}{\cancel{2} \times 2} = \dfrac{9}{2} = 4\dfrac{1}{2}$

(c) $\dfrac{3}{4} \div 1\dfrac{1}{2} = \dfrac{3}{4} \div \dfrac{3}{2} = \dfrac{\cancel{3}}{4} \times \dfrac{2}{\cancel{3}} = \dfrac{1}{\cancel{2} \times 2} \times \dfrac{\cancel{2}}{1} = \dfrac{1}{2}$ *Check:* $1\dfrac{1}{2} \times \dfrac{1}{2} = \dfrac{3}{2} \times \dfrac{1}{2} = \dfrac{3}{4}$

(d) $1\dfrac{1}{2} \div \dfrac{2}{3} = \dfrac{3}{2} \div \dfrac{2}{3} = \dfrac{3}{2} \times \dfrac{3}{2} = \dfrac{9}{4} = 2\dfrac{1}{4}$ *Check:* $\dfrac{2}{3} \times 2\dfrac{1}{4} = \dfrac{2}{3} \times \dfrac{9}{4} = \dfrac{\cancel{2}}{\cancel{3}} \times \dfrac{\cancel{3} \times 3}{\cancel{2} \times 2} = \dfrac{3}{2} = 1\dfrac{1}{2}$

(e) $2\dfrac{1}{2} \div 1\dfrac{1}{2} = \dfrac{5}{2} \div \dfrac{3}{2} = \dfrac{5}{\cancel{2}} \times \dfrac{\cancel{2}}{3} = \dfrac{5}{3} = 1\dfrac{2}{3}$ *Check:* $1\dfrac{1}{2} \times 1\dfrac{2}{3} = \dfrac{3}{2} \times \dfrac{5}{3} = \dfrac{\cancel{3}}{2} \times \dfrac{5}{\cancel{3}} = \dfrac{5}{2} = 2\dfrac{1}{2}$

(f) $3\dfrac{3}{4} \div 4\dfrac{1}{2} = \dfrac{15}{4} \div \dfrac{9}{2} = \dfrac{15}{4} \times \dfrac{2}{9} = \dfrac{\cancel{3} \times 5}{\cancel{2} \times 2} \times \dfrac{\cancel{2}}{\cancel{3} \times 3} = \dfrac{5}{6}$ *Check:* $4\dfrac{1}{2} \times \dfrac{5}{6} = \dfrac{9}{2} \times \dfrac{5}{6} = \dfrac{15}{4} = 3\dfrac{3}{4}$

(g) $9 \div 2\frac{1}{4} = \frac{9}{1} \div \frac{9}{4} = \frac{\cancel{9}}{1} \times \frac{4}{\cancel{9}} = \frac{4}{1} = 4$ *Check:* $2\frac{1}{4} \times 4 = \frac{9}{\cancel{4}} \times \frac{\cancel{4}}{1} = \frac{9}{1} = 9$

(h) $4\frac{1}{4} \div 6 = \frac{17}{4} \div \frac{6}{1} = \frac{17}{4} \times \frac{1}{6} = \frac{17}{24}$ *Check:* $6 \times \frac{17}{24} = \frac{6}{1} \times \frac{17}{24} = \frac{\cancel{6}}{1} \times \frac{17}{4 \times \cancel{6}} = \frac{17}{4} = 4\frac{1}{4}$

(i) $2\frac{5}{8} \div 3\frac{1}{2} = \frac{21}{8} \div \frac{7}{2} = \frac{21}{8} \times \frac{2}{7} = \frac{3 \times \cancel{7}}{\cancel{2} \times 4} \times \frac{\cancel{2}}{\cancel{7}} = \frac{3}{4}$ *Check:* $3\frac{1}{2} \times \frac{3}{4} = \frac{7}{2} \times \frac{3}{4} = \frac{21}{8} = 2\frac{5}{8}$

(j) $4\frac{1}{2} \div \frac{3}{4} = \frac{9}{2} \div \frac{3}{4} = \frac{9}{2} \times \frac{4}{3} = \frac{\cancel{3} \times 3}{\cancel{2}} \times \frac{\cancel{2} \times 2}{\cancel{3}} = \frac{6}{1} = 6$ *Check:* $\frac{3}{4} \times 6 = \frac{3}{4} \times \frac{6}{1} = \frac{9}{2} = 4\frac{1}{2}$

(k) $0 \div 5\frac{3}{4} = 0$ (0 divided by any nonzero number is 0.)   *Check:* $5\frac{3}{4} \times 0 = 0$

(l) $6\frac{7}{8} \div 1 = 6\frac{7}{8}$ (Any number divided by 1 equals that number.)   *Check:* $1 \times 6\frac{7}{8} = 6\frac{7}{8}$

PROBLEM 5-11 Decide what to do (multiply or divide with fractions) and then solve the following problems:

(a) A certain store has 40 radios on sale. Before noon $\frac{3}{4}$ of the radios on sale were sold. How many radios on sale were sold before noon?

(b) Joe bought $\frac{3}{4}$ dozen eggs. He used $\frac{1}{2}$ of them. What fractional part of the eggs did Joe use?

(c) Frieda wants to run $\frac{3}{4}$ mile. The track she runs on is $\frac{1}{8}$ mile long. How many laps around the track will Frieda need to run?

(d) Each jar of juice contains $\frac{3}{4}$ quart. How many jars are needed to make 18 quarts of juice?

(e) There are 18 vehicles in a certain parking lot. $\frac{2}{3}$ of the vehicles are cars. How many cars are in the parking lot?

(f) Bill sleeps $\frac{1}{3}$ of each day. There are 365 days in a normal year. How many days is Bill's total sleeping time during a normal year?

(g) James took a test with 100 questions on it. He got $\frac{9}{10}$ of the questions correct. How many questions did James get wrong?

(h) There are 27 bones in the human hand and wrist. This is $\frac{9}{10}$ of the number of bones in the human arm. How many bones in the human arm are not in the hand and wrist together?

Solution: Recall that to find a fractional part of an amount, you multiply; and to find an amount given a fractional part of it, you divide [see Example 5-20]:

(a) multiply: $\frac{3}{4} \times 40 = \frac{3}{4} \times \frac{40}{1} = 30$ (radios) **(b)** multiply: $\frac{1}{2} \times \frac{3}{4} = \frac{3}{8}$ (dozen)

(c) divide: $\frac{3}{4} \div \frac{1}{8} = \frac{3}{4} \times \frac{8}{1} = 6$ (laps) **(d)** divide: $18 \div \frac{3}{4} = \frac{18}{1} \times \frac{4}{3} = 24$ (jars)

(e) multiply: $\frac{2}{3} \times 18 = \frac{2}{3} \times \frac{18}{1} = 12$ (cars) **(f)** multiply: $\frac{1}{3} \times 365 = \frac{1}{3} \times \frac{365}{1} = 121\frac{2}{3}$ (days)

(g) multiply and then subtract: $\frac{9}{10} \times 100 = \frac{9}{10} \times \frac{100}{1} = 90$; $100 - 90 = 10$ (questions)

(h) divide and then subtract: $27 \div \frac{9}{10} = \frac{27}{1} \times \frac{10}{9} = 30$; $30 - 27 = 3$ (bones)

PROBLEM 5-12 Decide what to do (multiply or divide with mixed numbers) and then solve the following problems:

(a) Nancy drove $78\frac{3}{4}$ miles while using $3\frac{1}{2}$ gallons of gas. How many miles per gallon (mpg) did she get?

(b) Bob drove $78\frac{3}{4}$ miles while getting $17\frac{1}{2}$ mpg. How much gas did he use?

(c) John used $6\frac{1}{2}$ gallons of gas while getting $18\frac{1}{2}$ mpg. How far did John drive?

(d) Tina drove $192\frac{1}{2}$ miles at a constant speed in $3\frac{1}{2}$ hours. What was her constant speed?

(e) Felipe drove at a constant speed of 60 miles per hour (mph) for $4\frac{1}{4}$ hours. How far did he drive?

(f) Lynn drove $178\frac{3}{4}$ miles at a constant speed of 65 mph. How long did she drive?

(g) Kelley weighs $127\frac{1}{2}$ pounds. This is $2\frac{1}{2}$ times her daughter's weight. How much does Kelley's daughter weigh?

(h) Roger is $1\frac{3}{4}$ times his son's height. His son is $3\frac{1}{2}$ feet tall. How tall is Roger?

Solution: Recall that to solve a word problem containing mixed numbers, it is usually helpful to first solve a similar problem containing easier whole numbers [see Example 5-21]:

(a) divide: $78\frac{3}{4} \div 3\frac{1}{2} = 22\frac{1}{2}$ (mpg)

(b) divide: $78\frac{3}{4} \div 17\frac{1}{2} = 4\frac{1}{2}$ (gallons)

(c) multiply: $18\frac{1}{2} \times 6\frac{1}{2} = 120\frac{1}{4}$ (miles)

(d) divide: $192\frac{1}{2} \div 3\frac{1}{2} = 55$ (mph)

(e) multiply: $60 \times 4\frac{1}{4} = 255$ (miles)

(f) divide: $178\frac{3}{4} \div 65 = 2\frac{3}{4}$ (hours)

(g) divide: $127\frac{1}{2} \div 2\frac{1}{2} = 51$ (pounds)

(h) multiply: $1\frac{3}{4} \times 3\frac{1}{2} = 6\frac{1}{8}$ (feet)

Supplementary Exercises

PROBLEM 5-13   Multiply basic facts given horizontal form:

Row **(a)**	2×2	9×4	5×8	9×2	1×7	2×4	2×6	2×3	5×0	9×3
Row **(b)**	6×6	0×9	5×2	1×2	6×3	7×1	2×1	6×4	7×7	3×0
Row **(c)**	0×0	4×9	4×1	8×7	8×3	7×5	3×7	3×1	3×9	8×2
Row **(d)**	7×2	5×4	6×5	4×8	8×1	1×5	0×3	7×9	1×9	2×7
Row **(e)**	0×7	7×8	9×0	2×9	1×6	2×8	9×6	3×8	6×7	0×6
Row **(f)**	5×1	1×1	5×9	1×0	3×4	4×0	2×5	7×4	0×2	4×3
Row **(g)**	6×0	8×6	6×9	4×4	9×7	5×6	4×6	1×3	9×5	3×5
Row **(h)**	8×9	1×4	0×4	6×8	4×5	5×3	7×3	8×5	8×8	6×1
Row **(i)**	3×6	0×1	3×3	7×0	3×2	8×0	9×9	0×8	8×4	9×1
Row **(j)**	2×0	1×8	4×7	7×6	6×2	5×5	5×7	9×8	0×5	4×2

PROBLEM 5-14   Multiply with fractions and mixed numbers: **(a)** $\frac{3}{4} \times \frac{2}{3}$ **(b)** $\frac{1}{2} \times \frac{1}{4}$
(c) $\frac{2}{3} \times 2\frac{1}{4}$ **(d)** $2\frac{1}{2} \times \frac{4}{5}$ **(e)** $1\frac{3}{4} \times 2\frac{1}{8}$ **(f)** $1\frac{1}{2} \times 2\frac{1}{2}$ **(g)** $\frac{5}{8} \times \frac{1}{4}$ **(h)** $\frac{5}{6} \times \frac{2}{3}$ **(i)** $\frac{3}{8} \times 1\frac{1}{3}$
(j) $2\frac{1}{4} \times \frac{1}{3}$ **(k)** $1\frac{1}{2} \times \frac{3}{4} \times 2\frac{1}{3} \times 4$ **(l)** $\frac{5}{8} \times 2\frac{1}{2} \times \frac{3}{10} \times 2$ **(m)** $1\frac{1}{2} \times 1\frac{1}{3} \times 1\frac{1}{4} \times 1\frac{1}{5}$
(n) $\frac{3}{8} \times 1\frac{1}{3} \times \frac{4}{5} \times 1\frac{7}{10} \times 0 \times 5\frac{7}{10}$ **(o)** $\frac{1}{2} \times 1\frac{1}{2} \times \frac{3}{4} \times 1\frac{1}{3} \times \frac{4}{5} \times 1\frac{1}{4} \times \frac{5}{6} \times 1\frac{1}{5}$

PROBLEM 5-15   Find the reciprocal of each number: **(a)** $\frac{1}{4}$ **(b)** $\frac{1}{3}$ **(c)** 2   **(d)** 1   **(e)** $1\frac{1}{2}$
(f) $1\frac{1}{3}$ **(g)** $\frac{2}{5}$ **(h)** $\frac{0}{8}$ **(i)** 0   **(j)** 5   **(k)** $1\frac{1}{5}$ **(l)** $2\frac{1}{4}$ **(m)** $\frac{7}{10}$ **(n)** $\frac{39}{100}$ **(o)** $4\frac{7}{12}$

PROBLEM 5-16   Divide with fractions and mixed numbers: **(a)** $\frac{3}{4} \div \frac{1}{2}$ **(b)** $\frac{2}{3} \div \frac{1}{2}$ **(c)** $\frac{5}{8} \div \frac{5}{8}$
(d) $\frac{9}{2} \div \frac{7}{10}$ **(e)** $\frac{1}{3} \div 2$ **(f)** $3 \div \frac{2}{5}$ **(g)** $3\frac{1}{2} \div \frac{1}{2}$ **(h)** $\frac{1}{4} \div 2\frac{1}{2}$ **(i)** $1\frac{1}{2} \div 3$
(j) $2 \div 2\frac{3}{4}$ **(k)** $1\frac{1}{5} \div 1\frac{1}{3}$ **(l)** $2\frac{1}{2} \div 2\frac{1}{4}$ **(m)** $\frac{7}{12} \div \frac{5}{12}$ **(n)** $1\frac{1}{4} \div 1\frac{3}{4}$ **(o)** $\frac{5}{18} \div 1\frac{1}{9}$

PROBLEM 5-17   Solve word problems containing fractions and mixed numbers ($\times$, $\div$):

(a) Using $\frac{1}{4}$ ounce of medication, how much medication will there be in each of 8 equal doses?

(b) How many $\frac{1}{4}$-ounce doses can be made from 8 ounces of medication?

(c) Gloria was supposed to run for $1\frac{1}{2}$ hours each day. Today she only ran $\frac{3}{4}$ of her required time. How many hours did Gloria run today?

(d) Jack ran $1\frac{1}{2}$ hours today. He was supposed to run $1\frac{3}{4}$ hours. What fractional part of the time that Jack was supposed to run did he run?

(e) Cherie only ran $\frac{1}{2}$ of the time she was supposed to have run today. She ran for $\frac{3}{4}$ hour in all. How long was Cherie supposed to run?

(f) Only $\frac{1}{9}$ of the height of an iceberg is above the water's surface. If an iceberg is 216 feet from top to bottom, how high does it extend above the water's surface?

(g) If an iceberg is 342 feet from top to bottom, how far below the water's surface does it extend? [See problem **(f)**.]

(h) What fractional part of the iceberg extends below the water's surface? [See problem **(f)**.]

(i) How much of each ingredient is needed for a double recipe:
$\frac{1}{2}$ cup water
$\frac{1}{3}$ cup onions
$\frac{1}{4}$ cup celery
$\frac{1}{2}$ teaspoon salt
$\frac{1}{8}$ teaspoon pepper
$1\frac{1}{2}$ tablespoons flour
2 pounds hamburger meat

(j) How much of each ingredient is needed for one half of the recipe?
4 cups flour
3 tablespoons baking powder
1 teaspoon salt
$\frac{1}{2}$ teaspoon baking soda
$\frac{2}{3}$ cup shortening
$1\frac{1}{2}$ cup buttermilk

(k) Michael earns $18 per hour. How much per hour does he earn working overtime at time and one-half?

(l) Patti earns $21 working overtime at time and one-half. How much is her regular hourly wage?

(m) Jim needs $\frac{1}{3}$ pound of meat to make each hamburger. He has 6 pounds of meat. How many hamburgers can Jim make?

(n) Robin needs $\frac{1}{3}$ pound of meat to make each hamburger. She wants to make 6 hamburgers. How much meat will Robin need?

Answers to Supplementary Exercises

(5-13)

Row (a)	4	36	40	18	7	8	12	6	0	27
Row (b)	36	0	10	2	18	7	2	24	49	0
Row (c)	0	36	4	56	24	35	21	3	27	16
Row (d)	14	20	30	32	8	5	0	63	9	14
Row (e)	0	56	0	18	6	16	54	24	42	0
Row (f)	5	1	45	0	12	0	10	28	0	12
Row (g)	0	48	54	16	63	30	24	3	45	15
Row (h)	72	4	0	48	20	15	21	40	64	6
Row (i)	18	0	9	0	6	0	81	0	32	9
Row (j)	0	8	28	42	12	25	35	72	0	8

(5-14) **(a)** $\frac{1}{2}$ **(b)** $\frac{1}{8}$ **(c)** $1\frac{1}{2}$ **(d)** 2 **(e)** $3\frac{23}{32}$ **(f)** $3\frac{3}{4}$ **(g)** $\frac{5}{32}$ **(h)** $\frac{5}{9}$ **(i)** $\frac{1}{2}$ **(j)** $\frac{3}{4}$
(k) $10\frac{1}{2}$ **(l)** $\frac{15}{16}$ **(m)** 3 **(n)** 0 **(o)** $\frac{3}{4}$

(5-15) **(a)** $\frac{4}{1}$ **(b)** $\frac{3}{1}$ **(c)** $\frac{1}{2}$ **(d)** $\frac{1}{1}$ **(e)** $\frac{2}{3}$ **(f)** $\frac{3}{4}$ **(g)** $\frac{5}{2}$ **(h)** not defined
(i) not defined **(j)** $\frac{1}{5}$ **(k)** $\frac{5}{6}$ **(l)** $\frac{4}{9}$ **(m)** $\frac{10}{7}$ **(n)** $\frac{100}{39}$ **(o)** $\frac{12}{55}$

(5-16) **(a)** $1\frac{1}{2}$ **(b)** $1\frac{1}{3}$ **(c)** 1 **(d)** 0 **(e)** $\frac{1}{6}$ **(f)** $7\frac{1}{2}$ **(g)** 7 **(h)** $\frac{1}{10}$ **(i)** $\frac{1}{2}$
(j) $\frac{8}{11}$ **(k)** $\frac{9}{10}$ **(l)** $1\frac{1}{9}$ **(m)** $1\frac{2}{5}$ **(n)** 1 **(o)** $\frac{1}{4}$

(5-17) **(a)** $\frac{1}{32}$ ounce **(b)** 32 $\frac{1}{4}$-ounce doses **(c)** $1\frac{1}{8}$ hour **(d)** $\frac{6}{7}$ of the time **(e)** $1\frac{1}{2}$ hours
(f) 24 feet **(g)** 304 feet **(h)** $\frac{8}{9}$ of the iceberg **(i)** 1 cup water, $\frac{2}{3}$ cup onion, $\frac{1}{2}$ cup celery,
1 teaspoon salt, $\frac{1}{4}$ teaspoon pepper, 3 tablespoons flour, 4 pounds of hamburger meat
(j) 2 cups flour, $1\frac{1}{2}$ tablespoons baking powder, $\frac{1}{2}$ teaspoon salt, $\frac{1}{4}$ teaspoon baking soda,
$\frac{1}{3}$ cup shortening, $\frac{3}{4}$ cup buttermilk **(k)** $27 **(l)** $14 **(m)** 18 hamburgers **(n)** 2 pounds

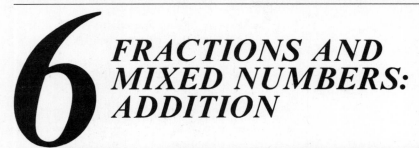

6 FRACTIONS AND MIXED NUMBERS: ADDITION

THIS CHAPTER IS ABOUT

☑ **Adding Like Fractions**
☑ **Finding the Least Common Denominator (LCD)**
☑ **Building Up Fractions**
☑ **Adding Unlike Fractions**
☑ **Adding with Mixed Numbers**

6-1. Adding Like Fractions

To join two or more fractional amounts together, you **add fractions.**

EXAMPLE 6-1: Rename the following fractional amounts as addition: $\frac{2}{8}$ joined together with $\frac{3}{8}$

Solution: $\frac{2}{8}$ joined together with $\frac{3}{8} = \frac{2}{8} + \frac{3}{8}$ ⟵ addition

A. Add two like fractions.

Fractions with the same denominator are called **like fractions.**

EXAMPLE 6-2: Write two like fractions.

Solution: $\frac{2}{8}$ and $\frac{3}{8}$ are two like fractions because their denominators are the same: $8 = 8$

To add like fractions, you can use the following rule:

Addition Rule for Like Fractions

If $\dfrac{a}{c}$ and $\dfrac{b}{c}$ are any two like fractions, then $\dfrac{a}{c} + \dfrac{b}{c} = \dfrac{a + b}{c}$.

Note: To add two like fractions, you add the two numerators and then write the same denominator.

EXAMPLE 6-3: Add $\frac{2}{8} + \frac{3}{8}$.

Solution: $\dfrac{2}{8} + \dfrac{3}{8} = \dfrac{2 + 3}{8}$ Add the two numerators.
Write the same denominator.

$\qquad\qquad = \dfrac{5}{8}$ Add in the numerator: $2 + 3 = 5$

The following example can help you understand addition of two like fractions.

EXAMPLE 6-4: Translate addition of like fractions to words and pictures.

Solution

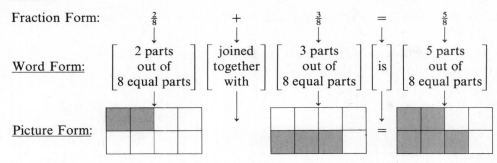

Fraction Form: $\frac{2}{8}$ $+$ $\frac{3}{8}$ $=$ $\frac{5}{8}$

Word Form: [2 parts out of 8 equal parts] [joined together with] [3 parts out of 8 equal parts] [is] [5 parts out of 8 equal parts]

Picture Form:

Caution: Always simplify a fraction sum when possible.

EXAMPLE 6-5: Add $\frac{3}{12} + \frac{5}{12}$.

Solution: $\dfrac{3}{12} + \dfrac{5}{12} = \dfrac{3 + 5}{12}$ Add like fractions.

$$= \frac{8}{12}$$

$$= \frac{2 \times \cancel{4}}{3 \times \cancel{4}}$$ Simplify [see Section 4.3.].

$$= \frac{2}{3} \longleftarrow \text{ simplest form}$$

B. Add more than two like fractions.

To add more than two like fractions, you add all the numerators and write the same denominator.

EXAMPLE 6-6: Add $\frac{3}{4} + \frac{1}{4} + \frac{5}{4} + \frac{1}{4}$.

Solution: $\dfrac{3}{4} + \dfrac{1}{4} + \dfrac{5}{4} + \dfrac{1}{4} = \dfrac{3 + 1 + 5 + 1}{4}$ Add all the numerators.
Write the same denominator.

$$= \frac{10}{4}$$ Add in the numerator.

$$= \frac{\cancel{2} \times 5}{\cancel{2} \times 2}$$ Simplify.

$$= \frac{5}{2}$$

$$= 2\tfrac{1}{2} \longleftarrow \text{ simplest form}$$

6-2. Finding the Least Common Denominator (LCD)

Fractions with different denominators are called **unlike fractions.** To add unlike fractions, the first step is to find the **least common denominator (LCD).** The least common denominator for two or more fractions is the smallest nonzero whole number that all of the denominators divide into **evenly** (with a zero remainder).

A. Find the LCD for two fractions.

When the smaller denominator divides the larger denominator evenly, the larger denominator is the LCD.

EXAMPLE 6-7: Find the LCD for the following fractions: $\frac{3}{4}$ and $\frac{5}{8}$

Solution: The LCD for $\frac{3}{4}$ and $\frac{5}{8}$ is 8 because 8 is the smallest nonzero whole number that both 4 and 8 will divide into evenly.

Note: In Example 6-7, the larger denominator 8 is the LCD because 4 divides 8 evenly: $8 \div 4 = 2$ R0, or just 2.

When the denominators do not share any common prime factors, the product of the denominators is the LCD.

EXAMPLE 6-8: Find the LCD for the following fractions: $\frac{2}{3}$ and $\frac{3}{4}$

Solution: The LCD for $\frac{2}{3}$ and $\frac{3}{4}$ is 12 because 12 (3×4) is the smallest nonzero whole number that both 3 and 4 will divide into evenly.

Note: In Example 6-8, you can multiply the denominators 3 and 4 to find the LCD 12 because 3 and 4 do not share a common prime factor: $3 = 1 \times 3$ and $4 = 1 \times 4$ or 2×2.

If you have trouble finding the LCD for two fractions, you can use the following method:

Factoring Method for Finding the LCD

1. Factor each denominator as a product of primes.
2. Identify the greatest number of times that each prime factor occurs in any single factorization from Step 1.
3. Write the LCD as a product of the factors found in Step 2.

EXAMPLE 6-9: Find the LCD for the following fractions: $\frac{3}{10}$ and $\frac{7}{12}$

Solution:
$$10 = 2 \times \boxed{5}$$
$$12 = \boxed{2 \times 2} \times \boxed{3}$$
$$\text{LCD} = 2 \times 2 \times 3 \times 5$$
$$= 60 \longleftarrow \text{LCD for } \tfrac{3}{10} \text{ and } \tfrac{7}{12}$$

Factor each denominator as a product of primes.

Circle the greatest number of times that each different prime factor occurs in any single factorization.

Write the LCD as a product of the circled factors.

Note 1: In Example 6-9, the LCD from $\frac{3}{10}$ and $\frac{7}{12}$ is 60 because 60 is the smallest nonzero whole number that both 10 and 12 will divide into evenly.

Note 2: In Example 6-9, there are two 2s in the LCD because the greatest number of times that 2 occurs as a factor in any single factorization is twice in $2 \times 2 \times 3$ (2 occurs only once in 2×5).

The following shortcut can save you time and effort in finding the LCD for two unlike fractions in almost every practical application.

Shortcut: To find the LCD for two fractions when the difference between their denominators divides each denominator evenly:
1. Find the difference between the denominators.
2. Divide the difference found in Step 1 into one of the denominators.
3. Multiply the quotient from Step 2 times the other denominator to get the LCD.

EXAMPLE 6-10: Find the LCD for $\frac{3}{10}$ and $\frac{7}{12}$ using a shortcut.

Solution: The LCD for $\frac{3}{10}$ and $\frac{7}{12}$ is 60 because

$$12 - 10 = 2 \longrightarrow 2\overline{)12} \quad 6 \times 10 = 60 \qquad \text{or} \qquad 12 - 10 = 2 \longrightarrow 2\overline{)10} \quad 5 \times 12 = 60$$

B. **Find the LCD for more than two fractions.**

When each smaller denominator divides the larger denominator evenly, the larger denominator is the LCD.

EXAMPLE 6-11: Find the LCD for the following fractions: $\frac{1}{6}$, $\frac{5}{12}$, and $\frac{3}{4}$

Solution: The LCD for $\frac{1}{6}$, $\frac{5}{12}$, and $\frac{3}{4}$ is 12 because 12 is the smallest nonzero whole number that 6, 12, and 4 will divide into evenly.

Note: In Example 6-11, the larger denominator 12 is the LCD because 4, 6, and 12 all divide into 12 evenly: $12 \div 4 = 3$, $12 \div 6 = 2$, and $12 \div 12 = 1$.

When no two denominators share a common prime factor, the product of all the denominators is the LCD.

EXAMPLE 6-12: Find the LCD for the following fractions: $\frac{1}{2}$, $\frac{2}{3}$, and $\frac{3}{5}$

Solution: The LCD for $\frac{1}{2}$, $\frac{2}{3}$, and $\frac{3}{5}$ is 30 because 30 ($2 \times 3 \times 5$) is the smallest nonzero whole number that 2, 3, and 5 will all divide into evenly.

Note: In Example 6-12, you can multiply the denominators, 2, 3, and 5 together to find the LCD because no two denominators share a common prime factor: $2 = 1 \times 2$, $3 = 1 \times 3$, and $5 = 1 \times 5$.

When you have trouble finding the LCD for more than two fractions, you can always use the Factoring Method for Finding the LCD presented in Example 6-9.

EXAMPLE 6-13: Find the LCD for the following fractions: $\frac{7}{18}$, $\frac{5}{6}$, and $\frac{3}{4}$

Solution: $18 = 2 \times \boxed{3 \times 3}$ Factor each denominator as a product of primes.

$\ 6 = 2 \times 3$ Circle the greatest number of times that each
$\ 4 = \boxed{2 \times 2}$ different prime factor occurs in any single
                 factorization.

$\text{LCD} = 2 \times 2 \times 3 \times 3$ Write the LCD as a product of the circled primes.

$\phantom{\text{LCD}}\ = 4 \times 9$

$\phantom{\text{LCD}}\ = 36 \longleftarrow \text{LCD for } \frac{7}{18}, \frac{5}{6}, \text{ and } \frac{3}{4}$

Note: In Example 6-13, the LCD for $\frac{7}{18}$, $\frac{5}{6}$, and $\frac{3}{4}$ is 36 because 36 is the smallest nonzero whole number that 18, 6, and 4 will all divide into evenly.

6-3. Building Up Fractions

Recall the **Fundamental Rule for Fractions from Chapter 4:**

If *a*, *b*, and *c* are whole numbers ($b \neq 0$ and $c \neq 0$), then $\dfrac{a}{b} = \dfrac{a \times c}{b \times c}$.

EXAMPLE 6-14: Show that $\dfrac{2}{3} = \dfrac{2 \times 4}{3 \times 4}$.

Solution: $\dfrac{2}{3} = \dfrac{2}{3} \times 1$ *Think:* Any number times one is that number.

$\phantom{\dfrac{2}{3}}\ = \dfrac{2}{3} \times \dfrac{4}{4}$ Substitute: $1 = \frac{4}{4}$

$\phantom{\dfrac{2}{3}}\ = \dfrac{2 \times 4}{3 \times 4}$ Multiply fractions.

A. Build up to get higher terms.

Recall that to add unlike fractions, the first step is to find the least common denominator (LCD). The second step is to **build up to get like fractions** using the LCD. To build up a given fraction as an equal fraction with a given denominator, you first compare denominators to find the correct **building factor.**

EXAMPLE 6-15: Build up $\frac{2}{3}$ as an equal fraction with a denominator of 12.

Solution: $\frac{2}{3} = \frac{?}{12}$ *Think:* 3 times what building factor is 12?

$3 \times 4 = 12$ *Think:* The correct building factor is 4
                                 because: $3 \times 4 = 12$.

$\frac{2}{3} = \frac{2 \times 4}{3 \times 4}$ Use the Fundamental Rule for Fractions.

$= \frac{8}{12} \longleftarrow \frac{2}{3}$ built up as an equal fraction with a denominator of 12

Note: In Example 6-15, to build up $\frac{2}{3}$ as an equal fraction with a denominator of 12, you multiply both the numerator and denominator of $\frac{2}{3}$ by the building factor 4.

B. Build up using the LCD.

To rename two unlike fractions as equivalent like fractions using the LCD, you first find the LCD and then build up to get like fractions using the LCD.

EXAMPLE 6-16: Rename the following fractions as equivalent like fractions using the LCD: $\frac{1}{2}$ and $\frac{2}{3}$

Solution: The LCD of $\frac{1}{2}$ and $\frac{2}{3}$ is 6.

$\frac{1}{2} = \frac{?}{6} \longleftarrow 2 \times \mathbf{3} = 6$ means the building factor for $\frac{1}{2}$ is 3

$\frac{2}{3} = \frac{?}{6} \longleftarrow 3 \times \mathbf{2} = 6$ means the building factor for $\frac{2}{3}$ is 2

$\frac{1}{2} = \frac{1 \times 3}{2 \times 3} = \frac{3}{6} \longleftarrow \frac{1}{2}$ built up as an equal fraction using the LCD 6

$\frac{2}{3} = \frac{2 \times 2}{3 \times 2} = \frac{4}{6} \longleftarrow \frac{2}{3}$ built up as an equal fraction using the LCD 6

$\frac{1}{2}$ and $\frac{2}{3}$ renamed as equivalent like fractions using the LCD are $\frac{3}{6}$ and $\frac{4}{6}$ respectively.

To rename more than two unlike fractions as equivalent like fractions using the LCD, you rename as with two unlike fractions.

EXAMPLE 6-17: Rename the following fractions as equivalent like fractions using the LCD: $\frac{5}{6}, \frac{1}{3}$, and $\frac{7}{12}$

Solution: The LCD of $\frac{5}{6}, \frac{1}{3}$, and $\frac{7}{12}$ is 12.

$\frac{5}{6} = \frac{?}{12} \longleftarrow 6 \times \mathbf{2} = 12$ means the building factor for $\frac{5}{6}$ is 2

$\frac{1}{3} = \frac{?}{12} \longleftarrow 3 \times \mathbf{4} = 12$ means the building factor for $\frac{1}{3}$ is 4

$$\frac{7}{12} = \frac{7}{12} \longleftarrow \text{ no building factor needed because 12 is the LCD}$$

$$\frac{5}{6} = \frac{5 \times 2}{6 \times 2} = \frac{10}{12} \longleftarrow \frac{5}{6} \text{ built up as an equal fraction using the LCD 12}$$

$$\frac{1}{3} = \frac{1 \times 4}{3 \times 4} = \frac{4}{12} \longleftarrow \frac{1}{3} \text{ built up as an equal fraction using the LCD 12}$$

$$\frac{7}{12} = \frac{7}{12} \longleftarrow \frac{7}{12} \text{ is already a fraction with the LCD 12 for a denominator.}$$

$\frac{5}{6}, \frac{1}{3}$, and $\frac{7}{12}$ renamed as equivalent fractions using the LCD are $\frac{10}{12}, \frac{4}{12}$, and $\frac{7}{12}$ respectively.

6-4. Adding Unlike Fractions

To add unlike fractions, you can use the following rules:

Addition Rules for Unlike Fractions

1. Find the LCD for the unlike fractions [see Section 6-2].
2. Build up to get like fractions using the LCD from Step 1 [see Section 6-3].
3. Substitute the like fractions for the original unlike fractions.
4. Add the like fractions from Step 3 [see Section 6-1].
5. Simplify the sum from Step 4 when possible [see Section 4-3].

Note: To add two or more unlike fractions, you first rename as like fractions using the LCD and then add the like fractions.

A. Add two unlike fractions.

To add two unlike fractions, you use the Addition Rules for Unlike Fractions.

EXAMPLE 6-18: Add $\frac{1}{2} + \frac{2}{3}$.

Solution: The LCD for $\frac{1}{2}$ and $\frac{2}{3}$ is 6.   Find the LCD.

$$\frac{1}{2} = \frac{1 \times 3}{2 \times 3} = \frac{3}{6} \longleftarrow$$ Build up to get like fractions using the LCD.

$$\text{LCD}$$

$$\frac{2}{3} = \frac{2 \times 2}{3 \times 2} = \frac{4}{6} \longleftarrow$$

$$\frac{1}{2} + \frac{2}{3} = \frac{3}{6} + \frac{4}{6} \longleftarrow \text{ like fractions}$$ Substitute the like fractions for the original unlike fractions.

$$= \frac{3 + 4}{6}$$ Add like fractions as before [see Example 6-1].

$$= \frac{7}{6}$$

$$= 1\frac{1}{6}$$ Simplify [see Section 4-3].

Note: $\frac{1}{2} + \frac{2}{3} = 1\frac{1}{6} \longleftarrow$ simplest form

B. Add more than two unlike fractions.

To add more than two unlike fractions, you use the Addition Rules for Unlike Fractions.

EXAMPLE 6-19: Add $\frac{11}{15} + \frac{3}{5} + \frac{1}{6}$.

Solution: The LCD for $\frac{11}{15}$, $\frac{3}{5}$, and $\frac{1}{6}$ is 30.   Find the LCD.

$$\frac{11}{15} = \frac{11 \times 2}{15 \times 2} = \frac{22}{30}$$ Build up to get like fractions.

$$\frac{3}{5} = \frac{3 \times 6}{5 \times 6} = \frac{18}{30} \quad \text{LCD}$$

$$\frac{1}{6} = \frac{1 \times 5}{6 \times 5} = \frac{5}{30}$$

$$\frac{11}{15} + \frac{3}{5} + \frac{1}{6} = \frac{22}{30} + \frac{18}{30} + \frac{5}{30} \longleftarrow \text{like fractions}$$ Substitute.

$$= \frac{22 + 18 + 5}{30}$$ Add like fractions.

$$= \frac{45}{30}$$

$$= \frac{3 \times 15}{2 \times 15}$$ Simplify.

$$= \frac{3}{2}$$

$$= 1\tfrac{1}{2}$$

Note: $\frac{11}{15} + \frac{3}{5} + \frac{1}{6} = 1\tfrac{1}{2}$ simplest form

6-5. Adding with Mixed Numbers

To add two or more mixed numbers, you can use the following rules:

Addition Rules for Mixed Numbers

(a) Write vertical addition form.
(b) Build up to get like-fraction parts when necessary.
(c) Add the like-fraction parts to find the fraction part of the sum.
(d) Add the whole-number parts to find the whole number part of the sum.
(e) Simplify the sum from Steps **(c)** and **(d)** when possible.

A. Add mixed numbers and whole numbers.

To add a mixed number to one or more whole numbers, you add the whole numbers and then write the same fraction part.

EXAMPLE 6-20: Add $5\tfrac{1}{2} + 3$.

Solution: $5\tfrac{1}{2} + 3 = 8\tfrac{1}{2}$ Add the whole-number parts: $5 + 3 = 8$
                                    Write the same fraction part.

Note: $5\tfrac{1}{2} + 3 = 8\tfrac{1}{2}$ because $5\tfrac{1}{2} + 3 = 5 + \tfrac{1}{2} + 3$

$$= (5 + 3) + \tfrac{1}{2}$$

$$= 8 + \tfrac{1}{2}$$

$$= 8\tfrac{1}{2}$$

B. Add mixed numbers and fractions.

To add a mixed number and one or more fractions, you add the fractions and then write the same whole-number part.

EXAMPLE 6-21: Add $\frac{5}{6} + 2\frac{1}{3}$.

Solution: $\frac{5}{6} + 2\frac{1}{3} = \frac{5}{6} + 2 + \frac{1}{3}$ *Think:* The LCD of $\frac{5}{6}$ and $\frac{1}{3}$ is 6.

$\quad\quad\quad\quad = 2 + (\frac{5}{6} + \frac{2}{6})$ Build up to get like fractions using the LCD.

$\quad\quad\quad\quad = 2 + \frac{7}{6}$ Add like fractions.

$\quad\quad\quad\quad = 2 + 1\frac{1}{6}$ Simplify.

$\quad\quad\quad\quad = 2 + 1 + \frac{1}{6}$

$\quad\quad\quad\quad = 3 + \frac{1}{6}$

$\quad\quad\quad\quad = 3\frac{1}{6}$

Note: $\frac{5}{6} + 2\frac{1}{3} = 3\frac{1}{6}$ ⟵ simplest form

C. Add with mixed numbers.

To add two or more mixed numbers, you first write the mixed numbers in vertical form, then add the fraction parts, and then add the whole-number parts.

Note: It is usually easier to add mixed numbers in vertical form.

EXAMPLE 6-22: Add $2\frac{3}{4} + 3\frac{1}{2}$.

Solution: The LCD for $\frac{3}{4}$ and $\frac{1}{2}$ is 4.

$$\begin{array}{l} 2\frac{3}{4} = 2\frac{3}{4} \\ +3\frac{1}{2} = 3\frac{2}{4} \\ \hline 5\frac{5}{4} = 5 + \frac{5}{4} \end{array}$$ > LCD

Write in vertical form.
Build up to get like-fraction parts.
Add the like-fraction parts: $\frac{3}{4} + \frac{2}{4} = \frac{5}{4}$
Add the whole-number parts: $2 + 3 = 5$

$\quad\quad\quad = 5 + 1\frac{1}{4}$ Simplify.

$\quad\quad\quad = 6\frac{1}{4}$

Note: $2\frac{3}{4} + 3\frac{1}{2} = 6\frac{1}{4}$ ⟵ simplest form

RAISE YOUR GRADES
Can you . . . ?

☑ identify like fractions
☑ add two or more like fractions using the Addition Rule for Like Fractions
☑ translate the addition of like fractions to words and pictures
☑ simplify a fractional sum when possible
☑ find the LCD for two or more unlike fractions
   (a) when all the smaller denominators divide the largest denominator evenly
   (b) when no two denominators share a common prime factor
   (c) using the Factoring Method for Finding the LCD
   (d) using a shortcut when the difference between the denominators divides each denominator evenly
☑ build up unlike fractions to get like fractions using the LCD
☑ add two or more unlike fractions using the Addition Rules for Unlike Fractions
☑ add with mixed numbers
☑ add mixed numbers and whole numbers
☑ add mixed numbers and fractions

SUMMARY

1. Addition is used to join two or more fractional amounts together.
2. Fractions with the same denominators are called like fractions.
3. To add like fractions, you add all the numerators together and then write the same denominators:

$$\frac{a}{c} + \frac{b}{c} = \frac{a + b}{c}$$

4. Always simplify a fractional sum when possible.
5. Fractions with different denominators are called unlike fractions.
6. The least common denominator (LCD) for two or more unlike fractions is the smallest nonzero whole number that all of the denominators divide into evenly.
7. When all of the smaller denominators divide the larger denominator evenly, the larger denominator is the LCD.
8. When the denominators share a common prime factor, the product of all the denominators is the LCD.
9. If you have trouble finding the LCD for two or more fractions, you can use the following method:
   Factoring Method for Finding the LCD
   (a) Factor each denominator as a product of primes.
   (b) Identify the greatest number of times that each prime factor occurs in any single factorization from Step (a).
   (c) Write the LCD as a product of the factors from Step (b).
10. To find the LCD for two fractions when the difference between their denominators divides each denominator evenly:
    (a) Find the difference between the denominators.
    (b) Divide the difference found in Step (a) into one of the denominators.
    (c) Multiply the quotient from Step (b) times the other denominator to get the LCD.
11. To build up two or more unlike fractions as equal fractions with a given common denominator, you multiply both the numerator and denominator of each original fraction by the correct building factor.
12. To rename two or more unlike fractions as equal fractions using the LCD, you first find the LCD and then build up to get like fractions using the LCD.
13. To add unlike fractions, you can use the following rules:
    Addition Rules for Unlike Fractions
    (a) Find the LCD for the unlike fractions.
    (b) Build up to get like fractions using the LCD from Step (a).
    (c) Substitute the like fractions for the original unlike fractions.
    (d) Add the like fractions from Step (c) [see Section 6-1.].
    (e) Simplify the sum from Step (d) when possible [see Section 4-3].
14. To add two or more mixed numbers, you can use the following rules:
    Addition Rules for Mixed Numbers
    (a) Write vertical addition form.
    (b) Build up to get like-fraction parts when necessary.
    (c) Add the like-fraction parts to find the fraction part of the sum.
    (d) Add the whole number parts to find the whole number part of the sum.
    (e) Simplify the sum from Steps (c) and (d) when possible.
15. To add a mixed number and a whole number, you add the whole numbers and then write the same fraction part.
16. To add a mixed number and a fraction, you add the fractions and then write the same whole-number part.

SOLVED PROBLEMS

PROBLEM 6-1   Add two like fractions: **(a)** $\frac{1}{2} + \frac{1}{2}$ **(b)** $\frac{1}{2} + \frac{3}{2}$ **(c)** $\frac{1}{4} + \frac{1}{4}$ **(d)** $\frac{3}{4} + \frac{1}{4}$
(e) $\frac{3}{4} + \frac{3}{4}$ **(f)** $\frac{1}{8} + \frac{1}{8}$ **(g)** $\frac{3}{8} + \frac{1}{8}$ **(h)** $\frac{3}{8} + \frac{5}{8}$ **(i)** $\frac{5}{8} + \frac{5}{8}$ **(j)** $\frac{1}{10} + \frac{1}{10}$ **(k)** $\frac{3}{10} + \frac{1}{10}$
(l) $\frac{1}{10} + \frac{7}{10}$ **(m)** $\frac{7}{10} + \frac{7}{10}$ **(n)** $\frac{1}{12} + \frac{1}{12}$ **(o)** $\frac{5}{12} + \frac{1}{12}$ **(p)** $\frac{1}{12} + \frac{7}{12}$ **(q)** $\frac{1}{5} + \frac{1}{5}$
(r) $\frac{2}{5} + \frac{2}{5}$ **(s)** $\frac{3}{100} + \frac{7}{100}$ **(t)** $\frac{19}{100} + \frac{31}{100}$

Solution: Recall that to add two like fractions, you add the numerators and then write the same denominator [see Examples 6-3 and 6-5]:

(a) $\dfrac{1}{2} + \dfrac{1}{2} = \dfrac{1+1}{2} = \dfrac{2}{2} = 1$ **(b)** $\dfrac{1}{2} + \dfrac{3}{2} = \dfrac{1+3}{2} = \dfrac{4}{2} = 2$

(c) $\dfrac{1}{4} + \dfrac{1}{4} = \dfrac{1+1}{4} = \dfrac{2}{4} = \dfrac{1}{2}$ **(d)** $\dfrac{3}{4} + \dfrac{1}{4} = \dfrac{3+1}{4} = \dfrac{4}{4} = 1$ **(e)** $\dfrac{3}{4} + \dfrac{3}{4} = \dfrac{3+3}{4} = \dfrac{6}{4} = \dfrac{3}{2} = 1\dfrac{1}{2}$

(f) $\dfrac{1}{8} + \dfrac{1}{8} = \dfrac{1+1}{8} = \dfrac{2}{8} = \dfrac{1}{4}$ **(g)** $\dfrac{3}{8} + \dfrac{1}{8} = \dfrac{3+1}{8} = \dfrac{4}{8} = \dfrac{1}{2}$ **(h)** $\dfrac{3}{8} + \dfrac{5}{8} = \dfrac{3+5}{8} = \dfrac{8}{8} = 1$

(i) $\dfrac{5}{8} + \dfrac{5}{8} = \dfrac{5+5}{8} = \dfrac{10}{8} = \dfrac{5}{4} = 1\dfrac{1}{4}$ **(j)** $\dfrac{1}{10} + \dfrac{1}{10} = \dfrac{1+1}{10} = \dfrac{2}{10} = \dfrac{1}{5}$

(k) $\dfrac{3}{10} + \dfrac{1}{10} = \dfrac{3+1}{10} = \dfrac{4}{10} = \dfrac{2}{5}$ **(l)** $\dfrac{1}{10} + \dfrac{7}{10} = \dfrac{1+7}{10} = \dfrac{8}{10} = \dfrac{4}{5}$

(m) $\dfrac{7}{10} + \dfrac{7}{10} = \dfrac{7+7}{10} = \dfrac{14}{10} = \dfrac{7}{5} = 1\dfrac{2}{5}$ **(n)** $\dfrac{1}{12} + \dfrac{1}{12} = \dfrac{1+1}{12} = \dfrac{2}{12} = \dfrac{1}{6}$

(o) $\dfrac{5}{12} + \dfrac{1}{12} = \dfrac{5+1}{12} = \dfrac{6}{12} = \dfrac{1}{2}$ **(p)** $\dfrac{1}{12} + \dfrac{7}{12} = \dfrac{1+7}{12} = \dfrac{8}{12} = \dfrac{2}{3}$ **(q)** $\dfrac{1}{5} + \dfrac{1}{5} = \dfrac{1+1}{5} = \dfrac{2}{5}$

(r) $\dfrac{2}{5} + \dfrac{2}{5} = \dfrac{2+2}{5} = \dfrac{4}{5}$ **(s)** $\dfrac{3}{100} + \dfrac{7}{100} = \dfrac{3+7}{100} = \dfrac{10}{100} = \dfrac{1}{10}$

(t) $\dfrac{19}{100} + \dfrac{31}{100} = \dfrac{19+31}{100} = \dfrac{50}{100} = \dfrac{1}{2}$

PROBLEM 6-2   Add more than two like fractions: **(a)** $\frac{1}{2} + \frac{1}{2} + \frac{1}{2}$ **(b)** $\frac{1}{3} + \frac{1}{3} + \frac{1}{3}$
(c) $\frac{1}{4} + \frac{1}{4} + \frac{1}{4}$ **(d)** $\frac{1}{5} + \frac{2}{5} + \frac{1}{5}$ **(e)** $\frac{1}{6} + \frac{5}{6} + \frac{5}{6}$ **(f)** $\frac{2}{7} + \frac{1}{7} + \frac{3}{7}$ **(g)** $\frac{5}{8} + \frac{1}{8} + \frac{3}{8}$
(h) $\frac{7}{9} + \frac{8}{9} + \frac{5}{9}$ **(i)** $\frac{1}{10} + \frac{3}{10} + \frac{7}{10}$ **(j)** $\frac{5}{12} + \frac{7}{12} + \frac{1}{12}$ **(k)** $\frac{3}{20} + \frac{1}{20} + \frac{7}{20}$ **(l)** $\frac{8}{25} + \frac{11}{25} + \frac{21}{25}$
(m) $\frac{49}{100} + \frac{79}{100} + \frac{91}{100}$ **(n)** $\frac{1}{3} + \frac{2}{3} + \frac{5}{3} + \frac{1}{3}$ **(o)** $\frac{3}{4} + \frac{3}{4} + \frac{5}{4} + \frac{1}{4}$ **(p)** $\frac{3}{8} + \frac{7}{8} + \frac{5}{8} + \frac{7}{8}$
(q) $\frac{3}{10} + \frac{7}{10} + \frac{11}{10} + \frac{1}{10}$ **(r)** $\frac{1}{9} + \frac{2}{9} + \frac{5}{9} + \frac{3}{9}$ **(s)** $\frac{1}{6} + \frac{2}{6} + \frac{3}{6} + \frac{4}{6} + \frac{5}{6}$ **(t)** $\frac{7}{12} + \frac{11}{12} + \frac{5}{12} + \frac{11}{12} + \frac{7}{12}$

Solution: Recall that to add more than two like fractions, you add all of the numerators and then write the same denominator [see Example 6-6]:

(a) $\dfrac{1}{2} + \dfrac{1}{2} + \dfrac{1}{2} = \dfrac{1+1+1}{2} = \dfrac{3}{2} = 1\dfrac{1}{2}$ **(b)** $\dfrac{1}{3} + \dfrac{1}{3} + \dfrac{1}{3} = \dfrac{1+1+1}{3} = \dfrac{3}{3} = 1$

(c) $\dfrac{1}{4} + \dfrac{1}{4} + \dfrac{1}{4} = \dfrac{1+1+1}{4} = \dfrac{3}{4}$ **(d)** $\dfrac{1}{5} + \dfrac{2}{5} + \dfrac{1}{5} = \dfrac{1+2+1}{5} = \dfrac{4}{5}$

(e) $\dfrac{1}{6} + \dfrac{5}{6} + \dfrac{5}{6} = \dfrac{1+5+5}{6} = \dfrac{11}{6} = 1\dfrac{5}{6}$ **(f)** $\dfrac{2}{7} + \dfrac{1}{7} + \dfrac{3}{7} = \dfrac{2+1+3}{7} = \dfrac{6}{7}$

(g) $\dfrac{5}{8} + \dfrac{1}{8} + \dfrac{3}{8} = \dfrac{5+1+3}{8} = \dfrac{9}{8} = 1\dfrac{1}{8}$ **(h)** $\dfrac{7}{9} + \dfrac{8}{9} + \dfrac{5}{9} = \dfrac{7+8+5}{9} = \dfrac{20}{9} = 2\dfrac{2}{9}$

(i) $\frac{1}{10} + \frac{3}{10} + \frac{7}{10} = \frac{1+3+7}{10} = \frac{11}{10} = 1\frac{1}{10}$ **(j)** $\frac{5}{12} + \frac{7}{12} + \frac{1}{12} = \frac{5+7+1}{12} = \frac{13}{12} = 1\frac{1}{12}$

(k) $\frac{3}{20} + \frac{1}{20} + \frac{7}{20} = \frac{3+1+7}{20} = \frac{11}{20}$ **(l)** $\frac{8}{25} + \frac{11}{25} + \frac{21}{25} = \frac{8+11+21}{25} = \frac{40}{25} = \frac{8}{5} = 1\frac{3}{5}$

(m) $\frac{49}{100} + \frac{79}{100} + \frac{91}{100} = \frac{49+79+91}{100} = \frac{219}{100} = 2\frac{19}{100}$

(n) $\frac{1}{3} + \frac{2}{3} + \frac{5}{3} + \frac{1}{3} = \frac{1+2+5+1}{3} = \frac{9}{3} = \frac{3}{1} = 3$

(o) $\frac{3}{4} + \frac{3}{4} + \frac{5}{4} + \frac{1}{4} = \frac{3+3+5+1}{4} = \frac{12}{4} = \frac{3}{1} = 3$

(p) $\frac{3}{8} + \frac{7}{8} + \frac{5}{8} + \frac{7}{8} = \frac{3+7+5+7}{8} = \frac{22}{8} = \frac{11}{4} = 2\frac{3}{4}$

(q) $\frac{3}{10} + \frac{7}{10} + \frac{11}{10} + \frac{1}{10} = \frac{3+7+11+1}{10} = \frac{22}{10} = \frac{11}{5} = 2\frac{1}{5}$

(r) $\frac{1}{9} + \frac{2}{9} + \frac{5}{9} + \frac{1}{9} = \frac{1+2+5+1}{9} = \frac{9}{9} = 1$

(s) $\frac{1}{6} + \frac{2}{6} + \frac{3}{6} + \frac{4}{6} + \frac{5}{6} = \frac{1+2+3+4+5}{6} = \frac{15}{6} = \frac{5}{2} = 2\frac{1}{2}$

(t) $\frac{7}{12} + \frac{11}{12} + \frac{5}{12} + \frac{11}{12} + \frac{7}{12} = \frac{7+11+5+11+7}{12} = \frac{41}{12} = 3\frac{5}{12}$

PROBLEM 6-3   Find the LCD for two unlike fractions when the smaller denominator divides the larger denominator evenly: **(a)** $\frac{1}{2}$ and $\frac{3}{4}$ **(b)** $\frac{1}{2}$ and $\frac{1}{6}$ **(c)** $\frac{5}{8}$ and $\frac{1}{2}$ **(d)** $\frac{3}{10}$ and $\frac{1}{2}$ **(e)** $\frac{1}{3}$ and $\frac{5}{6}$ **(f)** $\frac{2}{3}$ and $\frac{7}{9}$ **(g)** $\frac{5}{12}$ and $\frac{2}{3}$ **(h)** $\frac{1}{4}$ and $\frac{7}{8}$ **(i)** $\frac{3}{4}$ and $\frac{7}{12}$ **(j)** $\frac{1}{5}$ and $\frac{7}{10}$

Solution: Recall that when the smaller denominator divides the larger denominator evenly, the larger denominator is the LCD [see Example 6-7]:

(a) The LCD for $\frac{1}{2}$ and $\frac{3}{4}$ is 4 because 2 divides 4 evenly $(4 \div 2 = 2)$.
(b) The LCD for $\frac{1}{2}$ and $\frac{1}{6}$ is 6 because 2 divides 6 evenly $(6 \div 2 = 3)$.
(c) The LCD for $\frac{5}{8}$ and $\frac{1}{2}$ is 8 because 2 divides 8 evenly $(8 \div 2 = 4)$.
(d) The LCD for $\frac{3}{10}$ and $\frac{1}{2}$ is 10 because 2 divides 10 evenly $(10 \div 2 = 5)$.
(e) The LCD for $\frac{1}{3}$ and $\frac{5}{6}$ is 6 because 3 divides 6 evenly $(6 \div 3 = 2)$.
(f) The LCD for $\frac{2}{3}$ and $\frac{7}{9}$ is 9 because 3 divides 9 evenly $(9 \div 3 = 3)$.
(g) The LCD for $\frac{5}{12}$ and $\frac{2}{3}$ is 12 because 3 divides 12 evenly $(12 \div 3 = 4)$.
(h) The LCD for $\frac{1}{4}$ and $\frac{7}{8}$ is 8 because 4 divides 8 evenly $(8 \div 4 = 2)$.
(i) The LCD for $\frac{3}{4}$ and $\frac{7}{12}$ is 12 because 4 divides 12 evenly $(12 \div 4 = 3)$.
(j) The LCD for $\frac{1}{5}$ and $\frac{7}{10}$ is 10 because 5 divides 10 evenly $(10 \div 5 = 2)$.

PROBLEM 6-4   Find the LCD for two fractions when the denominators do not share a common prime factor: **(a)** $\frac{1}{2}$ and $\frac{2}{3}$ **(b)** $\frac{1}{2}$ and $\frac{3}{5}$ **(c)** $\frac{5}{7}$ and $\frac{1}{2}$ **(d)** $\frac{2}{9}$ and $\frac{1}{2}$ **(e)** $\frac{1}{3}$ and $\frac{3}{4}$ **(f)** $\frac{2}{3}$ and $\frac{1}{5}$ **(g)** $\frac{1}{4}$ and $\frac{4}{5}$ **(h)** $\frac{1}{9}$ and $\frac{3}{4}$ **(i)** $\frac{3}{5}$ and $\frac{5}{6}$ **(j)** $\frac{1}{8}$ and $\frac{2}{5}$

Solution: Recall that when two denominators do not share a common prime factor, the product of the denominators is the LCD [see Example 6-4]:

(a) The LCD of $\frac{1}{2}$ and $\frac{2}{3}$ is 6 (2×3) because 2 and 3 do not share a common prime factor.
(b) The LCD of $\frac{1}{2}$ and $\frac{3}{5}$ is 10 (2×5) because 2 and 5 do not share a common prime factor.
(c) The LCD of $\frac{5}{7}$ and $\frac{1}{2}$ is 14 (7×2) because 7 and 2 do not share a common prime factor.
(d) The LCD of $\frac{2}{9}$ and $\frac{1}{2}$ is 18 (9×2) because 9 and 2 do not share a common prime factor.

(e) The LCD of $\frac{1}{3}$ and $\frac{3}{4}$ is 12 (3 × 4) because 3 and 4 do not share a common prime factor.

(f) The LCD of $\frac{2}{3}$ and $\frac{1}{5}$ is 15 (3 × 5) because 3 and 5 do not share a common prime factor.

(g) The LCD of $\frac{1}{4}$ and $\frac{4}{5}$ is 20 (4 × 5) because 4 and 5 do not share a common prime factor.

(h) The LCD of $\frac{1}{9}$ and $\frac{3}{4}$ is 36 (9 × 4) because 9 and 4 do not share a common prime factor.

(i) The LCD of $\frac{3}{5}$ and $\frac{5}{6}$ is 30 (5 × 6) because 5 and 6 do not share a common prime factor.

(j) The LCD of $\frac{1}{8}$ and $\frac{2}{5}$ is 40 (8 × 5) because 8 and 5 do not share a common prime factor.

PROBLEM 6-5   Find the LCD of two fractions when the difference between the denominators divides each denominator evenly: (a) $\frac{3}{4}$ and $\frac{5}{6}$ (b) $\frac{3}{8}$ and $\frac{1}{6}$ (c) $\frac{1}{8}$ and $\frac{7}{10}$ (d) $\frac{1}{12}$ and $\frac{5}{8}$ (e) $\frac{7}{15}$ and $\frac{3}{10}$ (f) $\frac{2}{9}$ and $\frac{5}{12}$ (g) $\frac{5}{9}$ and $\frac{5}{6}$ (h) $\frac{7}{12}$ and $\frac{1}{14}$ (i) $\frac{11}{15}$ and $\frac{5}{12}$ (j) $\frac{5}{12}$ and $\frac{7}{18}$

Solution: Recall that when the difference between the two denominators divides each denominator evenly, you divide the difference into one of the denominators and then multiply that quotient times the other denominator to get the LCD [see Example 6-10]:

(a) The LCD for $\frac{3}{4}$ and $\frac{5}{6}$ is 12 because: $6 - 4 = 2$ and $2\overline{)6}$ $3 \times 4 = 12$ or $2\overline{)4}$ $2 \times 6 = 12$

(b) The LCD for $\frac{3}{8}$ and $\frac{1}{6}$ is 24 because: $8 - 6 = 2$ and $2\overline{)8}$ $4 \times 6 = 24$ or $2\overline{)6}$ $3 \times 8 = 24$

(c) The LCD for $\frac{1}{8}$ and $\frac{7}{10}$ is 40 because: $10 - 8 = 2$ and $2\overline{)10}$ $5 \times 8 = 40$ or $2\overline{)8}$ $4 \times 10 = 40$

(d) The LCD for $\frac{1}{12}$ and $\frac{5}{8}$ is 24 because: $12 - 8 = 4$ and $4\overline{)12}$ $3 \times 8 = 24$ or $4\overline{)8}$ $2 \times 12 = 24$

(e) The LCD for $\frac{7}{15}$ and $\frac{3}{10}$ is 30 because: $15 - 10 = 5$ and $5\overline{)15}$ $3 \times 10 = 30$ or $5\overline{)10}$ $2 \times 15 = 30$

(f) The LCD for $\frac{2}{9}$ and $\frac{5}{12}$ is 36 because: $12 - 9 = 3$ and $3\overline{)12}$ $4 \times 9 = 36$ or $3\overline{)9}$ $3 \times 12 = 36$

(g) The LCD for $\frac{5}{9}$ and $\frac{5}{6}$ is 18 because: $9 - 6 = 3$ and $3\overline{)9}$ $3 \times 6 = 18$ or $3\overline{)6}$ $2 \times 9 = 18$

(h) The LCD for $\frac{7}{12}$ and $\frac{1}{14}$ is 84 because: $14 - 12 = 2$ and $2\overline{)12}$ $6 \times 14 = 84$ or $2\overline{)14}$ $7 \times 12 = 84$

(i) The LCD for $\frac{11}{15}$ and $\frac{5}{12}$ is 60 because: $15 - 12 = 3$ and $3\overline{)15}$ $5 \times 12 = 60$ or $3\overline{)12}$ $4 \times 15 = 60$

(j) The LCD for $\frac{5}{12}$ and $\frac{7}{18}$ is 36 because: $18 - 12 = 6$ and $6\overline{)12}$ $2 \times 18 = 36$ or $6\overline{)18}$ $3 \times 12 = 36$

PROBLEM 6-6   Find the LCD of two fractions using the Factoring Method for finding the LCD:

(a) $\frac{1}{6}$ and $\frac{3}{10}$ (b) $\frac{1}{10}$ and $\frac{3}{4}$ (c) $\frac{1}{20}$ and $\frac{11}{12}$ (d) $\frac{5}{12}$ and $\frac{1}{21}$ (e) $\frac{4}{15}$ and $\frac{2}{9}$ (f) $\frac{3}{14}$ and $\frac{5}{6}$ (g) $\frac{3}{8}$ and $\frac{9}{14}$ (h) $\frac{1}{6}$ and $\frac{7}{15}$ (i) $\frac{17}{18}$ and $\frac{5}{8}$ (j) $\frac{7}{10}$ and $\frac{11}{18}$

Solution: Recall that when you have trouble finding the LCD, you can use the Factoring Method for finding the LCD [see Example 6-9]:

(a) The LCD for $\frac{1}{6}$ and $\frac{3}{10}$ is 30 because: $6 = \textcircled{2} \times \textcircled{3}$ and $10 = 2 \times \textcircled{5}$ means:
   LCD $= 2 \times 3 \times 5 = 30$.

(b) The LCD for $\frac{1}{10}$ and $\frac{3}{4}$ is 20 because: $10 = 2 \times \textcircled{5}$ and $4 = \boxed{2 \times 2}$ means:
   LCD $= 5 \times 2 \times 2 = 20$.

(c) The LCD for $\frac{1}{20}$ and $\frac{11}{12}$ is 60 because: $20 = \boxed{2 \times 2} \times \textcircled{5}$ and $12 = 2 \times 2 \times \textcircled{3}$ means:
   LCD $= 2 \times 2 \times 5 \times 3 = 60$.

(d) The LCD for $\frac{5}{12}$ and $\frac{1}{21}$ is 84 because: $12 = \boxed{2 \times 2} \times \textcircled{3}$ and $21 = 3 \times \textcircled{7}$ means:
   LCD $= 2 \times 2 \times 3 \times 7 = 84$.

(e) The LCD for $\frac{4}{15}$ and $\frac{2}{9}$ is 45 because: $15 = 3 \times \boxed{5}$ and $9 = \boxed{3 \times 3}$ means:
LCD = $5 \times 3 \times 3 = 45$.

(f) The LCD for $\frac{3}{14}$ and $\frac{5}{6}$ is 42 because: $14 = \boxed{2} \times \boxed{7}$ and $6 = 2 \times \boxed{3}$ means:
LCD = $2 \times 7 \times 3 = 42$.

(g) The LCD for $\frac{3}{8}$ and $\frac{9}{14}$ is 56 because: $8 = \boxed{2 \times 2 \times 2}$ and $14 = 2 \times \boxed{7}$ means:
LCD = $2 \times 2 \times 2 \times 7 = 56$.

(h) The LCD for $\frac{1}{6}$ and $\frac{7}{15}$ is 30 because: $6 = \boxed{2} \times \boxed{3}$ and $15 = 3 \times \boxed{5}$ means:
LCD = $2 \times 3 \times 5 = 30$.

(i) The LCD for $\frac{17}{18}$ and $\frac{5}{8}$ is 72 because: $18 = 2 \times \boxed{3 \times 3}$ and $8 = \boxed{2 \times 2 \times 2}$ means:
LCD = $3 \times 3 \times 2 \times 2 \times 2 = 72$.

(j) The LCD for $\frac{7}{10}$ and $\frac{11}{18}$ is 90 because: $10 = \boxed{2} \times \boxed{5}$ and $18 = 2 \times \boxed{3 \times 3}$ means:
LCD = $2 \times 5 \times 3 \times 3 = 90$.

PROBLEM 6-7 Find the LCD for more than two fractions: **(a)** $\frac{1}{2}$, $\frac{3}{4}$, and $\frac{5}{8}$ **(b)** $\frac{5}{6}$, $\frac{2}{3}$, and $\frac{7}{12}$
(c) $\frac{3}{10}$, $\frac{1}{2}$, and $\frac{2}{5}$ **(d)** $\frac{11}{25}$, $\frac{49}{100}$, and $\frac{3}{20}$ **(e)** $\frac{1}{2}$, $\frac{2}{3}$, and $\frac{1}{5}$ **(f)** $\frac{1}{3}$, $\frac{1}{4}$, and $\frac{3}{5}$
(g) $\frac{7}{8}$, $\frac{4}{5}$, and $\frac{2}{3}$ **(h)** $\frac{4}{5}$, $\frac{1}{3}$, and $\frac{6}{7}$ **(i)** $\frac{1}{2}$, $\frac{3}{4}$, and $\frac{1}{6}$ **(j)** $\frac{1}{4}$, $\frac{7}{10}$, and $\frac{5}{6}$
(k) $\frac{1}{12}$, $\frac{3}{8}$, and $\frac{5}{9}$ **(l)** $\frac{5}{12}$, $\frac{1}{10}$, and $\frac{7}{18}$ **(m)** $\frac{3}{16}$, $\frac{1}{2}$, $\frac{5}{8}$, and $\frac{3}{4}$ **(n)** $\frac{2}{3}$, $\frac{7}{12}$, $\frac{1}{4}$, and $\frac{1}{6}$
(o) $\frac{1}{3}$, $\frac{2}{5}$, $\frac{1}{2}$, and $\frac{1}{7}$ **(p)** $\frac{1}{2}$, $\frac{9}{11}$, $\frac{2}{3}$, and $\frac{3}{5}$ **(q)** $\frac{3}{4}$, $\frac{5}{6}$, $\frac{8}{9}$, and $\frac{4}{15}$ **(r)** $\frac{1}{15}$, $\frac{11}{12}$, $\frac{5}{18}$, and $\frac{3}{16}$

Solution: Recall that to find the LCD for more than two fractions you find the smallest nonzero whole number that each denominator will divide into evenly [see Examples 6-11, 6-12, and 6-13]:

(a) The LCD for $\frac{1}{2}$, $\frac{3}{4}$, and $\frac{5}{8}$ is 8 because 2, 4, and 8 all divide 8 evenly.

(b) The LCD for $\frac{5}{6}$, $\frac{2}{3}$, and $\frac{7}{12}$ is 12 because 3, 6, and 12 all divide 12 evenly.

(c) The LCD for $\frac{3}{10}$, $\frac{1}{2}$, and $\frac{2}{5}$ is 10 because 2, 5, and 10 all divide 10 evenly.

(d) The LCD for $\frac{11}{25}$, $\frac{49}{100}$, and $\frac{3}{20}$ is 100 because 20, 25, and 100 all divide 100 evenly.

(e) The LCD for $\frac{1}{2}$, $\frac{2}{3}$, and $\frac{1}{5}$ is 30 ($2 \times 3 \times 5$) because no two of 2, 3, and 5 share a common prime factor.

(f) The LCD for $\frac{1}{3}$, $\frac{1}{4}$, and $\frac{3}{5}$ is 60 ($3 \times 4 \times 5$) because no two of 3, 4, and 5 share a common prime factor.

(g) The LCD for $\frac{7}{8}$, $\frac{4}{5}$, and $\frac{2}{3}$ is 120 ($3 \times 5 \times 8$) because no two of 3, 5, and 8 share a common prime factor.

(h) The LCD for $\frac{4}{5}$, $\frac{1}{3}$, and $\frac{6}{7}$ is 105 ($3 \times 5 \times 7$) because no two of 3, 5, and 7 share a common prime factor.

(i) The LCD for $\frac{1}{2}$, $\frac{3}{4}$, and $\frac{1}{6}$ is 12 because: $2 = 2$, $4 = \boxed{2 \times 2}$, $6 = 2 \times \boxed{3}$, and $2 \times 2 \times 3 = 12$.

(j) The LCD for $\frac{1}{4}$, $\frac{7}{10}$, and $\frac{5}{6}$ is 60 because:
$4 = \boxed{2 \times 2}$, $10 = 2 \times \boxed{5}$, $6 = 2 \times \boxed{3}$, and $2 \times 2 \times 5 \times 3 = 60$.

(k) The LCD for $\frac{1}{12}$, $\frac{3}{8}$, and $\frac{5}{9}$ is 72 because:
$12 = 2 \times 2 \times 3$, $8 = \boxed{2 \times 2 \times 2}$, $9 = \boxed{3 \times 3}$, and $2 \times 2 \times 2 \times 3 \times 3 = 72$.

(l) The LCD for $\frac{5}{12}$, $\frac{1}{10}$, and $\frac{7}{18}$ is 180 because:
$12 = \boxed{2 \times 2} \times 3$, $10 = 2 \times \boxed{5}$, $18 = 2 \times \boxed{3 \times 3}$, and $2 \times 2 \times 5 \times 3 \times 3 = 180$.

(m) The LCD for $\frac{3}{16}$, $\frac{1}{2}$, $\frac{5}{8}$, and $\frac{3}{4}$ is 16 because 2, 4, 8, and 16 all divide 16 evenly.

(n) The LCD for $\frac{2}{3}$, $\frac{7}{12}$, $\frac{1}{4}$, and $\frac{1}{6}$ is 12 because 3, 4, 6, and 12 all divide 12 evenly.

(o) The LCD for $\frac{1}{3}$, $\frac{2}{5}$, $\frac{1}{2}$, and $\frac{1}{7}$ is 210 ($2 \times 3 \times 5 \times 7$) because no two of 2, 3, 5, and 7 share a common prime factor.

(p) The LCD for $\frac{1}{2}$, $\frac{9}{11}$, $\frac{2}{3}$, and $\frac{3}{5}$ is 330 ($2 \times 3 \times 5 \times 11$) because no two of 2, 3, 5, and 11 share a common prime factor.

(q) The LCD for $\frac{3}{4}$, $\frac{5}{6}$, $\frac{8}{9}$, and $\frac{4}{15}$ is 180 because:
$4 = \boxed{2 \times 2}$, $6 = 2 \times 3$, $9 = \boxed{3 \times 3}$, $15 = 3 \times \boxed{5}$ and $2 \times 2 \times 3 \times 3 \times 5 = 180$.

(r) The LCD for $\frac{1}{15}$, $\frac{11}{12}$, $\frac{5}{18}$, and $\frac{3}{16}$ is 720 because: $15 = 3 \times \boxed{5}$, $12 = 2 \times 2 \times 3$, $18 = 2 \times \boxed{3 \times 3}$,
$16 = \boxed{2 \times 2 \times 2 \times 2}$, and $2 \times 2 \times 2 \times 2 \times 3 \times 3 \times 5 = 720$.

PROBLEM 6-8   Build up each fraction as an equal fraction with the given denominator:

(a) $\frac{1}{2} = \frac{?}{6}$ (b) $\frac{1}{2} = \frac{?}{8}$ (c) $\frac{1}{3} = \frac{?}{9}$ (d) $\frac{2}{3} = \frac{?}{15}$ (e) $\frac{1}{4} = \frac{?}{8}$ (f) $\frac{3}{4} = \frac{?}{16}$

(g) $\frac{1}{5} = \frac{?}{15}$ (h) $\frac{2}{5} = \frac{?}{20}$ (i) $\frac{1}{6} = \frac{?}{30}$ (j) $\frac{5}{6} = \frac{?}{12}$ (k) $\frac{1}{8} = \frac{?}{40}$ (l) $\frac{3}{8} = \frac{?}{24}$

(m) $\frac{1}{9} = \frac{?}{36}$ (n) $\frac{2}{9} = \frac{?}{45}$ (o) $\frac{1}{10} = \frac{?}{60}$ (p) $\frac{3}{10} = \frac{?}{100}$ (q) $\frac{1}{12} = \frac{?}{24}$ (r) $\frac{5}{12} = \frac{?}{36}$

Solution: Recall that to build up a given fraction as an equal fraction with a given denominator, you first compare denominators to find the correct building factor and then multiply both the numerator and denominator by that building factor [see Examples 6-14 and 6-15]:

(a)   The building factor is 3 ($2 \times \mathbf{3} = 6$) and $\dfrac{1}{2} = \dfrac{1 \times 3}{2 \times 3} = \dfrac{3}{6}$.

(b)   The building factor is 4 ($2 \times \mathbf{4} = 8$) and $\dfrac{1}{2} = \dfrac{1 \times 4}{2 \times 4} = \dfrac{4}{8}$.

(c)   The building factor is 3 ($3 \times \mathbf{3} = 9$) and $\dfrac{1}{3} = \dfrac{1 \times 3}{3 \times 3} = \dfrac{3}{9}$.

(d)   The building factor is 5 ($3 \times \mathbf{5} = 15$) and $\dfrac{2}{3} = \dfrac{2 \times 5}{3 \times 5} = \dfrac{10}{15}$.

(e)   The building factor is 2 ($4 \times \mathbf{2} = 8$) and $\dfrac{1}{4} = \dfrac{1 \times 2}{4 \times 2} = \dfrac{2}{8}$.

(f)   The building factor is 4 ($4 \times \mathbf{4} = 16$) and $\dfrac{3}{4} = \dfrac{3 \times 4}{4 \times 4} = \dfrac{12}{16}$.

(g)   The building factor is 3 ($5 \times \mathbf{3} = 15$) and $\dfrac{1}{5} = \dfrac{1 \times 3}{5 \times 3} = \dfrac{3}{15}$.

(h)   The building factor is 4 ($5 \times \mathbf{4} = 20$) and $\dfrac{2}{5} = \dfrac{2 \times 4}{5 \times 4} = \dfrac{8}{20}$.

(i)   The building factor is 5 ($6 \times \mathbf{5} = 30$) and $\dfrac{1}{6} = \dfrac{1 \times 5}{6 \times 5} = \dfrac{5}{30}$.

(j)   The building factor is 2 ($6 \times \mathbf{2} = 12$) and $\dfrac{5}{6} = \dfrac{5 \times 2}{6 \times 2} = \dfrac{10}{12}$.

(k)   The building factor is 5 ($8 \times \mathbf{5} = 40$) and $\dfrac{1}{8} = \dfrac{1 \times 5}{8 \times 5} = \dfrac{5}{40}$.

(l)   The building factor is 3 ($8 \times \mathbf{3} = 24$) and $\dfrac{3}{8} = \dfrac{3 \times 3}{8 \times 3} = \dfrac{9}{24}$.

(m)   The building factor is 4 ($9 \times \mathbf{4} = 36$) and $\dfrac{1}{9} = \dfrac{1 \times 4}{9 \times 4} = \dfrac{4}{36}$.

(n)   The building factor is 5 ($9 \times \mathbf{5} = 45$) and $\dfrac{2}{9} = \dfrac{2 \times 5}{9 \times 5} = \dfrac{10}{45}$.

(o)   The building factor is 6 ($10 \times \mathbf{6} = 60$) and $\dfrac{1}{10} = \dfrac{1 \times 6}{10 \times 6} = \dfrac{6}{60}$.

(p)   The building factor is 10 ($10 \times \mathbf{10} = 100$) and $\dfrac{3}{10} = \dfrac{3 \times 10}{10 \times 10} = \dfrac{30}{100}$.

(q)   The building factor is 2 ($12 \times \mathbf{2} = 24$) and $\dfrac{1}{12} = \dfrac{1 \times 2}{12 \times 2} = \dfrac{2}{24}$.

(r)   The building factor is 3 ($12 \times \mathbf{3} = 36$) and $\dfrac{5}{12} = \dfrac{5 \times 3}{12 \times 3} = \dfrac{15}{36}$.

PROBLEM 6-9   Rename unlike fractions as equivalent like fractions using the LCD:

(a) $\frac{1}{2}$ and $\frac{3}{4}$ **(b)** $\frac{2}{3}$ and $\frac{5}{6}$ **(c)** $\frac{5}{8}$ and $\frac{1}{4}$ **(d)** $\frac{1}{2}$ and $\frac{1}{3}$ **(e)** $\frac{2}{3}$ and $\frac{3}{4}$ **(f)** $\frac{4}{5}$ and $\frac{1}{2}$
(g) $\frac{1}{4}$ and $\frac{1}{6}$ **(h)** $\frac{5}{6}$ and $\frac{7}{8}$ **(i)** $\frac{1}{10}$ and $\frac{3}{8}$ **(j)** $\frac{3}{4}$ and $\frac{3}{10}$ **(k)** $\frac{1}{12}$ and $\frac{7}{10}$ **(l)** $\frac{3}{8}$ and $\frac{5}{12}$
(m) $\frac{1}{4}, \frac{7}{8}$ and $\frac{1}{2}$ **(n)** $\frac{2}{3}, \frac{1}{2}$, and $\frac{3}{5}$ **(o)** $\frac{1}{10}, \frac{3}{8}$, and $\frac{7}{12}$ **(p)** $\frac{23}{25}, \frac{49}{100}$, and $\frac{19}{20}$
(q) $\frac{15}{16}, \frac{3}{4}, \frac{5}{8}$, and $\frac{1}{2}$ **(r)** $\frac{11}{12}, \frac{7}{18}, \frac{13}{15}$, and $\frac{9}{20}$

Solution: Recall that to rename unlike fractions as like fractions using the LCD, you first find the LCD and then build up to get like fractions using the LCD [see Examples 6-16 and 6-17]:

(a) The LCD of $\frac{1}{2}$ and $\frac{3}{4}$ is 4: $\frac{1}{2} = \frac{1 \times 2}{2 \times 2} = \frac{2}{4}$ and $\frac{3}{4} = \frac{3}{4}$.

(b) The LCD of $\frac{2}{3}$ and $\frac{5}{6}$ is 6: $\frac{2}{3} = \frac{2 \times 2}{3 \times 2} = \frac{4}{6}$ and $\frac{5}{6} = \frac{5}{6}$.

(c) The LCD of $\frac{5}{8}$ and $\frac{1}{4}$ is 8: $\frac{5}{8} = \frac{5}{8}$ and $\frac{1}{4} = \frac{1 \times 2}{4 \times 2} = \frac{2}{8}$.

(d) The LCD of $\frac{1}{2}$ and $\frac{1}{3}$ is 6: $\frac{1}{2} = \frac{1 \times 3}{2 \times 3} = \frac{3}{6}$ and $\frac{1}{3} = \frac{1 \times 2}{3 \times 2} = \frac{2}{6}$.

(e) The LCD of $\frac{2}{3}$ and $\frac{3}{4}$ is 12: $\frac{2}{3} = \frac{2 \times 4}{3 \times 4} = \frac{8}{12}$ and $\frac{3}{4} = \frac{3 \times 3}{4 \times 3} = \frac{9}{12}$.

(f) The LCD of $\frac{4}{5}$ and $\frac{1}{2}$ is 10: $\frac{4}{5} = \frac{4 \times 2}{5 \times 2} = \frac{8}{10}$ and $\frac{1}{2} = \frac{1 \times 5}{2 \times 5} = \frac{5}{10}$.

(g) The LCD of $\frac{1}{4}$ and $\frac{1}{6}$ is 12: $\frac{1}{4} = \frac{1 \times 3}{4 \times 3} = \frac{3}{12}$ and $\frac{1}{6} = \frac{1 \times 2}{6 \times 2} = \frac{2}{12}$.

(h) The LCD of $\frac{5}{6}$ and $\frac{7}{8}$ is 24: $\frac{5}{6} = \frac{5 \times 4}{6 \times 4} = \frac{20}{24}$ and $\frac{7}{8} = \frac{7 \times 3}{8 \times 3} = \frac{21}{24}$.

(i) The LCD of $\frac{1}{10}$ and $\frac{3}{8}$ is 40: $\frac{1}{10} = \frac{1 \times 4}{10 \times 4} = \frac{4}{40}$ and $\frac{3}{8} = \frac{3 \times 5}{8 \times 5} = \frac{15}{40}$.

(j) The LCD of $\frac{3}{4}$ and $\frac{3}{10}$ is 20: $\frac{3}{4} = \frac{3 \times 5}{4 \times 5} = \frac{15}{20}$ and $\frac{3}{10} = \frac{3 \times 2}{10 \times 2} = \frac{6}{20}$.

(k) The LCD of $\frac{1}{12}$ and $\frac{7}{10}$ is 60: $\frac{1}{12} = \frac{1 \times 5}{12 \times 5} = \frac{5}{60}$ and $\frac{7}{10} = \frac{7 \times 6}{10 \times 6} = \frac{42}{60}$.

(l) The LCD of $\frac{3}{8}$ and $\frac{5}{12}$ is 24: $\frac{3}{8} = \frac{3 \times 3}{8 \times 3} = \frac{9}{24}$ and $\frac{5}{12} = \frac{5 \times 2}{12 \times 2} = \frac{10}{24}$.

(m) The LCD of $\frac{1}{4}, \frac{7}{8}$, and $\frac{1}{2}$ is 8: $\frac{1}{4} = \frac{1 \times 2}{4 \times 2} = \frac{2}{8}, \frac{7}{8} = \frac{7}{8}$, and $\frac{1}{2} = \frac{1 \times 4}{2 \times 4} = \frac{4}{8}$.

(n) The LCD of $\frac{2}{3}, \frac{1}{2}$, and $\frac{3}{5}$ is 30: $\frac{2}{3} = \frac{2 \times 10}{3 \times 10} = \frac{20}{30}, \frac{1}{2} = \frac{1 \times 15}{2 \times 15} = \frac{15}{30}$, and $\frac{3}{5} = \frac{3 \times 6}{5 \times 6} = \frac{18}{30}$.

(o) The LCD of $\frac{1}{10}, \frac{3}{8}$, and $\frac{7}{12}$ is 120:

$$\frac{1}{10} = \frac{1 \times 12}{10 \times 12} = \frac{12}{120}, \frac{3}{8} = \frac{3 \times 15}{8 \times 15} = \frac{45}{120}, \text{ and } \frac{7}{12} = \frac{7 \times 10}{12 \times 10} = \frac{70}{120}.$$

(p) The LCD of $\frac{23}{25}, \frac{49}{100}$, and $\frac{19}{20}$ is 100:

$$\frac{23}{25} = \frac{23 \times 4}{25 \times 4} = \frac{92}{100}, \frac{49}{100} = \frac{49}{100}, \text{ and } \frac{19}{20} = \frac{19 \times 5}{20 \times 5} = \frac{95}{100}.$$

(q) The LCD of $\frac{15}{16}, \frac{3}{4}, \frac{5}{8}$, and $\frac{1}{2}$ is 16:

$$\frac{15}{16} = \frac{15}{16}, \frac{3}{4} = \frac{3 \times 4}{4 \times 4} = \frac{12}{16}, \frac{5}{8} = \frac{5 \times 2}{8 \times 2} = \frac{10}{16}, \text{ and } \frac{1}{2} = \frac{1 \times 8}{2 \times 8} = \frac{8}{16}.$$

(r) The LCD of $\frac{11}{12}, \frac{7}{18}, \frac{13}{15}$, and $\frac{9}{20}$ is 180:

$$\frac{11}{12} = \frac{11 \times 15}{12 \times 15} = \frac{165}{180}, \frac{7}{18} = \frac{7 \times 10}{18 \times 10} = \frac{70}{180}, \frac{13}{15} = \frac{13 \times 12}{15 \times 12} = \frac{156}{180}, \text{ and } \frac{9}{20} = \frac{9 \times 9}{20 \times 9} = \frac{81}{180}.$$

PROBLEM 6-10   Add two unlike fractions: **(a)** $\frac{1}{2} + \frac{1}{4}$ **(b)** $\frac{1}{4} + \frac{1}{8}$ **(c)** $\frac{1}{6} + \frac{1}{3}$
(d) $\frac{1}{12} + \frac{2}{3}$ **(e)** $\frac{3}{4} + \frac{5}{12}$ **(f)** $\frac{1}{2} + \frac{1}{3}$ **(g)** $\frac{1}{3} + \frac{1}{4}$ **(h)** $\frac{2}{5} + \frac{1}{2}$ **(i)** $\frac{3}{5} + \frac{1}{3}$
(j) $\frac{3}{4} + \frac{4}{5}$ **(k)** $\frac{1}{4} + \frac{1}{6}$ **(l)** $\frac{1}{6} + \frac{3}{8}$ **(m)** $\frac{3}{10} + \frac{1}{8}$ **(n)** $\frac{5}{12} + \frac{3}{8}$ **(o)** $\frac{7}{10} + \frac{11}{15}$
(p) $\frac{1}{6} + \frac{1}{10}$ **(q)** $\frac{1}{4} + \frac{3}{10}$ **(r)** $\frac{1}{15} + \frac{5}{8}$ **(s)** $\frac{7}{20} + \frac{7}{12}$ **(t)** $\frac{5}{9} + \frac{11}{15}$

Solution: Recall that to add two unlike fractions, first rename as equivalent like fractions using the LCD and then add the like fractions [see Example 6-18]:

(a) The LCD for $\frac{1}{2}$ and $\frac{1}{4}$ is 4: $\frac{1}{2} + \frac{1}{4} = \frac{1 \times 2}{2 \times 2} + \frac{1}{4} = \frac{2}{4} + \frac{1}{4} = \frac{2+1}{4} = \frac{3}{4}.$

(b) The LCD for $\frac{1}{4}$ and $\frac{1}{8}$ is 8: $\frac{1}{4} + \frac{1}{8} = \frac{1 \times 2}{4 \times 2} + \frac{1}{8} = \frac{2}{8} + \frac{1}{8} = \frac{3}{8}.$

(c) The LCD for $\frac{1}{6}$ and $\frac{1}{3}$ is 6: $\frac{1}{6} + \frac{1}{3} = \frac{1}{6} + \frac{1 \times 2}{3 \times 2} = \frac{1}{6} + \frac{2}{6} = \frac{3}{6} = \frac{1 \times \cancel{3}}{2 \times \cancel{3}} = \frac{1}{2}.$

(d) The LCD for $\frac{1}{12}$ and $\frac{2}{3}$ is 12: $\frac{1}{12} + \frac{2}{3} = \frac{1}{12} + \frac{2 \times 4}{3 \times 4} = \frac{1}{12} + \frac{8}{12} = \frac{9}{12} = \frac{\cancel{3} \times 3}{\cancel{3} \times 4} = \frac{3}{4}.$

(e) The LCD for $\frac{3}{4}$ and $\frac{5}{12}$ is 12: $\frac{3}{4} + \frac{5}{12} = \frac{3 \times 3}{4 \times 3} + \frac{5}{12} = \frac{9}{12} + \frac{5}{12} = \frac{14}{12} = \frac{\cancel{2} \times 7}{\cancel{2} \times 6} = \frac{7}{6} = 1\frac{1}{6}.$

(f) The LCD for $\frac{1}{2}$ and $\frac{1}{3}$ is 6: $\frac{1}{2} + \frac{1}{3} = \frac{1 \times 3}{2 \times 3} + \frac{1 \times 2}{3 \times 2} = \frac{3}{6} + \frac{2}{6} = \frac{5}{6}.$

(g) The LCD for $\frac{1}{3}$ and $\frac{1}{4}$ is 12: $\frac{1}{3} + \frac{1}{4} = \frac{1 \times 4}{3 \times 4} + \frac{1 \times 3}{4 \times 3} = \frac{4}{12} + \frac{3}{12} = \frac{7}{12}.$

(h) The LCD for $\frac{2}{5}$ and $\frac{1}{2}$ is 10: $\frac{2}{5} + \frac{1}{2} = \frac{2 \times 2}{5 \times 2} + \frac{1 \times 5}{2 \times 5} = \frac{4}{10} + \frac{5}{10} = \frac{9}{10}.$

(i) The LCD for $\frac{3}{5}$ and $\frac{1}{3}$ is 15: $\frac{3}{5} + \frac{1}{3} = \frac{3 \times 3}{5 \times 3} + \frac{1 \times 5}{3 \times 5} = \frac{9}{15} + \frac{5}{15} = \frac{14}{15}.$

(j) The LCD for $\frac{3}{4}$ and $\frac{4}{5}$ is 20: $\frac{3}{4} + \frac{4}{5} = \frac{3 \times 5}{4 \times 5} + \frac{4 \times 4}{5 \times 4} = \frac{15}{20} + \frac{16}{20} = \frac{31}{20} = 1\frac{11}{20}.$

(k) The LCD for $\frac{1}{4}$ and $\frac{1}{6}$ is 12: $\frac{1}{4} + \frac{1}{6} = \frac{1 \times 3}{4 \times 3} + \frac{1 \times 2}{6 \times 2} = \frac{3}{12} + \frac{2}{12} = \frac{5}{12}.$

(l) The LCD for $\frac{1}{6}$ and $\frac{3}{8}$ is 24: $\frac{1}{6} + \frac{3}{8} = \frac{1 \times 4}{6 \times 4} + \frac{3 \times 3}{8 \times 3} = \frac{4}{24} + \frac{9}{24} = \frac{13}{24}.$

(m) The LCD for $\frac{3}{10}$ and $\frac{1}{8}$ is 40: $\frac{3}{10} + \frac{1}{8} = \frac{3 \times 4}{10 \times 4} + \frac{1 \times 5}{8 \times 5} = \frac{12}{40} + \frac{5}{40} = \frac{17}{40}.$

(n) The LCD for $\frac{5}{12}$ and $\frac{3}{8}$ is 24: $\frac{5}{12} + \frac{3}{8} = \frac{5 \times 2}{12 \times 2} + \frac{3 \times 3}{8 \times 3} = \frac{10}{24} + \frac{9}{24} = \frac{19}{24}.$

(o) The LCD for $\frac{7}{10}$ and $\frac{11}{15}$ is 30: $\frac{7}{10} + \frac{11}{15} = \frac{7 \times 3}{10 \times 3} + \frac{11 \times 2}{15 \times 2} = \frac{21}{30} + \frac{22}{30} = \frac{43}{30} = 1\frac{13}{30}.$

(p) The LCD for $\frac{1}{6}$ and $\frac{1}{10}$ is 30: $\frac{1}{6} + \frac{1}{10} = \frac{1 \times 5}{6 \times 5} + \frac{1 \times 3}{10 \times 3} = \frac{5}{30} + \frac{3}{30} = \frac{8}{30} = \frac{\cancel{2} \times 4}{\cancel{2} \times 15} = \frac{4}{15}.$

(q) The LCD for $\frac{1}{4}$ and $\frac{3}{10}$ is 20: $\frac{1}{4} + \frac{3}{10} = \frac{1 \times 5}{4 \times 5} + \frac{3 \times 2}{10 \times 2} = \frac{5}{20} + \frac{6}{20} = \frac{11}{20}.$

(r) The LCD for $\frac{1}{15}$ and $\frac{5}{8}$ is 120: $\frac{1}{15} + \frac{5}{8} = \frac{1 \times 8}{15 \times 8} + \frac{5 \times 15}{8 \times 15} = \frac{8}{120} + \frac{75}{120} = \frac{83}{120}.$

(s) The LCD for $\frac{7}{20}$ and $\frac{7}{12}$ is 60: $\frac{7}{20} + \frac{7}{12} = \frac{7 \times 3}{20 \times 3} + \frac{7 \times 5}{12 \times 5} = \frac{21}{60} + \frac{35}{60} = \frac{56}{60} = \frac{\cancel{4} \times 14}{\cancel{4} \times 15} = \frac{14}{15}.$

(t) The LCD for $\frac{5}{9}$ and $\frac{11}{15}$ is 45: $\frac{5}{9} + \frac{11}{15} = \frac{5 \times 5}{9 \times 5} + \frac{11 \times 3}{15 \times 3} = \frac{25}{45} + \frac{33}{45} = \frac{58}{45} = 1\frac{13}{45}.$

PROBLEM 6-11 Add more than two unlike fractions: **(a)** $\frac{1}{4} + \frac{1}{2} + \frac{1}{8}$ **(b)** $\frac{5}{6} + \frac{1}{2} + \frac{2}{3}$
(c) $\frac{5}{12} + \frac{3}{4} + \frac{1}{3}$ **(d)** $\frac{1}{2} + \frac{7}{10} + \frac{4}{5}$ **(e)** $\frac{1}{3} + \frac{1}{2} + \frac{1}{5}$ **(f)** $\frac{3}{4} + \frac{4}{5} + \frac{2}{3}$ **(g)** $\frac{3}{8} + \frac{5}{6} + \frac{3}{10}$
(h) $\frac{7}{12} + \frac{9}{10} + \frac{5}{8}$ **(i)** $\frac{1}{12} + \frac{1}{2} + \frac{1}{4} + \frac{1}{6}$ **(j)** $\frac{7}{8} + \frac{5}{16} + \frac{3}{4} + \frac{1}{2}$

Solution: To add more than two unlike fractions, you proceed in the same way as you did with two unlike fractions [see Example 6-19]:

(a) The LCD for $\frac{1}{4}, \frac{1}{2},$ and $\frac{1}{8}$ is 8: $\frac{1}{4} + \frac{1}{2} + \frac{1}{8} = \frac{1 \times 2}{4 \times 2} + \frac{1 \times 4}{2 \times 4} + \frac{1}{8} = \frac{2}{8} + \frac{4}{8} + \frac{1}{8} = \frac{7}{8}.$

(b) The LCD for $\frac{5}{6}, \frac{1}{2},$ and $\frac{2}{3}$ is 6:

$$\frac{5}{6} + \frac{1}{2} + \frac{2}{3} = \frac{5}{6} + \frac{1 \times 3}{2 \times 3} + \frac{2 \times 2}{3 \times 2} = \frac{5}{6} + \frac{3}{6} + \frac{4}{6} = \frac{12}{6} = \frac{2}{1} = 2.$$

(c) The LCD for $\frac{5}{12}, \frac{3}{4},$ and $\frac{1}{3}$ is 12:

$$\frac{5}{12} + \frac{3}{4} + \frac{1}{3} = \frac{5}{12} + \frac{3 \times 3}{4 \times 3} + \frac{1 \times 4}{3 \times 4} = \frac{5}{12} + \frac{9}{12} + \frac{4}{12} = \frac{18}{12} = \frac{3}{2} = 1\frac{1}{2}.$$

(d) The LCD for $\frac{1}{2}, \frac{7}{10},$ and $\frac{4}{5}$ is 10:

$$\frac{1}{2} + \frac{7}{10} + \frac{4}{5} = \frac{1 \times 5}{2 \times 5} + \frac{7}{10} + \frac{4 \times 2}{5 \times 2} = \frac{5}{10} + \frac{7}{10} + \frac{8}{10} = \frac{20}{10} = \frac{2}{1} = 2.$$

(e) The LCD for $\frac{1}{3}, \frac{1}{2},$ and $\frac{1}{5}$ is 30:

$$\frac{1}{3} + \frac{1}{2} + \frac{1}{5} = \frac{1 \times 10}{3 \times 10} + \frac{1 \times 15}{2 \times 15} + \frac{1 \times 6}{5 \times 6} = \frac{10}{30} + \frac{15}{30} + \frac{6}{30} = \frac{31}{30} = 1\frac{1}{30}.$$

(f) The LCD for $\frac{3}{4}, \frac{4}{5},$ and $\frac{2}{3}$ is 60:

$$\frac{3}{4} + \frac{4}{5} + \frac{2}{3} = \frac{3 \times 15}{4 \times 15} + \frac{4 \times 12}{5 \times 12} + \frac{2 \times 20}{3 \times 20} = \frac{45}{60} + \frac{48}{60} + \frac{40}{60} = \frac{133}{60} = 2\frac{13}{60}.$$

(g) The LCD for $\frac{3}{8}, \frac{5}{6},$ and $\frac{3}{10}$ is 120:

$$\frac{3}{8} + \frac{5}{6} + \frac{3}{10} = \frac{3 \times 15}{8 \times 15} + \frac{5 \times 20}{6 \times 20} + \frac{3 \times 12}{10 \times 12} = \frac{45}{120} + \frac{100}{120} + \frac{36}{120} = \frac{181}{120} = 1\frac{61}{120}.$$

(h) The LCD for $\frac{7}{12}$, $\frac{9}{10}$, and $\frac{5}{8}$ is 120:

$$\frac{7}{12} + \frac{9}{10} + \frac{5}{8} = \frac{7 \times 10}{12 \times 10} + \frac{9 \times 12}{10 \times 12} + \frac{5 \times 15}{8 \times 15} = \frac{70}{120} + \frac{108}{120} + \frac{75}{120} = \frac{253}{120} = 2\frac{13}{120}.$$

(i) The LCD for $\frac{1}{12}$, $\frac{1}{2}$, $\frac{1}{4}$, and $\frac{1}{6}$ is 12:

$$\frac{1}{12} + \frac{1}{2} + \frac{1}{4} + \frac{1}{6} = \frac{1}{12} + \frac{1 \times 6}{2 \times 6} + \frac{1 \times 3}{4 \times 3} + \frac{1 \times 2}{6 \times 2} = \frac{1}{12} + \frac{6}{12} + \frac{3}{12} + \frac{2}{12} = \frac{12}{12} = 1.$$

(j) The LCD for $\frac{7}{8}$, $\frac{5}{16}$, $\frac{3}{4}$, and $\frac{1}{2}$ is 16:

$$\frac{7}{8} + \frac{5}{16} + \frac{3}{4} + \frac{1}{2} = \frac{7 \times 2}{8 \times 2} + \frac{5}{16} + \frac{3 \times 4}{4 \times 4} + \frac{1 \times 8}{2 \times 8} = \frac{14}{16} + \frac{5}{16} + \frac{12}{16} + \frac{8}{16} = \frac{39}{16} = 2\frac{7}{16}.$$

PROBLEM 6-12   Add mixed numbers and whole numbers: **(a)** $2 + 5\frac{1}{3}$ **(b)** $1 + 3\frac{3}{4}$
(c) $4\frac{1}{5} + 8$ **(d)** $6\frac{5}{8} + 9$ **(e)** $4 + 1\frac{2}{3}$ **(f)** $7 + 2\frac{5}{6}$ **(g)** $1\frac{1}{4} + 2$ **(h)** $5\frac{1}{2} + 1$ **(i)** $3 + 3\frac{2}{5}$
(j) $3 + 5\frac{1}{8}$ **(k)** $2\frac{7}{10} + 4$ **(l)** $4\frac{1}{6} + 3$ **(m)** $4 + 5\frac{3}{10} + 2$ **(n)** $1\frac{3}{5} + 3 + 6$
(o) $5 + 8\frac{7}{8} + 2 + 7$ **(p)** $5\frac{9}{10} + 8 + 7 + 6$

Solution: Recall that to add a mixed number to one or more whole numbers, you add the whole numbers and then write the same fraction part [see Example 6-20]:

(a) $2 + 5\frac{1}{3} = 7\frac{1}{3}$ **(b)** $1 + 3\frac{3}{4} = 4\frac{3}{4}$ **(c)** $4\frac{1}{5} + 8 = 12\frac{1}{5}$ **(d)** $6\frac{5}{8} + 9 = 15\frac{5}{8}$
(e) $4 + 1\frac{2}{3} = 5\frac{2}{3}$ **(f)** $7 + 2\frac{5}{6} = 9\frac{5}{6}$ **(g)** $1\frac{1}{4} + 2 = 3\frac{1}{4}$ **(h)** $5\frac{1}{2} + 1 = 6\frac{1}{2}$ **(i)** $3 + 3\frac{2}{5} = 6\frac{2}{5}$
(j) $3 + 5\frac{1}{8} = 8\frac{1}{8}$ **(k)** $2\frac{7}{10} + 4 = 6\frac{7}{10}$ **(l)** $4\frac{1}{6} + 3 = 7\frac{1}{6}$ **(m)** $4 + 5\frac{3}{10} + 2 = 11\frac{3}{10}$
(n) $1\frac{3}{5} + 3 + 6 = 10\frac{3}{5}$ **(o)** $5 + 8\frac{7}{8} + 2 + 7 = 22\frac{7}{8}$ **(p)** $5\frac{9}{10} + 8 + 7 + 6 = 26\frac{9}{10}$

PROBLEM 6-13   Add mixed numbers and fractions: **(a)** $\frac{1}{4} + 3\frac{1}{4}$ **(b)** $\frac{1}{8} + 9\frac{3}{8}$ **(c)** $8\frac{1}{2} + \frac{1}{2}$
(d) $1\frac{1}{3} + \frac{2}{3}$ **(e)** $\frac{1}{2} + 7\frac{1}{4}$ **(f)** $\frac{1}{6} + 2\frac{1}{3}$ **(g)** $4\frac{5}{8} + \frac{3}{4}$ **(h)** $6\frac{4}{5} + \frac{9}{10}$ **(i)** $\frac{1}{2} + 5\frac{1}{3}$
(j) $\frac{1}{3} + 3\frac{3}{4}$ **(k)** $7\frac{5}{8} + \frac{2}{3}$ **(l)** $1\frac{2}{3} + \frac{4}{5}$ **(m)** $\frac{5}{6} + 5\frac{1}{10}$ **(n)** $\frac{3}{10} + 4\frac{1}{4}$ **(o)** $\frac{11}{12} + 9\frac{7}{9}$
(p) $6\frac{7}{12} + \frac{8}{15}$ **(q)** $2\frac{5}{6} + \frac{1}{2} + \frac{2}{3}$ **(r)** $\frac{3}{5} + 8\frac{1}{2} + \frac{2}{3}$ **(s)** $\frac{1}{8} + \frac{1}{2} + 3\frac{3}{4} + \frac{5}{16}$ **(t)** $\frac{5}{8} + \frac{1}{10} + \frac{5}{12} + 5\frac{14}{15}$

Solution: Recall that to add a mixed number and one or more fractions, you add the fractions and then write the same whole-number part [see Example 6-21]:

(a) $3\frac{1}{4}$
$\underline{+ \ \frac{1}{4}}$
$3\frac{2}{4} = 3\frac{1}{2}$

(b) $9\frac{3}{8}$
$\underline{+ \ \frac{1}{8}}$
$9\frac{4}{8} = 9\frac{1}{2}$

(c) $8\frac{1}{2}$
$\underline{+ \ \frac{1}{2}}$
$8\frac{2}{2} = 8 + \frac{2}{2} = 8 + 1 = 9$

(d) $1\frac{1}{3}$
$\underline{+ \ \frac{2}{3}}$
$1\frac{3}{3} = 1 + \frac{3}{3} = 1 + 1 = 2$

(e) $7\frac{1}{4} = 7\frac{1}{4}$
$\underline{+ \ \frac{1}{2} = \ \frac{2}{4}}$
$7\frac{3}{4}$

(f) $2\frac{1}{3} = 2\frac{2}{6}$
$\underline{+ \ \frac{1}{6} = \ \frac{1}{6}}$
$2\frac{3}{6} = 2\frac{1}{2}$

(g) $4\frac{5}{8} = 4\frac{5}{8}$
$\underline{+ \ \frac{3}{4} = \ \frac{6}{8}}$
$4\frac{11}{8} = 4 + \frac{11}{8} = 4 + 1\frac{3}{8} = 5\frac{3}{8}$

(h) $6\frac{4}{5} = 6\frac{8}{10}$
$\underline{+ \ \frac{9}{10} = \ \frac{9}{10}}$
$6\frac{17}{10} = 6 + \frac{17}{10} = 6 + 1\frac{7}{10} = 7\frac{7}{10}$

(i) $5\frac{1}{3} = 5\frac{2}{6}$
$\underline{+ \ \frac{1}{2} = \ \frac{3}{6}}$
$5\frac{5}{6}$

(j) $3\frac{3}{4} = 3\frac{9}{12}$
$\underline{+ \ \frac{1}{3} = \ \frac{4}{12}}$
$3\frac{13}{12} = 3 + \frac{13}{12} = 3 + 1\frac{1}{12} = 4\frac{1}{12}$

(k) $7\frac{5}{8} = 7\frac{15}{24}$
$\underline{+ \ \frac{2}{3} = \ \frac{16}{24}}$
$7\frac{31}{24} = 7 + \frac{31}{24} = 7 + 1\frac{7}{24} = 8\frac{7}{24}$

(l) $1\frac{2}{3} = 1\frac{10}{15}$
$\underline{+ \ \frac{4}{5} = \ \frac{12}{15}}$
$1\frac{22}{15} = 1 + \frac{22}{15} = 1 + 1\frac{7}{15} = 2\frac{7}{15}$

(m) $5\frac{1}{10} = 5\frac{3}{30}$
$\underline{+ \ \frac{5}{6} = \ \frac{25}{30}}$
$5\frac{28}{30} = 5\frac{14}{15}$

(n) $\quad 4\frac{1}{4} = 4\frac{5}{20}$
$\quad +\frac{3}{10} = \ \frac{6}{20}$
$\quad\quad\quad\quad\ \ 4\frac{11}{20}$

(o) $\quad 9\frac{7}{9} = 9\frac{28}{36}$
$\quad +\frac{11}{12} = \ \frac{33}{36}$
$\quad\quad\quad 9\frac{61}{36} = 9 + \frac{61}{36} = 9 + 1\frac{25}{36} = 10\frac{25}{36}$

(p) $\quad 6\frac{7}{12} = 6\frac{35}{60}$
$\quad +\ \frac{8}{15} = \ \frac{32}{60}$
$\quad\quad\quad 6\frac{67}{60} = 6 + \frac{67}{60} = 6 + 1\frac{7}{60} = 7\frac{7}{60}$

(q) $\quad 2\frac{5}{6} = 2\frac{5}{6}$
$\quad\quad \frac{1}{2} = \ \frac{3}{6}$
$\quad +\ \frac{2}{3} = \ \frac{4}{6}$
$\quad\quad\quad 2\frac{12}{6} = 2 + \frac{12}{6} = 2 + \frac{2}{1} = 4$

(r) $\quad 8\frac{1}{2} = 8\frac{15}{30}$
$\quad\quad \frac{3}{5} = \ \frac{18}{30}$
$\quad +\ \frac{2}{3} = \ \frac{20}{30}$
$\quad\quad\quad 8\frac{53}{30} = 8 + \frac{53}{30} = 8 + 1\frac{23}{30} = 9\frac{23}{30}$

(s) $\quad 3\frac{3}{4} = 3\frac{12}{16}$
$\quad\quad \frac{1}{8} = \ \frac{2}{16}$
$\quad\quad \frac{1}{2} = \ \frac{8}{16}$
$\quad +\ \frac{5}{16} = \ \frac{5}{16}$
$\quad\quad\quad 3\frac{27}{16} = 3 + \frac{27}{16} = 3 + 1\frac{11}{16} = 4\frac{11}{16}$

(t) $\quad 5\frac{14}{15} = 5\frac{112}{120}$
$\quad\quad \frac{5}{8} = \ \frac{75}{120}$
$\quad\quad \frac{1}{10} = \ \frac{12}{120}$
$\quad +\ \frac{5}{12} = \ \frac{50}{120}$
$\quad\quad\quad 5\frac{249}{120} = 5\frac{83}{40} = 5 + \frac{83}{40} = 5 + 2\frac{3}{40} = 7\frac{3}{40}$

PROBLEM 6-14 Add mixed numbers: **(a)** $1\frac{1}{6} + 1\frac{1}{6}$ **(b)** $2\frac{1}{3} + 3\frac{1}{3}$ **(c)** $5\frac{3}{4} + 2\frac{1}{4}$ **(d)** $3\frac{5}{8} + 5\frac{7}{8}$ **(e)** $2\frac{1}{4} + 2\frac{1}{2}$ **(f)** $3\frac{1}{3} + 4\frac{1}{6}$ **(g)** $3\frac{3}{4} + 3\frac{5}{8}$ **(h)** $4\frac{1}{2} + 5\frac{1}{3}$ **(i)** $2\frac{1}{3} + 4\frac{1}{4}$ **(j)** $4\frac{3}{4} + 4\frac{4}{5}$ **(k)** $1\frac{1}{4} + 5\frac{1}{6}$ **(l)** $4\frac{1}{8} + 1\frac{5}{6}$ **(m)** $5\frac{9}{10} + 2\frac{7}{8}$ **(n)** $1\frac{1}{6} + 3\frac{1}{10}$ **(o)** $7\frac{7}{10} + 9\frac{1}{4}$ **(p)** $8\frac{8}{9} + 3\frac{11}{12}$ **(q)** $3\frac{1}{2} + 1\frac{1}{2} + 2\frac{1}{2}$ **(r)** $4\frac{1}{2} + 3\frac{3}{5} + 5\frac{7}{10}$ **(s)** $2\frac{11}{16} + 1\frac{5}{8} + 3\frac{1}{2} + 8\frac{3}{4}$ **(t)** $5\frac{1}{2} + 1\frac{4}{5} + 3\frac{1}{3} + 4\frac{3}{4}$

Solution: Recall that to add two or more mixed numbers, you first write the numbers in vertical form, then add the fraction parts, and then add the whole-number parts [see Example 6-22]:

(a) $\quad 1\frac{1}{6}$
$\quad +1\frac{1}{6}$
$\quad\ \ 2\frac{2}{6} = 2\frac{1}{3}$

(b) $\quad 2\frac{1}{3}$
$\quad +3\frac{1}{3}$
$\quad\ \ 5\frac{2}{3}$

(c) $\quad 5\frac{3}{4}$
$\quad +2\frac{1}{4}$
$\quad\ \ 7\frac{4}{4} = 7 + \frac{4}{4} = 7 + 1 = 8$

(d) $\quad 3\frac{5}{8}$
$\quad +5\frac{7}{8}$
$\quad\ \ 8\frac{12}{8} = 8\frac{3}{2} = 8 + \frac{3}{2} = 8 + 1\frac{1}{2} = 9\frac{1}{2}$

(e) $\quad 2\frac{1}{4} = 2\frac{1}{4}$
$\quad +2\frac{1}{2} = 2\frac{2}{4}$
$\quad\quad\quad\ \ 4\frac{3}{4}$

(f) $\quad 3\frac{1}{3} = 3\frac{2}{6}$
$\quad +4\frac{1}{6} = 4\frac{1}{6}$
$\quad\quad\quad\ \ 7\frac{3}{6} = 7\frac{1}{2}$

(g) $\quad 3\frac{3}{4} = 3\frac{6}{8}$
$\quad +3\frac{5}{8} = 3\frac{5}{8}$
$\quad\quad\quad\ \ 6\frac{11}{8} = 6 + \frac{11}{8} = 6 + 1\frac{3}{8} = 7\frac{3}{8}$

(h) $\quad 4\frac{1}{2} = 4\frac{3}{6}$
$\quad +5\frac{1}{3} = 5\frac{2}{6}$
$\quad\quad\quad\ \ 9\frac{5}{6}$

(i) $\quad 2\frac{1}{3} = 2\frac{4}{12}$
$\quad +4\frac{1}{4} = 4\frac{3}{12}$
$\quad\quad\quad\ \ 6\frac{7}{12}$

(j) $\quad 4\frac{3}{4} = 4\frac{15}{20}$
$\quad +4\frac{4}{5} = 4\frac{16}{20}$
$\quad\quad\quad\ \ 8\frac{31}{20} = 8 + \frac{31}{20} = 8 + 1\frac{11}{20} = 9\frac{11}{20}$

(k) $\quad 1\frac{1}{4} = 1\frac{3}{12}$
$\quad +5\frac{1}{6} = 5\frac{2}{12}$
$\quad\quad\quad\ \ 6\frac{5}{12}$

(l) $\quad 4\frac{1}{8} = 4\frac{3}{24}$
$\quad +1\frac{5}{6} = 1\frac{20}{24}$
$\quad\quad\quad\ \ 5\frac{23}{24}$

(m) $\quad 5\frac{9}{10} = 5\frac{36}{40}$
$\quad +2\frac{7}{8} = 2\frac{35}{40}$
$\quad\quad\quad\ \ 7\frac{71}{40} = 7 + \frac{71}{40} = 7 + 1\frac{31}{40} = 8\frac{31}{40}$

(n) $\quad 1\frac{1}{6} = 1\frac{5}{30}$
$\quad +3\frac{1}{10} = 3\frac{3}{30}$
$\quad\quad\quad\ \ 4\frac{8}{30} = 4\frac{4}{15}$

(o) $\quad 7\frac{7}{10} = \ 7\frac{14}{20}$
$\quad +9\frac{1}{4} = \ 9\frac{5}{20}$
$\quad\quad\quad\quad\ 16\frac{19}{20}$

(p) $\quad 8\frac{8}{9} = \ 8\frac{32}{36}$
$\quad +3\frac{11}{12} = \ 3\frac{33}{36}$
$\quad\quad\quad\quad\ 11\frac{65}{36} = 11 + \frac{65}{36} = 11 + 1\frac{29}{36} = 12\frac{29}{36}$

(q) $\quad 3\frac{1}{2}$
$\quad\quad 1\frac{1}{2}$
$\quad +2\frac{1}{2}$
$\quad\ \ 6\frac{3}{2} = 6 + \frac{3}{2} = 6 + 1\frac{1}{2} = 7\frac{1}{2}$

(r) $\quad 4\frac{1}{2} = \ 4\frac{5}{10}$
$\quad\quad 3\frac{3}{5} = \ 3\frac{6}{10}$
$\quad +5\frac{7}{10} = \ 5\frac{7}{10}$
$\quad\quad\quad\quad 12\frac{18}{10} = 12\frac{9}{5} = 12 + \frac{9}{5} = 12 + 1\frac{4}{5} = 13\frac{4}{5}$

(s) $2\frac{11}{16} = 2\frac{11}{16}$
$\quad\quad 1\frac{5}{8} = 1\frac{10}{16}$
$\quad\quad 3\frac{1}{2} = 3\frac{8}{16}$
$\quad\quad +8\frac{3}{4} = 8\frac{12}{16}$
$\quad\quad\quad\quad 14\frac{41}{16} = 14 + \frac{41}{16} = 14 + 2\frac{9}{16} = 16\frac{9}{16}$

(t) $5\frac{1}{2} = 5\frac{30}{60}$
$\quad\quad 1\frac{4}{5} = 1\frac{48}{60}$
$\quad\quad 3\frac{2}{3} = 3\frac{40}{60}$
$\quad\quad +4\frac{3}{4} = 4\frac{45}{60}$
$\quad\quad\quad\quad 13\frac{163}{60} = 13 + \frac{163}{60} = 13 + 2\frac{43}{60} = 15\frac{43}{60}$

Supplementary Exercises

PROBLEM 6-15 Add basic addition facts in horizontal form:

Row (a)	$9 + 3$	$3 + 0$	$8 + 2$	$2 + 7$	$0 + 6$	$5 + 1$	$3 + 5$	$6 + 1$	$9 + 1$	$4 + 2$
Row (b)	$1 + 9$	$7 + 7$	$3 + 9$	$5 + 0$	$6 + 7$	$1 + 1$	$9 + 5$	$8 + 8$	$8 + 4$	$0 + 5$
Row (c)	$2 + 3$	$6 + 4$	$3 + 1$	$7 + 9$	$0 + 0$	$5 + 9$	$1 + 3$	$8 + 5$	$0 + 8$	$9 + 8$
Row (d)	$2 + 6$	$2 + 1$	$3 + 7$	$0 + 3$	$9 + 6$	$1 + 0$	$4 + 6$	$7 + 3$	$9 + 9$	$5 + 7$
Row (e)	$2 + 4$	$7 + 1$	$7 + 5$	$1 + 5$	$2 + 8$	$3 + 4$	$5 + 6$	$5 + 3$	$8 + 0$	$5 + 5$
Row (f)	$1 + 7$	$6 + 3$	$8 + 3$	$8 + 1$	$1 + 6$	$4 + 0$	$9 + 7$	$4 + 5$	$3 + 2$	$6 + 2$
Row (g)	$9 + 2$	$1 + 2$	$8 + 7$	$4 + 8$	$2 + 9$	$0 + 2$	$4 + 4$	$6 + 8$	$7 + 0$	$7 + 6$
Row (h)	$5 + 8$	$5 + 2$	$4 + 1$	$6 + 5$	$9 + 0$	$2 + 5$	$6 + 9$	$0 + 4$	$3 + 3$	$4 + 7$
Row (i)	$9 + 4$	$0 + 9$	$4 + 9$	$5 + 4$	$7 + 8$	$7 + 4$	$8 + 6$	$1 + 4$	$0 + 1$	$1 + 8$
Row (j)	$2 + 2$	$6 + 6$	$3 + 8$	$7 + 2$	$0 + 7$	$4 + 3$	$6 + 0$	$8 + 9$	$3 + 6$	$2 + 0$

PROBLEM 6-16 Add like fractions: (a) $\frac{1}{8} + \frac{5}{8}$ (b) $\frac{3}{10} + \frac{3}{10}$ (c) $\frac{1}{12} + \frac{11}{12}$ (d) $\frac{1}{5} + \frac{2}{5}$
(e) $\frac{1}{6} + \frac{1}{6}$ (f) $\frac{1}{3} + \frac{1}{3}$ (g) $\frac{7}{8} + \frac{1}{8}$ (h) $\frac{3}{10} + \frac{7}{10}$ (i) $\frac{5}{12} + \frac{5}{12}$ (j) $\frac{3}{5} + \frac{1}{5}$ (k) $\frac{1}{6} + \frac{5}{6}$
(l) $\frac{2}{3} + \frac{2}{3}$ (m) $\frac{3}{8} + \frac{7}{8}$ (n) $\frac{7}{12} + \frac{5}{12}$ (o) $\frac{2}{5} + \frac{3}{5}$ (p) $\frac{5}{6} + \frac{5}{6}$ (q) $\frac{1}{3} + \frac{2}{3}$ (r) $\frac{7}{8} + \frac{7}{8}$
(s) $\frac{5}{12} + \frac{11}{12} + \frac{7}{12}$ (t) $\frac{3}{5} + \frac{4}{5} + \frac{1}{5}$ (u) $\frac{1}{2} + \frac{1}{2} + \frac{1}{2}$ (v) $\frac{1}{4} + \frac{3}{4} + \frac{3}{4}$ (w) $\frac{3}{8} + \frac{5}{8} + \frac{1}{8} + \frac{7}{8}$
(x) $\frac{5}{10} + \frac{7}{10} + \frac{9}{10} + \frac{9}{10}$ (y) $\frac{2}{3} + \frac{1}{3} + \frac{1}{3} + \frac{1}{3}$ (z) $\frac{1}{6} + \frac{5}{6} + \frac{5}{6} + \frac{5}{6}$

PROBLEM 6-17 Find the LCD for unlike fractions without using the Factoring Method for finding the LCD: (a) $\frac{1}{2}$ and $\frac{5}{12}$ (b) $\frac{2}{3}$ and $\frac{7}{15}$ (c) $\frac{3}{4}$ and $\frac{5}{16}$ (d) $\frac{2}{5}$ and $\frac{1}{15}$ (e) $\frac{1}{6}$ and $\frac{1}{12}$
(f) $\frac{3}{8}$ and $\frac{5}{16}$ (g) $\frac{1}{3}$ and $\frac{2}{7}$ (h) $\frac{1}{4}$ and $\frac{1}{3}$ (i) $\frac{3}{4}$ and $\frac{5}{7}$ (j) $\frac{2}{3}$ and $\frac{1}{7}$ (k) $\frac{3}{8}$ and $\frac{1}{9}$
(l) $\frac{1}{4}$ and $\frac{5}{9}$ (m) $\frac{5}{6}$ and $\frac{7}{8}$ (n) $\frac{1}{8}$ and $\frac{7}{10}$ (o) $\frac{3}{10}$ and $\frac{7}{12}$ (p) $\frac{5}{9}$ and $\frac{11}{15}$ (q) $\frac{9}{10}$ and $\frac{11}{12}$
(r) $\frac{5}{12}$ and $\frac{13}{18}$ (s) $\frac{3}{4}, \frac{1}{2}$, and $\frac{5}{8}$ (t) $\frac{2}{3}, \frac{1}{6}$, and $\frac{1}{2}$ (u) $\frac{2}{3}, \frac{3}{5}$, and $\frac{1}{2}$ (v) $\frac{3}{4}, \frac{4}{5}$, and $\frac{1}{3}$
(w) $\frac{5}{6}, \frac{3}{8}$, and $\frac{7}{10}$ (x) $\frac{3}{10}, \frac{5}{18}$, and $\frac{9}{8}$ (y) $\frac{7}{18}, \frac{11}{12}$, and $\frac{10}{9}$ (z) $\frac{1}{24}, \frac{7}{18}$, and $\frac{15}{16}$

PROBLEM 6-18 Find the LCD for unlike fractions using the Factoring Method for finding the LCD: (a) $\frac{1}{4}$ and $\frac{7}{10}$ (b) $\frac{1}{6}$ and $\frac{3}{14}$ (c) $\frac{5}{8}$ and $\frac{9}{14}$ (d) $\frac{2}{9}$ and $\frac{1}{15}$ (e) $\frac{7}{16}$ and $\frac{4}{9}$
(f) $\frac{3}{10}$ and $\frac{1}{14}$ (g) $\frac{1}{16}$ and $\frac{1}{10}$ (h) $\frac{3}{10}$ and $\frac{5}{18}$ (i) $\frac{1}{12}$ and $\frac{11}{20}$ (j) $\frac{5}{12}$ and $\frac{11}{21}$ (k) $\frac{1}{12}$ and $\frac{9}{22}$
(l) $\frac{5}{14}$ and $\frac{17}{18}$ (m) $\frac{19}{20}$ and $\frac{1}{14}$ (n) $\frac{13}{14}$ and $\frac{17}{21}$ (o) $\frac{23}{24}$ and $\frac{13}{14}$ (p) $\frac{15}{16}$ and $\frac{20}{21}$
(q) $\frac{17}{18}$ and $\frac{3}{25}$ (r) $\frac{1}{4}, \frac{1}{6}$, and $\frac{1}{8}$ (s) $\frac{2}{3}, \frac{5}{8}$, and $\frac{1}{6}$ (t) $\frac{3}{4}, \frac{7}{8}$, and $\frac{4}{9}$ (u) $\frac{5}{6}, \frac{1}{8}$, and $\frac{3}{10}$
(v) $\frac{7}{12}, \frac{5}{8}$, and $\frac{5}{6}$ (w) $\frac{7}{10}, \frac{11}{12}$, and $\frac{14}{15}$ (x) $\frac{1}{12}, \frac{1}{18}$, and $\frac{1}{10}$ (y) $\frac{3}{20}, \frac{5}{18}$, and $\frac{11}{16}$ (z) $\frac{23}{24}, \frac{13}{18}$, and $\frac{14}{15}$

PROBLEM 6-19 Rename unlike fractions as equivalent like fractions using the LCD: (a) $\frac{1}{2}$ and $\frac{3}{8}$
(b) $\frac{5}{6}$ and $\frac{1}{2}$ (c) $\frac{1}{4}$ and $\frac{5}{12}$ (d) $\frac{3}{5}$ and $\frac{7}{10}$ (e) $\frac{1}{2}$ and $\frac{1}{4}$ (f) $\frac{2}{3}$ and $\frac{5}{9}$ (g) $\frac{1}{3}$ and $\frac{3}{5}$

(h) $\frac{3}{4}$ and $\frac{1}{5}$ **(i)** $\frac{5}{6}$ and $\frac{2}{5}$ **(j)** $\frac{4}{5}$ and $\frac{5}{8}$ **(k)** $\frac{1}{2}$ and $\frac{2}{3}$ **(l)** $\frac{1}{3}$ and $\frac{7}{8}$ **(m)** $\frac{5}{12}$ and $\frac{7}{18}$
(n) $\frac{1}{15}$ and $\frac{11}{12}$ **(o)** $\frac{4}{15}$ and $\frac{3}{20}$ **(p)** $\frac{17}{18}$ and $\frac{11}{15}$ **(q)** $\frac{7}{12}$ and $\frac{9}{16}$ **(r)** $\frac{13}{16}$ and $\frac{15}{18}$
(s) $\frac{5}{6}, \frac{1}{3}$, and $\frac{1}{2}$ **(t)** $\frac{3}{4}, \frac{5}{6}$, and $\frac{7}{12}$ **(u)** $\frac{1}{4}, \frac{3}{4}$, and $\frac{4}{5}$ **(v)** $\frac{1}{2}, \frac{3}{7}$, and $\frac{2}{5}$ **(w)** $\frac{7}{10}, \frac{5}{12}$, and $\frac{1}{9}$
(x) $\frac{3}{8}, \frac{2}{15}$, and $\frac{7}{12}$ **(y)** $\frac{15}{16}, \frac{1}{4}, \frac{1}{2}$, and $\frac{7}{8}$ **(z)** $\frac{5}{18}, \frac{1}{2}, \frac{2}{3}$, and $\frac{5}{6}$

PROBLEM 6-20   Add unlike fractions: **(a)** $\frac{1}{2} + \frac{5}{6}$ **(b)** $\frac{3}{8} + \frac{1}{2}$ **(c)** $\frac{2}{3} + \frac{7}{9}$ **(d)** $\frac{3}{4} + \frac{5}{16}$
(e) $\frac{1}{5} + \frac{7}{10}$ **(f)** $\frac{5}{12} + \frac{1}{6}$ **(g)** $\frac{1}{2} + \frac{3}{7}$ **(h)** $\frac{2}{5} + \frac{5}{8}$ **(i)** $\frac{1}{4} + \frac{4}{5}$ **(j)** $\frac{4}{7} + \frac{1}{5}$ **(k)** $\frac{1}{6} + \frac{3}{5}$
(l) $\frac{7}{8} + \frac{1}{3}$ **(m)** $\frac{5}{6} + \frac{1}{9}$ **(n)** $\frac{1}{12} + \frac{11}{14}$ **(o)** $\frac{15}{16} + \frac{7}{12}$ **(p)** $\frac{1}{12} + \frac{8}{15}$ **(q)** $\frac{5}{18} + \frac{11}{12}$
(r) $\frac{14}{15} + \frac{1}{18}$ **(s)** $\frac{2}{15} + \frac{3}{5} + \frac{1}{3}$ **(t)** $\frac{7}{12} + \frac{5}{6} + \frac{1}{2}$ **(u)** $\frac{1}{2} + \frac{1}{3} + \frac{1}{5}$ **(v)** $\frac{5}{6} + \frac{7}{8} + \frac{9}{10}$
(w) $\frac{1}{8} + \frac{3}{16} + \frac{1}{4} + \frac{1}{2}$ **(x)** $\frac{5}{6} + \frac{1}{2} + \frac{5}{12} + \frac{1}{3}$ **(y)** $\frac{3}{10} + \frac{1}{12} + \frac{2}{15} + \frac{5}{18}$ **(z)** $\frac{11}{20} + \frac{13}{15} + \frac{7}{8} + \frac{7}{9}$

PROBLEM 6-21   Add with mixed numbers: **(a)** $2 + 3\frac{2}{3}$ **(b)** $5 + 8\frac{1}{2}$ **(c)** $3\frac{1}{4} + 4$
(d) $1\frac{1}{5} + 2$ **(e)** $3\frac{3}{4} + 1$ **(f)** $3 + 9\frac{5}{6}$ **(g)** $\frac{1}{2} + 4\frac{1}{2}$ **(h)** $\frac{1}{3} + 9\frac{1}{3}$ **(i)** $8\frac{3}{4} + \frac{3}{4}$
(j) $7\frac{5}{8} + \frac{7}{8}$ **(k)** $\frac{1}{3} + 6\frac{1}{6}$ **(l)** $\frac{3}{4} + 5\frac{5}{8}$ **(m)** $4\frac{1}{2} + \frac{3}{4}$ **(n)** $3\frac{1}{2} + 1\frac{2}{3}$ **(o)** $3\frac{3}{4} + 5\frac{3}{5}$
(p) $4\frac{1}{6} + 1\frac{5}{8}$ **(q)** $5\frac{3}{10} + 4\frac{7}{12}$ **(r)** $6\frac{5}{15} + 5\frac{4}{20}$ **(s)** $2 + \frac{1}{3} + 8\frac{5}{9}$ **(t)** $\frac{1}{12} + 9 + 1\frac{1}{3}$
(u) $3 + 4\frac{4}{8} + \frac{1}{9}$ **(v)** $\frac{7}{12} + 5\frac{1}{6} + 2$ **(w)** $6\frac{3}{12} + 7 + \frac{7}{18}$ **(x)** $8\frac{2}{3} + \frac{4}{5} + 2$
(y) $9\frac{1}{2} + \frac{2}{3} + 3 + 5\frac{5}{6}$ **(z)** $\frac{1}{8} + 6\frac{7}{10} + \frac{1}{12} + 8$

Answers to Supplementary Exercises

(6-15)

Row										
(a)	12	3	10	9	6	6	8	7	10	6
(b)	10	14	12	5	13	2	14	16	12	5
(c)	5	10	4	16	0	14	4	13	8	17
(d)	8	3	10	3	15	1	10	10	18	12
(e)	6	8	12	6	10	7	11	8	8	10
(f)	8	9	11	9	7	4	16	9	5	8
(g)	11	3	15	12	11	2	8	14	7	13
(h)	13	7	5	11	9	7	15	4	6	11
(i)	13	9	13	9	15	11	14	5	1	9
(j)	4	12	11	9	7	7	6	17	9	2

(6-16)   **(a)** $\frac{3}{4}$ **(b)** $\frac{3}{5}$ **(c)** 1   **(d)** $\frac{3}{5}$ **(e)** $\frac{1}{3}$ **(f)** $\frac{2}{3}$ **(g)** 1   **(h)** 1   **(i)** $\frac{5}{6}$ **(j)** $\frac{4}{5}$
(k) 1   **(l)** $1\frac{1}{3}$ **(m)** $1\frac{1}{4}$ **(n)** 1   **(o)** 1   **(p)** $1\frac{2}{3}$ **(q)** 1   **(r)** $1\frac{3}{4}$ **(s)** $1\frac{11}{12}$ **(t)** $1\frac{3}{5}$
(u) $1\frac{1}{2}$ **(v)** $1\frac{3}{4}$ **(w)** 2   **(x)** 3   **(y)** $1\frac{2}{3}$ **(z)** $2\frac{2}{3}$

(6-17)   **(a)** 12   **(b)** 15   **(c)** 16   **(d)** 15   **(e)** 12   **(f)** 16   **(g)** 21   **(h)** 12   **(i)** 28
(j) 35   **(k)** 72   **(l)** 36   **(m)** 24   **(n)** 40   **(o)** 60   **(p)** 45   **(q)** 60   **(r)** 36   **(s)** 8
(t) 6   **(u)** 30   **(v)** 60   **(w)** 120   **(x)** 360   **(y)** 36   **(z)** 144

(6-18)   **(a)** 20   **(b)** 42   **(c)** 56   **(d)** 45   **(e)** 144   **(f)** 70   **(g)** 80   **(h)** 90   **(i)** 60
(j) 84   **(k)** 132   **(l)** 126   **(m)** 140   **(n)** 42   **(o)** 168   **(p)** 336   **(q)** 450   **(r)** 24
(s) 24   **(t)** 72   **(u)** 120   **(v)** 24   **(w)** 60   **(x)** 180   **(y)** 720   **(z)** 360

(6-19)   **(a)** $\frac{4}{8}$ and $\frac{3}{8}$ **(b)** $\frac{5}{6}$ and $\frac{3}{6}$ **(c)** $\frac{3}{12}$ and $\frac{5}{12}$ **(d)** $\frac{6}{10}$ and $\frac{7}{10}$ **(e)** $\frac{2}{4}$ and $\frac{1}{4}$ **(f)** $\frac{6}{9}$ and $\frac{5}{9}$
(g) $\frac{5}{15}$ and $\frac{9}{15}$ **(h)** $\frac{15}{20}$ and $\frac{4}{20}$ **(i)** $\frac{25}{30}$ and $\frac{12}{30}$ **(j)** $\frac{32}{40}$ and $\frac{25}{40}$ **(k)** $\frac{3}{6}$ and $\frac{4}{6}$ **(l)** $\frac{8}{24}$ and $\frac{21}{24}$
(m) $\frac{15}{36}$ and $\frac{14}{36}$ **(n)** $\frac{4}{60}$ and $\frac{55}{60}$ **(o)** $\frac{16}{60}$ and $\frac{9}{60}$ **(p)** $\frac{85}{90}$ and $\frac{66}{90}$ **(q)** $\frac{28}{48}$ and $\frac{27}{48}$ **(r)** $\frac{117}{144}$ and $\frac{120}{144}$
(s) $\frac{5}{6}, \frac{2}{6}$, and $\frac{3}{6}$ **(t)** $\frac{9}{12}, \frac{10}{12}$, and $\frac{7}{12}$ **(u)** $\frac{15}{60}, \frac{40}{60}$, and $\frac{48}{60}$ **(v)** $\frac{35}{70}, \frac{30}{70}$, and $\frac{28}{70}$
(w) $\frac{126}{180}, \frac{75}{180}$, and $\frac{20}{180}$ **(x)** $\frac{45}{120}, \frac{16}{120}$, and $\frac{70}{120}$ **(y)** $\frac{15}{16}, \frac{4}{16}, \frac{8}{16}$, and $\frac{14}{16}$ **(z)** $\frac{5}{18}, \frac{9}{18}, \frac{12}{18}$, and $\frac{15}{18}$

(6-20) (a) $1\frac{1}{3}$ (b) $\frac{7}{8}$ (c) $1\frac{4}{9}$ (d) $1\frac{1}{16}$ (e) $\frac{9}{10}$ (f) $\frac{7}{12}$ (g) $\frac{13}{14}$ (h) $1\frac{1}{40}$ (i) $1\frac{1}{20}$
(j) $\frac{27}{35}$ (k) $\frac{23}{30}$ (l) $1\frac{5}{24}$ (m) $\frac{17}{18}$ (n) $\frac{73}{84}$ (o) $12\frac{25}{48}$ (p) $\frac{37}{60}$ (q) $1\frac{7}{36}$ (r) $\frac{89}{90}$
(s) $1\frac{1}{15}$ (t) $1\frac{11}{12}$ (u) $1\frac{1}{30}$ (v) $2\frac{73}{120}$ (w) $1\frac{1}{16}$ (x) $2\frac{1}{12}$ (y) $\frac{143}{180}$ (z) $3\frac{25}{360}$

(6-21) (a) $5\frac{2}{3}$ (b) $13\frac{1}{2}$ (c) $7\frac{1}{4}$ (d) $3\frac{1}{5}$ (e) $4\frac{3}{4}$ (f) $12\frac{5}{6}$ (g) 5 (h) $9\frac{2}{3}$ (i) $9\frac{1}{2}$
(j) $8\frac{1}{2}$ (k) $6\frac{1}{2}$ (l) $6\frac{3}{8}$ (m) $5\frac{1}{4}$ (n) $5\frac{1}{6}$ (o) $9\frac{7}{20}$ (p) $5\frac{19}{24}$ (q) $9\frac{53}{60}$ (r) $11\frac{8}{15}$
(s) $10\frac{8}{9}$ (t) $10\frac{5}{12}$ (u) $7\frac{11}{18}$ (v) $7\frac{3}{4}$ (w) $13\frac{19}{36}$ (x) $11\frac{7}{15}$ (y) 19 (z) $14\frac{109}{120}$

7 FRACTIONS AND MIXED NUMBERS: SUBTRACTION

THIS CHAPTER IS ABOUT

☑ **Subtracting Like Fractions**
☑ **Subtracting Unlike Fractions**
☑ **Subtracting with Mixed Numbers**
☑ **Renaming to Subtract**
☑ **Solving Word Problems Containing Fractions and Mixed Numbers (+, −, ×, ÷)**

7-1. Subtracting Like Fractions

To take one fractional amount away from another fractional amount, you **subtract fractions.**

EXAMPLE 7-1: Rename the following as subtraction: $\frac{5}{8}$ take away $\frac{2}{8}$

Solution: $\frac{5}{8}$ take away $\frac{2}{8} = \frac{5}{8} - \frac{2}{8}$ ⟵ subtraction

To subtract like fractions, you can use the following rule:

Subtraction Rule for Like Fractions

If $\frac{a}{c}$ and $\frac{b}{c}$ are any two like fractions, then $\frac{a}{c} - \frac{b}{c} = \frac{a - b}{c}$.

Note: To subtract like fractions, you subtract the numerators and then write the same denominator.

EXAMPLE 7-2: Subtract the following fractions: $\frac{5}{8} - \frac{2}{8}$

Solution: $\frac{5}{8} - \frac{2}{8} = \frac{5 - 2}{8}$ Subtract the numerators.
                                       Write the same denominator.

$\qquad\qquad = \frac{3}{8}$ Subtract in the numerator: $5 - 2 = 3$

To check a subtraction answer, you add the proposed difference to the original subtrahend to see if you get the original minuend.

EXAMPLE 7-3: Check the following subtraction problem: $\frac{5}{8} - \frac{2}{8} = \frac{3}{8}$

Solution: $\frac{5}{8} - \frac{2}{8} = \frac{3}{8}$ because $\frac{3}{8} + \frac{2}{8} = \frac{5}{8}$ ⟵ $\frac{3}{8}$ checks

proposed difference
original subtrahend
original minuend

The following example can help you understand subtraction of like fractions.

EXAMPLE 7-4: Translate subtraction of like fractions to words and pictures.

Solution

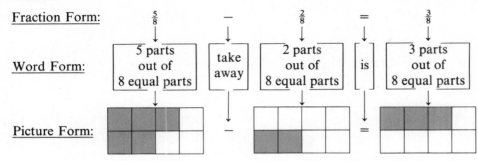

Caution: Always simplify a fractional difference when possible [see Section 4-3].

EXAMPLE 7-5: Subtract the following fractions: $\frac{11}{12} - \frac{5}{12}$

Solution: $\dfrac{11}{12} - \dfrac{5}{12} = \dfrac{11 - 5}{12}$ Subtract like fractions.

$$= \frac{6}{12}$$

$$= \frac{1 \times \cancel{6}}{2 \times \cancel{6}}$$ Simplify [see Section 4-3.]

$$= \frac{1}{2} \longleftarrow \text{ simplest form}$$

7-2. Subtracting Unlike Fractions

To subtract unlike fractions, you can use the following method:

Subtracting Rules for Unlike Fractions

1. Find the LCD for the unlike fractions [see Section 6-2].
2. Build up to get like fractions using the LCD from Step 1 [see Section 6-3].
3. Substitute the like fractions from Step 2 for the original unlike fractions.
4. Subtract the like fractions from Step 3 [see Section 7-2].
5. Simplify the difference from Step 4 when possible [see Section 4-3].
6. Check by adding the simplest form of the proposed difference from Step 4 or 5 to the original subtrahend to see if you get the original minuend [see Example 7-3].

Note: To subtract unlike fractions, you first rename as equivalent like fractions using the LCD and then subtract the like fractions.

EXAMPLE 7-6: Subtract the following fractions: $\frac{11}{12} - \frac{1}{6}$

Solution: The LCD for $\frac{11}{12}$ and $\frac{1}{6}$ is 12.   Find the LCD.

$$\frac{11}{12} = \frac{11}{12}$$ Build up to get like fractions using the LCD.

$$\frac{1}{6} = \frac{1 \times 2}{6 \times 2} = \frac{2}{12}$$ LCD

$$\frac{11}{12} - \frac{1}{6} = \frac{11}{12} - \frac{2}{12} \longleftarrow \text{ like fractions}$$

Substitute the like fractions for the original unlike fractions.

$$= \frac{11 - 2}{12}$$

Subtract the like fractions.

$$= \frac{9}{12}$$

$$= \frac{\not{3} \times 3}{\not{3} \times 4}$$

Simplify.

$$= \frac{3}{4} \longleftarrow \text{ proposed difference}$$

$$\frac{3}{4} + \frac{1}{6} = \frac{9}{12} + \frac{2}{12}$$

Check by adding the proposed difference to the original subtrahend to see if you get the original minuend.

$$= \frac{9 + 2}{12}$$

$$= \frac{11}{12} \longleftarrow \frac{3}{4} \text{ checks}$$

Note: $\frac{11}{12} - \frac{1}{6} = \frac{3}{4} \longleftarrow$ simplest form

7-3. Subtracting with Mixed Numbers

A. Subtract a whole number from a mixed number.

To subtract a whole number from a mixed number, you subtract the whole numbers and then write the same fraction part.

EXAMPLE 7-7: Subtract the following whole number from a mixed number: $7\frac{2}{3} - 3$

Solution: $7\frac{2}{3} - 3 = 4\frac{2}{3}$ or
$$\begin{array}{r} 7\frac{2}{3} \\ -3 \\ \hline 4\frac{2}{3} \end{array}$$

Subtract the whole numbers.
Write the same fractional part.

Note: $7\frac{2}{3} - 3 = 4\frac{2}{3}$ because $7\frac{2}{3} - 3 = 7 + \frac{2}{3} - 3$
$$= (7 - 3) + \frac{2}{3}$$
$$= 4 + \frac{2}{3}$$
$$= 4\frac{2}{3}$$

B. Subtract a fraction from a mixed number.

To subtract a fraction from a mixed number, you first write the numbers in vertical form, then subtract the fractions, and then write the same whole-number part.

EXAMPLE 7-8: Subtract the following fraction from a mixed number: $5\frac{2}{3} - \frac{1}{2}$

Solution:
$$\begin{array}{r} 5\frac{2}{3} = 5\frac{4}{6} \\ -\frac{1}{2} = \frac{3}{6} \\ \hline 5\frac{1}{6} \end{array}$$

Write in vertical subtraction form.
Build up like fractions using the LCD 6.
Subtract like fractions and write the same whole-number part.

$$\begin{array}{r} 5\frac{1}{6} = 5\frac{1}{6} \\ +\frac{1}{2} = \frac{3}{6} \\ \hline 5\frac{4}{6} = 5\frac{2}{3} \longleftarrow 5\frac{1}{6} \text{ checks} \end{array}$$

Check the proposed difference.

Note: $5\frac{2}{3} - \frac{1}{2} = 5\frac{1}{6} \longleftarrow$ simplest form

C. Subtract mixed numbers.

To subtract two mixed numbers, you first write the numbers in vertical form, then subtract the fraction parts, and then subtract the whole-number parts.

EXAMPLE 7-9: Subtract the following mixed numbers: $9\frac{5}{8} - 2\frac{1}{6}$

$$4 \times 6 = 24$$

Solution: The LCD for $\frac{5}{8}$ and $\frac{1}{6}$ is 24.   *Think:* $8 - 6 = 2$ and $2\overline{)8}$

$$9\frac{5}{8} = 9\frac{15}{24}$$
$$-2\frac{1}{6} = 2\frac{4}{24} \qquad \text{LCD}$$
$$7\frac{11}{24} \longleftarrow \text{proposed difference}$$

Write in vertical form.
Build up to get like fraction parts.
Subtract the like fraction parts: $\frac{15}{24} - \frac{4}{24} = \frac{11}{24}$
Subtract the whole-number parts: $9 - 2 = 7$

$$7\frac{11}{24} = 7\frac{11}{24}$$
$$+2\frac{1}{6} = 2\frac{4}{24}$$
$$9\frac{15}{24} = 9 + \frac{15}{24}$$

Check by adding the proposed difference to the original subtrahend to see if you get the original minuend.

$$= 9 + \frac{\cancel{3} \times 5}{\cancel{3} \times 8}$$

$$= 9 + \frac{5}{8}$$

$$= 9\frac{5}{8} \longleftarrow 7\frac{11}{24} \text{ checks}$$

Note: $9\frac{5}{8} - 2\frac{1}{6} = 7\frac{11}{24} \longleftarrow$ simplest form

7-4. Renaming to Subtract

A. Rename to subtract from a whole number.

To subtract a fraction or a mixed number from a whole number, you first rename the whole number as an equal mixed number so that the fraction parts are like fractions.

EXAMPLE 7-10: Subtract the following mixed number from a whole number: $8 - 2\frac{3}{4}$

missing fraction part

Solution:
$$8 = 7\frac{4}{4}$$
$$-2\frac{3}{4} = 2\frac{3}{4} \qquad \text{like fractions}$$
$$5\frac{1}{4} \longleftarrow \text{proposed difference}$$

Think: $8 = 7 + 1 = 7 + \frac{4}{4} = 7\frac{4}{4}$

$$5\frac{1}{4}$$
$$+2\frac{3}{4}$$
$$7\frac{4}{4} = 7 + \frac{4}{4}$$

Check the proposed difference.

$$= 7 + 1$$
$$= 8 \longleftarrow 5\frac{1}{4} \text{ checks}$$

Recall that if you are subtracting a whole number from a mixed number, you subtract the whole number and then write the same fraction part [see Example 7-7]. However, you should never move a fraction in the mixed number to the whole number and then subtract using this method. Your answer will always be wrong if you do this.

Caution:

$8 - 2\frac{3}{4} \neq 8\frac{3}{4} - 2$.   So $8\frac{3}{4} - 2 = (8 - 2) + \frac{3}{4} = 6 + \frac{3}{4} = 6\frac{3}{4} \longleftarrow$ wrong answer [see Example 7-7].

$8 - 2\frac{3}{4} = 7\frac{4}{4} - 2\frac{3}{4} = 5\frac{3}{4} \longleftarrow$ correct answer [see Example 7-10].

Always rename the whole number as an equal mixed number when subtracting a mixed number from a whole number.

To avoid making an error when subtracting whole numbers and fractions or mixed numbers, you should always use vertical subtraction form.

EXAMPLE 7-11: Subtract the following whole numbers and fractions: (a) $5 - \frac{1}{2}$ (b) $6 - 3\frac{5}{8}$
(c) $6\frac{5}{8} - 3$

Solution: (a)
$$5 = 4\frac{2}{2}$$
$$\frac{-\frac{1}{2} = \frac{1}{2}}{4\frac{1}{2}}$$

(b)
$$6 = 5\frac{8}{8}$$
$$\frac{-3\frac{5}{8} = 3\frac{5}{8}}{2\frac{3}{8}}$$

(c)
$$6\frac{5}{8}$$
$$\frac{-3}{3\frac{5}{8}}$$ Check as before.

B. Rename to subtract from a mixed number.

To subtract two mixed numbers when the minuend of the fraction part is too small, you first rename the minuend to get an equal mixed number with a larger fraction part.

EXAMPLE 7-12: Subtract the following mixed numbers: $9\frac{1}{6} - 1\frac{5}{6}$

too
small   more sixths
↓   ↓

Solution: $9\frac{1}{6} = 8\frac{7}{6}$ *Think:* $9\frac{1}{6} = 9 + \frac{1}{6} = 8 + 1 + \frac{1}{6} = 8 + \frac{6}{6} + \frac{1}{6} = 8 + \frac{7}{6} = 8\frac{7}{6}$

$$\frac{-1\frac{5}{6} = 1\frac{5}{6}}{7\frac{2}{6}}$$ or: $9\frac{1}{6} = 8\frac{7}{6}$ (The new numerator is the sum of the old numerator and denominator.)

with arrows: $1 + 6$ and $9 - 1$

$7\frac{2}{6} = 7\frac{1}{3}$ ⟵ proposed difference   *Think:* $\frac{2}{6} = \frac{\cancel{2} \times 1}{\cancel{2} \times 3} = \frac{1}{3}$

$$7\frac{1}{3} = 7\frac{2}{6}$$ Check the proposed difference.
$$\frac{+1\frac{5}{6} = 1\frac{5}{6}}{8\frac{7}{6} = 8 + 1\frac{1}{6}}$$
$$= 9\frac{1}{6} \longleftarrow 7\frac{1}{3} \text{ checks}$$

Caution: To avoid making an error when subtracting a fraction from a mixed number, you should always use vertical subtraction form.

EXAMPLE 7-13: Subtract the following fraction from a mixed number: $5\frac{1}{2} - \frac{3}{4}$

too small   more fourths
↓   ↓

Solution: $5\frac{1}{2} = 5\frac{2}{4} = 4\frac{6}{4}$ *Think:* $5\frac{2}{4} = 5 + \frac{2}{4} = 4 + 1 + \frac{2}{4} = 4 + \frac{4}{4} + \frac{2}{4} = 4 + \frac{6}{4} = 4\frac{6}{4}$

$$\frac{-\frac{3}{4} = \frac{3}{4} = \frac{3}{4}}{4\frac{3}{4}} \longleftarrow \text{ proposed difference}$$

$$4\frac{3}{4}$$ Check the proposed difference.
$$\frac{+\frac{3}{4}}{4\frac{6}{4} = 4 + \frac{3}{2}}$$
$$= 4 + 1\frac{1}{2}$$
$$= 5\frac{1}{2} \longleftarrow 4\frac{3}{4} \text{ checks}$$

7-5. Solving Word Problems Containing Fractions and Mixed Numbers (+, −, ×, ÷)

A. Solve word problems containing fractions (+, −, ×, ÷).

To solve addition word problems containing fractions, you find how much two or more fractional parts are all together.

To solve subtraction word problems containing fractions, you find
(a) how much is left;
(b) how much more is needed;
(c) the difference between two fractional parts.

To solve multiplication word problems containing fractions, you find
(a) a fractional part of a given amount;
(b) how much two or more equal fractional parts are all together;
(c) the product of one fractional part times another fractional part.

To solve division word problems containing fractions, you find
(a) an amount given a fractional part of it;
(b) how many equal fractional parts there are in all;
(c) how much is in each equal fractional part;
(d) how many times larger one fractional part is than another fractional part.

EXAMPLE 7-14: Solve the following word problem containing fractions:

1. *Identify:* Lillian used $\frac{1}{2}$ of the paint. ⟶ Circle each fact.
George used $\frac{1}{4}$ of the paint.
(a) How much of the paint did they ⟶ Underline the questions.
use together? **(b)** How much
of the paint was left?

2. *Understand:* **(a)** The first part of the problem asks you to join $\frac{1}{2}$ and $\frac{1}{4}$ together.

(b) The second part of the problem asks you to subtract the sum of $\frac{1}{4}$ and $\frac{1}{2}$ from the total amount of paint.
To take a fractional part away, you **subtract.**

3. *Decide:* To join fractional parts together, you **add.**

4. *Compute:* $\frac{1}{2} + \frac{1}{4} = \frac{2}{4} + \frac{1}{4} = \frac{3}{4}$
$1 - \frac{3}{4} = \frac{4}{4} - \frac{3}{4} = \frac{1}{4}$

5. *Interpret:* $\frac{3}{4}$ means they used $\frac{3}{4}$ of the paint.
$\frac{1}{4}$ means $\frac{1}{4}$ of the paint is left.

6. *Check:* Is $\frac{3}{4}$ of the paint (the amount used) and $\frac{1}{4}$ of the paint (the amount left) equal to all of the paint? Yes: $\frac{3}{4} + \frac{1}{4} = \frac{4}{4} = 1$ ⟵ the whole amount of the paint

B. Solve word problems containing mixed numbers (+, −, ×, ÷).

To solve a word problem containing mixed numbers, it is usually helpful to first solve a similar problem containing easier whole numbers.

EXAMPLE 7-15: Solve the following word problem containing mixed numbers:

1. *Identify:* Margaret needs $3\frac{1}{3}$ yards of material ⟶ Circle each fact.
for each curtain panel. She needs
$3\frac{3}{4}$ panels in all. How much material ⟵ Underline the question.
should Margaret buy?

2. *Think:* Margaret needs **3** yards of material for each curtain panel. She needs **4** panels in all. How much material should Margaret buy?

Use a similar problem containing easier whole numbers.

3. *Understand:*   The question asks you to join 4 equal amounts of 3 together.

4. *Decide:*   To join equal amounts together, you **multiply.**

5. *Compute
using whole
numbers:* $4 \times 3 = 12$ ⟵ approximate answer

6. *Compute
using mixed
numbers:* $3\frac{3}{4} \times 3\frac{1}{3} = \frac{15}{4} \times \frac{10}{3}$

$$= \frac{\cancel{3} \times 5}{\cancel{2} \times 2} \times \frac{\cancel{2} \times 5}{\cancel{3}}$$

Replace the easier whole numbers with the original mixed numbers and compute as before.

$$= \frac{5}{2} \times \frac{5}{1}$$

$$= \frac{25}{2}$$

$$= 12\frac{1}{2}$$ ⟵ exact answer

7. *Interpret:* $12\frac{1}{2}$ means that $3\frac{3}{4}$ panels at $3\frac{1}{3}$ yards of material each requires **$12\frac{1}{2}$ yards of material.**

RAISE YOUR GRADES

Can you . . . ?

☑ subtract like fractions using the Subtraction Rule for Like Fractions
☑ check a proposed difference using addition
☑ translate subtraction of like fractions into words and pictures
☑ simplify a fractional difference when possible
☑ subtract unlike fractions using the Subtraction Rule for Unlike Fractions
☑ subtract a whole number from a mixed number
☑ subtract a fraction from a mixed number
☑ subtract two mixed numbers
☑ rename to subtract a fraction or mixed number from a whole number
☑ rename to subtract a fraction or mixed number from a mixed number when the minuend of the fraction part is too small
☑ solve word problems containing fractions
☑ solve word problems containing mixed numbers

SUMMARY

1. Subtraction is used to take one fractional amount away from another fractional amount.
2. To subtract like fractions, you subtract the numerators and then write the same denominator:
Subtraction Rule for Like Fractions
If $\frac{a}{c}$ and $\frac{b}{c}$ are any two like fractions, then $\frac{a}{c} - \frac{b}{c} = \frac{a - b}{c}$.
3. To check a subtraction answer, you add the proposed difference to the original subtrahend to see if you get the original minuend.
4. Always simplify a fractional difference when possible.
5. To subtract unlike fractions, you can use the following method:
Subtraction Rules for Unlike Fractions:
(a) Find the LCD for the unlike fractions.
(b) Build up to get like fractions using the LCD from Step (a).
(c) Substitute the like fractions from Step (b) for the original unlike fractions.
(d) Subtract the like fractions from Step (c).

(e) Simplify the difference from Step **(d)** when possible.

(f) Check by adding the simplest form of the proposed difference from Step **(d)** or **(e)** to the original subtrahend to see if you get the original minuend.

6. To subtract whole numbers, mixed numbers, or fractions, you should use vertical subtraction form to avoid making an error.

7. To subtract a whole number from a mixed number, you subtract the whole number and then write the same fraction part.

8. To subtract a fraction from a mixed number, you subtract the fractions and then write the same whole-number part.

9. To subtract two mixed numbers, you subtract the fraction parts, and then subtract the whole-number parts.

10. To subtract a fraction or a mixed number from a whole number, you first rename the whole number as an equal mixed number so that the fraction parts are like fractions.

11. To subtract a fraction or a mixed number from another mixed number whose minuend is too small, you first rename the minuend of the fraction part to get an equal mixed number with a larger fraction part.

12. To solve addition word problems containing fractions, you find how much two or more fractional parts are all together.

13. To solve subtraction word problems containing fractions, you find
    (a) how much is left;
    (b) how much more is needed;
    (c) the difference between two fractional parts.

14. To solve multiplication problems containing fractions, you find
    (a) a fractional part of a given amount;
    (b) how much two or more equal fractional parts are all together;
    (c) the product of one fractional part times another fractional part.

15. To solve division word problems containing fractions, you find
    (a) an amount given a fractional part of it;
    (b) how many equal fractional parts there are in all;
    (c) how much is in each equal fractional part;
    (d) how many times larger one fractional part is than another fractional part.

16. To solve a word problem containing mixed numbers, it is usually helpful to first solve a similar problem containing easier whole numbers.

SOLVED PROBLEMS

PROBLEM 7-1   Subtract like fractions and check: **(a)** $\frac{2}{3} - \frac{1}{3}$ **(b)** $\frac{7}{8} - \frac{1}{8}$ **(c)** $\frac{1}{2} - \frac{1}{2}$
(d) $\frac{9}{10} - \frac{3}{10}$ **(e)** $\frac{3}{4} - \frac{1}{4}$ **(f)** $\frac{11}{12} - \frac{1}{12}$ **(g)** $\frac{4}{5} - \frac{3}{5}$ **(h)** $\frac{8}{9} - \frac{4}{9}$ **(i)** $\frac{5}{6} - \frac{1}{6}$ **(j)** $\frac{6}{7} - \frac{3}{7}$
(k) $\frac{13}{18} - \frac{11}{18}$ **(l)** $\frac{19}{20} - \frac{3}{20}$ **(m)** $\frac{43}{50} - \frac{7}{50}$ **(n)** $\frac{24}{25} - \frac{1}{25}$ **(o)** $\frac{69}{75} - \frac{7}{75}$ **(p)** $\frac{9}{16} - \frac{5}{16}$
(q) $\frac{29}{32} - \frac{1}{32}$ **(r)** $\frac{53}{64} - \frac{9}{64}$

Solution: Recall that to subtract like fractions, you subtract the numerators and then write the same denominator [see Examples 7-2, 7-3, and 7-5]:

(a) $\dfrac{2}{3} - \dfrac{1}{3} = \dfrac{2-1}{3} = \dfrac{1}{3}$ *Check:* $\dfrac{1}{3} + \dfrac{1}{3} = \dfrac{2}{3}$

(b) $\dfrac{7}{8} - \dfrac{1}{8} = \dfrac{7-1}{8} = \dfrac{6}{8} = \dfrac{3}{4}$ *Check:* $\dfrac{3}{4} + \dfrac{1}{8} = \dfrac{6}{8} + \dfrac{1}{8} = \dfrac{7}{8}$

(c) $\dfrac{1}{2} - \dfrac{1}{2} = \dfrac{1-1}{2} = \dfrac{0}{2} = 0$ *Check:* $0 + \dfrac{1}{2} = \dfrac{1}{2}$

(d) $\dfrac{9}{10} - \dfrac{3}{10} = \dfrac{9-3}{10} = \dfrac{6}{10} = \dfrac{3}{5}$ *Check:* $\dfrac{3}{5} + \dfrac{3}{10} = \dfrac{6}{10} + \dfrac{3}{10} = \dfrac{9}{10}$

(e) $\dfrac{3}{4} - \dfrac{1}{4} = \dfrac{3-1}{4} = \dfrac{2}{4} = \dfrac{1}{2}$ *Check:* $\dfrac{1}{2} + \dfrac{1}{4} = \dfrac{2}{4} + \dfrac{1}{4} = \dfrac{3}{4}$

(f) $\dfrac{11}{12} - \dfrac{1}{12} = \dfrac{11-1}{12} = \dfrac{10}{12} = \dfrac{5}{6}$ *Check:* $\dfrac{5}{6} + \dfrac{1}{12} = \dfrac{10}{12} + \dfrac{1}{12} = \dfrac{11}{12}$

(g) $\dfrac{4}{5} - \dfrac{3}{5} = \dfrac{4-3}{5} = \dfrac{1}{5}$ *Check:* $\dfrac{1}{5} + \dfrac{3}{5} = \dfrac{4}{5}$

(h) $\dfrac{8}{9} - \dfrac{4}{9} = \dfrac{8-4}{9} = \dfrac{4}{9}$ *Check:* $\dfrac{4}{9} + \dfrac{4}{9} = \dfrac{8}{9}$

(i) $\dfrac{5}{6} - \dfrac{1}{6} = \dfrac{5-1}{6} = \dfrac{4}{6} = \dfrac{2}{3}$ *Check:* $\dfrac{2}{3} + \dfrac{1}{6} = \dfrac{4}{6} + \dfrac{1}{6} = \dfrac{5}{6}$

(j) $\dfrac{6}{7} - \dfrac{3}{7} = \dfrac{6-3}{7} = \dfrac{3}{7}$ *Check:* $\dfrac{3}{7} + \dfrac{3}{7} = \dfrac{6}{7}$

(k) $\dfrac{13}{18} - \dfrac{11}{18} = \dfrac{13-11}{18} = \dfrac{2}{18} = \dfrac{1}{9}$ *Check:* $\dfrac{1}{9} + \dfrac{11}{18} = \dfrac{2}{18} + \dfrac{11}{18} = \dfrac{13}{18}$

(l) $\dfrac{19}{20} - \dfrac{3}{20} = \dfrac{19-3}{20} = \dfrac{16}{20} = \dfrac{4}{5}$ *Check:* $\dfrac{4}{5} + \dfrac{3}{20} = \dfrac{16}{20} + \dfrac{3}{20} = \dfrac{19}{20}$

(m) $\dfrac{43}{50} - \dfrac{7}{50} = \dfrac{43-7}{50} = \dfrac{36}{50} = \dfrac{18}{25}$ *Check:* $\dfrac{18}{25} + \dfrac{7}{50} = \dfrac{36}{50} + \dfrac{7}{50} = \dfrac{43}{50}$

(n) $\dfrac{24}{25} - \dfrac{1}{25} = \dfrac{24-1}{25} = \dfrac{23}{25}$ *Check:* $\dfrac{23}{25} + \dfrac{1}{25} = \dfrac{24}{25}$

(o) $\dfrac{69}{75} - \dfrac{7}{75} = \dfrac{69-7}{75} = \dfrac{62}{75}$ *Check:* $\dfrac{62}{75} + \dfrac{7}{75} = \dfrac{69}{75}$

(p) $\dfrac{9}{16} - \dfrac{5}{16} = \dfrac{9-5}{16} = \dfrac{4}{16} = \dfrac{1}{4}$ *Check:* $\dfrac{1}{4} + \dfrac{5}{16} = \dfrac{4}{16} + \dfrac{5}{16} = \dfrac{9}{16}$

(q) $\dfrac{29}{32} - \dfrac{1}{32} = \dfrac{29-1}{32} = \dfrac{28}{32} = \dfrac{7}{8}$ *Check:* $\dfrac{7}{8} + \dfrac{1}{32} = \dfrac{28}{32} + \dfrac{1}{32} = \dfrac{29}{32}$

(r) $\dfrac{53}{64} - \dfrac{9}{64} = \dfrac{53-9}{64} = \dfrac{44}{64} = \dfrac{11}{16}$ *Check:* $\dfrac{11}{16} + \dfrac{9}{64} = \dfrac{44}{64} + \dfrac{9}{64} = \dfrac{53}{64}$

PROBLEM 7-2 Subtract unlike fractions and check: (a) $\frac{3}{4} - \frac{1}{2}$ (b) $\frac{1}{2} - \frac{1}{6}$ (c) $\frac{5}{8} - \frac{1}{4}$

(d) $\frac{4}{5} - \frac{3}{10}$ (e) $\frac{1}{2} - \frac{1}{3}$ (f) $\frac{3}{4} - \frac{1}{3}$ (g) $\frac{3}{5} - \frac{1}{2}$ (h) $\frac{1}{3} - \frac{1}{5}$ (i) $\frac{1}{4} - \frac{1}{6}$ (j) $\frac{5}{6} - \frac{5}{8}$

(k) $\frac{9}{10} - \frac{7}{12}$ (l) $\frac{11}{12} - \frac{1}{9}$ (m) $\frac{1}{4} - \frac{1}{10}$ (n) $\frac{7}{10} - \frac{1}{6}$ (o) $\frac{9}{10} - \frac{5}{14}$ (p) $\frac{5}{12} - \frac{3}{20}$

(q) $\frac{1}{12} - \frac{1}{18}$ (r) $\frac{24}{25} - \frac{3}{20}$

Solution: Recall that to subtract unlike fractions, first rename as equivalent like fractions using the LCD and then subtract the like fractions [see Example 7-6]:

(a) The LCD for $\frac{3}{4}$ and $\frac{1}{2}$ is 4: $\frac{3}{4} - \frac{1}{2} = \frac{3}{4} - \frac{2}{4} = \frac{1}{4}$ *Check:* $\frac{1}{4} + \frac{1}{2} = \frac{1}{4} + \frac{2}{4} = \frac{3}{4}$

(b) The LCD for $\frac{1}{2}$ and $\frac{1}{6}$ is 6: $\frac{1}{2} - \frac{1}{6} = \frac{3}{6} - \frac{1}{6} = \frac{2}{6} = \frac{1}{3}$ *Check:* $\frac{1}{3} + \frac{1}{6} = \frac{2}{6} + \frac{1}{6} = \frac{3}{6} = \frac{1}{2}$

(c) The LCD for $\frac{5}{8}$ and $\frac{1}{4}$ is 8: $\frac{5}{8} - \frac{1}{4} = \frac{5}{8} - \frac{2}{8} = \frac{3}{8}$ *Check:* $\frac{3}{8} + \frac{1}{4} = \frac{3}{8} + \frac{2}{8} = \frac{5}{8}$

(d) The LCD for $\frac{4}{5}$ and $\frac{3}{10}$ is 10: $\frac{4}{5} - \frac{3}{10} = \frac{8}{10} - \frac{3}{10} = \frac{5}{10} = \frac{1}{2}$ *Check:* $\frac{1}{2} + \frac{3}{10} = \frac{5}{10} + \frac{3}{10} = \frac{8}{10} = \frac{4}{5}$

(e) The LCD for $\frac{1}{2}$ and $\frac{1}{3}$ is 6: $\frac{1}{2} - \frac{1}{3} = \frac{3}{6} - \frac{2}{6} = \frac{1}{6}$ *Check:* $\frac{1}{6} + \frac{1}{3} = \frac{1}{6} + \frac{2}{6} = \frac{3}{6} = \frac{1}{2}$

(f) The LCD for $\frac{3}{4}$ and $\frac{1}{3}$ is 12: $\frac{3}{4} - \frac{1}{3} = \frac{9}{12} - \frac{4}{12} = \frac{5}{12}$ *Check:* $\frac{5}{12} + \frac{1}{3} = \frac{5}{12} + \frac{4}{12} = \frac{9}{12} = \frac{3}{4}$

(g) The LCD for $\frac{3}{5}$ and $\frac{1}{2}$ is 10: $\frac{3}{5} - \frac{1}{2} = \frac{6}{10} - \frac{5}{10} = \frac{1}{10}$ *Check:* $\frac{1}{10} + \frac{1}{2} = \frac{1}{10} + \frac{5}{10} = \frac{6}{10} = \frac{3}{5}$

(h) The LCD for $\frac{1}{3}$ and $\frac{1}{5}$ is 15: $\frac{1}{3} - \frac{1}{5} = \frac{5}{15} - \frac{3}{15} = \frac{2}{15}$ *Check:* $\frac{2}{15} + \frac{1}{5} = \frac{2}{15} + \frac{3}{15} = \frac{5}{15} = \frac{1}{3}$

(i) The LCD for $\frac{1}{4}$ and $\frac{1}{6}$ is 12: $\frac{1}{4} - \frac{1}{6} = \frac{3}{12} - \frac{2}{12} = \frac{1}{12}$ *Check:* $\frac{1}{12} + \frac{1}{6} = \frac{1}{12} + \frac{2}{12} = \frac{3}{12} = \frac{1}{4}$

(j) The LCD for $\frac{5}{6}$ and $\frac{5}{8}$ is 24: $\frac{5}{6} - \frac{5}{8} = \frac{20}{24} - \frac{15}{24} = \frac{5}{24}$ *Check:* $\frac{5}{24} + \frac{5}{8} = \frac{5}{24} + \frac{15}{24} = \frac{20}{24} = \frac{5}{6}$

(k) The LCD for $\frac{9}{10}$ and $\frac{7}{12}$ is 60: $\frac{9}{10} - \frac{7}{12} = \frac{54}{60} - \frac{35}{60} = \frac{19}{60}$ *Check:* $\frac{19}{60} + \frac{7}{12} = \frac{19}{60} + \frac{35}{60} = \frac{54}{60} = \frac{9}{10}$

(l) The LCD for $\frac{11}{12}$ and $\frac{1}{9}$ is 36: $\frac{11}{12} - \frac{1}{9} = \frac{33}{36} - \frac{4}{36} = \frac{29}{36}$ *Check:* $\frac{29}{36} + \frac{1}{9} = \frac{29}{36} + \frac{4}{36} = \frac{33}{36} = \frac{11}{12}$

(m) The LCD for $\frac{1}{4}$ and $\frac{1}{10}$ is 20: $\frac{1}{4} - \frac{1}{10} = \frac{5}{20} - \frac{2}{20} = \frac{3}{20}$ *Check:* $\frac{3}{20} + \frac{1}{10} = \frac{3}{20} + \frac{2}{20} = \frac{5}{20} = \frac{1}{4}$

(n) The LCD for $\frac{7}{10}$ and $\frac{1}{6}$ is 30: $\frac{7}{10} - \frac{1}{6} = \frac{21}{30} - \frac{5}{30} = \frac{16}{30} = \frac{8}{15}$ *Check:* $\frac{8}{15} + \frac{1}{6} = \frac{16}{30} + \frac{5}{30} = \frac{7}{10}$

(o) The LCD for $\frac{9}{10}$ and $\frac{5}{14}$ is 70: $\frac{9}{10} - \frac{5}{14} = \frac{63}{70} - \frac{25}{70} = \frac{38}{70} = \frac{19}{35}$ *Check:* $\frac{19}{35} + \frac{5}{14} = \frac{38}{70} + \frac{25}{70} = \frac{9}{10}$

(p) The LCD for $\frac{5}{12}$ and $\frac{3}{20}$ is 60: $\frac{5}{12} - \frac{3}{20} = \frac{25}{60} - \frac{9}{60} = \frac{16}{60} = \frac{4}{15}$ *Check:* $\frac{4}{15} + \frac{3}{20} = \frac{16}{60} + \frac{9}{60} = \frac{5}{12}$

(q) The LCD for $\frac{1}{12}$ and $\frac{1}{18}$ is 36: $\frac{1}{12} - \frac{1}{18} = \frac{3}{36} - \frac{2}{36} = \frac{1}{36}$ *Check:* $\frac{1}{36} + \frac{1}{18} = \frac{1}{36} + \frac{2}{36} = \frac{3}{36} = \frac{1}{12}$

(r) The LCD for $\frac{24}{25}$ and $\frac{3}{20}$ is 100: $\frac{24}{25} - \frac{3}{20} = \frac{96}{100} - \frac{15}{100} = \frac{81}{100}$ *Check:* $\frac{81}{100} + \frac{3}{20} = \frac{81}{100} + \frac{15}{100} = \frac{24}{25}$

PROBLEM 7-3 Subtract whole numbers from mixed numbers: **(a)** $7\frac{3}{8} - 2$ **(b)** $8\frac{1}{4} - 1$
(c) $4\frac{1}{2} - 3$ **(d)** $6\frac{2}{3} - 5$ **(e)** $9\frac{1}{5} - 4$ **(f)** $1\frac{1}{6} - 1$ **(g)** $2\frac{4}{9} - 1$ **(h)** $3\frac{3}{10} - 1$
(i) $8\frac{11}{12} - 5$ **(j)** $7\frac{3}{4} - 6$ **(k)** $9\frac{1}{3} - 9$ **(l)** $6\frac{5}{6} - 1$ **(m)** $11\frac{2}{5} - 2$ **(n)** $23\frac{7}{10} - 15$
(o) $5\frac{2}{9} - 1$ **(p)** $15\frac{5}{12} - 11$ **(q)** $10\frac{1}{10} - 1$ **(r)** $8\frac{4}{5} - 8$

Solution: Recall that to subtract a whole number from a mixed number, you subtract the whole numbers and then write the fraction part [see Example 7-7]:

(a) $\begin{array}{r} 7\frac{3}{8} \\ -2 \\ \hline 5\frac{3}{8} \end{array}$ **(b)** $\begin{array}{r} 8\frac{1}{4} \\ -1 \\ \hline 7\frac{1}{4} \end{array}$ **(c)** $\begin{array}{r} 4\frac{1}{2} \\ -3 \\ \hline 1\frac{1}{2} \end{array}$ **(d)** $\begin{array}{r} 6\frac{2}{3} \\ -5 \\ \hline 1\frac{2}{3} \end{array}$ **(e)** $\begin{array}{r} 9\frac{1}{5} \\ -4 \\ \hline 5\frac{1}{5} \end{array}$ **(f)** $\begin{array}{r} 1\frac{1}{6} \\ -1 \\ \hline \frac{1}{6} \end{array}$ **(g)** $\begin{array}{r} 2\frac{4}{9} \\ -1 \\ \hline 1\frac{4}{9} \end{array}$ **(h)** $\begin{array}{r} 3\frac{3}{10} \\ -1 \\ \hline 2\frac{3}{10} \end{array}$ **(i)** $\begin{array}{r} 8\frac{11}{12} \\ -5 \\ \hline 3\frac{11}{12} \end{array}$

(j) $\begin{array}{r} 7\frac{3}{4} \\ -6 \\ \hline 1\frac{3}{4} \end{array}$ **(k)** $\begin{array}{r} 9\frac{1}{3} \\ -9 \\ \hline \frac{1}{3} \end{array}$ **(l)** $\begin{array}{r} 6\frac{5}{6} \\ -1 \\ \hline 5\frac{5}{6} \end{array}$ **(m)** $\begin{array}{r} 11\frac{2}{5} \\ -2 \\ \hline 9\frac{2}{5} \end{array}$ **(n)** $\begin{array}{r} 23\frac{7}{10} \\ -15 \\ \hline 8\frac{7}{10} \end{array}$ **(o)** $\begin{array}{r} 5\frac{2}{9} \\ -1 \\ \hline 4\frac{2}{9} \end{array}$ **(p)** $\begin{array}{r} 15\frac{5}{12} \\ -11 \\ \hline 4\frac{5}{12} \end{array}$ **(q)** $\begin{array}{r} 10\frac{1}{10} \\ -1 \\ \hline 9\frac{1}{10} \end{array}$ **(r)** $\begin{array}{r} 8\frac{4}{5} \\ -8 \\ \hline \frac{4}{5} \end{array}$

PROBLEM 7-4 Subtract fractions from mixed numbers: **(a)** $7\frac{5}{6} - \frac{1}{6}$ **(b)** $2\frac{7}{8} - \frac{1}{8}$ **(c)** $9\frac{3}{4} - \frac{1}{4}$
(d) $1\frac{9}{10} - \frac{3}{10}$ **(e)** $8\frac{2}{3} - \frac{1}{3}$ **(f)** $3\frac{11}{12} - \frac{1}{12}$ **(g)** $5\frac{1}{2} - \frac{1}{2}$ **(h)** $4\frac{4}{5} - \frac{1}{5}$ **(i)** $6\frac{1}{2} - \frac{3}{8}$
(j) $12\frac{9}{10} - \frac{1}{2}$ **(k)** $5\frac{5}{6} - \frac{1}{5}$ **(l)** $11\frac{3}{4} - \frac{2}{5}$ **(m)** $5\frac{7}{10} - \frac{2}{15}$ **(n)** $5\frac{13}{15} - \frac{5}{12}$ **(o)** $3\frac{11}{12} - \frac{7}{16}$
(p) $10\frac{1}{10} - \frac{1}{16}$ **(q)** $7\frac{11}{16} - \frac{3}{20}$ **(r)** $8\frac{17}{18} - \frac{7}{20}$

Solution: Recall that to subtract a fraction from a mixed number, you first write the numbers in vertical form, then subtract fractions, and then write the same whole-number part [see Example 7-8]:

(a) $\begin{array}{r} 7\frac{5}{6} \\ -\frac{1}{6} \\ \hline 7\frac{4}{6} = 7\frac{2}{3} \end{array}$ **(b)** $\begin{array}{r} 2\frac{7}{8} \\ -\frac{1}{8} \\ \hline 2\frac{6}{8} = 2\frac{3}{4} \end{array}$ **(c)** $\begin{array}{r} 9\frac{3}{4} \\ -\frac{1}{4} \\ \hline 9\frac{2}{4} = 9\frac{1}{2} \end{array}$ **(d)** $\begin{array}{r} 1\frac{9}{10} \\ -\frac{3}{10} \\ \hline 1\frac{6}{10} = 1\frac{3}{5} \end{array}$ **(e)** $\begin{array}{r} 8\frac{2}{3} \\ -\frac{1}{3} \\ \hline 8\frac{1}{3} \end{array}$ **(f)** $\begin{array}{r} 3\frac{11}{12} \\ -\frac{1}{12} \\ \hline 3\frac{10}{12} = 3\frac{5}{6} \end{array}$

(g) $\begin{array}{r} 5\frac{1}{1} \\ -\frac{1}{2} \\ \hline 5 \end{array}$ **(h)** $\begin{array}{r} 4\frac{4}{5} \\ -\frac{1}{5} \\ \hline 4\frac{3}{5} \end{array}$ **(i)** $\begin{array}{r} 6\frac{1}{2} = 6\frac{4}{8} \\ -\frac{3}{8} = \frac{3}{8} \\ \hline 6\frac{1}{8} \end{array}$ **(j)** $\begin{array}{r} 12\frac{9}{10} = 12\frac{9}{10} \\ -\frac{1}{2} = \frac{5}{10} \\ \hline 12\frac{4}{10} = 12\frac{2}{5} \end{array}$ **(k)** $\begin{array}{r} 5\frac{5}{6} = 5\frac{25}{30} \\ -\frac{1}{5} = \frac{6}{30} \\ \hline 5\frac{19}{30} \end{array}$

(l) $\begin{array}{r} 11\frac{3}{4} = 11\frac{15}{20} \\ -\frac{2}{5} = \frac{8}{20} \\ \hline 11\frac{7}{20} \end{array}$ **(m)** $\begin{array}{r} 5\frac{7}{10} = 5\frac{21}{30} \\ -\frac{2}{15} = \frac{4}{30} \\ \hline 5\frac{17}{30} \end{array}$ **(n)** $\begin{array}{r} 5\frac{13}{15} = 5\frac{52}{60} \\ -\frac{5}{12} = \frac{25}{60} \\ \hline 5\frac{27}{60} = 5\frac{9}{20} \end{array}$ **(o)** $\begin{array}{r} 3\frac{11}{12} = 3\frac{44}{48} \\ -\frac{7}{16} = \frac{21}{48} \\ \hline 3\frac{23}{48} \end{array}$

(p) $\begin{array}{r} 10\frac{1}{10} = 10\frac{8}{80} \\ -\frac{1}{16} = \frac{5}{80} \\ \hline 10\frac{3}{80} \end{array}$ **(q)** $\begin{array}{r} 7\frac{11}{16} = 7\frac{55}{80} \\ -\frac{3}{20} = \frac{12}{80} \\ \hline 7\frac{43}{80} \end{array}$ **(r)** $\begin{array}{r} 8\frac{17}{18} = 8\frac{170}{180} \\ -\frac{7}{20} = \frac{63}{180} \\ \hline 8\frac{107}{180} \end{array}$

PROBLEM 7-5 Subtract two mixed numbers: **(a)** $9\frac{3}{4} - 5\frac{1}{4}$ **(b)** $2\frac{2}{3} - 1\frac{1}{3}$ **(c)** $5\frac{11}{12} - 5\frac{7}{12}$
(d) $8\frac{1}{2} - 7\frac{1}{2}$ **(e)** $6\frac{5}{6} - 1\frac{1}{6}$ **(f)** $4\frac{3}{8} - 3\frac{1}{8}$ **(g)** $8\frac{7}{10} - 8\frac{3}{10}$ **(h)** $7\frac{2}{5} - 2\frac{1}{5}$ **(i)** $6\frac{1}{2} - 4\frac{1}{4}$
(j) $10\frac{11}{12} - 2\frac{1}{3}$ **(k)** $12\frac{2}{3} - 1\frac{1}{2}$ **(l)** $20\frac{3}{4} - 5\frac{1}{3}$ **(m)** $6\frac{1}{4} - 3\frac{1}{6}$ **(n)** $1\frac{9}{10} - 1\frac{1}{8}$
(o) $12\frac{1}{4} - 8\frac{1}{10}$ **(p)** $15\frac{8}{9} - 3\frac{1}{12}$ **(q)** $10\frac{5}{16} - 9\frac{7}{32}$ **(r)** $1\frac{11}{16} - 1\frac{3}{8}$

Solution: Recall that to subtract two mixed numbers, you first write the numbers in vertical form, then subtract the fraction parts, and then subtract the whole-number parts [see Example 7-9]:

(a) $9\frac{3}{4}$
$\underline{-5\frac{1}{4}}$
$4\frac{2}{4} = 4\frac{1}{2}$

(b) $2\frac{2}{3}$
$\underline{-1\frac{1}{3}}$
$1\frac{1}{3}$

(c) $5\frac{11}{12}$
$\underline{-5\frac{7}{12}}$
$\frac{4}{12} = \frac{1}{3}$

(d) $8\frac{1}{2}$
$\underline{-7\frac{1}{2}}$
1

(e) $6\frac{5}{6}$
$\underline{-1\frac{1}{6}}$
$5\frac{4}{6} = 5\frac{2}{3}$

(f) $4\frac{3}{8}$
$\underline{-3\frac{1}{8}}$
$1\frac{2}{8} = 1\frac{1}{4}$

(g) $8\frac{7}{10}$
$\underline{-8\frac{3}{10}}$
$\frac{4}{10} = \frac{2}{5}$

(h) $7\frac{2}{5}$
$\underline{-2\frac{1}{5}}$
$5\frac{1}{5}$

(i) $6\frac{1}{2} = 6\frac{2}{4}$
$\underline{-4\frac{1}{4} = 4\frac{1}{4}}$
$2\frac{1}{4}$

(j) $10\frac{11}{12} = 10\frac{11}{12}$
$\underline{- 2\frac{1}{3} = 2\frac{4}{12}}$
$8\frac{7}{12}$

(k) $12\frac{2}{3} = 12\frac{4}{6}$
$\underline{- 1\frac{1}{2} = 1\frac{3}{6}}$
$11\frac{1}{6}$

(l) $20\frac{3}{4} = 20\frac{9}{12}$
$\underline{- 5\frac{1}{3} = 5\frac{4}{12}}$
$15\frac{5}{12}$

(m) $6\frac{1}{4} = 6\frac{3}{12}$
$\underline{-3\frac{1}{6} = 3\frac{2}{12}}$
$3\frac{1}{12}$

(n) $1\frac{9}{10} = 1\frac{36}{40}$
$\underline{-1\frac{1}{8} = 1\frac{5}{40}}$
$\frac{31}{40}$

(o) $12\frac{1}{4} = 12\frac{5}{20}$
$\underline{- 8\frac{1}{10} = 8\frac{2}{20}}$
$4\frac{3}{20}$

(p) $15\frac{8}{9} = 15\frac{32}{36}$
$\underline{- 3\frac{1}{12} = 3\frac{3}{36}}$
$12\frac{29}{36}$

(q) $10\frac{5}{16} = 10\frac{10}{32}$
$\underline{- 9\frac{7}{32} = 9\frac{7}{32}}$
$1\frac{3}{32}$

(r) $1\frac{11}{16} = 1\frac{11}{16}$
$\underline{-1\frac{3}{8} = 1\frac{6}{16}}$
$\frac{5}{16}$

PROBLEM 7-6   Subtract mixed numbers from whole numbers: (a) $8 - 2\frac{1}{2}$ (b) $3 - 1\frac{3}{4}$
(c) $12 - 3\frac{7}{8}$ (d) $9 - 1\frac{1}{3}$ (e) $2 - 1\frac{5}{6}$ (f) $10 - 5\frac{4}{5}$ (g) $7 - 4\frac{3}{10}$ (h) $4 - 3\frac{1}{4}$
(i) $15 - 1\frac{1}{8}$ (j) $6 - 2\frac{2}{3}$ (k) $4 - 2\frac{1}{6}$ (l) $20 - 3\frac{1}{5}$ (m) $5 - 1\frac{7}{10}$ (n) $3 - 2\frac{11}{12}$
(o) $18 - 11\frac{3}{5}$ (p) $11 - 2\frac{7}{12}$ (q) $19 - 5\frac{7}{18}$ (r) $12 - 10\frac{3}{20}$

Solution: Recall that to subtract a mixed number from a whole number, you first rename the whole number as an equal mixed number so that the fraction parts are like fractions [see Example 7-10]:

(a) $8 = 7\frac{2}{2}$
$\underline{-2\frac{1}{2} = 2\frac{1}{2}}$
$5\frac{1}{2}$

(b) $3 = 2\frac{4}{4}$
$\underline{-1\frac{3}{4} = 1\frac{3}{4}}$
$1\frac{1}{4}$

(c) $12 = 11\frac{8}{8}$
$\underline{- 3\frac{7}{8} = 3\frac{7}{8}}$
$8\frac{1}{8}$

(d) $9 = 8\frac{3}{3}$
$\underline{-1\frac{1}{3} = 1\frac{1}{3}}$
$7\frac{2}{3}$

(e) $2 = 1\frac{6}{6}$
$\underline{-1\frac{5}{6} = 1\frac{5}{6}}$
$\frac{1}{6}$

(f) $10 = 9\frac{5}{5}$
$\underline{- 5\frac{4}{5} = 5\frac{4}{5}}$
$4\frac{1}{5}$

(g) $7 = 6\frac{10}{10}$
$\underline{-4\frac{3}{10} = 4\frac{3}{10}}$
$2\frac{7}{10}$

(h) $4 = 3\frac{4}{4}$
$\underline{-3\frac{1}{4} = 3\frac{1}{4}}$
$\frac{3}{4}$

(i) $15 = 14\frac{8}{8}$
$\underline{- 1\frac{1}{8} = 1\frac{1}{8}}$
$13\frac{7}{8}$

(j) $6 = 5\frac{3}{3}$
$\underline{-2\frac{2}{3} = 2\frac{2}{3}}$
$3\frac{1}{3}$

(k) $4 = 3\frac{6}{6}$
$\underline{-2\frac{1}{6} = 2\frac{1}{6}}$
$1\frac{5}{6}$

(l) $20 = 19\frac{5}{5}$
$\underline{- 3\frac{1}{5} = 3\frac{1}{5}}$
$16\frac{4}{5}$

(m) $5 = 4\frac{10}{10}$
$\underline{-1\frac{7}{10} = 1\frac{7}{10}}$
$3\frac{3}{10}$

(n) $3 = 2\frac{12}{12}$
$\underline{-2\frac{11}{12} = 2\frac{11}{12}}$
$\frac{1}{12}$

(o) $18 = 17\frac{5}{5}$
$\underline{-11\frac{3}{5} = 11\frac{3}{5}}$
$6\frac{2}{5}$

(p) $11 = 10\frac{12}{12}$
$\underline{- 2\frac{7}{12} = 2\frac{7}{12}}$
$8\frac{5}{12}$

(q) $19 = 18\frac{18}{18}$
$\underline{- 5\frac{7}{18} = 5\frac{7}{18}}$
$13\frac{11}{18}$

(r) $12 = 11\frac{20}{20}$
$\underline{-10\frac{3}{20} = 10\frac{3}{20}}$
$1\frac{17}{20}$

PROBLEM 7-7   Subtract fractions from whole numbers: (a) $2 - \frac{3}{8}$ (b) $10 - \frac{1}{5}$ (c) $7 - \frac{1}{10}$
(d) $1 - \frac{1}{12}$ (e) $12 - \frac{1}{2}$ (f) $9 - \frac{5}{8}$ (g) $3 - \frac{5}{12}$ (h) $13 - \frac{1}{3}$ (i) $8 - \frac{3}{4}$ (j) $4 - \frac{2}{3}$
(k) $15 - \frac{1}{6}$ (l) $7 - \frac{1}{8}$ (m) $5 - \frac{1}{5}$ (n) $16 - \frac{9}{16}$ (o) $6 - \frac{9}{10}$ (p) $17 - \frac{5}{32}$
(q) $20 - \frac{7}{8}$ (r) $1 - \frac{5}{6}$

Solution: Recall that to subtract a fraction from a whole number, you first rename the whole number as an equal mixed number so that the fractions are like fractions [see Example 7-11]:

(a) $2 = 1\frac{8}{8}$
$\underline{-\frac{3}{8} = \frac{3}{8}}$
$1\frac{5}{8}$

(b) $10 = 9\frac{5}{5}$
$\underline{- \frac{1}{5} = \frac{1}{5}}$
$9\frac{4}{5}$

(c) $7 = 6\frac{10}{10}$
$\underline{- \frac{1}{10} = \frac{1}{10}}$
$6\frac{9}{10}$

(d) $1 = \frac{12}{12}$
$\underline{-\frac{1}{12} = \frac{1}{12}}$
$\frac{11}{12}$

(e) $12 = 11\frac{2}{2}$
$\underline{- \frac{1}{2} = \frac{1}{2}}$
$11\frac{1}{2}$

(f) $9 = 8\frac{8}{8}$

$-\frac{5}{8} = \frac{5}{8}$

$\overline{\qquad 8\frac{3}{8}}$

(g) $3 = 2\frac{12}{12}$

$-\frac{5}{12} = \frac{5}{12}$

$\overline{\qquad 2\frac{7}{12}}$

(h) $13 = 12\frac{3}{3}$

$-\frac{1}{3} = \frac{1}{3}$

$\overline{\qquad 12\frac{2}{3}}$

(i) $8 = 7\frac{4}{4}$

$-\frac{3}{4} = \frac{3}{4}$

$\overline{\qquad 7\frac{1}{4}}$

(j) $4 = 3\frac{3}{3}$

$-\frac{2}{3} = \frac{2}{3}$

$\overline{\qquad 3\frac{1}{3}}$

(k) $15 = 14\frac{6}{6}$

$-\frac{1}{6} = \frac{1}{6}$

$\overline{\qquad 14\frac{5}{6}}$

(l) $7 = 6\frac{8}{8}$

$-\frac{1}{8} = \frac{1}{8}$

$\overline{\qquad 6\frac{7}{8}}$

(m) $5 = 4\frac{5}{5}$

$-\frac{1}{5} = \frac{1}{5}$

$\overline{\qquad 4\frac{4}{5}}$

(n) $16 = 15\frac{16}{16}$

$-\frac{9}{16} = \frac{9}{16}$

$\overline{\qquad 15\frac{7}{16}}$

(o) $6 = 5\frac{10}{10}$

$-\frac{9}{10} = \frac{9}{10}$

$\overline{\qquad 5\frac{1}{10}}$

(p) $17 = 16\frac{32}{32}$

$-\frac{5}{32} = \frac{5}{32}$

$\overline{\qquad 16\frac{27}{32}}$

(q) $20 = 19\frac{8}{8}$

$-\frac{7}{8} = \frac{7}{8}$

$\overline{\qquad 19\frac{1}{8}}$

(r) $1 = \frac{6}{6}$

$-\frac{5}{6} = \frac{5}{6}$

$\overline{\qquad \frac{1}{6}}$

PROBLEM 7-8   Subtract two mixed numbers when the fraction part of the minuend is too small:

(a) $8\frac{1}{3} - 3\frac{2}{3}$ **(b)** $9\frac{2}{5} - 2\frac{3}{5}$ **(c)** $7\frac{1}{8} - 4\frac{7}{8}$ **(d)** $10\frac{1}{4} - 1\frac{3}{4}$ **(e)** $6\frac{1}{6} - 5\frac{5}{6}$ **(f)** $8\frac{3}{10} - 2\frac{7}{10}$
(g) $9\frac{1}{12} - 1\frac{11}{12}$ **(h)** $7\frac{4}{9} - 3\frac{5}{9}$ **(i)** $11\frac{1}{2} - 1\frac{1}{4}$ **(j)** $6\frac{1}{5} - 4\frac{3}{10}$ **(k)** $7\frac{1}{2} - 2\frac{2}{3}$ **(l)** $8\frac{2}{5} - 1\frac{1}{2}$
(m) $6\frac{1}{8} - 3\frac{5}{12}$ **(n)** $9\frac{1}{4} - 3\frac{5}{6}$ **(o)** $5\frac{7}{10} - 4\frac{13}{16}$ **(p)** $12\frac{3}{4} - 5\frac{9}{10}$ **(q)** $10\frac{3}{16} - 2\frac{9}{32}$
(r) $2\frac{1}{8} - 1\frac{3}{16}$

Solution: Recall that to subtract two mixed numbers when the minuend of the fraction part is too small, you first rename the minuend to get an equal mixed number with a larger fraction part [see Example 7-12]:

(a) $8\frac{1}{3} = 7\frac{4}{3}$

$-3\frac{2}{3} = 3\frac{2}{3}$

$\overline{\qquad 4\frac{2}{3}}$

(b) $9\frac{2}{5} = 8\frac{7}{5}$

$-2\frac{3}{5} = 2\frac{3}{5}$

$\overline{\qquad 6\frac{4}{5}}$

(c) $7\frac{1}{8} = 6\frac{9}{8}$

$-4\frac{7}{8} = 4\frac{7}{8}$

$\overline{\qquad 2\frac{2}{8} = 2\frac{1}{4}}$

(d) $10\frac{1}{4} = 9\frac{5}{4}$

$- 1\frac{3}{4} = 1\frac{3}{4}$

$\overline{\qquad 8\frac{2}{4} = 8\frac{1}{2}}$

(e) $6\frac{1}{6} = 5\frac{7}{6}$

$-5\frac{5}{6} = 5\frac{5}{6}$

$\overline{\qquad \frac{2}{6} = \frac{1}{3}}$

(f) $8\frac{3}{10} = 7\frac{13}{10}$

$-2\frac{7}{10} = 2\frac{7}{10}$

$\overline{\qquad 5\frac{6}{10} = 5\frac{3}{5}}$

(g) $9\frac{1}{12} = 8\frac{13}{12}$

$-1\frac{11}{12} = 1\frac{11}{12}$

$\overline{\qquad 7\frac{2}{12} = 7\frac{1}{6}}$

(h) $7\frac{4}{9} = 6\frac{13}{9}$

$-3\frac{5}{9} = 3\frac{5}{9}$

$\overline{\qquad 3\frac{8}{9}}$

(i) $11\frac{1}{2} = 11\frac{2}{4} = 10\frac{6}{4}$

$- 1\frac{3}{4} = 1\frac{3}{4} = 1\frac{3}{4}$

$\overline{\qquad 9\frac{3}{4}}$

(j) $6\frac{1}{5} = 6\frac{2}{10} = 5\frac{12}{10}$

$-4\frac{3}{10} = 4\frac{3}{10} = 4\frac{3}{10}$

$\overline{\qquad 1\frac{9}{10}}$

(k) $7\frac{1}{2} = 7\frac{3}{6} = 6\frac{9}{6}$

$-2\frac{2}{3} = 2\frac{4}{6} = 2\frac{4}{6}$

$\overline{\qquad 4\frac{5}{6}}$

(l) $8\frac{2}{5} = 8\frac{4}{10} = 7\frac{14}{10}$

$-1\frac{1}{2} = 1\frac{5}{10} = 1\frac{5}{10}$

$\overline{\qquad 6\frac{9}{10}}$

(m) $6\frac{1}{8} = 6\frac{3}{24} = 5\frac{27}{24}$

$-3\frac{5}{12} = 3\frac{10}{24} = 3\frac{10}{24}$

$\overline{\qquad 2\frac{17}{24}}$

(n) $9\frac{1}{4} = 9\frac{3}{12} = 8\frac{15}{12}$

$-3\frac{5}{6} = 3\frac{10}{12} = 3\frac{10}{12}$

$\overline{\qquad 5\frac{5}{12}}$

(o) $5\frac{7}{10} = 5\frac{56}{80} = 4\frac{136}{80}$

$-4\frac{13}{16} = 4\frac{65}{80} = 4\frac{65}{80}$

$\overline{\qquad \frac{71}{80}}$

(p) $12\frac{3}{4} = 12\frac{15}{20} = 11\frac{35}{20}$

$- 5\frac{9}{10} = 5\frac{18}{20} = 5\frac{18}{20}$

$\overline{\qquad 6\frac{17}{20}}$

(q) $10\frac{3}{16} = 10\frac{6}{32} = 9\frac{38}{32}$

$- 2\frac{9}{32} = 2\frac{9}{32} = 2\frac{9}{32}$

$\overline{\qquad 7\frac{29}{32}}$

(r) $2\frac{1}{8} = 2\frac{2}{16} = 1\frac{18}{16}$

$-1\frac{3}{16} = 1\frac{3}{16} = 1\frac{3}{16}$

$\overline{\qquad \frac{15}{16}}$

PROBLEM 7-9   Subtract fractions from mixed numbers when the minuend of the fraction part is too small: **(a)** $2\frac{2}{9} - \frac{7}{9}$ **(b)** $5\frac{1}{3} - \frac{2}{3}$ **(c)** $1\frac{3}{8} - \frac{5}{8}$ **(d)** $9\frac{1}{10} - \frac{3}{10}$ **(e)** $3\frac{1}{4} - \frac{3}{4}$ **(f)** $8\frac{1}{5} - \frac{4}{5}$
(g) $7\frac{3}{7} - \frac{5}{7}$ **(h)** $1\frac{5}{12} - \frac{7}{12}$ **(i)** $4\frac{1}{2} - \frac{5}{6}$ **(j)** $6\frac{1}{8} - \frac{1}{4}$ **(k)** $10\frac{1}{4} - \frac{1}{3}$ **(l)** $12\frac{1}{3} - \frac{2}{5}$
(m) $2\frac{1}{8} - \frac{1}{6}$ **(n)** $5\frac{1}{10} - \frac{11}{12}$ **(o)** $15\frac{5}{16} - \frac{11}{20}$ **(p)** $9\frac{1}{6} - \frac{9}{10}$ **(q)** $2\frac{5}{16} - \frac{7}{8}$ **(r)** $1\frac{1}{32} - \frac{5}{8}$

Solution: Recall that to subtract a fraction from a mixed number when the minuend of the fraction part is too small, you first rename the mixed number as an equal mixed number with a larger fraction part [see Example 7-13]:

(a) $2\frac{2}{9} = 1\frac{11}{9}$

$-\frac{7}{9} \qquad \frac{7}{9}$

$\overline{\qquad 1\frac{4}{9}}$

(b) $5\frac{1}{3} = 4\frac{4}{3}$

$-\frac{2}{3} \qquad \frac{2}{3}$

$\overline{\qquad 4\frac{2}{3}}$

(c) $1\frac{3}{8} = \frac{11}{8}$

$-\frac{5}{8} = \frac{5}{8}$

$\overline{\qquad \frac{6}{8} = \frac{3}{4}}$

(d) $9\frac{1}{10} = 8\frac{11}{10}$
$\underline{-\ \frac{3}{10} = \ \frac{3}{10}}$
$\qquad 8\frac{8}{10} = 8\frac{4}{5}$

(e) $3\frac{1}{4} = 2\frac{5}{4}$
$\underline{-\ \frac{3}{4} = \ \frac{3}{4}}$
$\qquad 2\frac{2}{4} = 2\frac{1}{2}$

(f) $8\frac{1}{5} = 7\frac{6}{5}$
$\underline{-\ \frac{4}{5} = \ \frac{4}{5}}$
$\qquad 7\frac{2}{5}$

(g) $7\frac{3}{7} = 6\frac{10}{7}$
$\underline{-\ \frac{5}{7} = \ \frac{5}{7}}$
$\qquad 6\frac{5}{7}$

(h) $1\frac{5}{12} = \frac{17}{12}$
$\underline{-\ \frac{7}{12} = \ \frac{7}{12}}$
$\qquad \frac{10}{12} = \frac{5}{6}$

(i) $4\frac{1}{2} = 4\frac{3}{6} = 3\frac{9}{6}$
$\underline{-\ \frac{5}{6} = \ \frac{5}{6} = \ \frac{5}{6}}$
$\qquad\qquad 3\frac{4}{6} = 3\frac{2}{3}$

(j) $6\frac{1}{8} = 6\frac{1}{8} = 5\frac{9}{8}$
$\underline{-\ \frac{1}{4} = \ \frac{2}{8} = \ \frac{2}{8}}$
$\qquad\qquad 5\frac{7}{8}$

(k) $10\frac{1}{4} = 10\frac{3}{12} = 9\frac{15}{12}$
$\underline{-\ \frac{1}{3} = \ \frac{4}{12} = \ \frac{4}{12}}$
$\qquad\qquad\quad 9\frac{11}{12}$

(l) $12\frac{1}{3} = 12\frac{5}{15} = 11\frac{20}{15}$
$\underline{-\ \frac{2}{5} = \ \frac{6}{15} = \ \frac{6}{15}}$
$\qquad\qquad\quad 11\frac{14}{15}$

(m) $2\frac{1}{8} = 2\frac{3}{24} = 1\frac{27}{24}$
$\underline{-\ \frac{1}{6} = \ \frac{4}{24} = \ \frac{4}{24}}$
$\qquad\qquad\quad 1\frac{23}{24}$

(n) $5\frac{1}{10} = 5\frac{6}{60} = 4\frac{66}{60}$
$\underline{-\ \frac{11}{12} = \ \frac{55}{60} = \ \frac{55}{60}}$
$\qquad\qquad\quad 4\frac{11}{60}$

(o) $15\frac{5}{16} = 15\frac{25}{80} = 14\frac{105}{80}$
$\underline{-\ \frac{11}{20} = \ \frac{44}{80} = \ \frac{44}{80}}$
$\qquad\qquad\quad 14\frac{61}{80}$

(p) $9\frac{1}{6} = 9\frac{5}{30} = 8\frac{35}{30}$
$\underline{-\ \frac{9}{10} = \ \frac{27}{30} = \ \frac{27}{30}}$
$\qquad\qquad 8\frac{8}{30} = 8\frac{4}{15}$

(q) $2\frac{5}{16} = 2\frac{5}{16} = 1\frac{21}{16}$
$\underline{-\ \frac{7}{8} = \ \frac{14}{16} = \ \frac{14}{16}}$
$\qquad\qquad\quad 1\frac{7}{16}$

(r) $1\frac{1}{32} = 1\frac{1}{32} = \frac{33}{32}$
$\underline{-\ \frac{5}{8} = \ \frac{20}{32} = \ \frac{20}{32}}$
$\qquad\qquad\quad \frac{13}{32}$

PROBLEM 7-10 Decide what to do (add, subtract, multiply, or divide with fractions) and then solve the following problems:

(a) A weekend is $\frac{2}{7}$ of a week. What fractional part of a week are the weekdays, not including the weekend?

(b) Each lap around the track is $\frac{1}{4}$ mile. How many laps must be completed to run 5 miles?

(c) George spent $\frac{1}{2}$ hour studying math and $\frac{3}{4}$ hour studying English. How long did he study in all?

(d) Pat practices the piano $\frac{3}{4}$ hour each day. How long does she practice in a 7-day week?

(e) A recipe calls for $\frac{3}{4}$ cup sugar. Richard has only $\frac{1}{3}$ cup on hand. How much more sugar is needed?

(f) Mom gave Tom $\frac{2}{3}$ of the pie. He ate $\frac{1}{2}$ of that portion. What fractional part of the whole pie did he eat?

(g) Greg studied for $\frac{1}{2}$ hour. This was only $\frac{1}{3}$ of the time he should have studied. How long should Greg have studied?

(h) Bobbi paid $\frac{1}{4}$ of her monthly pay for rent and $\frac{1}{3}$ for food. What part of her monthly pay did she pay for rent and food together?

Solution: Recall that to join parts together, you add; to take one part away from another part, you subtract; to join equal parts together, you multiply; and to separate an amount into equal parts, you divide [see Examples 7-14 and 7-15]:

(a) subtract: whole week $\longrightarrow$ 1 or $\frac{7}{7}$; $\frac{7}{7} - \frac{2}{7} = \frac{5}{7}$ (of a week)

(b) divide: $5 \div \frac{1}{4} = \frac{5}{1} \div \frac{1}{4} = \frac{5}{1} \times \frac{4}{1} = \frac{20}{1} = 20$ (laps)

(c) add: $\frac{1}{2} + \frac{3}{4} = \frac{2}{4} + \frac{3}{4} = \frac{5}{4} = 1\frac{1}{4}$ (hours)

(d) multiply: $\frac{3}{4} \times 7 = \frac{3}{4} \times \frac{7}{1} = \frac{21}{4} = 5\frac{1}{4}$ (hours)

(e) subtract: $\frac{3}{4} - \frac{1}{3} = \frac{9}{12} - \frac{4}{12} = \frac{5}{12}$ (of a cup)

(f) multiply: $\frac{1}{2} \times \frac{2}{3} = \frac{1}{3}$ (of the pie)

(g) divide: $\frac{1}{2} \div \frac{1}{3} = \frac{1}{2} \times \frac{3}{1} = \frac{3}{2} = 1\frac{1}{2}$ (hours)

(h) add: $\frac{1}{4} + \frac{1}{3} = \frac{3}{12} + \frac{4}{12} = \frac{7}{12}$ (of her monthly pay)

PROBLEM 7-11 Decide what to do (add, subtract, multiply, or divide with mixed numbers) and then solve the following problems:

(a) Amy swam $8\frac{1}{2}$ laps. Bill swam $2\frac{3}{4}$ laps. How many more laps did Amy swim than Bill?

(b) Aaron ran $3\frac{1}{2}$ miles and then walked $4\frac{3}{4}$ miles. How far did he run and walk all together?

(c) If Ann takes $3\frac{1}{2}$ minutes to run each lap around the track, how long will it take her to run 45 laps?

(d) If Alfred takes $3\frac{1}{2}$ minutes to run each lap around the track, how many laps will he complete in 45 minutes?

(e) If each book weighs $1\frac{1}{3}$ pounds, how much will 100 books weigh?

(f) Brant types at $72\frac{1}{2}$ words per minute. About how long will it take him to type a report of 1000 words?

(g) Emil is $62\frac{1}{2}$ inches tall. Last year he was $60\frac{3}{4}$ inches tall. How much did Emil grow since last year?

(h) Brice was a $5\frac{1}{2}$-pound baby. His twin sister weighed $6\frac{3}{4}$ pounds at birth. How much was the combined weight of the twins at birth?

Solution: Recall that to join amounts together, you add; to take one amount away from another, you subtract; to join equal amounts together, you multiply; and to separate into equal amounts, you divide [see Examples 7-14 and 7-15]:

(a) subtract: $8\frac{1}{2} - 2\frac{3}{4} = 5\frac{3}{4}$ (laps)   **(b)** add: $3\frac{1}{2} + 4\frac{3}{4} = 8\frac{1}{4}$ (miles)

(c) multiply: $3\frac{1}{2} \times 45 = 157\frac{1}{2}$ (minutes)   **(d)** divide: $45 \div 3\frac{1}{2} = 12\frac{6}{7}$ or 12 (complete laps)

(e) multiply: $1\frac{1}{3} \times 100 = 133\frac{1}{3}$ (pounds)   **(f)** divide: $1000 \div 72\frac{1}{2} = 13\frac{23}{29}$ or about 14 (minutes)

(g) subtract: $62\frac{1}{2} - 60\frac{3}{4} = 1\frac{3}{4}$ (inches)   **(h)** add: $5\frac{1}{2} + 6\frac{3}{4} = 12\frac{1}{4}$ (pounds)

Supplementary Exercises

PROBLEM 7-12   Subtract basic facts given horizontal form:

Row **(a)** $4-2$ $2-0$ $14-8$ $12-7$ $10-6$ $12-9$ $7-2$ $13-6$ $5-2$ $11-5$
Row **(b)** $8-1$ $14-6$ $3-2$ $10-5$ $9-2$ $9-5$ $3-0$ $6-1$ $16-9$ $6-3$
Row **(c)** $11-7$ $7-4$ $12-6$ $3-1$ $2-2$ $15-6$ $8-7$ $3-3$ $15-7$ $10-2$
Row **(d)** $10-4$ $12-5$ $12-3$ $11-2$ $16-7$ $13-5$ $11-3$ $14-7$ $9-3$ $8-6$
Row **(e)** $8-4$ $4-4$ $9-9$ $4-3$ $17-9$ $9-4$ $11-6$ $18-9$ $11-4$ $14-9$
Row **(f)** $10-3$ $13-9$ $9-1$ $7-5$ $9-7$ $5-3$ $7-7$ $8-2$ $5-4$ $17-8$
Row **(g)** $6-2$ $8-5$ $8-0$ $10-7$ $9-6$ $13-8$ $15-8$ $13-4$ $0-0$ $2-1$
Row **(h)** $13-7$ $8-3$ $12-4$ $1-0$ $7-1$ $6-4$ $7-0$ $14-5$ $10-1$ $5-0$
Row **(i)** $4-1$ $1-1$ $6-6$ $6-0$ $7-3$ $16-8$ $5-1$ $8-8$ $5-5$ $10-9$
Row **(j)** $4-0$ $7-6$ $6-5$ $15-9$ $12-8$ $11-8$ $10-8$ $9-8$ $11-9$ $9-0$

PROBLEM 7-13   Subtract fractions: **(a)** $\frac{3}{5} - \frac{1}{5}$ **(b)** $\frac{4}{3} - \frac{1}{3}$ **(c)** $\frac{1}{6} - \frac{1}{6}$ **(d)** $\frac{5}{2} - \frac{1}{2}$
(e) $\frac{5}{8} - \frac{3}{8}$ **(f)** $\frac{5}{4} - \frac{3}{4}$ **(g)** $\frac{7}{10} - \frac{3}{10}$ **(h)** $\frac{7}{12} - \frac{1}{12}$ **(i)** $\frac{7}{32} - \frac{3}{32}$ **(j)** $\frac{11}{16} - \frac{9}{16}$
(k) $\frac{1}{2} - \frac{1}{8}$ **(l)** $\frac{9}{10} - \frac{1}{2}$ **(m)** $\frac{3}{4} - \frac{5}{12}$ **(n)** $\frac{7}{8} - \frac{1}{16}$ **(o)** $\frac{4}{5} - \frac{1}{4}$ **(p)** $\frac{1}{2} - \frac{2}{7}$ **(q)** $\frac{5}{8} - \frac{3}{10}$
(r) $\frac{11}{12} - \frac{5}{8}$ **(s)** $\frac{7}{10} - \frac{1}{12}$ **(t)** $\frac{14}{15} - \frac{5}{12}$ **(u)** $\frac{9}{10} - \frac{11}{16}$ **(v)** $\frac{11}{20} - \frac{13}{30}$ **(w)** $\frac{15}{16} - \frac{11}{24}$
(x) $\frac{11}{18} - \frac{11}{24}$ **(y)** $\frac{9}{16} - \frac{5}{32}$ **(z)** $\frac{11}{32} - \frac{1}{8}$

PROBLEM 7-14   Subtract with mixed numbers: **(a)** $5\frac{3}{4} - 3$ **(b)** $8\frac{2}{3} - 1$ **(c)** $3\frac{5}{6} - 2$
(d) $4\frac{7}{10} - 1$ **(e)** $1\frac{2}{3} - 1$ **(f)** $6\frac{4}{5} - 2$ **(g)** $2\frac{5}{9} - 1$ **(h)** $7\frac{1}{2} - 7$ **(i)** $9\frac{3}{10} - \frac{1}{10}$
(j) $1\frac{5}{8} - \frac{1}{2}$ **(k)** $8\frac{3}{4} - \frac{1}{2}$ **(l)** $2\frac{2}{3} - \frac{1}{6}$ **(m)** $7\frac{1}{2} - \frac{1}{3}$ **(n)** $3\frac{2}{3} - \frac{1}{4}$ **(o)** $6\frac{5}{6} - \frac{3}{10}$
(p) $4\frac{5}{12} - \frac{1}{5}$ **(q)** $5\frac{1}{4} - 4\frac{1}{4}$ **(r)** $6\frac{7}{8} - 3\frac{1}{8}$ **(s)** $10\frac{1}{2} - 2\frac{1}{6}$ **(t)** $15\frac{2}{3} - 4\frac{5}{12}$ **(u)** $12\frac{1}{2} - 1\frac{2}{5}$
(v) $8\frac{4}{5} - 5\frac{1}{4}$ **(w)** $9\frac{3}{4} - 2\frac{1}{10}$ **(x)** $11\frac{11}{12} - 8\frac{7}{18}$ **(y)** $3\frac{11}{20} - 3\frac{5}{12}$ **(z)** $1\frac{7}{8} - 1\frac{3}{16}$

PROBLEM 7-15   Subtract fractions and mixed numbers from whole numbers: **(a)** $5 - 1\frac{1}{2}$
(b) $12 - 2\frac{1}{3}$ **(c)** $2 - 1\frac{1}{4}$ **(d)** $9 - 5\frac{1}{8}$ **(e)** $15 - 6\frac{5}{6}$ **(f)** $16 - 2\frac{4}{5}$ **(g)** $3 - 2\frac{7}{10}$
(h) $11 - 1\frac{5}{12}$ **(i)** $7 - 5\frac{3}{7}$ **(j)** $4 - 1\frac{7}{9}$ **(k)** $15 - 3\frac{5}{16}$ **(l)** $8 - 5\frac{3}{32}$ **(m)** $6 - 5\frac{1}{5}$
(n) $10 - 3\frac{3}{8}$ **(o)** $2 - \frac{2}{3}$ **(p)** $5 - \frac{3}{4}$ **(q)** $10 - \frac{5}{8}$ **(r)** $3 - \frac{1}{6}$ **(s)** $7 - \frac{2}{5}$ **(t)** $9 - \frac{3}{10}$
(u) $6 - \frac{1}{12}$ **(v)** $1 - \frac{1}{16}$ **(w)** $4 - \frac{29}{32}$ **(x)** $6 - \frac{7}{8}$ **(y)** $14 - \frac{1}{10}$ **(z)** $13 - \frac{11}{12}$

PROBLEM 7-16   Subtract fractions and mixed numbers from mixed numbers when the minuend of the fraction part is too small: **(a)** $8\frac{1}{4} - 5\frac{3}{4}$ **(b)** $7\frac{1}{8} - 2\frac{7}{8}$ **(c)** $9\frac{1}{3} - 8\frac{2}{3}$ **(d)** $3\frac{1}{5} - 2\frac{2}{5}$
(e) $6\frac{1}{3} - 1\frac{1}{2}$ **(f)** $10\frac{1}{4} - 2\frac{1}{3}$ **(g)** $11\frac{1}{4} - 3\frac{1}{2}$ **(h)** $6\frac{1}{3} - 1\frac{5}{6}$ **(i)** $8\frac{3}{4} - 2\frac{5}{6}$ **(j)** $3\frac{5}{8} - 1\frac{15}{16}$
(k) $5\frac{3}{10} - 1\frac{3}{4}$ **(l)** $12\frac{1}{16} - 3\frac{3}{16}$ **(m)** $5\frac{1}{2} - 4\frac{3}{4}$ **(n)** $2\frac{1}{6} - 1\frac{5}{6}$ **(o)** $9\frac{1}{2} - \frac{7}{10}$ **(p)** $1\frac{1}{4} - \frac{3}{4}$
(q) $8\frac{1}{3} - \frac{5}{8}$ **(r)** $2\frac{1}{6} - \frac{1}{3}$ **(s)** $7\frac{2}{5} - \frac{3}{5}$ **(t)** $3\frac{3}{10} - \frac{7}{10}$ **(u)** $6\frac{3}{8} - \frac{1}{2}$ **(v)** $4\frac{1}{12} - \frac{1}{10}$
(w) $5\frac{5}{12} - \frac{3}{4}$ **(x)** $10\frac{1}{16} - \frac{1}{4}$ **(y)** $12\frac{3}{16} - \frac{5}{8}$ **(z)** $1\frac{1}{2} - \frac{3}{4}$

PROBLEM 7-17   Solve word problems containing fractions and mixed numbers $(+, -, \times, \div)$:

(a) How many $7\frac{1}{2}$-inch boards can be cut from a 22-foot piece of stock?

(b) A board is cut into 20 equal pieces so that each piece is $3\frac{1}{2}$ inches in length. How long was the board before it was cut?

(c) A bedroom is $8\frac{1}{2}$ feet by $9\frac{1}{4}$ feet. How much baseboard is needed to go around the room if there is one doorway that is $3\frac{1}{2}$ feet wide and 7 feet high?

(d) The walls of the room in problem **(c)** are 8 feet high. Wall paper is $1\frac{1}{2}$ feet wide and the length of each roll is 20 feet. How many rolls are needed to paper the room using only whole 8-feet strips except for the doorway?

(e) What is the difference between the width and length of an $8\frac{1}{2} \times 11$ inch sheet of paper?

(f) It takes $\frac{1}{2}$ pound of meat for each serving of meat loaf. **(a)** How much meat is needed for $7\frac{1}{2}$ servings? **(b)** How many complete servings can be made from $7\frac{1}{4}$ pounds of meat?

(g) Stella worked $4\frac{1}{2}$ hours Monday, $5\frac{1}{4}$ hours Tuesday, $3\frac{3}{4}$ hours Wednesday, $6\frac{1}{2}$ hours Thursday, and $7\frac{1}{4}$ hours Friday. How many hours did she work in all?

(h) Ralph purchased $3\frac{1}{2}$ pounds of potatoes for 49¢ per pound and $5\frac{1}{4}$ pounds of onions at 19¢ per pound. How much will the potatoes and onions cost? (*Hint:* Stores always round up to the next nearest cent.)

Answers to Supplementary Exercises

(7-12)

Row **(a)**	2	2	6	5	4	3	5	7	3	6
Row **(b)**	7	8	1	5	7	4	3	5	7	3
Row **(c)**	4	3	6	2	0	9	1	0	8	8
Row **(d)**	6	7	9	9	9	8	8	7	6	2
Row **(e)**	4	0	0	1	8	5	5	9	7	5
Row **(f)**	7	4	8	2	2	2	0	6	1	9
Row **(g)**	4	3	8	3	3	5	7	9	0	1
Row **(h)**	6	5	8	1	6	2	7	9	9	5
Row **(i)**	3	0	0	6	4	8	4	0	0	1
Row **(j)**	4	1	1	6	4	3	2	1	2	9

(7-13)   **(a)** $\frac{2}{5}$ **(b)** 1   **(c)** 0   **(d)** 2   **(e)** $\frac{1}{4}$ **(f)** $\frac{1}{2}$ **(g)** $\frac{2}{5}$ **(h)** $\frac{1}{2}$ **(i)** $\frac{1}{8}$ **(j)** $\frac{1}{8}$
(k) $\frac{3}{8}$ **(l)** $\frac{2}{5}$ **(m)** $\frac{1}{3}$ **(n)** $\frac{13}{16}$ **(o)** $\frac{11}{20}$ **(p)** $\frac{3}{14}$ **(q)** $\frac{13}{40}$ **(r)** $\frac{7}{24}$ **(s)** $\frac{37}{60}$ **(t)** $\frac{31}{60}$
(u) $\frac{17}{80}$ **(v)** $\frac{7}{60}$ **(w)** $\frac{23}{48}$ **(x)** $\frac{11}{72}$ **(y)** $\frac{13}{32}$ **(z)** $\frac{7}{32}$

(7-14)   **(a)** $2\frac{3}{4}$ **(b)** $7\frac{2}{3}$ **(c)** $1\frac{5}{6}$ **(d)** $3\frac{7}{10}$ **(e)** $\frac{2}{3}$ **(f)** $4\frac{4}{5}$ **(g)** $1\frac{5}{9}$ **(h)** $\frac{1}{2}$ **(i)** $9\frac{1}{5}$
(j) $1\frac{1}{8}$ **(k)** $8\frac{1}{4}$ **(l)** $2\frac{1}{2}$ **(m)** $7\frac{1}{6}$ **(n)** $3\frac{5}{12}$ **(o)** $6\frac{8}{15}$ **(p)** $4\frac{13}{60}$ **(q)** 1   **(r)** $3\frac{3}{4}$
(s) $8\frac{1}{3}$ **(t)** $11\frac{1}{4}$ **(u)** $11\frac{1}{10}$ **(v)** $3\frac{11}{20}$ **(w)** $7\frac{13}{20}$ **(x)** $3\frac{19}{36}$ **(y)** $\frac{2}{15}$ **(z)** $\frac{11}{16}$

(7-15)   **(a)** $3\frac{1}{2}$ **(b)** $9\frac{2}{3}$ **(c)** $\frac{3}{4}$ **(d)** $3\frac{7}{8}$ **(e)** $8\frac{1}{6}$ **(f)** $13\frac{1}{5}$ **(g)** $\frac{3}{10}$ **(h)** $9\frac{7}{12}$ **(i)** $1\frac{4}{7}$
(j) $2\frac{2}{9}$ **(k)** $11\frac{11}{16}$ **(l)** $2\frac{27}{32}$ **(m)** $\frac{4}{5}$ **(n)** $6\frac{5}{8}$ **(o)** $1\frac{1}{3}$ **(p)** $4\frac{1}{4}$ **(q)** $9\frac{3}{8}$ **(r)** $2\frac{5}{6}$
(s) $6\frac{3}{5}$ **(t)** $8\frac{7}{10}$ **(u)** $5\frac{11}{12}$ **(v)** $\frac{15}{16}$ **(w)** $3\frac{3}{32}$ **(x)** $5\frac{1}{8}$ **(y)** $13\frac{9}{10}$ **(z)** $12\frac{1}{12}$

(7-16)   **(a)** $2\frac{1}{2}$ **(b)** $4\frac{1}{4}$ **(c)** $\frac{2}{3}$ **(d)** $\frac{4}{5}$ **(e)** $4\frac{5}{6}$ **(f)** $7\frac{11}{12}$ **(g)** $7\frac{3}{4}$ **(h)** $4\frac{1}{2}$ **(i)** $5\frac{11}{12}$
(j) $1\frac{11}{16}$ **(k)** $3\frac{11}{20}$ **(l)** $8\frac{7}{8}$ **(m)** $\frac{3}{4}$ **(n)** $\frac{1}{3}$ **(o)** $8\frac{4}{5}$ **(p)** $\frac{1}{2}$ **(q)** $7\frac{17}{24}$ **(r)** $1\frac{5}{6}$
(s) $6\frac{4}{5}$ **(t)** $2\frac{3}{5}$ **(u)** $5\frac{7}{8}$ **(v)** $3\frac{59}{60}$ **(w)** $4\frac{2}{3}$ **(x)** $9\frac{13}{16}$ **(y)** $11\frac{9}{16}$ **(z)** $\frac{3}{4}$

(7-17)   **(a)** 2 $7\frac{1}{2}$-inch boards   **(b)** 70 inches   **(c)** 32 feet   **(d)** 11 rolls   **(e)** $2\frac{1}{2}$ inches
(f) **(a)** $3\frac{3}{4}$ pounds **(b)** 14 complete servings   **(g)** $27\frac{1}{4}$ hours   **(h)** $2.72

8 DECIMALS: ADDITION AND SUBTRACTION

THIS CHAPTER IS ABOUT

☑ **Using Place Value with Decimals**
☑ **Renaming with Decimals**
☑ **Adding with Decimals**
☑ **Subtracting with Decimals**
☑ **Solving Word Problems Containing Decimals (+, −)**

8-1. Using Place Value with Decimals

A. Identify digits in decimals.

The numbers used to measure things accurately are called **decimal numbers** or just **decimals.**

EXAMPLE 8-1: List all the decimals using the digits 0, 2, and 5 exactly once in each number.

Solution: The decimals containing 0, 2, and 5 exactly once are 52.0, 25.0, 50.2, 20.5, 05.2 or 5.20, 5.02, 02.5 or 2.50, 2.05, 0.52, and 0.25.

Every decimal has a **whole-number part** and a **decimal-fraction part** separated by a **decimal point.**

EXAMPLE 8-2: Identify the three parts of 0.25.

Solution: In 0.25: 0 is the whole-number part.
25 is the decimal-fraction part.
. is the decimal point.

To help locate the correct digit in a decimal given the digit's place-value, you can use a place-value chart.

EXAMPLE 8-3: In 1.732051, draw a line under the digit that is in the thousandths place.

Solution

Decimal Place-Value Chart
(from millions to millionths)

Whole-Number Part							Decimal-Fraction Part					
millions	hundred thousands	ten thousands	thousands	hundreds	tens	ones	tenths	hundredths	thousandths	ten thousandths	hundred thousandths	millionths
						1	7	3	<u>2</u>	0	5	1

← place-value names

Draw a line under the 2 thousandths.

134

To help identify the correct place-value name for a given digit, you can use a place-value chart.

EXAMPLE 8-4: Write the place-value name for the underlined digit in the following decimal: 3.1<u>4</u>1592

Solution

Whole-Number Part							Decimal-Fraction Part					
millions	hundred thousands	ten thousands	thousands	hundreds	tens	ones	tenths	hundredths	thousandths	ten thousandths	hundred thousandths	millionths
						3 .	1	<u>4</u> ↑	1	5	9	2

The place-value name of the underlined digit is hundredths.

Once the place-value name for a given digit is identified, you can write the value of that digit.

EXAMPLE 8-5: In 235.980176, what is the value of each of the following digits? **(a)** 9   **(b)** 8   **(c)** 0   **(d)** 1   **(e)** 7   **(f)** 6

Solution: In 235.980176: **(a)** the value of 9 is 9 tenths (or 0.9);
 (b) the value of 8 is 8 hundredths (or 0.08);
 (c) the value of 0 is 0 thousandths (or 0);
 (d) the value of 1 is 1 ten thousandths (or 0.0001);
 (e) the value of 7 is 7 hundred thousandths (or 0.00007);
 (f) the value of 6 is 6 millionths (or 0.000006).

B. Write decimals.

To write the **decimal-word name** for a given decimal number, you write the word *and* for the decimal point when the whole-number part is not zero.

EXAMPLE 8-6: Write the decimal-word name for the following decimal number: 2.5

Solution: Read or write 2.5 as "two and five tenths" or "2 and 5 tenths."

Note: "2 and 5 tenths" is often referred to as the **short decimal-word name** for 2.5.

To write the correct short decimal-word name for a given decimal number, you always name the whole number, then name the decimal-fraction part with the place-value name of the last digit on the right.

EXAMPLE 8-7: Write the short decimal-word name for the following decimal numbers:
(a) 1.7   **(b)** 1.73   **(c)** 1.732   **(d)** 1.7320   **(e)** 1.73205   **(f)** 1.732051

Solution: Read or write: **(a)** 1.7 as "1 and 7 tenths;"
 (b) 1.73 as "1 and 73 hundredths;"
 (c) 1.732 as "1 and 732 thousandths;"
 (d) 1.7320 as "1 and 7320 thousandths;"
 (e) 1.73205 as "1 and 73,205 thousandths;"
 (f) 1.732051 as "1 and 732,051 millionths."

To write the correct decimal number given a short decimal-word name, you write the last digit of the decimal-fraction part so that it has the same value as the given short decimal-word name.

EXAMPLE 8-8: Write the decimal number for the following short decimal-word name:
3 and 75 thousandths

Whole-Number Part							Decimal-Fraction Part					
millions	hundred thousands	ten thousands	thousands	hundreds	tens	ones	tenths	hundredths	thousandths	ten thousandths	hundred thousandths	millionths
						3 .	0	7	5			

write a
zero

and

"75 thousandths" means that
the last digit 5 must go
in the thousandths place.

When the whole-number part of a decimal number is zero, the decimal is called a **decimal fraction.**

EXAMPLE 8-9: Write decimal fractions for the following decimal-word names: **(a)** 2 tenths
(b) 3 hundredths **(c)** 7 thousandths **(d)** 15 ten thousandths **(e)** 9 hundred thousandths
(f) 235 millionths

Solution: The decimal fraction for: **(a)** 2 tenths is 0.2;
(b) 3 hundredths is 0.03;
(c) 7 thousandths is 0.007;
(d) 15 ten thousandths is 0.0015;
(e) 9 hundred thousandths is 0.00009;
(f) 235 millionths is 0.000235.

Note: A zero (0) is usually written for the whole-number part of a decimal fraction.

EXAMPLE 8-10: Which is the preferred form of 3 tenths: **(a)** .3 **(b)** 0.3

Solution: **(b)** 0.3 is the preferred form of 3 tenths because the zero is included in the whole-number part. .2 is a correct form of 3 tenths but not the preferred form.

C. Compare decimals.

To **compare two decimals,** you determine which decimal is greater (larger) or which decimal is less (smaller) or that the two decimals are equal.

Two decimals can always be compared using one of the following symbols:

> (is greater than)
< (is less than)
= (is equal to)

To compare two decimals, you may find it helpful to write one decimal under the other while aligning the decimal points and like place values.

EXAMPLE 8-11: Compare the following decimals: **(a)** 0.09   **(b)** 0.1

Whole-Number Part							Decimal-Fraction Part					
millions	hundred thousands	ten thousands	thousands	hundreds	tens	ones	tenths	hundredths	thousandths	ten thousandths	hundred thousandths	millionths
						0	0	9				
						0	1					

    same   different (0 tenths < 1 tenths)

Write one number under the other while aligning the decimal points and like place values.

Solution: **(a)** 0.09 is less than 0.1 (0.09 < 0.1), or 0.1 is greater than 0.09 (0.1 > 0.09).

Note: The inequality symbols < and > should always point at the smaller number.

D. Order decimals.

To **order two or more decimals,** you list them from smallest to largest or from largest to smallest.

EXAMPLE 8-12: Order the following decimals from smallest to largest: 9.99, 15.29, 83.00, 7.05, 15.30, 9.49, 5.29, 83.01, 10.00, 15.28

Solution: 5.2 9 ←—— smallest decimal
         7.0 5
         9.4 9
         9.9 9
     1 0.0 0
     1 5.2 8
     1 5.2 9
     1 5.3 0
     8 3.0 0
     8 3.0 1 ←—— largest decimal

Note: To order two or more decimals from largest to smallest, you just reverse the smallest to largest order.

EXAMPLE 8-13: Order the following decimals from largest to smallest: 138.49, 138.50, 138.51, 39.49, 37.99, 38.00, 8.50, 30.00, 8.49, 18.51, 0.99, 0.09, 0.1

Solution: 1 3 8.5 1 ←—— largest decimal
      1 3 8.5 0
      1 3 8.4 9
        3 9.4 9
        3 8.0 0
        3 7.9 9
        3 0.0 0
        1 8.5 1
          8.5 0
          8.4 9
          0.9 9
          0.1
          0.0 9 ←—— smallest decimal

E. **Round numbers.**

When an exact answer is not needed for a problem using decimals, you may want to make the computation easier and **round decimals** before computing. The place to which you round the decimals will depend on how accurate the answer needs to be.

If the digit to the right of the digit to be rounded is 5 or more, then **round up** as follows:
1. Increase the digit to be rounded by 1.
2. If the rounded digit is in the whole-number part, replace the digits to the right of the rounded digit with zeros in the whole-number part.
3. Omit all the digits to the right of the rounded digit in the decimal-fraction part.

EXAMPLE 8-14: Round the following decimal to the nearest thousandth: 3.14159

$$\text{thousandths} \downarrow$$

Solution: 3.14159 = 3.14159   Draw a line under the digit to be rounded.

= 3.14159   *Think:* Is the digit to the right 5 or more? Yes: 5 = 5.

≈ 3.142   Then round up: Increase 1 by 1 to get 2.
                                    Omit 5 and 9.

Note: To the nearest thousandth, 3.14159 ≈ 3.142.

If the digit to the right of the digit to be rounded is less than 5, then **round down** as follows:
1. Leave the digit to be rounded the same.
2. If the rounded digit is in the whole-number part, replace the digits to the right of the rounded digit by zeros in the whole-number part.
3. Omit all the digits to the right of the rounded digit in the decimal-fraction part.

EXAMPLE 8-15: Round the following decimal to the nearest ten: 963.25

Solution: 963.25 = 963.25   Draw a line under the digit to be rounded.

= 963.25   *Think:* Is the digit to the right 5 or more? No: 3 < 5

≈ 960   Then round down: Leave 6 the same.
                                    Replace 3 with a zero.
                                    Omit 2 and 5.

Note: To the nearest ten, 963.25 ≈ 960.

8-2. Renaming with Decimals

A. **Rename whole numbers as decimals.**

Every whole number can be renamed as an equal decimal by writing the decimal point following the whole number and then writing one or more zeros following the decimal point.

EXAMPLE 8-16: Rename the following whole numbers as decimals: **(a)** 8   **(b)** 10   **(c)** 258

Solution

whole number   decimals
      ↓
(a)   8 =   8.0 or   8.00 or   8.000 (and so on)
(b) 10 =   10.0 or 10.00 or 10.000 (and so on)
(c) 258 = 258.0 or 258.00 or 258.000 (and so on)

Note: When the decimal-fraction part of a decimal is zero, the decimal can always be renamed as an equal whole number.

EXAMPLE 8-17: Rename the following decimals as equal whole numbers: **(a)** 5.0   **(b)** 23.00 **(c)** 1.000

Solution

decimal   whole number

(a) 5.0   = 5
(b) 23.00 = 23
(c) 1.000 = 1

B. Rename decimals as fractions and mixed numbers.

To rename a decimal-fraction as an equal proper fraction, you first write the value of the decimal-fraction, and then rename using tenths, hundredths, thousandths, $\cdots$, respectively.

EXAMPLE 8-18: Rename the following decimal-fractions as equal proper fractions in simplest form: **(a)** 0.5   **(b)** 0.25   **(c)** 0.001

Solution

value   fraction simplest form

(a) 0.5   =   5 tenths   = $\frac{5}{10}$ = $\frac{1}{2}$
(b) 0.25 = 25 hundredths = $\frac{25}{100}$ = $\frac{1}{4}$
(c) 0.001 = 1 thousandth = $\frac{1}{1000}$

Note: A value of tenths, hundredths, thousandths, $\cdots$ means the denominator of the proper fraction is 10, 100, 1000, $\cdots$, respectively.

When a fraction has a denominator of 10, or 100, or 1000, $\cdots$, you can rename that proper fraction as a equal decimal fraction using tenths, hundredths, thousandths, $\cdots$, respectively.

EXAMPLE 8-19: Rename the following fractions as equal decimals: **(a)** $\frac{7}{10}$ **(b)** $\frac{3}{100}$ **(c)** $\frac{581}{1000}$

Solution

value   decimal

(a) $\frac{7}{10}$ =   7 tenths   = 0.7
(b) $\frac{3}{100}$ 3 hundredths = 0.03
(c) $\frac{581}{1000}$ = 581 thousandths = 0.581

Note: When a fraction has a denominator that is not 10, 100, or 1000, $\cdots$, you can rename that fraction as an equal decimal also. The method for renaming such a fraction as an equal decimal is presented in Section 9-6.

When the whole-number part of a decimal is not zero, the decimal can be renamed as an equal mixed number using the short decimal-word name.

EXAMPLE 8-20: Rename the following decimals as equal mixed numbers in simplest form: **(a)** 2.8 **(b)** 49.05   **(c)** 8.003

Solution

	decimal-word name	proper fraction	mixed number	simplest form
(a) 2.8 =	2 and 8 tenths	= 2 and $\frac{8}{10}$	= $2\frac{8}{10}$	= $2\frac{4}{5}$
(b) 49.05 =	49 and 5 hundredths	= 49 and $\frac{5}{100}$	= $49\frac{5}{100}$	= $49\frac{1}{20}$
(c) 8.003 =	8 and 3 thousandths	= 8 and $\frac{3}{1000}$	= $8\frac{3}{1000}$	

When the whole-number part of a decimal is not zero, the decimal can also be renamed as an equal improper fraction by first using place values.

EXAMPLE 8-21: Rename as equal improper fractions in lowest terms: **(a)** 2.8   **(b)** 49.05 **(c)** 8.003

Solution

$$\begin{array}{cccc} & \text{place} & \text{improper} & \text{lowest} \\ & \text{value} & \text{fraction} & \text{terms} \\ & \downarrow & \downarrow & \downarrow \end{array}$$

(a) 2.8 $=$ 28 tenths $= \frac{28}{10} = \frac{14}{5}$

(b) 49.05 $=$ 4905 hundredths $= \frac{4905}{100} = \frac{981}{20}$

(c) 8.003 $=$ 8003 thousandths $= \frac{8003}{1000}$

C. Rename decimals to get more decimal places.

Each digit in the decimal-fraction part of a decimal counts as one **decimal place.**

EXAMPLE 8-22: Find the number of decimal places in the following decimals: **(a)** 0.70   **(b)** 21.5 **(c)** 0.001

Solution

(a) 0.70 has 2 decimal places because there are 2 digits in the decimal-fraction part.
(b) 21.5 has 1 decimal place because there is 1 digit in the decimal-fraction part.
(c) 0.001 has 3 decimal places because there are 3 digits in the decimal-fraction part.

To rename a decimal as an equal decimal containing more decimal places, you can add the necessary number of zeros on the right-hand side of the decimal-fraction part.

EXAMPLE 8-23: Rename 1.5 as an equal decimal with **(a)** 2 decimal places;
(b) 3 decimal places;
(c) 4 decimal places.

Solution: 1.5 = 1.50 ⟵── 2 decimal places
1.5 = 1.500 ⟵── 3 decimal places
1.5 = 1.5000 ⟵── 4 decimal places

Note: Any number of zeros can be written on the right-hand side of the decimal-fraction part of a decimal without changing the value of that decimal.

8-3. Adding with Decimals

A. Add decimals.

To join two or more decimal amounts together, you **add decimals.**

To add decimals, you can first rename as equal fractions.

EXAMPLE 8-24: Add the following decimals using fractions: 0.3 + 0.4

Solution: 0.3 + 0.4 = 3 tenths + 4 tenths   Rename as equal fractions.

$\qquad\qquad = \frac{3}{10} + \frac{4}{10}$ ⟵── equal fractions

$\qquad\qquad = \frac{7}{10}$ Add fractions.

$\qquad\qquad = 7$ tenths   Rename as an equal decimal.

$\qquad\qquad = 0.7$ ⟵── sum of 0.3 + 0.4

Note: It is much easier to add decimals by first writing in vertical form, and then adding as you would for whole numbers, remembering to write the decimal point in the answer.

EXAMPLE 8-25: Add the following decimals in vertical form: 0.3 + 0.4

Solution

1. Write in vertical form.

$$
\begin{array}{r}
0.3 \\
+0.4 \\
\hline
\end{array}
$$

↑

align
decimal
points

2. Add as you would for whole numbers.

$$
\begin{array}{r}
0.3 \\
+0.4 \\
\hline
0\ 7
\end{array}
$$

↑ ↑

3 + 4 = 7
0 + 0 = 0

3. Write the decimal point in the answer.

$$
\begin{array}{r}
0.3 \\
+0.4 \\
\hline
0.7
\end{array}
$$

↑

write the decimal
point directly
below the other
decimal points

When decimals do not have the same number of decimal places, you will usually find it easier to add if you first rename to get an equal number of decimal places in each decimal.

EXAMPLE 8-26: Add the following decimals: 0.5 + 2.73 + 18.2 + 1.125

Solution:

$$
\begin{array}{r}
{}^{1\ \ 1} \\
0.5\ 0\ 0 \longleftarrow 0.5 = 0.500 \\
2.7\ 3\ 0 \longleftarrow 2.73 = 2.730 \\
1\ 8.2\ 0\ 0 \longleftarrow 18.2 = 18.200 \\
+\ \ 1.1\ 2\ 5 \\
\hline
2\ 2.5\ 5\ 5
\end{array}
$$

Write in vertical form.
Rename to get an equal number of decimal places.
Add as you would for whole numbers.

Write the decimal point in the answer.

B. Add decimals and whole numbers.

To add decimals and whole numbers, you first rename each whole number as an equal decimal with the same number of decimal places as the given decimal and then add as before.

EXAMPLE 8-27: Add the following whole number and decimal: 258 + 63.49

Solution:

$$
\begin{array}{r}
{}^{1\ \ 1} \\
2\ 5\ 8.0\ 0 \longleftarrow 258 = 258. = 258.00 \\
+\ \ 6\ 3.4\ 9 \\
\hline
3\ 2\ 1.4\ 9
\end{array}
$$

Rename to get the same
number of decimal places.

C. Add with money.

To **add with money,** you first rename each dollar-and-cent amount as an equal decimal with 2 decimal places and then add as before.

EXAMPLE 8-28: Add the following amounts of money: $54.62 + $15.76 + 87¢ + $9 + 15¢ + 9¢

Solution:

$$
\begin{array}{r}
{}^{2\ \ 2\ \ 2} \\
\$5\ 4.6\ 2 \\
1\ 5.7\ 6 \\
0.8\ 7 \longleftarrow 87¢ = 87 \text{ hundredths} = \$0.87 \\
9.0\ 0 \longleftarrow \$9 = \$9.00 \ \ \ \ \ \ \ \ = \$9.00 \\
0.1\ 5 \longleftarrow 15¢ = 15 \text{ hundredths} = \$0.15 \\
0.0\ 9 \longleftarrow 9¢ = 9 \text{ hundredths} = \$0.09 \\
\hline
\$8\ 0.4\ 9
\end{array}
$$

8-4. Subtracting with Decimals

A. Subtract decimals.

To take one decimal amount away from another, you **subtract decimals.**

To subtract decimals, you can first rename the decimals as equal fractions.

EXAMPLE 8-29: Subtract the following decimals using fractions: $0.7 - 0.3$

Solution: $0.7 - 0.3 = 7$ tenths $- 3$ tenths $\qquad$ Rename as equal fractions.

$\qquad\qquad = \frac{7}{10} - \frac{3}{10}$ ←— like fractions

$\qquad\qquad = \frac{4}{10}$ $\qquad\qquad\qquad\qquad$ Subtract like fractions.

$\qquad\qquad = 4$ tenths $\qquad\qquad\qquad$ Rename as an equal decimal.

$\qquad\qquad = 0.4$ ←— difference of 0.7 and 0.4

Note: It is much easier to subtract decimals by first writing in vertical form, then subtracting as you would for whole numbers, remembering to write the decimal point in the answer.

EXAMPLE 8-30: Subtract the following decimals in vertical form: $0.7 - 0.4$

Solution

1. Write in vertical form. **2.** Subtract as you would for whole numbers. **3.** Write the decimal point in the answer.

$$\begin{array}{r} 0.7 \\ -0.4 \\ \hline \end{array}$$
↑
align
decimal
points

$$\begin{array}{r} 0.7 \\ -0.4 \\ \hline 0\ 3 \end{array}$$
↑ ↑
$7 - 4 = 3$
$0 - 0 = 0$

$$\begin{array}{r} 0.7 \\ -0.4 \\ \hline 0.3 \end{array}$$
↑
write the decimal
point directly
below the other
decimal points

When the minuend has more decimal places than the subtrahend, you will usually find it easier to subtract if you first rename the subtrahend to get an equal number of decimal places in both decimals.

EXAMPLE 8-31: Subtract the following decimals in vertical form: $25.375 - 4.8$

Solution:
$$\begin{array}{r} {}^{4}2\overset{1}{\cancel{5}}.3\ 7\ 5 \\ -\ \ 4.8\ 0\ 0 \\ \hline 2\ 0.5\ 7\ 5 \end{array}$$ ←— $4.8 = 4.800$

Write in vertical form.
Rename to get an equal number of decimal places.
Subtract as you would for whole numbers.
Write the decimal point in the answer.

Caution: When the subtrahend has more decimal places than the minuend, you must rename to get an equal number of decimal places in order to subtract.

EXAMPLE 8-32: Subtract the following decimals in vertical form: $45.2 - 9.125$

Solution:
$$\begin{array}{r} {}^{3}\overset{1}{\cancel{4}}\ 5.\overset{1}{2}\overset{9}{\cancel{0}}\ \overset{1}{0} \\ -\ \ 9.1\ 2\ 5 \\ \hline 3\ 6.0\ 7\ 5 \end{array}$$ ←— $45.2 = 45.200$

Rename to get the same number of decimal places in order to subtract.

B. Subtract decimals and whole numbers.

To subtract a whole number from a decimal, you must first rename the whole number as an equal decimal with the same number of decimal places as the given decimal.

EXAMPLE 8-33: Subtract the following decimal and whole number: $185.36 - 78$

Solution:
$$\begin{array}{r} 1\ \overset{7}{\cancel{8}}\ \overset{1}{5}.3\ 6 \\ -\ \ 7\ 8.0\ 0 \\ \hline 1\ 0\ 7.3\ 6 \end{array}$$ ←— $78 = 78.00$

Rename to get the same number of decimal places.

Caution: To subtract a decimal from a whole number, you must first rename the whole number as an equal decimal with the same number of decimal places as the given decimal.

EXAMPLE 8-34: Subtract the following whole number and decimal: $5 - 0.008$

Solution:
$$\begin{array}{r} {}^{4}{}^{9}{}^{9}{}^{1} \\ 5.\!\!\!\!\diagup 0\!\!\!\!\diagup 0\!\!\!\!\diagup 0 \\ -0.0\ 0\ 8 \\ \hline 4.9\ 9\ 2 \end{array}$$
$\longleftarrow 5 = 5.000$ Rename to get the same number of decimal places in order to subtract.

C. Subtract with money.

To **subtract with money,** you first rename each dollar-and-cent amount as an equal decimal with 2 decimal places and then subtract as before.

EXAMPLE 8-35: Subtract the following amounts of money: $\$18 - 5\cent$

Solution:
$$\begin{array}{r} {}^{7}{}^{9}{}^{1} \\ \$1\ 8.\!\!\!\!\diagup 0\!\!\!\!\diagup 0 \\ -\quad 0.0\ 5 \\ \hline \$1\ 7.9\ 5 \end{array}$$
$\longleftarrow \$18 = \$18.00$
$\longleftarrow 5\cent = 5\text{ hundredths} = 0.05$

8-5. Solving Word Problems Containing Decimals $(+, -)$

To solve addition word problems containing decimals, you find how much two or more amounts are all together.

To solve subtraction word problems containing decimals, you find
(a) how much is left;
(b) how much more is needed;
(c) the difference between two amounts.

EXAMPLE 8-36: Solve the following word problems using addition or subtraction $(+, -)$

1. *Identify:*	Ginger had $25. She found another 59¢. How much did Ginger have then?	**1.** *Identify:*	Eric had $25. He spent 59¢. of it. How much did Eric have then?
2. *Understand:*	The question asks you to join amounts of $25 and 59¢ together.	**2.** *Understand:*	The question asks you to take the amount of 59¢ away from $25.
3. *Decide:*	To join amounts together, you **add.**	**3.** *Decide:*	To take an amount away, you **subtract.**

4. *Compute:*
$$\begin{array}{r} \$2\ 5.0\ 0 \\ +\quad 0.5\ 9 \\ \hline \$2\ 5.5\ 9 \end{array}$$

4. *Compute:*
$$\begin{array}{r} {}^{4}{}^{9}{}^{1} \\ \$2\ 5.\!\!\!\!\diagup 0\!\!\!\!\diagup 0 \\ -\quad 0.5\ 9 \\ \hline \$2\ 4.4\ 1 \end{array}$$

5. *Interpret:* Ginger then had **$25.59.**

5. *Interpret:* Eric then had **$24.41.**

RAISE YOUR GRADES

Can you . . . ?

☑ identify the three parts of a decimal number
☑ locate the correct digit in a decimal given the digit's place-value name
☑ identify the correct place-value name of a given digit in a decimal
☑ write the correct value of a given digit in a decimal
☑ write the short decimal-word name for a given decimal
☑ write the correct decimal number given a short decimal-word name
☑ write the preferred form of a decimal fraction
☑ compare two decimals using the symbols $<$, $>$, or $=$

☑ order two or more decimals from smallest to largest or from largest to smallest
☑ round a decimal up or down as required
☑ rename a whole number as an equal decimal
☑ rename a decimal as an equal whole number when its decimal-fraction part is zero
☑ rename a decimal fraction as an equal proper fraction
☑ rename a fraction with a denominator of 10, or 100, or 1000, · · · as an equal decimal fraction
☑ rename a decimal as an equal mixed number or improper fraction when the whole-number part is not zero
☑ identify the number of decimal places in a decimal number
☑ rename a decimal as an equal decimal containing more decimal places than in the given decimal
☑ add two or more decimals
☑ add decimals and whole numbers
☑ add with money
☑ subtract decimals
☑ subtract decimals and whole numbers
☑ subtract with money
☑ solve word problems containing decimals using addition or subtraction

SUMMARY

1. The numbers used to measure things accurately are called decimal numbers or just decimals.
2. Every decimal has a whole-number part and a decimal-fraction part separated by a decimal point.
3. To help locate the correct digit in a decimal given the digit's place-value name, you can use the following place-value chart:

Whole-Number Part							Decimal-Fraction Part					
millions	hundred thousands	ten thousands	thousands	hundreds	tens	ones	tenths	hundredths	thousandths	ten thousandths	hundred thousandths	millionths

4. To help identify the correct place-value name for a given digit, you can use a place-value chart.
5. Once the place-value name for a given digit is identified, you can write the value of that digit.
6. To write the decimal-word name for a given decimal number, you write the word *and* for the decimal point when the whole number part is not zero.
7. To write the correct short decimal-word name for a given decimal number, you always name the whole number, then name the decimal-fractional part with the place-value name of the last digit on the right.
8. To write the correct decimal number given a short decimal-word name, you write the last digit of the decimal-fraction part so that it has the same value as the given short decimal-word name.
9. When the whole-number part of a decimal is zero, the decimal is called a decimal fraction.
10. A zero (0) is usually written for the whole-number part of a decimal fraction.
11. To compare two decimals, you determine which decimal is greater (larger) or which decimal is less (smaller) or that the two decimals are equal.
12. Two decimals can always be compared using either <, >, or =.
13. The inequality symbols < and > should always point at the smaller number.

14. To order two or more decimals, you list them from smallest to largest or from largest to smallest.
15. To round a decimal when the digit to the right of the digit to be rounded is 5 or more, you round up as follows:
  (a) Increase the digit to be rounded by 1.
  (b) If the rounded digit is in the whole-number part, replace digits to the right of the rounded digit with zeros in the whole-number part.
  (c) Omit all the digits to the right of the rounded digit in the decimal-fraction part.
16. To round a decimal when the digit to the right of the digit to be rounded is less than 5, you round down as follows:
  (a) Leave the digit to be rounded the same.
  (b) If the rounded digit is in the whole number part, replace the digits to the right of the rounded digit with zeros in the whole-number part.
  (c) Omit all the digits to the right of the rounded digit in the decimal-fraction part.
17. Every whole number can be renamed as an equal decimal ($8 = 8.0 = 8.00 = \cdots$).
18. When the decimal-fraction part of a decimal is zero, the decimal can always be renamed as an equal whole number ($8.0 = 8$).
19. To rename a decimal-fraction as an equal proper fraction, you first write the value of the decimal fraction and then rename using tenths, hundredths, thousandths, $\cdots$, so that the denominator of the fraction is 10, 100, 1000, $\cdots$, respectively.
20. When a fraction has a denominator of 10, 100, 1000, $\cdots$, you can rename that fraction as an equal decimal fraction using tenths, hundredths, thousandths, $\cdots$, respectively.
21. When a fraction has a denominator that is not 10, 100, 1000, $\cdots$, you can rename that fraction as an equal decimal using the methods shown in Section 9-6.
22. When the whole-number part of a decimal is not zero, the decimal can be renamed as an equal mixed number or improper fraction.
23. In a decimal number, each digit in the decimal-fraction part counts as one decimal place.
24. To rename a decimal as an equal decimal containing more decimal places, you can add the necessary number of zeros on the right-hand side of the decimal-fraction part without changing the value of the original decimal.
25. To join two or more decimal amounts together, you add decimals.
26. To take one decimal amount away from another, you subtract decimals.
27. To add or subtract decimals, you
  (a) write in vertical form while aligning the decimal points and like values;
  (b) rename to get an equal number of decimal places as necessary;
  (c) add or subtract as you would for whole numbers;
  (d) write the decimal point in the answer directly below the other decimal points.
28. To add or subtract decimals and whole numbers, you first rename each whole number as an equal decimal with the same number of decimal places as the given decimal.
29. To add or subtract with money, you first rename each dollar-and-cent amount as an equal decimal with 2 decimal places then add or subtract as before.
30. To solve addition word problems containing decimals, you find how much two or more amounts are all together.
31. To solve subtraction word problems containing decimals, you find
  (a) how much is left;
  (b) how much more is needed;
  (c) the difference between two amounts.

SOLVED PROBLEMS

PROBLEM 8-1   Draw a line under the digit in each decimal that has the given value:

(a) hundredths: 5813.490267
(b) millionths: 4871.032965
(c) thousandths: 1026.589736
(d) tenths: 6105.478329
(e) hundred thousandths: 7580.613294
(f) hundreds: 4259.836107
(g) thousands: 2768.105934
(h) ten thousandths: 8315.069274

Solution: Recall that to help locate the correct digit in a decimal, you can use a place-value chart [see Example 8-3]:

(a) 5813.4<u>9</u>0267   (b) 4871.03296<u>5</u>   (c) 1026.589<u>7</u>36   (d) 6105.<u>4</u>78329
(e) 7580.6132<u>9</u>4   (f) 4<u>2</u>59.836107   (g) <u>2</u>768.105934   (h) 8315.069<u>2</u>74

PROBLEM 8-2   Write the place-value name for each underlined digit:

(a) 3587.01924<u>6</u>   (b) 5813.<u>6</u>49072   (c) 1705.8<u>9</u>2364   (d) 7258.39614<u>0</u>
(e) 6278.30<u>9</u>145   (f) 2340.785<u>1</u>69   (g) 486<u>5</u>.079132   (h) 9<u>0</u>75.481263

Solution: Recall that to help identify the correct place-value name for a given digit, you can use a place-value chart [see Example 8-4]:

(a) hundred thousandths   (b) tenths   (c) hundredths   (d) millionths   (e) thousandths
(f) ten thousandths   (g) ones   (h) hundreds

PROBLEM 8-3   Write the value of 3 in each decimal:

(a) 5281.637094   (b) 8157.024693   (c) 2879.145306   (d) 7283.156940
(e) 4170.329568   (f) 1235.049876   (g) 6248.013579   (h) 9257.018436

Solution: Recall that to write the value for a given digit, you first identify the place-value name of that digit [see Example 8-5]:

(a) 3 hundredths or 0.03   (b) 3 millionths or 0.000003   (c) 3 ten thousandths or 0.0003
(d) 3 ones or 3   (e) 3 tenths or 0.3   (f) 3 tens or 30   (g) 3 thousandths or 0.003
(h) 3 hundred thousandths or 0.00003

PROBLEM 8-4   Write the short decimal-word name for each decimal number:

(a) 1.5   (b) 3.25   (c) 25.075   (d) 187.0015   (e) 0.125   (f) 0.05   (g) 0.000001   (h) 0.00279

Solution: Recall that to write the short decimal-word name for a given decimal number, you write *and* for the decimal point when the whole-number part is not zero and then name the whole number and the decimal-fraction part with the place-value name of the last digit on the right [see Examples 8-6 and 8-7]:

(a) 1 and 5 tenths   (b) 3 and 25 hundredths   (c) 25 and 75 thousandths
(d) 187 and 15 ten thousandths   (e) 125 thousandths   (f) 5 hundredths   (g) 1 millionth
(h) 279 hundred thousandths

PROBLEM 8-5   Write the decimal number for each short decimal-word name:
(a) 1 and 375 thousandths   (b) 2 and 3 tenths   (c) ninety-nine and 9 hundredths
(d) 3 and 2 ten thousandths   (e) 5 millionths   (f) 36 hundred thousandths
(g) 2 and 5 tenths   (h) 3259 hundredths

Solution: Recall that to write the correct decimal number for a short decimal-word name, you write the last digit of the decimal-fraction part so that it has the same value as the given short decimal-word name [see Example 8-8]:

(a) 1.375   (b) 2.3   (c) 99.09   (d) 3.0002   (e) 0.000005   (f) 0.00036   (g) 2.5
(h) 32.59

PROBLEM 8-6   Which is the smallest decimal in the following pairs of decimals? (a) 0.9 and 1.0
(b) 0.05 and 0.5   (c) 1.03 and 0.4   (d) 25.0 and 25   (e) 49 and 48.9
(f) 0.625 and 0.75   (g) 3.14 and 3.141   (h) 2.25 and 3.99

Solution: Recall that to compare two decimals, you may find it helpful to write one decimal under the other while aligning the decimal points and like place values [see Example 8-11]:

(a) 0.9   (b) 0.05   (c) 0.4   (d) 25.0 = 25   (e) 48.9   (f) 0.625   (g) 3.14   (h) 2.25

PROBLEM 8-7 List each group of decimals from smallest to largest: **(a)** 0.5, 0.8, 0.1
(b) 0.01, 0.1, 0.001   **(c)** 1.5, 0.5, 1.0   **(d)** 2.5, 0.75, 1.99   **(e)** 0.25, 0.125, 0.75, 0.375
(f) 5.49, 6.25, 4.99, 12.00   **(g)** 0.125384, 0.125834, 0.125348, 0.125483, 0.125843, 0.125438

Solution: Recall that to order numbers from smallest to largest, you write the smallest number first, then the next smallest, and so on, until you have listed all the given numbers [see Example 8-12]:

(a) 0.1, 0.5, 0.8   **(b)** 0.001, 0.01, 0.1   **(c)** 0.5, 1.0, 1.5   **(d)** 0.75, 1.99, 2.5
(e) 0.125, 0.25, 0.375, 0.75   **(f)** 4.99, 5.49, 6.25, 12.00
(g) 0.125348, 0.125384, 0.125438, 0.125483, 0.125834, 0.125843

PROBLEM 8-8 Round each decimal to the given place:

(a) 0.25 to the nearest tenth
(c) 1.82539 to the nearest thousandth
(e) 0.851246 to the nearest hundred thousandth
(g) 4582.1675 to the nearest hundred
(i) 8213.92 to the nearest thousand

(b) 0.138 to the nearest hundredth
(d) 0.238504 to the nearest ten thousandth
(f) 0.834 to the nearest tenth
(h) 3487.139 to the nearest whole number
(j) 8056.2 to the nearest ten

Solution: Recall that if the digit on the right of the digit to be rounded is 5 or more, you round up; if it is less than 5, you round down [see Examples 8-14 and 8-15]: **(a)** 0.3   **(b)** 0.14   **(c)** 1.825
(d) 0.2385   **(e)** 0.85125   **(f)** 0.8   **(g)** 4600   **(h)** 3487   **(i)** 8000   **(j)** 8060

PROBLEM 8-9 Rename each whole number as an equal decimal:
(a) 5   **(b)** 26   **(c)** 0   **(d)** 1

Solution: Recall that to rename a whole number as an equal decimal, you write the decimal point following the whole number and then write one or more zeros following the decimal point [see Example 8-16]:

(a) $5 = 5.0 = 5.00 = 5.000 = \cdots$ **(b)** $26 = 26.0 = 26.00 = 26.000 = \cdots$
(c) $0 = 0.0 = 0.00 = 0.000 = \cdots$ **(d)** $1 = 1.0 = 1.00 = 1.000 = \cdots$

PROBLEM 8-10 Rename each decimal as an equal whole number: **(a)** 2.0   **(b)** 5.000
(c) 1.00   **(d)** 0.0000   **(e)** 20.0   **(f)** 89.00   **(g)** 125.000   **(h)** 10.0000

Solution: Recall that when the decimal-fraction part of a decimal is zero, the decimal can always be renamed as an equal whole number [see Example 8-17]: **(a)** $2.\cancel{0} = 2. = 2$ **(b)** $5.\cancel{000} = 5. = 5$
(c) $1.\cancel{00} = 1. = 1$ **(d)** $0.\cancel{0000} = 0. = 0$ **(e)** $20.\cancel{0} = 20. = 20$ **(f)** $89.\cancel{00} = 89. = 89$
(g) $125.\cancel{000} = 125. = 125$ **(h)** $10.\cancel{0000} = 10. = 10$

PROBLEM 8-11 Rename each decimal fraction as an equal proper fraction in simplest form:
(a) 0.1   **(b)** 0.7   **(c)** 0.4   **(d)** 0.8   **(e)** 0.75   **(f)** 0.01   **(g)** 0.125   **(h)** 0.025
(i) 0.0625

Solution: Recall that to help rename a decimal-fraction as an equal proper fraction, you first write the value of the decimal fraction, and then rename using tenths, hundredths, thousandths, $\cdots$, respectively [see Example 8-18]:

(a) $0.1 = 1 \text{ tenth} = \dfrac{1}{10}$

(b) $0.7 = 7 \text{ tenths} = \dfrac{7}{10}$

(c) $0.4 = 4 \text{ tenths} = \dfrac{4}{10} = \dfrac{\cancel{2} \times 2}{\cancel{2} \times 5} = \dfrac{2}{5}$

(d) $0.8 = 8 \text{ tenths} = \dfrac{8}{10} = \dfrac{\cancel{2} \times 4}{\cancel{2} \times 5} = \dfrac{4}{5}$

(e) $0.75 = 75 \text{ hundredths} = \dfrac{75}{100} = \dfrac{3 \times \cancel{25}}{4 \times \cancel{25}} = \dfrac{3}{4}$

(f) $0.01 = 1 \text{ hundredth} = \dfrac{1}{100}$

(g) $0.125 = 125 \text{ thousandths} = \dfrac{125}{1000} = \dfrac{1 \times \cancel{125}}{8 \times \cancel{125}} = \dfrac{1}{8}$

(h) $0.025 = 25 \text{ thousandths} = \dfrac{25}{1000} = \dfrac{1 \times 25}{40 \times 25} = \dfrac{1}{40}$

(i) $0.0625 = 625 \text{ ten thousandths} = \dfrac{625}{10,000} = \dfrac{1 \times 625}{16 \times 625} = \dfrac{1}{16}$

PROBLEM 8-12 Rename each fraction as an equal decimal: **(a)** $\frac{3}{10}$ **(b)** $\frac{9}{10}$ **(c)** $\frac{11}{10}$ **(d)** $\frac{23}{10}$ **(e)** $\frac{9}{100}$ **(f)** $\frac{99}{100}$ **(g)** $\frac{7}{1000}$

Solution: Recall that when a fraction has a denominator of 10, 100, 1000, $\cdots$, you can rename that fraction as an equal decimal using tenths, hundredths, thousandths, $\cdots$, respectively [see Example 8-19]:

(a) $\frac{3}{10} = 3 \text{ tenths} = 0.3$ **(b)** $\frac{9}{10} = 9 \text{ tenths} = 0.9$ **(c)** $\frac{11}{10} = 11 \text{ tenths} = 1.1$
(d) $\frac{23}{10} = 23 \text{ tenths} = 2.3$ **(e)** $\frac{9}{100} = 9 \text{ hundredths} = 0.09$ **(f)** $\frac{99}{100} = 99 \text{ hundredths} = 0.99$
(g) $\frac{7}{1000} = 7 \text{ thousandths} = 0.007$

PROBLEM 8-13 Rename each decimal as an equal mixed number in simplest form: **(a)** 5.3 **(b)** 2.9 **(c)** 1.5 **(d)** 3.2 **(e)** 4.25 **(f)** 8.03 **(g)** 6.375 **(h)** 10.075 **(i)** 1.3125

Solution: Recall that when the whole-number part of a decimal is not zero, the decimal can be renamed as an equal mixed number using the short decimal-word name [see Example 8-20]:

(a) $5.3 = 5 \text{ and } 3 \text{ tenths} = 5 \text{ and } \frac{3}{10} = 5\frac{3}{10}$

(b) $2.9 = 2 \text{ and } 9 \text{ tenths} = 2 \text{ and } \frac{9}{10} = 2\frac{9}{10}$

(c) $1.5 = 1 \text{ and } 5 \text{ tenths} = 1 \text{ and } \frac{5}{10} = 1\frac{5}{10} = 1\frac{1}{2}$

(d) $3.2 = 3 \text{ and } 2 \text{ tenths} = 3 \text{ and } \frac{2}{10} = 3\frac{2}{10} = 3\frac{1}{5}$

(e) $4.25 = 4 \text{ and } 25 \text{ hundredths} = 4 \text{ and } \frac{25}{100} = 4\frac{1}{4}$

(f) $8.03 = 8 \text{ and } 3 \text{ hundredths} = 8 \text{ and } \frac{3}{100} = 8\frac{3}{100}$

(g) $6.375 = 6 \text{ and } 375 \text{ thousandths} = 6 \text{ and } \frac{375}{1000} = 6\frac{3}{8}$

(h) $10.075 = 10 \text{ and } 75 \text{ thousandths} = 10 \text{ and } \frac{75}{1000} = 10\frac{75}{1000} = 10\frac{3}{40}$

(i) $1.3125 = 1 \text{ and } 3125 \text{ ten thousandths} = 1 \text{ and } \frac{3125}{10,000} = 1\frac{3125}{10,000} = 1\frac{5}{16}$

PROBLEM 8-14 Rename each decimal as an equal improper fraction in lowest terms: **(a)** 1.6 **(b)** 2.5 **(c)** 3.625 **(d)** 4.875 **(e)** 1.75 **(f)** 2.03 **(g)** 3.25 **(h)** 2.6875

Solution: Recall that when the whole-number part of a decimal is not zero, the decimal can be renamed as an equal improper fraction using place values [see Example 8-21]:

(a) $1.6 = 16 \text{ tenths} = \dfrac{16}{10} = \dfrac{2 \times 8}{2 \times 5} = \dfrac{8}{5}$ **(b)** $2.5 = 25 \text{ tenths} = \dfrac{25}{10} = \dfrac{5 \times 5}{2 \times 5} = \dfrac{5}{2}$

(c) $3.625 = 3625 \text{ thousandths} = \dfrac{3625}{1000} = \dfrac{29 \times 125}{8 \times 125} = \dfrac{29}{8}$

(d) $4.875 = 4875 \text{ thousandths} = \dfrac{4875}{1000} = \dfrac{39 \times 125}{8 \times 125} = \dfrac{39}{8}$

(e) $1.75 = 175 \text{ hundreds} = \dfrac{175}{100} = \dfrac{7 \times 25}{4 \times 25} = \dfrac{7}{4}$ **(f)** $2.03 = 203 \text{ hundreds} = \dfrac{203}{100}$

(g) $3.25 = 325 \text{ hundredths} = \dfrac{325}{100} = \dfrac{13 \times 25}{4 \times 25} = \dfrac{13}{4}$

(h) $2.6875 = 26,875 \text{ ten thousandths} = \dfrac{26,875}{10,000} = \dfrac{43 \times 625}{16 \times 625} = \dfrac{43}{16}$

PROBLEM 8-15 Write the number of decimal places in each decimal number: **(a)** 1.3 **(b)** 5.2 **(c)** 6.39 **(d)** 2.05 **(e)** 5.60 **(f)** 8.00 **(g)** 3.125 **(h)** 6.0300

Solution: Recall that in a decimal number, each digit in the decimal-fraction part counts as one decimal place [see Example 8-22]:

(a) 1.3 has 1 decimal place because there is 1 digit in the decimal-fraction part.
(b) 5.2 has 1 decimal place because there is 1 digit in the decimal-fraction part.
(c) 6.39 has 2 decimal places because there are 2 digits in the decimal-fraction part.
(d) 2.05 has 2 decimal places because there are 2 digits in the decimal-fraction part.
(e) 5.60 has 2 decimal places because there are 2 digits in the decimal-fraction part.
(f) 8.00 has 2 decimal places because there are 2 digits in the decimal-fraction part.
(g) 3.125 has 3 decimal places because there are 3 digits in the decimal-fraction part.
(h) 6.0300 has 4 decimal places because there are 4 digits in the decimal-fraction part.

PROBLEM 8-16 Rename each number as an equal decimal with the given number of decimal places:

(a) 3 with 1 decimal place **(b)** 5 with 2 decimal places **(c)** 2 with 3 decimal places
(d) 9 with 4 decimal places **(e)** 2.9 with 2 decimal places **(f)** 0.1 with 3 decimal places
(g) 10.37 with 5 decimal places **(h)** 5.07 with 4 decimal places

Solution: Recall that to rename a decimal as an equal decimal containing more decimal places, you can add the necessary number of zeros on the right-hand side of the decimal-fraction part [see Example 8-23]:

(a) 3 = 3. = 3.0 **(b)** 5 = 5. = 5.00 **(c)** 2 = 2. = 2.000 **(d)** 9 = 9. = 9.0000
(e) 2.9 = 2.90 **(f)** 0.1 = 0.100 **(g)** 10.37 = 10.37000 **(h)** 5.07 = 5.0700

PROBLEM 8-17 Add decimals: **(a)** 1.5 + 0.8 **(b)** 8.4 + 7.9 **(c)** 12.49 + 5.89
(d) 16.785 + 0.792 **(e)** 0.85 + 10.6 **(f)** 51.6 + 8.925
(g) 15.49 + 8.79 + 8.49 + 5.99 + 25.69 **(h)** 0.812 + 0.05 + 0.1 + 0.725 + 0.3495 + 0.9

Solution: Recall that to add decimals, you first write in vertical form while aligning decimal points and like values, then rename to get an equal number of decimal places as needed. Add as you would for whole numbers, and then write the decimal point in the answer directly below the other decimal points [see Examples 8-24, 8-25, and 8-26]:

(a)
```
    1
  0.5
+0.8
─────
  1.3
```

(b)
```
      1
    8.4
+   7.9
──────
  1 6.3
```

(c)
```
    1 1
  1 2.4 9
+   5.8 9
────────
  1 8.3 8
```

(d)
```
      1 1
  1 6.7 8 5
+   0.7 9 2
──────────
  1 7.5 7 7
```

(e)
```
    1
    0.8 5
+1 0.6 0
────────
  1 1.4 5
```

(f)
```
    1 1
  5 1.6 0 0
+   8.9 2 5
──────────
  6 0.5 2 5
```

(g)
```
    3 3 4
  1 5.4 9
    8.7 9
    8.4 9
    5.9 9
+2 5.6 9
────────
  6 4.4 5
```

(h)
```
    2 1 1
  0.8 1 2 0
  0.0 5 0 0
  0.1 0 0 0
  0.7 2 5 0
  0.3 4 9 5
+0.9 0 0 0
──────────
  2.9 3 6 5
```

PROBLEM 8-18 Add decimals and whole numbers: **(a)** 3 + 0.1 **(b)** 0.5 + 2 **(c)** 8 + 1.2
(d) 3.5 + 1 **(e)** 4 + 1.25 **(f)** 7.49 + 12 **(g)** 5 + 0.25 **(h)** 0.125 + 3
(i) 90 + 127.39 **(j)** 689 + 25.185

Solution: Recall that to add decimals and whole numbers, you first rename each whole number as an equal decimal with the necessary number of decimal places and then add as before [see Example 8-28]:

(a)
```
  3.0  ⟵── 3 = 3. = 3.0
+0.1
─────
  3.1
```

(b)
```
  0.5
+2.0  ⟵── 2 = 2. = 2.0
─────
  2.5
```

(c)
```
  8.0  ⟵── 8 = 8. = 8.0
+1.2
─────
  9.2
```

(d)   3.5
      +1.0 ⟵ 1 = 1. = 1.0
      ‾‾‾‾‾
       4.5

(e)   4.0 0 ⟵ 4 = 4. = 4.00
      +1.2 5
      ‾‾‾‾‾‾‾
       5.2 5

(f)   7.4 9
      +1 2.0 0 ⟵ 12 = 12. = 12.00
      ‾‾‾‾‾‾‾‾‾
       1 9.4 9

(g)   5.0 0 ⟵ 5 = 5. = 5.00
      +0.2 5
      ‾‾‾‾‾‾‾
       5.2 5

(h)   0.1 2 5
      +3.0 0 0 ⟵ 3 = 3. = 3.000
      ‾‾‾‾‾‾‾‾‾
       3.1 2 5

(i)   9 0.0 0 ⟵ 90 = 90. = 90.00
      +1 2 7.3 9
      ‾‾‾‾‾‾‾‾‾‾‾
       2 1 7.3 9

(j)   1 1
          6 8 9.0 0 0 ⟵ 689 = 689. = 689.000
      +   2 5.1 8 5
      ‾‾‾‾‾‾‾‾‾‾‾‾‾
       7 1 4.1 8 5

PROBLEM 8-19   Add with money: **(a)** $15.85 + $4.19   **(b)** $6.37 + $25   **(c)** 87¢ + 37¢
(d) 79¢ + 95¢   **(e)** $8.75 + 27¢   **(f)** 85¢ + $15
(g) $8.21 + $4.75 + $3.99 + $45 + 27¢ + 32¢   **(h)** 15¢ + $5 + $8.42 + $125.38 + 87¢ + $59

Solution: Recall that to add with money, you first rename each dollar-and-cent amount as an equal decimal with 2 decimal places and then add as before [see Example 8-29]:

(a)   1 1 1
       $1 5.8 5
      +   4.1 9
      ‾‾‾‾‾‾‾‾‾
       $2 0.0 4

(b)   1
         $ 6.3 7
       + 2 5.0 0 ⟵ $25 = $25.00
      ‾‾‾‾‾‾‾‾‾‾‾
         $3 1.3 7

(c)   1 1
       $0.8 7 ⟵ 87¢ = $0.87
      + 0.3 7 ⟵ 37¢ = $0.37
      ‾‾‾‾‾‾‾
       $1.2 4

(d)   1 1
       $0.7 9 ⟵ 79¢ = $0.79
      + 0.9 5 ⟵ 95¢ = $0.95
      ‾‾‾‾‾‾‾
       $1.7 4

(e)   1 1
       $8.7 5
      + 0.2 7 ⟵ 27¢ = $0.27
      ‾‾‾‾‾‾‾
       $9.0 2

(f) $ 0.8 5 ⟵ 85¢ = $0.85
       + 1 5.0 0 ⟵ $15 = $15.00
      ‾‾‾‾‾‾‾‾‾‾‾
        $1 5.8 5

(g)   2 2 2
       $ 8.2 1
          4.7 5
          3.9 9
        4 5.0 0 ⟵ $45 = $45.00
          0.2 7 ⟵ 27¢ = $0.27
      +   0.3 2 ⟵ 32¢ = $0.32
      ‾‾‾‾‾‾‾‾‾‾
       $6 2.5 4

(h)   2 1 2
       $ 0.1 5 ⟵ 15¢ = $0.15
              5.0 0 ⟵ $5 = $5.00
              8.4 2
          1 2 5.3 8
              0.8 7 ⟵ 87¢ = $0.87
      +   5 9.0 0 ⟵ $59 = $59.00
      ‾‾‾‾‾‾‾‾‾‾‾‾‾
       $1 9 8.8 2

PROBLEM 8-20   Subtract decimals: **(a)** 0.9 − 0.1   **(b)** 8.3 − 3.7   **(c)** 258.79 − 168.85
(d) 4.1857 − 2.3462   **(e)** 4.15 − 0.9   **(f)** 18.125 − 5.3   **(g)** 2.1 − 1.89
(h) 35.75 − 0.3652

Solution: Recall that to subtract decimals, you first write in vertical form while aligning the decimal points and like values, then rename to get an equal number of decimal places as needed. Subtract as you would for whole numbers, and then write the decimal point in the answer directly below the other decimal points [see Examples 8-30, 8-31, 8-32, and 8-33]:

(a)   0.9
       −0.1
      ‾‾‾‾
       0.8

(b)   7 1
        8̸.3
       −3.7
      ‾‾‾‾
       4.6

(c)   1 14 17 1
        2̸ 5̸ 8̸.7 9
       −1 6 8.8 5
      ‾‾‾‾‾‾‾‾‾
         8 9.9 4

(d)   3 1   7 1
        4̸.1 8̸ 5 7
       −2.3 4 6 2
      ‾‾‾‾‾‾‾‾‾
        1.8 3 9 5

(e) $\overset{3}{\overset{1}{\cancel{4}}}.1\ 5$
      $-0.9\ 0$
      $\overline{3.2\ 5}$

(f) $1\ \overset{7}{\overset{1}{\cancel{8}}}.1\ 2\ 5$
      $-\quad 5.3\ 0\ 0$
      $\overline{1\ 2.8\ 2\ 5}$

(g) $\overset{1}{\cancel{2}}\overset{10}{.}\overset{1}{\cancel{1}}\ 0$
      $-1.8\ 9$
      $\overline{0.2\ 1}$

(h) $3\ 5.\overset{6}{\cancel{7}}\ \overset{14}{\cancel{5}}\ \overset{9}{\cancel{0}}\ \overset{1}{0}$
      $-\quad 0.3\ 6\ 5\ 2$
      $\overline{3\ 5.3\ 8\ 4\ 8}$

PROBLEM 8-21   Subtract decimals and whole numbers: **(a)** 5.25 − 3   **(b)** 9.5 − 6
(c) 28.49 − 15   **(d)** 629.029 − 587   **(e)** 5 − 0.1   **(f)** 3 − 1.85   **(g)** 23 − 7.49
(h) 500 − 8.205

Solution: Recall that to subtract decimals and whole numbers, you first rename the whole number as an equal decimal with the necessary number of decimal places and then subtract as before [see Examples 8-34 and 8-35]:

(a) $5.2\ 5$
         $-3.0\ 0$ ⟵ 3 = 3. = 3.00
         $\overline{2.2\ 5}$

(b) 9.5
         -6.0 ⟵ 6 = 6. = 6.0
         $\overline{3.5}$

(c) $2\ 8.4\ 9$
         $-1\ 5.0\ 0$ ⟵ 15 = 15. = 15.00
         $\overline{1\ 3.4\ 9}$

(d) $\overset{5}{\cancel{6}}\overset{1}{2}\ 9.0\ 2\ 9$
         $-5\ 8\ 7.0\ 0\ 0$ ⟵ 587 = 587. = 587.000
         $\overline{4\ 2.0\ 2\ 9}$

(e) $\overset{4}{\cancel{5}}\overset{1}{.}0$ ⟵ 5 = 5. = 5.0
         -0.1
         $\overline{4.9}$

(f) $\overset{2}{\cancel{3}}\overset{9}{.}\overset{1}{\cancel{0}}\ 0$ ⟵ 3 = 3. = 3.00
         $-1.8\ 5$
         $\overline{1.1\ 5}$

(g) $\overset{1}{\cancel{2}}\ \overset{12}{\cancel{3}}.\overset{9}{\cancel{0}}\ \overset{1}{0}$ ⟵ 23 = 23. = 23.00
         $-\quad 7.4\ 9$
         $\overline{1\ 5.5\ 1}$

(h) $\overset{4}{\cancel{5}}\ \overset{9}{\cancel{0}}\ \overset{9}{\cancel{0}}.\overset{9}{\cancel{0}}\ \overset{9}{\cancel{0}}\ \overset{1}{0}$ ⟵ 500 = 500. = 500.000
         $-\quad\quad 8.2\ 0\ 5$
         $\overline{4\ 9\ 1.7\ 9\ 5}$

PROBLEM 8-22   Subtract with money: **(a)** \$25 − \$17   **(b)** \$108 − \$57   **(c)** \$6.89 − \$3
(d) \$27.45 − \$19   **(e)** \$45 − \$8.45   **(f)** \$82 − \$37.99   **(g)** 85¢ − 32¢   **(h)** 91¢ − 17¢
(i) \$5.25 − 63¢   **(j)** \$14.23 − 97¢   **(k)** \$5 − 6¢   **(l)** \$15 − 63¢

Solution: Recall that to subtract with money, you first rename each dollar-and-cent amount as an equal decimal with 2 decimal places (when necessary) and then subtract as before [see Example 8-36]:

(a) $\$\overset{1}{\cancel{2}}\overset{1}{5}$
         $-\ 1\ 7$
         $\overline{\$\quad 8}$

(b) $\$\overset{0}{\cancel{1}}\overset{1}{0}\ 8$
         $-\quad 5\ 7$
         $\overline{\$\quad 5\ 1}$

(c) $\$6.8\ 9$
         $-\ 3.0\ 0$
         $\overline{\$3.8\ 9}$

(d) $\$\overset{1}{\cancel{2}}\overset{1}{7}.4\ 5$
         $-\ 1\ 9.0\ 0$
         $\overline{\$\quad 8.4\ 5}$

(e) $\$\overset{3}{\cancel{4}}\ \overset{14}{\cancel{5}}.\overset{9}{\cancel{0}}\ \overset{1}{0}$
         $-\quad 8.4\ 5$
         $\overline{\$3\ 6.5\ 5}$

(f) $\$\overset{7}{\cancel{8}}\ \overset{11}{\cancel{2}}.\overset{9}{\cancel{0}}\ \overset{1}{0}$
         $-\ 3\ 7.9\ 9$
         $\overline{\$4\ 4.0\ 1}$

(g) $8\ 5¢$
         $-3\ 2¢$
         $\overline{5\ 3¢}$

(h) $\overset{8}{\cancel{9}}\overset{1}{1}¢$
         $-1\ 7¢$
         $\overline{7\ 4¢}$

(i) $\$\overset{4}{\cancel{5}}\overset{1}{.}2\ 5$
         $-\ 0.6\ 3$ ⟵ 63¢ = \$0.63
         $\overline{\$4.6\ 2}$

(j) $\$1\ \overset{3}{\cancel{4}}.\overset{11}{\cancel{2}}\ \overset{1}{3}$
         $-\quad 0.9\ 7$ ⟵ 97¢ = \$0.97
         $\overline{\$1\ 3.2\ 6}$

(k) $\$5.\overset{4}{\cancel{0}}\ \overset{9}{\cancel{0}}\ \overset{1}{0}$
         $-\ 0.0\ 6$ ⟵ 6¢ = \$0.06
         $\overline{\$4.9\ 4}$

(l) $\$1\ 5.\overset{4}{\cancel{0}}\ \overset{9}{\cancel{0}}\ \overset{1}{0}$
         $-\quad 0.6\ 3$ ⟵ 63¢ = \$0.63
         $\overline{\$1\ 4.3\ 7}$

PROBLEM 8-23   Decide what to do (add or subtract with decimals) and then solve the following problems:

(a) A radio cost $50 plus $3.50 sales tax. What is the total cost of the radio?

(b) Hope had $50. She spent $3.50 on magazines. How much did Hope have then?

(c) Mark has $8.75. He needs $25 for tickets to a concert. How much more does Mark need?

(d) Martin has $82.79 and Jane has $53.29. What is the difference between their amounts?

(e) Otis paid $3.17 in tax on a purchase of $52.79. How much did Otis pay in all?

(f) Clara paid 39¢ tax for a total of $8.28. How much was the purchase amount?

(g) Louise earned $485.60 last week and $872.83 this week. She expects to earn $675 next week. How much more did Louise earn this week and last combined than she will earn next week?

(h) Burgess walked 1.5 km the first day, 2.75 km the second day, and 8.2 km the third day. He wants to walk 30 km for the week. How many more kilometers does Burgess need to walk after the third day?

Solution: Recall that to solve addition word problems containing decimals, you find how much two or more amounts are all together; to solve subtraction word problems containing decimals, you find how much is left, how much more is needed, or the difference between two amounts [see Example 8-37]:

(a) add: $50 + $3.50 = $53.50

(b) subtract: $50 − $3.50 = $46.50

(c) subtract: $25 − $8.75 = $16.25

(d) subtract: $82.79 − $53.29 = $29.50

(e) add: $3.17 + $52.79 = $55.96

(f) subtract: $8.28 − $0.39 = $7.89

(g) add and then subtract: $485.60 + $872.83 = $1358.43, and $1358.43 − $675 = $683.43

(h) add and then subtract: 1.5 + 2.75 + 8.2 = 12.45, and 30 − 12.45 = 17.55 (km)

Supplementary Exercises

PROBLEM 8-24   Draw a line under the digit in each decimal that has the given value:

(a) thousandths: 6582.197034

(b) tens: 3075.192864

(c) tenths: 4132.709865

(d) hundredths: 9827.651430

(e) hundreds: 7043.928165

(f) millionths: 5897.264013

(g) ten thousandths: 1087.649235

(h) hundred thousandths: 2540.738169

PROBLEM 8-25   Write the place-value name for each underlined digit:

(a) 6528.13<u>9</u>704

(b) 5764.12083<u>9</u>

(c) 4126.5<u>1</u>8093

(d) 3487.6591<u>0</u>2

(e) 9870.<u>4</u>31625

(f) 8213.597<u>6</u>04

(g) 208<u>7</u>.941365

(h) 7<u>5</u>01.249863

PROBLEM 8-26   Write the value of 2 in each decimal:

(a) 4123.587609

(b) 8570.624139

(c) 2658.901743

(d) 6409.218753

(e) 1587.604932

(f) 9413.850287

(g) 7502.613489

(h) 3415.098726

PROBLEM 8-27   Write the short decimal-word name for each decimal number: (a) 2.6   (b) 1.75   (c) 3.08   (d) 4.375   (e) 0.045   (f) 0.005   (g) 0.0015   (h) 0.0125

PROBLEM 8-28   Write the decimal number for each short decimal-word name: (a) 5 and 5 tenths   (b) 3 and 4 hundredths   (c) 6 and 125 thousandths   (d) 12 and 2 thousands   (e) 3 thousandths   (f) 75 millionths   (g) 815 tenths

PROBLEM 8-29   Write the largest decimal in the following pairs of decimals:

(a) 2.35 and 2.53   (b) 0.01 and 0.001   (c) 3.1 and 3.10   (d) 1.01 and 1.011   (e) 34.89 and 34.90   (f) 26.78 and 25.79   (g) 0.0101 and 0.01

PROBLEM 8-30   List each group of decimals from largest to smallest:

(a) 1.1, 1.111, 1.11   (b) 0.9, 0.99, 0.909   (c) 5.2, 2.5, 25.0, 0.25
(d) 3.1, 3.141, 3.14, 3.1415   (e) 3.025, 3.520, 3.052, 3.250, 3.502, 3.205
(f) 0.125, 12.5, 1.25, 125.0   (g) 1.414, 1, 1.732, 2.828, 2.646, 2, 2.449, 3, 2.236

PROBLEM 8-31   Round each decimal to the given place: (a) 5.8361 to the nearest hundredth
(b) 3.14 to the nearest whole number   (c) 0.81539 to the nearest thousandth
(d) 2.7506 to the nearest tenth   (e) 0.815649 to the nearest ten thousandth
(f) 0.2185 to the nearest hundredth   (g) 139.0005 to the nearest ten

PROBLEM 8-32   Rename each whole number as an equal decimal: (a) 4   (b) 9   (c) 0
(d) 7   (e) 15   (f) 18   (g) 135   (h) 1

PROBLEM 8-33   Rename each decimal as an equal whole number: (a) 5.0   (b) 1.0
(c) 0.00   (d) 4.00   (e) 15.00   (f) 7.000   (g) 8.0   (h) 3.0000

PROBLEM 8-34 Rename each decimal fraction as an equal proper fraction in simplest form: (a) 0.3
(b) 0.9   (c) 0.2   (d) 0.6   (e) 0.25   (f) 0.05   (g) 0.375   (h) 0.1875

PROBLEM 8-35   Rename each fraction as an equal decimal: (a) $\frac{1}{10}$ (b) $\frac{7}{10}$ (c) $\frac{29}{10}$
(d) $\frac{43}{10}$ (e) $\frac{1}{100}$ (f) $\frac{23}{100}$ (g) $\frac{859}{100}$ (h) $\frac{1}{1000}$

PROBLEM 8-36   Rename each decimal as an equal mixed number in simplest form and as an equal
improper fraction in lowest terms: (a) 2.9   (b) 1.3   (c) 3.75   (d) 6.50   (e) 9.125
(f) 7.375

PROBLEM 8-37   Rename each number as an equal decimal with the given number of decimal places:

(a) 2 with 1 decimal places   (b) 4 with 5 decimal places   (c) 8 with 3 decimal places
(d) 1 with 4 decimal places   (e) 1.5 with 2 decimal places   (f) 0.9 with 3 decimal places
(g) 18.25 with 4 decimal places

PROBLEM 8-38   Add basic facts in vertical form:

Row (a)	8 +9	1 +0	3 +3	1 +5	3 +0	0 +1	1 +4	8 +8	5 +5	8 +6
Row (b)	6 +8	3 +1	4 +6	9 +8	9 +2	1 +2	7 +3	6 +0	0 +3	2 +3
Row (c)	8 +1	7 +7	5 +1	4 +2	2 +0	8 +0	2 +7	6 +4	4 +8	7 +4
Row (d)	3 +7	0 +0	0 +7	0 +8	4 +4	0 +4	8 +3	9 +1	9 +4	6 +7
Row (e)	1 +1	3 +5	1 +6	0 +2	6 +2	2 +9	3 +8	5 +9	9 +5	1 +9
Row (f)	9 +0	5 +6	2 +1	4 +7	1 +3	4 +0	3 +6	6 +6	5 +2	6 +5

Row (g)

6	2	4	8	5	5	9	8	7	9
+9	+2	+5	+5	+0	+3	+6	+7	+9	+9

Row (h)

0	8	1	4	0	2	2	7	2	5
+6	+.1	+7	+3	+9	+4	+6	+8	+8	+8

Row (i)

5	4	6	0	4	5	6	3	8	9
+7	+9	+3	+5	+1	+4	+1	+9	+2	+3

Row (j)

3	2	8	7	3	7	7	7	9	7
+2	+5	+4	+2	+4	+1	+5	+0	+7	+6

PROBLEM 8-39 Add with decimals: **(a)** $1.125 + 3.375$ **(b)** $4 + 8.25$ **(c)** $5.1 + 16$
(d) $0.915 + 0.6$ **(e)** $1.4 + 18.35$ **(f)** $123 + 0.123$ **(g)** $14.58 + 8 + 0.06 + 12.1 + 125$

PROBLEM 8-40 Subtract basic facts in vertical form:

Row (a)

4	7	6	15	12	4	1	6	6	7
−0	−6	−5	− 9	− 8	−1	−1	−6	−0	−3

Row (b)

13	8	12	1	7	6	8	8	10	9
− 7	−3	− 4	−0	−1	−2	−5	−0	− 7	−6

Row (c)

10	13	9	7	9	8	4	9	4	17
− 3	− 9	−1	−5	−7	−4	−4	−9	−3	− 9

Row (d)

10	12	12	11	16	11	7	12	3	2
− 4	− 5	− 3	− 2	− 7	− 7	−4	− 6	−1	−2

Row (e)

8	14	3	10	9	4	2	14	12	10
−1	− 6	−2	− 5	−2	−2	−0	− 8	− 7	− 6

Row (f)

11	10	9	11	9	16	5	8	5	10
− 8	− 8	−8	− 9	−0	− 8	−1	−8	−5	− 9

Row (g)

6	7	14	10	5	13	15	13	0	2
−4	−0	− 5	− 1	−0	− 8	− 8	− 4	−0	−1

Row (h)

5	7	8	5	17	9	11	18	11	14
−3	−7	−2	−4	− 8	−4	− 6	− 9	− 4	− 9

Row (i)

13	11	14	9	8	15	8	3	15	10
− 5	− 3	− 7	−3	−6	− 6	−7	−3	− 7	− 2

Row (j)   9   3   6   16   6   12   7   13   5   11
          −5   −0   −1   − 9   −3   − 9   −2   − 6   −2   − 5

PROBLEM 8-41   Subtract with decimals: **(a)** 8.451 − 0.079   **(b)** 6.85 − 3   **(c)** 5 − 2.84
(d) 6.3 − 5.859   **(e)** 7.285 − 6.9   **(f)** 18.49 − 9.99   **(g)** 8 − 0.1   **(h)** 17.5 − 9

PROBLEM 8-42   Add and subtract with money: **(a)** $15.49 + $8.37   **(b)** $19 + $8.50
(c) $21.75 − $16   **(d)** $43 − $29   **(e)** $5 − 97¢   **(f)** 82¢ − 57¢   **(g)** $20 − $8.43
(h) $8.52 + 19.75 + 87¢ + $25 + 83¢ + 95.75 + $253

PROBLEM 8-43   Solve each word problem containing decimals using addition and/or subtraction:

(a) The sales tax on a $349.99 TV set is $17.50. How much is the total cost of the TV set?

(b) One color TV set is $499.99 and a second color TV set of another brand is $649.99. How much would be saved (not counting tax) by buying the cheaper set?

(c) Linda earned $850 this week. She paid $475 for her rent. How much did Linda have left after paying her rent?

(d) Gerry was left with $275 of his pay after paying his rent of $585. How much was Gerry's pay?

(e) Foster should weigh 20.8 kg more than Mary. Mary and Foster now weigh 54.4 kg and 90.7 kg respectively. How much weight will Foster need to lose to reach the desired level if Mary remains the same weight?

(f) Rita gave the clerk $20 for a $7.83 purchase. Then she used her change to purchase $3.99 worth of books. How much of the original $20 did Rita have then?

(g) Burton won $15 the first race, $83.50 the second race and lost $79 the 3rd race. How much ahead was Burton after the third race?

(h) Gail flew 353.8 km on Monday, 567.4 km on Tuesday, and 485.9 km on Wednesday. How far did she fly during the 3 days?

Answers to Supplementary Exercises

(8-24)   **(a)** 6582.197034   **(b)** 3075.192864   **(c)** 4132.709865   **(d)** 9827.651430
(e) 7043.928165   **(f)** 5897.264013   **(g)** 1087.649235   **(h)** 2540.738169

(8-25)   **(a)** thousandths   **(b)** millionths   **(c)** hundredths   **(d)** hundred thousandths
(e) tenths   **(f)** ten thousandths   **(g)** ones   **(h)** hundreds

(8-26)   **(a)** 20   **(b)** 0.02   **(c)** 2000   **(d)** 0.2   **(e)** 0.000002   **(f)** 0.0002   **(g)** 2
(h) 0.00002

(8-27)   **(a)** 2 and 6 tenths   **(b)** 1 and 75 hundredths   **(c)** 3 and 8 hundredths
(d) 4 and 375 thousandths   **(e)** 45 thousandths   **(f)** 5 thousandths
(g) 15 ten thousandths   **(h)** 125 ten thousandths

(8-28)   **(a)** 5.5   **(b)** 3.04   **(c)** 6.125   **(d)** 12.002   **(e)** 0.003   **(f)** 0.000075   **(g)** 81.5

(8-29)   **(a)** 2.53   **(b)** 0.01   **(c)** 3.1 = 3.10   **(d)** 1.011   **(e)** 34.90   **(f)** 26.78
(g) 0.0101

(8-30)   **(a)** 1.111, 1.11, 1.1   **(b)** 0.99, 0.909, 0.9   **(c)** 25.0, 5.2, 2.5, 0.25
(d) 3.1415, 3.141, 3.14, 3.1   **(e)** 3.520, 3.502, 3.250, 3.205, 3.052, 3.025
(f) 125.0, 12.5, 1.25, 0.125   **(g)** 3, 2.828, 2.646, 2.449, 2.236, 2, 1.732, 1.414, 1

(8-31) (a) 5.84 (b) 3 (c) 0.815 (d) 2.8 (e) 0.8156 (f) 0.22 (g) 140

(8-32) (a) 4.0 (b) 9.0 (c) 0.0 (d) 7.0 (e) 15.0 (f) 18.0 (g) 135.0 (h) 1.0

(8-33) (a) 5 (b) 1 (c) 0 (d) 4 (e) 15 (f) 7 (g) 8 (h) 3

(8-34) (a) $\frac{3}{10}$ (b) $\frac{9}{10}$ (c) $\frac{1}{5}$ (d) $\frac{3}{5}$ (e) $\frac{1}{4}$ (f) $\frac{1}{20}$ (g) $\frac{3}{8}$ (h) $\frac{3}{16}$

(8-35) (a) 0.1 (b) 0.7 (c) 2.9 (d) 4.3 (e) 0.01 (f) 0.23 (g) 8.59 (h) 0.001

(8-36) (a) $2\frac{9}{10}, \frac{29}{10}$ (b) $1\frac{3}{10}, \frac{13}{10}$ (c) $3\frac{3}{4}, \frac{15}{4}$ (d) $6\frac{1}{2}, \frac{13}{2}$ (e) $9\frac{1}{8}, \frac{73}{8}$ (f) $7\frac{3}{8}, \frac{59}{8}$

(8-37) (a) 2.0 (b) 4.00000 (c) 8.000 (d) 1.0000 (e) 1.50 (f) 0.900 (g) 18.2500

(8-38)

Row (a)	17	1	6	6	3	1	5	16	10	14
Row (b)	14	4	10	17	11	3	10	6	3	5
Row (c)	9	14	6	6	2	8	9	10	12	11
Row (d)	10	0	7	8	8	4	11	10	13	13
Row (e)	2	8	7	2	8	11	11	14	14	10
Row (f)	9	11	3	11	4	4	9	12	7	11
Row (g)	15	4	9	13	5	8	15	15	16	18
Row (h)	6	9	8	7	9	6	8	15	10	13
Row (i)	12	13	9	5	5	9	7	12	10	12
Row (j)	5	7	12	9	7	8	12	7	16	13

(8-39) (a) 4.5 (b) 12.25 (c) 21.1 (d) 1.515 (e) 19.75 (f) 123.123 (g) 159.74

(8-40)

Row (a)	4	1	1	6	4	3	0	0	6	4
Row (b)	6	5	8	1	6	4	3	8	3	3
Row (c)	7	4	8	2	2	4	0	0	1	8
Row (d)	6	7	9	9	9	4	3	6	2	0
Row (e)	7	8	1	5	7	2	2	6	5	4
Row (f)	3	2	1	2	9	8	4	0	0	1
Row (g)	2	7	9	9	5	5	7	9	0	1
Row (h)	2	0	6	1	9	5	5	9	7	5
Row (i)	8	8	7	6	2	9	1	0	8	8
Row (j)	4	3	5	7	3	3	5	7	3	6

(8-41) (a) 8.372 (b) 3.85 (c) 2.16 (d) 0.441 (e) 0.385 (f) 8.5
(g) 7.9 (h) 8.5

(8-42) (a) $23.86 (b) $27.50 (c) $5.75 (d) $14 (e) $4.03 (f) $0.25 or 25¢
(g) $11.57 (h) $403.72

(8-43) (a) $367.49 (b) $150 (c) $375 (d) $860 (e) 15.5 kg (f) $8.18
(g) $19.50 ahead (h) 1407.1 km

9 DECIMALS: MULTIPLICATION AND DIVISION

9-1. Multiplying with Decimals

A. Multiply a decimal by a whole number.

To join two or more equal decimal amounts together, you **multiply the decimal** amount by the appropriate whole number.

EXAMPLE 9-1: Write the following as multiplication: 3 equal amounts of 0.4

Solution: 3 equal amounts of 0.4 means $0.4 + 0.4 + 0.4$ or 3×0.4 ⟵ multiplication

To multiply a decimal by a whole number, you can first rename both as equal fractions.

EXAMPLE 9-2: Multiply the following decimal and whole number using fractions: 3×0.4

Solution: $3 \times 0.4 = 3 \times 4$ tenths   Rename as equal fractions.

$$= \frac{3}{1} \times \frac{4}{10} \longleftarrow \text{ equal fractions}$$

$$= \frac{3 \times 4}{1 \times 10} \qquad\qquad \text{Multiply fractions.}$$

$$= \frac{12}{10}$$

$$= 12 \text{ tenths} \qquad\qquad \text{Rename as an equal decimal.}$$

$$= 1.2 \longleftarrow 3 \times 0.4$$

It is much easier to multiply a decimal by a whole number by first writing in vertical form, multiplying as you would for whole numbers, and then writing the decimal point in the product.

EXAMPLE 9-3: Multiply the following whole number and decimal in vertical form: 3×0.4

Solution

1. Write in vertical form.

align right-hand digits
↓

```
  0.4
×   3
_____
```

2. Multiply as you would for whole numbers.
$3 \times 4 = 12$

```
  0.4
×   3
_____
  1 2
```

3. Write the decimal point in the product.

```
  0.4  ⟵ 1 decimal place
×   3  ⟵ 0 decimal places
_____
  1.2  ⟵ 1 (1 + 0) decimal places
```

Note: The number of decimal places in the product should equal the total number of decimal places contained in the two factors.

Multiplication Rule for Decimals

1. Write in vertical form while aligning the right-hand digits.
2. Multiply as you would for two whole numbers.
3. Put the same number of decimal places in the product as the total number of decimal places contained in both factors.

EXAMPLE 9-4: Multiply the following whole number and decimal: 25×3.49

Solution

1. Write in vertical form.

align right-hand digits
↓

```
  3.4 9
×   2 5
_____
```

2. Multiply as you would for whole numbers.
$25 \times 349 = 8725$

```
    1
  2 4
  3.4 9
×   2 5
_____
  1 7 4 5
  6 9 8
_____
  8 7 2 5
```

3. Write the decimal point in the product.

```
    1
  2 4
  3.4 9  ⟵ 2 decimal places
×   2 5  ⟵ 0 decimal places
_____
  1 7 4 5
  6 9 8
_____
  8 7.2 5  ⟵ 2 (2 + 0) decimal places
```

Note: When you multiply a decimal by a whole number, the product will always have the same number of decimal places as the decimal factor because a whole number always has 0 decimal places.

B. Multiply decimals.

When both factors are decimals, you must add the number of decimal places in each factor together to find the number of decimal places that belong in the product.

EXAMPLE 9-5: Multiply the following decimals: 5.14×23.6

Solution

1. Write in vertical form.

```
  2 3.6
×5.1 4
_____
```

2. Multiply as you would for whole numbers.
$514 \times 236 = 121304$

```
    1 3
    1 2
  2 3.6
×5.1 4
_____
    9 4 4
  2 3 6
1 1 8 0
_____
1 2 1 3 0 4
```

3. Write the decimal point in the product.

```
    1 3
    1 2
  2 3.6  ⟵ 1 decimal place
×5.1 4  ⟵ 2 decimal places
_____
    9 4 4
  2 3 6
1 1 8 0
_____
1 2 1.3 0 4  ⟵ 3 (1 + 2) decimal places
```

C. **Write zeros in the product.**

When the number of decimal places that belong in the product is greater than the number of digits in the product, you must write zeros in the product until there are enough digits to write the decimal point in the correct place.

EXAMPLE 9-6: Multiply the following decimals: 0.02×0.03

Solution

1. Write in vertical form.

2. Multiply as you would for whole numbers.
$2 \times 3 = 6$

3. Write the decimal point in the product

$$
\begin{array}{r}
0.0\ 3 \\
\times 0.0\ 2 \\
\hline
\end{array}
$$

$$
\begin{array}{r}
0.0\ 3 \\
\times 0.0\ 2 \\
\hline
6
\end{array}
$$

write zeros ⟶
as needed

$0.0\ 3$ ⟵ 2 decimal places
$\times 0.0\ 2$ ⟵ 2 decimal places
$0.0\ 0\ 0\ 6$ ⟵ 4 (2 + 2) decimal places

Note: In Example 9-6, 3 zeros were needed to write the decimal point in the product and a 4th zero was needed to show that the whole-number part of the product is zero.

9-2. **Multiplying by 10, 100, and 1000**

A. **Multiply by 10.**

To multiply a decimal by 10, you can proceed as shown in Examples 9-3 and 9-4.

EXAMPLE 9-7: Multiply the following decimal by 10: 7.49

Solution

1. Write in vertical form.

$$
\begin{array}{r}
7.4\ 9 \\
\times\ \ 1\ 0 \\
\hline
\end{array}
$$

2. Multiply as before.

$7.4\ 9$ ⟵ 2 decimal places
$\times\ \ 1\ 0$ ⟵ 0 decimal places
$7\ 4.9\ 0$ ⟵ 2 (2 + 0) decimal places

Shortcut: To multiply 7.49 by 10, you could have moved the decimal point 1 decimal place to the right: $10 \times 7.49 = 74.9$

Multiply by 10 Rule
To multiply by 10, you move the decimal point 1 decimal place to the right.

EXAMPLE 9-8: Multiply the following decimals by 10: **(a)** 1.89 **(b)** 3.2 **(c)** 0.25 **(d)** 5

Solution
(a) $10 \times 1.89 = 18.9 = 18.9$

(b) $10 \times 3.2\ \ = 32.\ \ = 32$

(c) $10 \times 0.25 = 02.5 = \ \ 2.5$

(d) $10 \times 5\ \ \ \ = 50.\ \ = 50$

B. **Multiply by 100.**

To multiply a decimal by 100, you can proceed as shown in Examples 9-3 and 9-4.

EXAMPLE 9-9: Multiply the following decimal by 100: 0.125

Solution

1. Write in vertical form.   **2.** Multiply as before.

$$
\begin{array}{r}
0.1\ 2\ 5 \\
\times\quad 1\ 0\ 0 \\
\hline
\end{array}
$$

$$
\begin{array}{r}
0.1\ 2\ 5 \longleftarrow 3 \text{ decimal places} \\
\times\qquad 1\ 0\ 0 \longleftarrow 0 \text{ decimal places} \\
\hline
1\ 2.5\ 0\ 0 \longleftarrow 3\ (3+0) \text{ decimal places}
\end{array}
$$

Shortcut: To multiply 0.125 by 100, you could have moved the decimal point 2 decimal places to the right: $100 \times 0.125 = 012.5 = 12.5$

Multiply by 100 Rule

To multiply by 100, you can move the decimal point 2 decimal places to the right.

EXAMPLE 9-10: Multiply the following decimals by 100: **(a)** 0.059   **(b)** 0.02   **(c)** 2.5   **(d)** 39

Solution

(a) $100 \times 0.059 = \quad 005.9 \quad = 5.9$

(b) $100 \times 0.02 = \quad 002. \quad = 2$

(c) $100 \times 2.5 = \quad 250. \quad = 250. \quad = 250$

(d) $100 \times 39 = 100 \times 39. = 3900. = 3900$

C. Multiply by 1000.

To multiply by 1000, you can proceed as shown in Examples 9-3 and 9-4.

EXAMPLE 9-11: Multiply the following decimal by 1000: 3.14159

Solution

1. Write in vertical form.   **2.** Multiply as before.

$$
\begin{array}{r}
3.1\ 4\ 1\ 5\ 9 \\
\times\qquad 1\ 0\ 0\ 0 \\
\hline
\end{array}
$$

$$
\begin{array}{r}
3.1\ 4\ 1\ 5\ 9 \longleftarrow 5 \text{ decimal places} \\
\times\qquad\qquad 1\ 0\ 0\ 0 \longleftarrow 0 \text{ decimal places} \\
\hline
3\ 1\ 4\ 1.5\ 9\ 0\ 0\ 0 \longleftarrow 5\ (5+0) \text{ decimal places}
\end{array}
$$

Shortcut: To multiply 3.14159 by 1000, you could have moved the decimal point 3 decimal places to the right: $1000 \times 3.14159 = 3141.59$

Multiply by 1000 Rule

To multiply by 1000, you move the decimal point 3 decimal places to the right.

EXAMPLE 9-12: Multiply the following numbers by 1000: **(a)** 0.2576   **(b)** 0.002   **(c)** 1.3 **(d)** 8

Solution

(a) $1000 \times 0.2576 = 0257.6 = 257.6$

(b) $1000 \times 0.002 = 0002. = 0002 = 2$

(c) $1000 \times 1.3 = 1300. = 1300$

(d) $1000 \times 8 = 8000. = 8000$

D. Multiply by powers of 10.

The whole numbers 10, 100, 1000, $\cdots$, are called **powers of 10.** The following rule is a basic rule for multiplying by 10, 100, 1000, or any power of 10.

Multiply by Powers of 10 Rule

To multiply by a power of 10, you move the decimal point 1 decimal place to the right for each zero in the power of 10.

EXAMPLE 9-13: Multiply 0.75 by **(a)** 10   **(b)** 100   **(c)** 1000   **(d)** 10,000.

Solution

(a) 10×0.75 $= 07.5$ $= 7.5$ One zero in 10 means 1 decimal place to the right.

(b) 100×0.75 $= 075.$ $= 75$ Two zeros in 100 means 2 decimal places to the right.

(c) 1000×0.75 $= 0750.$ $= 750$ Three zeros in 1000 means 3 decimal places to the right.

(d) $10,000 \times 0.75 = 07500.$ $= 7500$ Four zeros in 10,000 means 4 decimal places to the right.

9-3.   Dividing with Decimals

A. Divide a decimal by a whole number.

To separate a decimal amount into two or more equal amounts, you **divide the decimal** amount by the appropriate whole number.

EXAMPLE 9-14: Write the following as division: 1.2 separated into 3 equal amounts

Solution: 1.2 separated into 3 equal amounts means $1.2 \div 3$ ⟵ division

To divide a decimal by a whole number, you can first rename the decimal and whole number as equal fractions.

EXAMPLE 9-15: Divide the following decimal and whole number using fractions: $1.2 \div 3$

Solution: $1.2 \div 3 = 12$ tenths $\div 3$ Rename as equal fractions.

$$= \frac{12}{10} \div \frac{3}{1} \longleftarrow \text{equal fractions}$$

$$= \frac{12}{10} \times \frac{1}{3}$$ Divide fractions.

$$= \frac{\cancel{3} \times 4}{10} \times \frac{1}{\cancel{3}}$$

$$= \frac{4}{10}$$

$$= 4 \text{ tenths}$$ Rename as an equal decimal.

$$= 0.4 \longleftarrow 1.2 \div 3$$

It is much easier to divide a decimal by a whole number by first writing in division box form, dividing as you would for whole numbers, and then writing the decimal point in the quotient.

EXAMPLE 9-16: Divide the following decimal and whole number in division box form: $1.2 \div 3$

Solution

1. Write in division box form.

$$3\overline{)1.2}$$

2. Divide as you would for whole numbers.
$12 \div 3 = 4$

$$\begin{array}{r} 4 \\ 3\overline{)\ 1.2} \\ -1\ 2 \\ \hline 0 \end{array}$$

3. Write the decimal point in the quotient.

$$\begin{array}{r} 0.4 \\ 3\overline{)\ 1.2} \\ -1\ 2 \\ \hline 0 \end{array}$$ 1 decimal place

Note: When you divide a decimal by a whole number, the number of decimal places in the quotient should equal number of decimal places in the decimal dividend.

Recall: To check division, you multiply the proposed quotient by the original divisor to see if you get the original dividend.

EXAMPLE 9-17: Check the following division problem: $1.2 \div 3 = 0.4$

Solution: 0 . 4 ⟵ proposed quotient
             ×   3 ⟵ original divisor
         ‾‾‾‾‾‾‾‾
             1 . 2 ⟵ original dividend (0.4 checks)

B. Divide by a decimal.

To divide by a decimal, you can first rename the dividend and the divisor as equal fractions.

EXAMPLE 9-18: Divide the following decimals using fractions: $0.12 \div 0.3$

Solution: $0.12 \div 0.3 = 12$ hundredths $\div 3$ tenths   Rename as equal fractions.

$$= \frac{12}{100} \div \frac{3}{10} \longleftarrow \text{equal fractions}$$

$$= \frac{12}{100} \times \frac{10}{3} \qquad\qquad\qquad \text{Divide fractions.}$$

$$= \frac{\cancel{3} \times 4}{\cancel{10} \times 10} \times \frac{\cancel{10}}{\cancel{3}}$$

$$= \frac{4}{10}$$

$$= 4 \text{ tenths} \qquad\qquad\qquad \text{Rename as an equal decimal.}$$

$$= 0.4 \longleftarrow 0.12 \div 0.3$$

It is much easier to divide by a decimal in division box form.

Caution: To divide by a decimal in division box form, you must first make the divisor a whole number by multiplying both the divisor and dividend by the same power of 10.

EXAMPLE 9-19: Divide the following decimals in division box form: $0.12 \div 0.3$

Solution

1. Write in division box form.

$$0.3 \overline{)0.1\,2}$$

2. Make the divisor a whole number.

whole-number divisor
$$0.3 \overline{)0.1\,2} = 3 \overline{)1.2} \longleftarrow 10 \times 0.12$$
$$\uparrow$$
$$10 \times 0.3$$

3. Divide by a whole number.

$$\begin{array}{r} 0.4 \\ 3 \overline{)\,1.2} \\ -1\,2 \\ \hline 0 \end{array} \quad \text{1 decimal place}$$

4. Check.

         0 . 4 ⟵ proposed quotient
     ×   0 . 3 ⟵ original divisor
     ‾‾‾‾‾‾‾‾‾‾‾
     0 . 1 2 ⟵ original dividend (0.4 checks)

Caution: If the decimal divisor is multiplied by 10, 100, 1000, $\cdots$, to get a whole number divisor, then the dividend must also be multiplied by the same power of 10.

$$0.12 \div 0.3 = 1.2 \div 3 = 0.4$$

Note: Multiplying both the divisor and dividend by the same nonzero number will not change the quotient of the original problem.

Division Rule for Decimals
1. Write in division box form.
2. If the divisor is a decimal, then make the divisor a whole number by multiplying both the divisor and the dividend by the same power of 10.
3. Divide as you would for whole numbers.
4. Put the same number of decimal places in the quotient as in the decimal dividend from Step 2.

EXAMPLE 9-20: Divide the following decimals: $1.71 \div 0.025$
Solution

1. Write in division box form.

$$0.025\,\overline{)\,1.7\;1\;}$$

2. Make the divisor a whole number.

whole-number divisor

$$0.025\,\overline{)\,1.7\;1\;0\;} = 25\,\overline{)\,1\;7\;1\;0.} \longleftarrow 1000 \times 1.71$$

$$\uparrow$$
$$1000 \times 0.025$$

3. Divide as you would for whole numbers.

$$17100 \div 25 = 684$$

$$
\begin{array}{r}
6\;8\;4 \\
25\,\overline{)\,1\;7\;1\;0.0} \longleftarrow 1710 = 1710.0 \\
-1\;5\;0 \\
\hline
2\;1\;0 \\
-2\;0\;0 \\
\hline
1\;0\;0 \\
-1\;0\;0 \\
\hline
0
\end{array}
$$

4. Write the decimal point in the quotient.

$$
\begin{array}{r}
6\;\;8.4 \\
25\,\overline{)\,1\;7\;1\;0.0}
\end{array}
\quad\rangle\; \text{1 decimal place}
$$

Note: $0.025 \times 68.4 = 1.71$ (68.4 checks)

C. Write zeros in the quotient.

When the number of decimal places that belong in the quotient is greater than the number of digits in the quotient, you must write zeros in the quotient until there are enough digits to write the decimal point in the correct place.

EXAMPLE 9-21: Divide the following decimals $0.0001 \div 0.02$

Solution

1. Write in division box form.

$$0.02\,\overline{)\,0.0\;0\;0\;1\;}$$

2. Make the divisor a whole number.

whole-number divisor

$$0.02\,\overline{)\,0.0001\;} = 2\,\overline{)\,0.01} \longleftarrow 100 \times 0.0001$$

$$\uparrow$$
$$100 \times 0.02$$

3. Divide as you would for whole numbers

$$10 \div 2 = 5$$

$$
\begin{array}{r}
5 \\
2\,\overline{)\,0.0\;1\;0} \\
-1\;0 \\
\hline
0
\end{array}
$$

4. Write the decimal point in the quotient and write zeros as needed.

write zeros

$$
\begin{array}{r}
\overbrace{0.0\;0\;5} \\
2\,\overline{)\,0.0\;1\;0}
\end{array}
\quad\rangle\; \text{3 decimal places}
$$

Note: $0.0001 \div 0.02 = 0.01 \div 2 = 0.005$ and $0.02 \times 0.005 = 0.0001$ (0.005 checks)

D. **Divide and round the quotient.**

When you divide decimals and get too many digits in the quotient, you will usually want to round the quotient. The place to which you round the quotient will depend on how accurate your answer needs to be.

To divide and round the quotient to a given place, you must carry out the division to one place beyond the given place and then round the quotient (see Examples 8-14 and 8-15).

EXAMPLE 9-22: Divide and round the quotient to the nearest tenth in the following division problem: $0.23 \div 0.3$

Solution

1. Make the divisor a whole number.

2. Divide one place beyond the given place.

3. Round to the given place.

$$0.3 \overline{)0.2\,3} = 3\overline{)2.3}$$

hundredths place
$$3\overline{)\begin{array}{c} 0.7\ 6 \\ 2.3\ 0 \\ -2\ 1 \\ \hline 2\ 0 \\ -1\ 8 \\ \hline 2 \end{array}}$$

nearest tenth
$$3\overline{)\begin{array}{c} 0.7\ 6 \approx 0.8 \\ 2.3\ 0 \end{array}}$$

Note: $0.3 \times 0.8 = 0.24 \approx 0.23$ (0.8 checks)

9-4. Dividing by 10, 100, and 1000

A. **Divide by 10.**

To divide a decimal by 10, you can proceed as shown in Example 9-16.

EXAMPLE 9-23: Divide the following decimal by 10: 25.8

Solution

1. Write in division box form.   2. Divide as before.   3. Check.

$$10\overline{)2\,5.8}$$

$$10\overline{)\begin{array}{c} 2.5\ 8 \\ 2\ 5.8\ 0 \\ -2\ 0 \\ \hline 5\ 8 \\ -5\ 0 \\ \hline 8\ 0 \\ -8\ 0 \\ \hline 0 \end{array}} \begin{array}{l} \leftarrow \\ \swarrow \end{array} \begin{array}{l} \text{2 decimal} \\ \text{places} \end{array}$$

$10 \times 2.58 = 25.8$ (2.58 checks)

Note: To divide 25.8 by 10, you could have moved the decimal point 1 decimal place to the left: $25.8 \div 10 = 25.8 = 2.58$

Divide by 10 Rule

To divide by 10, you move the decimal point 1 decimal place to the left.

EXAMPLE 9-24: Divide the following numbers by 10: **(a)** 63   **(b)** 2.35   **(c)** 50   **(d)** 200

Solution

(a) $63 \div 10 = 63.0 \div 10 = 6.30 = 6.3$

(b) $2.35 \div 10 = .235 = 0.235$

(c) $50 \div 10 = 5.0 = 5. = 5$ or $5\emptyset \div 1\emptyset = 5 \div 1 = 5$

(d) $200 \div 10 = 20.0 = 20. = 20$ or $2\emptyset\emptyset \div 1\emptyset = 20 \div 1 = 20$

B. Divide by 100.

To divide a decimal by 100, you can proceed as shown in Example 9-16.

EXAMPLE 9-25: Divide the following decimal by 100: 2.5

Solution

1. Write in division box form. **2.** Divide as before. **3.** Check.

$$100\overline{)2.5}$$

$$\begin{array}{r} 0.0\ 2\ 5 \\ 100\overline{)\ 2.5\ 0\ 0} \\ -2\ 0\ 0 \\ \hline 5\ 0\ 0 \\ -5\ 0\ 0 \\ \hline 0 \end{array}$$

3 decimal places

$100 \times 0.025 = 2.5$ (0.025 checks)

Note: To divide 2.5 by 100, you could have moved the decimal point 2 decimal places to the left: $2.5 \div 100 = .0250 = 0.025$.

Divide by 100 Rule

To divide by 100, you move the decimal point 2 decimal places to the left.

EXAMPLE 9-26: Divide the following numbers by 100: **(a)** 275 **(b)** 37.5 **(c)** 40 **(d)** 600

Solution

(a) $275 \div 100 = 275.0 \div 100 = 2.750 = 2.75$

(b) $37.5 \div 100 = .375 = 0.375$

(c) $40 \div 100 = 40. \div 100 = .40 = 0.4$

(d) $600 \div 100 = 600. \div 100 = 6.00 = 6. = 6$ or $6\emptyset\emptyset \div 1\emptyset\emptyset = 6 \div 1 = 6$

C. Divide by 1000.

To divide a decimal by 1000, you can proceed as shown in Example 9-16.

EXAMPLE 9-27: Divide the following decimal by 1000 in division box form: 635.2

Solution

1. Write in division box form. **2.** Divide as before. **3.** Check.

$$1000\overline{)6\ 3\ 5.2}$$

$$\begin{array}{r} 0.6\ 3\ 5\ 2 \\ 1000\overline{)\ 6\ 3\ 5.2\ 0\ 0\ 0} \\ -6\ 0\ 0\ 0 \\ \hline 3\ 5\ 2\ 0 \\ -3\ 0\ 0\ 0 \\ \hline 5\ 2\ 0\ 0 \\ -5\ 0\ 0\ 0 \\ \hline 2\ 0\ 0\ 0 \\ -2\ 0\ 0\ 0 \\ \hline 0 \end{array}$$

4 decimal places

$1000 \times 0.6352 = 635.2$ (0.6352 checks)

Note: To divide 635.2 by 1000, you could have moved the decimal point 3 decimal places to the left: $635.2 \div 1000 = .6352 = 0.6352$

Divide by 1000 Rule

To divide by 1000, you move the decimal point 3 decimal places to the left.

EXAMPLE 9-28: Divide the following numbers by 1000: **(a)** 85.7   **(b)** 9   **(c)** 1280   **(d)** 500

Solution
(a) $85.7 \div 1000 = .0857 \qquad = 0.0857$

(b) $9 \div 1000 \quad = 9. \div 1000 \quad = .009 \quad = 0.009$

(c) $1280 \div 1000 = 1280. \div 1000 = 1.280 = 1.28$

(d) $5000 \div 1000 = 5000. \div 1000 = 5.000 = 5 \text{ or } 5000 \div 1000 = 5 \div 1 = 5$

D. Divide by powers of 10.

The following rule is a basic rule for dividing by 10, 100, 1000, or any power of 10.

Divide by Powers of 10 Rule
To divide by a power of 10, you move the decimal point 1 decimal place to the left for each zero in the power of 10.

EXAMPLE 9-29: Divide 3.8 by the following powers of 10: **(a)** 10   **(b)** 100   **(c)** 1000   **(d)** 10,000

Solution
(a) $3.8 \div 10 \qquad = .38 \qquad = 0.38$ One zero in 10 means 1 decimal place to the left.

(b) $3.8 \div 100 \qquad = .038 \qquad = 0.038$ Two zeros in 100 means 2 decimal places to the left.

(c) $3.8 \div 1000 \quad = .0038 \quad = 0.0038$ Three zeros in 1000 means 3 decimal places to the left.

(d) $3.8 \div 10,000 = .00038 = 0.00038$ Four zeros in 10,000 means 4 decimal places to the left.

9-5. Finding the Average (Mean)

A. Find the average (mean) of two or more given numbers.

To find the **average (mean)** of two or more given numbers:
1. Add the numbers.
2. Divide the sum from Step 1 by the number of given numbers.

EXAMPLE 9-30: Find the average (mean) of the following numbers: 82, 9.1, 36, 0.5, 1.3

Solution
1. Find the sum.   **2.** Divide the sum by the number of given numbers.   **3.** Check.

$$
\begin{array}{r}
1 \\
8\,2.0 \\
9.1 \\
3\,6.0 \\
0.5 \\
+\quad 1.3 \\
\hline
1\,2\,8.9
\end{array}
$$

$$
\begin{array}{r}
\;2\,5.7\,8 \leftarrow \text{average} \\
\text{number} \rightarrow 5\overline{)\,1\,2\,8.9\,0} \leftarrow \text{sum} \\
\text{of given} \quad -1\,0 \\
\text{numbers} \quad \overline{2\,8} \\
-2\,5 \\
\overline{3\,9} \\
-3\,5 \\
\overline{4\,0} \\
-4\,0 \\
\overline{0}
\end{array}
$$

$$
\begin{array}{r}
2\;3\;4 \\
2\,5.7\,8 \\
\times \qquad 5 \\
\hline
1\,2\,8.9\,0 \leftarrow \text{25.78 checks}
\end{array}
$$

Note: The average (mean) of 82, 9.1, 36, 0.5, and 1.3 is 25.78.

B. Find the average (mean) of a given number of addends, given the sum.

To find the **average (mean)** of a given number of addends when you know the sum, you divide the sum by the number of addends.

EXAMPLE 9-31: Find the average (mean) of 4 numbers whose sum is 0.1.

$$
\begin{array}{r}
0.0\ 2\ 5 \longleftarrow \text{average (mean)} \\
4\overline{)\ 0.1\ 0\ 0} \longleftarrow \text{sum} \qquad \text{Check as before.} \\
-\ \ \ 8 \\
\hline
2\ 0 \\
-2\ 0 \\
\hline
0
\end{array}
$$

Solution: number of addends $\longrightarrow$

9-6. Renaming Fractions as Decimals

To rename certain fractions as decimals, you can first rename the fraction with a denominator of 10, 100, 1000, $\cdots$, and then rename as a decimal, as in Example 8-19.

EXAMPLE 9-32: Rename $\frac{2}{5}$ as a decimal by renaming first as a fraction with a denominator of 10.

Solution: $\dfrac{2}{5} = \dfrac{2}{5} \times 1$ \qquad Multiply by 1.

$\qquad = \dfrac{2}{5} \times \dfrac{2}{2}$ \qquad *Think:* $1 = \dfrac{2}{2}$

$\qquad = \dfrac{2 \times 2}{5 \times 2}$ \qquad Multiply fractions.

$\qquad = \dfrac{4}{10}$

$\qquad = 4$ tenths \qquad Rename as an equal decimal [see Example 8-19].

$\qquad = 0.4 \longleftarrow \frac{2}{5}$

It is easier to rename a fraction as an equal decimal by first renaming the fraction as a division problem, and then dividing to get a decimal quotient.

EXAMPLE 9-33: Rename $\frac{2}{5}$ as an equal decimal using division.

Solution: $\frac{2}{5} = 2 \div 5$ or $5\overline{)2}$ \qquad Rename as division.

$\qquad 5\overline{)2.0}$ \qquad Rename the dividend as a decimal.

$$
\begin{array}{r}
0.4 \longleftarrow \frac{2}{5} \\
5\overline{)\ 2.0} \\
-2\ 0 \\
\hline
2\ 0
\end{array}
$$
Divide to get a decimal quotient.

The decimal 0.4 is called a **terminating decimal** because it ends in a given place (the tenths place). A decimal that does not end in any given place is called a **nonterminating decimal.**

EXAMPLE 9-34: Which of the following decimals is a nonterminating decimal: **(a)** 0.3 **(b)** 0.33 **(c)** 0.333 **(d)** 0.333 $\cdots$

Solution

(a) 0.3 is a terminating decimal because it ends in the tenths place.
(b) 0.33 is a terminating decimal because it ends in the hundredths place.
(c) 0.333 is a terminating decimal because it ends in the thousandths place.
(d) 0.333 $\cdots$ is a nonterminating decimal because the ellipsis symbol ($\cdots$) indicates that the 3s continue on forever.

A nonterminating decimal (like 0.333 $\cdots$) that repeats one or more digits forever is called a **repeating decimal.**

EXAMPLE 9-35: Which of the following decimals is a repeating decimal: **(a)** 1.234 $\cdots$ **(b)** 27.2727 $\cdots$

Solution

(a) 1.234 $\cdots$ is not a repeating decimal because the counting pattern shown goes on forever: 1.2345678910111213 $\cdots$

(b) 27.2727 $\cdots$ is a repeating decimal because the digits 2 and 7 repeat forever in the same way.

Repeating decimals (like 0.666 $\cdots$ and 27.2727 $\cdots$) are usually written with **bar notation** instead of ellipsis notation. To write a repeating decimal using bar notation, you draw the bar over each digit in the decimal fraction part that repeats.

EXAMPLE 9-36: Write the following repeating decimals using bar notation: **(a)** 0.666   **(b)** 27.2727 **(c)** 0.8333

Solution

(a)   0.666 $\cdots$ = $0.\overline{6}$

(b) 27.2727 $\cdots$ = $27.\overline{27}$ $\longrightarrow$ bar notation

(c)   0.8333 $\cdots$ = $0.8\overline{3}$

simplest form
$\downarrow$

Note: 0.666 $\cdots$ = $\overline{0.\overline{6}}$ = $0.6\overline{6}$ = $0.66\overline{6}$ = $\cdots$

Caution: 27.2727 $\cdots$ cannot be written as $\overline{27}$ $\longleftarrow$ wrong

because $\overline{27}$ = 27272727 $\cdots$

but 27.272727 $\cdots$ can be written as $27.\overline{27}$

Caution: Do not write 0.8333 $\cdots$ as $0.\overline{83}$ $\longleftarrow$ wrong

because $0.\overline{83}$ = 0.838383 $\cdots$

but $0.8\overline{3}$ = 0.8333 $\cdots$

Every fraction can be renamed as an equal terminating decimal or as an equal repeating decimal by dividing the denominator into the numerator to get the decimal quotient.

EXAMPLE 9-37: Rename the following fractions as equal decimals: **(a)** $\frac{1}{2}$ **(b)** $\frac{3}{4}$ **(c)** $\frac{5}{8}$ **(d)** $\frac{1}{3}$ **(e)** $\frac{5}{6}$

Solution

(a) $\frac{1}{2}$ = 1 ÷ 2 or $2\overline{)1}$:

$$\begin{array}{r} 0.5 \longleftarrow \frac{1}{2} \text{ as an equal decimal} \\ 2\overline{)\ 1.0} \\ \underline{-1\ 0} \\ 0 \end{array}$$

(b) $\frac{3}{4}$ = 3 ÷ 4 or $4\overline{)3}$:

$$\begin{array}{r} 0.7\ 5 \longleftarrow \frac{3}{4} \text{ as an equal decimal} \\ 4\overline{)\ 3.0\ 0} \\ \underline{-2\ 8} \\ 2\ 0 \\ \underline{-2\ 0} \\ 0 \end{array}$$

(c) $\frac{5}{8}$ = 5 ÷ 8 or $8\overline{)5}$:

$$\begin{array}{r} 0.6\ 2\ 5 \longleftarrow \frac{5}{8} \text{ as an equal decimal} \\ 8\overline{)\ 5.0\ 0\ 0} \\ \underline{-4\ 8} \\ 2\ 0 \\ \underline{-1\ 6} \\ 4\ 0 \\ \underline{-4\ 0} \\ 0 \end{array}$$

(d) $\frac{1}{3} = 1 \div 3$ or $3\overline{)1}$:

$$
\begin{array}{r}
0.3\ 3\ 3 \cdots = 0.\overline{3} \longleftarrow \tfrac{1}{3} \text{ as an equal decimal} \\
3\overline{)\ 1.0\ 0\ 0} \\
-\ 9 \\
\hline
1\diagdown 0 \\
-\ 9 \\
\hline
1\diagdown 0 \\
-\ 9 \\
\hline
1
\end{array}
$$

repeating 1s

(e) $\frac{5}{6} = 5 \div 6$ or $6\overline{)5}$:

$$
\begin{array}{r}
0.8\ 3\ 3\ 3 \cdots = 0.8\overline{3} \longleftarrow \tfrac{5}{6} \text{ as an equal decimal} \\
6\overline{)\ 5.0\ 0\ 0\ 0} \\
-4\ 8 \\
\hline
2\diagdown 0 \\
-1\ 8 \\
\hline
2\diagdown 0 \\
-1\ 8 \\
\hline
2
\end{array}
$$

repeating 2s

9-7. Solving Word Problems Containing Decimals ($+$, $-$, $\times$, $\div$)

To solve addition word problems containing decimals, you find how much two or more amounts are all together.

To solve subtraction word problems containing decimals, you find
(a) how much is left;
(b) how much more is needed;
(c) the difference between two amounts.

To solve multiplication word problems containing decimals, you find
(a) how much two or more equal amounts are all together;
(b) the product of one amount times another amount.

To solve division word problems containing decimals, you find
(a) how many equal amounts there are in all;
(b) how much is in each equal amount;
(c) how many times larger one amount is than another amount.

EXAMPLE 9-38: Solve each word problem using addition, subtraction, multiplication or division.

1. *Identify:*	Morris earns $8.25 per hour. How much does Morris earn in a 40-hour week?		**1.** *Identify:*	Rachel earns $390 for a 40-hour work week. How much does she earn per hour?
2. *Understand:*	The question asks you to find 40 equal amounts of 8.25.		**2.** *Understand:*	The question asks you to find how much is in each of 40 equal amounts.
3. *Decide:*	To join equal amounts together, you **multiply.**		**3.** *Decide:*	To find how much is in each equal amount, you **divide.**
4. *Compute:*	$40 \times 8.25 = 330$		**4.** *Compute:*	$390 \div 40 = 9.75$
5. *Interpret:*	330 means that in a 40-hour work week Morris earns **$330.**		**5.** *Interpret:*	9.75 means that Rachel's hourly pay for Rachel is **$9.75.**
6. *Check:*	Is $330 for 40 hours equal to $8.25 per hour? Yes: $330 \div 40 = 8.25$		**6.** *Check:*	Is 40 hours at $9.75 per hour equal to $390? Yes: $40 \times 9.75 = 390$

RAISE YOUR GRADES

Can you . . . ?

☑ multiply a decimal by a whole number
☑ determine how many decimal places belong in a decimal product
☑ multiply two or more decimals
☑ identify a power of 10
☑ multiply by a power of 10
☑ divide a decimal by a whole number
☑ rename a decimal divisor as a whole number divisor without changing the value of the quotient
☑ divide a decimal by a decimal divisor
☑ check a proposed quotient
☑ round a quotient to a given place
☑ divide by a power of 10
☑ find the average (mean) of two or more given numbers
☑ find the average (mean) of a given number of addends given the sum
☑ identify a terminating decimal
☑ identify a nonterminating decimal
☑ identify a repeating decimal
☑ write bar notation for a repeating decimal
☑ rename a fraction as a decimal using division
☑ solve word problems containing decimals that require addition, subtraction, multiplication, or division

SUMMARY

1. To join two or more equal decimal amounts together, you multiply the decimal amount by the appropriate whole number.
2. **Multiplication Rule for Decimals**
   (a) Write in vertical form while aligning the right-hand digits.
   (b) Multiply as you would for whole numbers.
   (c) Put the same number of decimal places in the product as the total number of decimal places contained in both factors.
3. When the number of decimal places that belong in the product is greater than the number of digits in the product, you must write zeros in the product until there are enough digits to write the decimal point in the correct place.
4. The whole numbers 10, 100, 1000, $\cdots$ are called powers of 10.
5. **Multiply by Powers of 10 Rule**
   To multiply by a power of 10, you move the decimal point 1 decimal place to the right for each zero in the power of 10.
6. To separate a decimal amount into two or more equal amounts, you divide the decimal amount by the appropriate whole number.
7. When you divide a decimal by a whole number, the number of decimal places in the quotient should equal the number of decimal places in the decimal dividend.
8. Multiplying both the divisor and dividend by the same nonzero number will not change the quotient of the original problem.
9. **Division Rule for Decimals**
   (a) Write in division box form.
   (b) If the divisor is a decimal, then make the divisor a whole number by multiplying both the divisor and the dividend by the same power of 10.
   (c) Divide as you would for whole numbers.
   (d) Put the same number of decimal places in the quotient as in the decimal dividend in Step (b).

10. When the number of decimal places that belong in the quotient is greater than the number of digits in the quotient, you must write zeros in the quotient until there are enough digits to write the decimal point in the correct place.
11. To check division, you multiply the proposed quotient by the original divisor to see if you get the original dividend.
12. **Divide by Powers of 10 Rule**
    To divide by a power of 10, you move the decimal point 1 decimal place to the left for each zero in the power of 10.
13. To find the average (mean) of two or more given numbers:
    (a) Add the numbers.
    (b) Divide the sum from Step (a) by the number of given numbers.
14. To find the average (mean) of a given number of addends given the sum, you divide the sum by the number of addends.
15. A decimal that ends in a given place is called a terminating decimal.
16. A decimal that does not end in any given place is called a nonterminating decimal.
17. A nonterminating decimal that repeats one or more digits forever is called a repeating decimal.
18. Repeating decimals are usually written using bar notation: $0.666\cdots = 0.\overline{6}$
19. Every fraction can be renamed as an equal terminating decimal or as an equal repeating decimal using division.
20. To solve addition word problems containing decimals, you find how much two or more amounts are all together.
21. To solve subtraction word problems containing decimals, you find
    (a) how much is left;
    (b) how much more is needed;
    (c) the difference between two amounts.
22. To solve multiplication word problems containing decimals, you find
    (a) how much two or more equal amounts are all together;
    (b) the product of one amount times another amount.
23. To solve division word problems containing decimals, you find
    (a) how many equal amounts there are in all;
    (b) how much is in each equal amount;
    (c) how many times larger one amount is than another amount.

SOLVED PROBLEMS

PROBLEM 9-1   Multiply decimals and whole numbers: **(a)** 3×0.2 **(b)** 0.1×4 **(c)** 6×0.8
(d) 0.9×7 **(e)** 2×0.5 **(f)** 0.4×5 **(g)** 3.45×3 **(h)** 2×0.75 **(i)** 12.495×25

Solution: Recall that to multiply a decimal by a whole number, you first write in vertical form while aligning the right-hand digits, then multiply as you would for whole numbers, and then write the decimal point in the product so that it has the same number of decimal places as the decimal factor [see Examples 9-3 and 9-4]:

(a)
$$
\begin{array}{r}
0.2 \\
\times\ \ 3 \\
\hline
0.6
\end{array}
$$
0.2 ⟵ 1 decimal place
× 3 ⟵ 0 decimal places
0.6 ⟵ 1 (1 + 0) decimal place

(b)
$$
\begin{array}{r}
0.1 \\
\times\ \ 4 \\
\hline
0.4
\end{array}
$$
0.1 ⟵ 1 decimal place
× 4
0.4

(c)
$$
\begin{array}{r}
0.8 \\
\times\ \ 6 \\
\hline
4.8
\end{array}
$$
0.8 ⟵ 1 decimal place
× 6
4.8

(d)
$$
\begin{array}{r}
0.9 \\
\times\ \ 7 \\
\hline
6.3
\end{array}
$$
0.9 ⟵ 1 decimal place
× 7
6.3

(e)
$$
\begin{array}{r}
0.5 \\
\times\ \ 2 \\
\hline
1.0 = 1
\end{array}
$$
0.5 ⟵ 1 decimal place
× 2
1.0 = 1

(f)
$$
\begin{array}{r}
0.4 \\
\times\ \ 5 \\
\hline
2.0 = 2
\end{array}
$$
0.4 ⟵ 1 decimal place
× 5
2.0 = 2

(g)
$$
\begin{array}{r}
3.4\ 5 \\
\times\quad\ \ 3 \\
\hline
1\ 0.3\ 5
\end{array}
$$
← 2 decimal places

(h)
$$
\begin{array}{r}
0.7\ 5 \\
\times\quad\ \ 2 \\
\hline
1.5\ 0 = 1.5
\end{array}
$$
← 2 decimal places

(i)
$$
\begin{array}{r}
1\ 2.4\ 9\ 5 \\
\times\quad\ \ 2\ 5 \\
\hline
6\ 2\ 4\ 7\ 5 \\
2\ 4\ 9\ 9\ 0\quad \\
\hline
3\ 1\ 2.3\ 7\ 5
\end{array}
$$
← 3 decimal places

PROBLEM 9-2   Multiply two decimals: **(a)** 0.3 × 0.6   **(b)** 0.4 × 0.7   **(c)** 0.5 × 0.2   **(d)** 0.8 × 0.5   **(e)** 3.49 × 0.9   **(f)** 0.1 × 1.25   **(g)** 5.25 × 0.35   **(h)** 2.15 × 8.03   **(i)** 2.03 × 0.125   **(j)** 5.621 × 2.12

Solution: Recall the Multiplication Rule for Two or More Decimals [see Example 9-5]:
1. Write in vertical form while aligning the right-hand digits.
2. Multiply as you would for two whole numbers.
3. Put the same number of decimal places in the product as the total number of decimal places contained in both factors.

(a)
$$
\begin{array}{r}
0.6 \\
\times 0.3 \\
\hline
0.1\ 8
\end{array}
$$
← 1 decimal place
← 1 decimal place
← 2 (1 + 1) decimal places

(b)
$$
\begin{array}{r}
0.7 \\
\times 0.4 \\
\hline
0.2\ 8
\end{array}
$$

(c)
$$
\begin{array}{r}
0.2 \\
\times 0.5 \\
\hline
0.1\ 0 = 0.1
\end{array}
$$

(d)
$$
\begin{array}{r}
0.5 \\
\times 0.8 \\
\hline
0.4\ 0 = 0.4
\end{array}
$$

(e)
$$
\begin{array}{r}
3.4\ 9 \\
\times\ \ 0.9 \\
\hline
3.1\ 4\ 1
\end{array}
$$
← 2 decimal places
← 1 decimal place
← 3 (2 + 1) decimal places

(f)
$$
\begin{array}{r}
1.2\ 5 \\
\times\ \ 0.1 \\
\hline
0.1\ 2\ 5
\end{array}
$$

(g)
$$
\begin{array}{r}
5.2\ 5 \\
\times 0.3\ 5 \\
\hline
2\ 6\ 2\ 5 \\
1\ 5\ 7\ 5\quad \\
\hline
1.8\ 3\ 7\ 5
\end{array}
$$
← 2 decimal places
← 2 decimal places
← 4 (2 + 2) decimal places

(h)
$$
\begin{array}{r}
2.1\ 5 \\
\times 8.0\ 3 \\
\hline
6\ 4\ 5 \\
1\ 7\ 2\ 0\ 0\quad \\
\hline
1\ 7.2\ 6\ 4\ 5
\end{array}
$$

(i)
$$
\begin{array}{r}
0.1\ 2\ 5 \\
\times\ \ 2.0\ 3 \\
\hline
3\ 7\ 5 \\
2\ 5\ 0\ 0\quad \\
\hline
0.2\ 5\ 3\ 7\ 5
\end{array}
$$
← 3 decimal places
← 2 decimal places
← 5 (3 + 2) decimal places

(j)
$$
\begin{array}{r}
5.6\ 2\ 1 \\
\times\ \ 2.1\ 2 \\
\hline
1\ 1\ 2\ 4\ 2 \\
5\ 6\ 2\ 1\quad \\
1\ 1\ 2\ 4\ 2\quad\ \ \\
\hline
1\ 1.9\ 1\ 6\ 5\ 2
\end{array}
$$

PROBLEM 9-3   Multiply decimals when zeros are needed in the product: **(a)** 0.3 × 0.1   **(b)** 0.1 × 0.1   **(c)** 0.4 × 0.02   **(d)** 0.04 × 0.5   **(e)** 0.06 × 0.01   **(f)** 0.05 × 0.04   **(g)** 0.003 × 0.03   **(h)** 0.25 × 0.038   **(i)** 0.002 × 0.002   **(j)** 0.014 × 0.215

Solution: Recall that when the number of decimal places that belong in the product are more than the number of digits in the product, you must write zeros in the product until there are enough digits to write the decimal point in the correct place [see Example 9-6]:

(a)
$$
\begin{array}{r}
0.1 \\
\times 0.3 \\
\hline
0.0\ 3
\end{array}
$$
← 1 decimal place
← 1 decimal place
← 2 (1 + 1) decimal places

write zeros

(b)
$$
\begin{array}{r}
0.1 \\
\times 0.1 \\
\hline
0.0\ 1
\end{array}
$$

(c)
$$
\begin{array}{r}
0.0\ 2 \\
\times\ \ 0.4 \\
\hline
0.0\ 0\ 8
\end{array}
$$
← 2 decimal places
← 1 decimal places
← 3 (2 + 1) decimal places

write zeros

(d)   0.0 4
  × 0.5
  ‾‾‾‾‾‾‾‾
  0.0 2 0 = 0.02

(e)   0.0 1 ⟵ 2 decimal places
  ×0.0 6 ⟵ 2 decimal places
  ‾‾‾‾‾‾‾‾
  0.0 0 0 6 ⟵ 4 (2 + 2) decimal places
       write zeros

(f)   0.0 4
  ×0.0 5
  ‾‾‾‾‾‾‾‾
  0.0 0 2 0 = 0.002

(g)   0.0 0 3 ⟵ 3 decimal places
  × 0.0 3 ⟵ 2 decimal places
  ‾‾‾‾‾‾‾‾
  0.0 0 0 0 9 ⟵ 5 (3 + 2) decimal places
      write zeros

(h)   0.0 3 8
  × 0.2 5
  ‾‾‾‾‾‾‾‾
    1 9 0
    7 6
  ‾‾‾‾‾‾‾‾
  0.0 0 9 5 0 = 0.0095

(i)   0.0 0 2 ⟵ 3 decimal places
  ×0.0 0 2 ⟵ 3 decimal places
  ‾‾‾‾‾‾‾‾
  0.0 0 0 0 0 4 ⟵ 6 (3 + 3) decimal places
      write zeros

(j)   0.2 1 5
  ×0.0 1 4
  ‾‾‾‾‾‾‾‾
    8 6 0
    2 1 5
  ‾‾‾‾‾‾‾‾
  0.0 0 3 0 1 0 = 0.00301

PROBLEM 9-4   Multiply by 10: **(a)** 10 × 4.25   **(b)** 10 × 13.125   **(c)** 10 × 1.5
(d) 10 × 423.8   **(e)** 10 × 0.75   **(f)** 10 × 9   **(g)** 10 × 0.02   **(h)** 10 × 8   **(i)** 10 × 50
(j) 10 × 600 [see Example 9-8]:

Solution: Recall that to multiply by 10, you move the decimal point 1 decimal place to the right:

(a) 10 × 4.25 = 42.5 = 42.5   **(b)** 10 × 13.125 = 131.25 = 131.25   **(c)** 10 × 1.5 = 15. = 15

(d) 10 × 423.8 = 4238. = 4238   **(e)** 10 × 0.75 = 07.5 = 7.5   **(f)** 10 × 9 = 90. = 90

(g) 10 × 0.02 = 00.2 = 0.2   **(h)** 10 × 8 = 80. = 80   **(i)** 10 × 50 = 500. = 500

(j) 10 × 600 = 6000. = 6000

PROBLEM 9-5   Multiply by 100: **(a)** 100 × 1.125   **(b)** 100 × 0.375   **(c)** 100 × 0.025
(d) 100 × 0.003   **(e)** 100 × 0.05   **(f)** 100 × 0.19   **(g)** 100 × 3.1   **(h)** 100 × 25.6
(i) 100 × 8   **(j)** 100 × 200

Solution: Recall that to multiply by 100, you move the decimal point 2 places to the right [see Example 9-10]:

(a) 100 × 1.125 = 112.5 = 112.5   **(b)** 100 × 0.375 = 037.5 = 37.5

(c) 100 × 0.025 = 002.5 = 2.5   **(d)** 100 × 0.003 = 000.3 = 0.3   **(e)** 100 × 0.05 = 005. = 5

(f) 100 × 0.19 = 019. = 19   **(g)** 100 × 3.1 = 100 × 3.10 = 310. = 310

(h) 100 × 25.6 = 100 × 25.60 = 2560   **(i)** 100 × 8 = 100 × 8.00 = 800

(j) 100 × 200 = 100 × 200.00 = 20,000

PROBLEM 9-6   Multiply by 1000: **(a)** 1000 × 5.8196   **(b)** 1000 × 0.2458
(c) 1000 × 0.0675   **(d)** 1000 × 0.0025   **(e)** 1000 × 0.0004   **(f)** 1000 × 0.009
(g) 1000 × 5.12   **(h)** 1000 × 3.4   **(i)** 1000 × 7   **(j)** 1000 × 800

Solution: Recall that to multiply by 1000, you move the decimal point 3 places to the right [see Example 9-12]:

(a) 1000 × 5.8196 = 5819.6 = 5819.6   **(b)** 1000 × 0.2458 = 0245.8 = 245.8

(c) 1000 × 0.0675 = 0067.5 = 67.5   **(d)** 1000 × 0.0025 = 0002.5 = 2.5

(e) 1000 × 0.0004 = 0000.4 = 0.4   **(f)** 1000 × 0.009 = 0009. = 9

(g) 1000 × 5.12 = 1000 × 5.120 = 5120. = 5120   **(h)** 1000 × 3.4 = 1000 × 3.400 = 3400

(i) 1000 × 7 = 1000 × 7.000 = 7000   **(j)** 1000 × 800.000 = 800,000

PROBLEM 9-7   Multiply by powers of 10: **(a)** 10×2.5 **(b)** 100×2.5 **(c)** 1000×2.5 **(d)** $10,000 \times 2.5$ **(e)** $100,000 \times 0.005$ **(f)** $1,000,000 \times 3$ **(g)** $10,000,000 \times 0.000002$ **(h)** $100,000,000 \times 800$ **(i)** $1,000,000,000 \times 1,000,000$ **(j)** $10,000 \times 0.0000075$

Solution: Recall that to multiply by a power of 10, you move the decimal point 1 decimal place to the right for each zero in the power of 10 [See Example 9-13]:

| 1 zero | right 1 place | | 2 zeros | right 2 places | | 3 zeros | right 3 places |

(a) $10 \times 2.5 = 25$ **(b)** $100 \times 2.5 = 250$ **(c)** $1000 \times 2.5 = 2500$

(d) $10,000 \times 2.5 = 25,000$ **(e)** $100,000 \times 0.005 = 500$

(f) $1,000,000 \times 3 = 3,000,000$ **(g)** $10,000,000 \times 0.000002 = 20$

| 8 zeros | 2 zeros | 10 (8 + 2) zeros |

(h) $100,000,000 \times 800 = 80,000,000,000$

| 9 zeros | 6 zeros | 15 (9 + 6) zeros |

(i) $1,000,000,000 \times 1,000,000 = 1,000,000,000,000,000$ **(j)** $10,000 \times 0.0000075 = 0.075$

PROBLEM 9-8   Divide decimals by whole numbers: **(a)** $1.0 \div 5$ **(b)** $0.18 \div 3$ **(c)** $0.03 \div 2$ **(d)** $0.0504 \div 6$ **(e)** $1.4 \div 20$ **(f)** $22.42 \div 3$ **(g)** $11.4 \div 9$ **(h)** $77.5 \div 31$ **(i)** $40.0 \div 400$ **(j)** $287.5 \div 230$

Solution: Recall that to divide a decimal by a whole number, you first write in division box form, divide as you would for whole numbers, then put the same number of decimal places in the quotient as in the decimal dividend [see Example 9-16]:

(a)
```
      0.2
  5 ) 1.0      ← 1 decimal place
     -1 0
        0
```

(b)
```
       0.0 6
  3 ) 0.1 8      ← 2 decimal places
     -  1 8
           0
```

(c)
```
       0.0 1 5
  2 ) 0.0 3 0      ← 3 decimal places
     -    2
          1 0
         -1 0
             0
```

(d)
```
       0.0 0 8 4
  6 ) 0.0 5 0 4      ← 4 decimal places
     -    4 8
            2 4
           -2 4
               0
```

(e)
```
       0.0 7
  20 ) 1.4 0
      -1 4 0
           0
```

(f)
```
        7.4 7 3 3 3 ··· = 7.47\overline{3}
  3 ) 2 2.4 2 0 0 0
     -2 1
        1 4
       -1 2
          2 2
         -2 1
            1 0
           - 9
              1 0
             - 9
                1 0
               - 9
                  1
```

(g)
```
        1.2 6 6 6 ··· = 1.2\overline{6}
  9 ) 1 1.4 0 0 0
     -  9
        2 4
       -1 8
          6 0
         -5 4
            6 0
           -5 4
              6 0
             -5 4
                6
```

$$
\begin{array}{r}
2.5 \\
\textbf{(h) } 31\overline{)\ 7\ 7.5} \\
-6\ 2 \\
\hline
1\ 5\ 5 \\
-1\ 5\ 5 \\
\hline
0
\end{array}
\qquad
\begin{array}{r}
0.1 \\
\textbf{(i) } 400\overline{)\ 4\ 0.0} \\
-4\ 0\ 0 \\
\hline
0
\end{array}
\qquad
\begin{array}{r}
1.2\ 5 \\
\textbf{(j) } 230\overline{)\ 2\ 8\ 7.5\ 0} \\
-2\ 3\ 0 \\
\hline
5\ 7\ 5 \\
-4\ 6\ 0 \\
\hline
1\ 1\ 5\ 0 \\
-1\ 1\ 5\ 0 \\
\hline
0
\end{array}
$$

PROBLEM 9-9 Divide decimals by decimals: **(a)** 0.18 ÷ 0.2 **(b)** 0.35 ÷ 0.5 **(c)** 0.36 ÷ 0.3 **(d)** 1.08 ÷ 0.4 **(e)** 0.15 ÷ 1.2 **(f)** 7.875 ÷ 2.1 **(g)** 0.6603 ÷ 0.31 **(h)** 1.1137 ÷ 0.43 **(i)** 0.045 ÷ 0.135 **(j)** 13.5 ÷ 2.025

Solution: Recall the Division Rule for Decimals [see Examples 9-19 and 9-20]:
1. Write in division box form.
2. If the division is a decimal, then make the divisor a whole number by multiplying both the divisor and the dividend by the same power of 10.
3. Divide as you would for whole numbers.
4. Put the same number of decimal places in the quotient as in the decimal dividend from Step 2.

$$
\textbf{(a) } 0.2\overline{)0.18} = 2\overline{)\ 1.8}
\begin{array}{r}
0.9 \\
\\
-1\ 8 \\
\hline
0
\end{array}
\qquad
\textbf{(b) } 0.5\overline{)0.35} = 5\overline{)\ 3.5}
\begin{array}{r}
0.7 \\
\\
-3\ 5 \\
\hline
0
\end{array}
\qquad
\textbf{(c) } 0.3\overline{)0.36} = 3\overline{)\ 3.6}
\begin{array}{r}
1.2 \\
\\
-3 \\
\hline
\cancel{0}\ 6 \\
-\ 6 \\
\hline
0
\end{array}
$$

$$
\textbf{(d) } 0.4\overline{)1.08} = 4\overline{)\ 1\ 0.8}
\begin{array}{r}
2.7 \\
\\
-\ 8 \\
\hline
2\ 8 \\
-2\ 8 \\
\hline
0
\end{array}
\qquad
\textbf{(e) } 1.2\overline{)0.15} = 12\overline{)\ 1.5\ 0\ 0}
\begin{array}{r}
0.1\ 2\ 5 \\
\\
-1\ 2 \\
\hline
3\ 0 \\
-2\ 4 \\
\hline
6\ 0 \\
-6\ 0 \\
\hline
0
\end{array}
\qquad
\textbf{(f) } 2.1\overline{)7.875} = 21\overline{)\ 7\ 8.7\ 5}
\begin{array}{r}
3.7\ 5 \\
\\
-6\ 3 \\
\hline
1\ 5\ 7 \\
-1\ 4\ 7 \\
\hline
1\ 0\ 5 \\
-1\ 0\ 5 \\
\hline
0
\end{array}
$$

$$
\textbf{(g) } 0.31\overline{)0.6603} = 31\overline{)\ 6\ 6.0\ 3}
\begin{array}{r}
2.1\ 3 \\
\\
-6\ 2 \\
\hline
4\ 0 \\
-3\ 1 \\
\hline
9\ 3 \\
-9\ 3 \\
\hline
0
\end{array}
\qquad
\textbf{(h) } 0.43\overline{)1.1137} = 43\overline{)\ 1\ 1\ 1.3\ 7}
\begin{array}{r}
2.5\ 9 \\
\\
-\ 8\ 6 \\
\hline
2\ 5\ 3 \\
-2\ 1\ 5 \\
\hline
3\ 8\ 7 \\
-3\ 8\ 7 \\
\hline
0
\end{array}
$$

$$
\textbf{(i) } 0.135\overline{)0.045} = 135\overline{)\ 4\ 5.0\ 0\ 0}
\begin{array}{r}
0.3\ 3\ 3\cdots = 0.\overline{3} \\
\\
-4\ 0\ 5 \\
\hline
4\ 5\ 0 \\
-4\ 0\ 5 \\
\hline
4\ 5\ 0 \\
-4\ 0\ 5 \\
\hline
4\ 5
\end{array}
$$

$$6.6\ 6\ 6\cdots = 6.\overline{6}$$

(j) $2.\underset{\smile}{025}\,\overline{)\,13.\underset{\smile}{5}}\,\,\,= 2025\,\overline{)\,1\ 3\ 5\ 0\ 0.0\ 0\ 0}$

$$\begin{array}{r} -1\ 2\ 1\ 5\ 0 \\ \hline 1\ 3\ 5\ 0\ 0 \\ -1\ 2\ 1\ 5\ 0 \\ \hline 1\ 3\ 5\ 0\ 0 \\ -1\ 2\ 1\ 5\ 0 \\ \hline 1\ 3\ 5\ 0\ 0 \\ -1\ 2\ 1\ 5\ 0 \\ \hline 0 \end{array}$$

PROBLEM 9-10   Divide when zeros are needed in the quotient: **(a)** $0.04 \div 2$ **(b)** $0.015 \div 3$
(c) $0.0008 \div 8$ **(d)** $0.0001 \div 5$ **(e)** $0.001 \div 0.4$ **(f)** $0.002 \div 0.3$ **(g)** $0.1 \div 2.5$
(h) $0.011 \div 3.3$ **(i)** $0.24064 \div 75.2$ **(j)** $0.001195 \div 0.239$

Solution: Recall that when the number of decimal places that belong in the quotient is greater than the number of digits in the quotient, you must write zeros in the quotient until there are enough digits to write the decimal point in the correct place [see Example 9-21]:

write zeros

(a) $2\,\overline{)\,0.0\ 4}$ quotient $\overset{\frown}{0.0}\,2$

$$\begin{array}{r} 0.0\ 2 \\ 2\,\overline{)\,0.0\ 4} \\ -\quad 4 \\ \hline 0 \end{array} \qquad \begin{array}{r} 0.0\ 0\ 5 \\ 3\,\overline{)\,0.0\ 1\ 5} \\ -\quad 1\ 5 \\ \hline 0 \end{array} \qquad \begin{array}{r} 0.0\ 0\ 0\ 1 \\ 8\,\overline{)\,0.0\ 0\ 0\ 8} \\ -\quad 8 \\ \hline 0 \end{array} \qquad \begin{array}{r} 0.0\ 0\ 0\ 0\ 2 \\ 5\,\overline{)\,0.0\ 0\ 0\ 1\ 0} \\ -\quad 1\ 0 \\ \hline 0 \end{array}$$

(b) — **(c)** — **(d)**

(e) $0.\underset{\smile}{4}\,\overline{)\,0.\underset{\smile}{0}01} = 4\,\overline{)\,0.0\ 1\ 0\ 0}$

$$\begin{array}{r} 0.0\ 0\ 2\ 5 \\ 4\,\overline{)\,0.0\ 1\ 0\ 0} \\ -\quad 8 \\ \hline 2\ 0 \\ -2\ 0 \\ \hline 0 \end{array}$$

(f) $0.\underset{\smile}{3}\,\overline{)\,0.\underset{\smile}{0}02} = 3\,\overline{)\,0.0\ 2\ 0\ 0\ 0}$

$$0.0\ 0\ 6\ 6\ 6\cdots = 0.00\overline{6}$$
$$\begin{array}{r} 3\,\overline{)\,0.0\ 2\ 0\ 0\ 0} \\ -\quad 1\ 8 \\ \hline 2\ 0 \\ -1\ 8 \\ \hline 2\ 0 \\ -1\ 8 \\ \hline 2 \end{array}$$

(g) $2.\underset{\smile}{5}\,\overline{)\,0.\underset{\smile}{1}} = 25\,\overline{)\,1.0\ 0}$

$$\begin{array}{r} 0.0\ 4 \\ 25\,\overline{)\,1.0\ 0} \\ -1\ 0\ 0 \\ \hline 0 \end{array}$$

(h) $3.\underset{\smile}{3}\,\overline{)\,0.\underset{\smile}{0}11} = 33\,\overline{)\,0.1\ 1\ 0\ 0\ 0}$

$$0.0\ 0\ 3\ 3\ 3\cdots = 0.00\overline{3}$$
$$\begin{array}{r} 33\,\overline{)\,0.1\ 1\ 0\ 0\ 0} \\ -\quad 9\ 9 \\ \hline 1\ 1\ 0 \\ -\ 9\ 9 \\ \hline 1\ 1\ 0 \\ -\ 9\ 9 \\ \hline 1\ 1 \end{array}$$

(i) $75.\underset{\smile}{2}\,\overline{)\,0.\underset{\smile}{24064}} = 752\,\overline{)\,2.4\ 0\ 6\ 4}$

$$\begin{array}{r} 0.0\ 0\ 3\ 2 \\ 752\,\overline{)\,2.4\ 0\ 6\ 4} \\ -2\ 2\ 5\ 6 \\ \hline 1\ 5\ 0\ 4 \\ -1\ 5\ 0\ 4 \\ \hline 0 \end{array}$$

(j) $0.\underset{\smile}{239}\,\overline{)\,0.\underset{\smile}{001195}} = 239\,\overline{)\,1.1\ 9\ 5}$

$$\begin{array}{r} 0.0\ 0\ 5 \\ 239\,\overline{)\,1.1\ 9\ 5} \\ -1\ 1\ 9\ 5 \\ \hline 0 \end{array}$$

PROBLEM 9-11   Divide and round the quotient to the given place value:

(a) $5 \div 0.3$ (nearest tenth)   **(b)** $0.1 \div 0.6$ (nearest tenth)
(c) $1.3 \div 0.07$ (nearest hundredth)   **(d)** $2 \div 0.09$ (nearest hundredth)

(e) $3.1 \div 1.2$ (nearest thousandth) **(f)** $0.1 \div 0.15$ (nearest thousandth)
(g) $4 \div 0.021$ (nearest whole number) **(h)** $35.9 \div 2.4$ (nearest whole number)
(i) $2.1395 \div 16.8$ (nearest tenth) **(j)** $5.03 \div 0.825$ (nearest tenth)

Solution: Recall that to divide and round to a given place, you carry out the division to one place beyond the given place and then round the quotient [see Example 9-22]:

(a)
$$
\begin{array}{r}
1\,6.6\,6 \approx 16.7 \\
0.3\overline{)5} = 3\overline{)\,5\,0.0\,0} \\
-3 \\
\hline
2\,0 \\
-1\,8 \\
\hline
2\,0 \\
-1\,8 \\
\hline
2\,0 \\
-1\,8 \\
\hline
2 \\
\end{array}
$$

(b)
$$
\begin{array}{r}
0.1\,6 \approx 0.2 \\
0.6\overline{)0.1} = 6\overline{)\,1.0\,0} \\
-\,6 \\
\hline
4\,0 \\
-3\,6 \\
\hline
4 \\
\end{array}
$$

(c)
$$
\begin{array}{r}
1\,8.5\,7\,1 \approx 18.57 \\
0.07\overline{)1.3} = 7\overline{)\,1\,3\,0.0\,0\,0} \\
-\,7 \\
\hline
6\,0 \\
-5\,6 \\
\hline
4\,0 \\
-3\,5 \\
\hline
5\,0 \\
-4\,9 \\
\hline
1\,0 \\
-\,7 \\
\hline
3 \\
\end{array}
$$

(d)
$$
\begin{array}{r}
2\,2.2\,2\,2 \approx 22.22 \\
0.09\overline{)2} = 9\overline{)\,2\,0\,0.0\,0\,0} \\
-1\,8 \\
\hline
2\,0 \\
-1\,8 \\
\hline
2\,0 \\
-1\,8 \\
\hline
2\,0 \\
-1\,8 \\
\hline
2\,0 \\
-1\,8 \\
\hline
2 \\
\end{array}
$$

(e)
$$
\begin{array}{r}
2.5\,8\,3\,3 \approx 2.583 \\
1.2\overline{)3.1} = 12\overline{)\,3\,1.0\,0\,0\,0} \\
-2\,4 \\
\hline
7\,0 \\
-6\,0 \\
\hline
1\,0\,0 \\
-\,9\,6 \\
\hline
4\,0 \\
-3\,6 \\
\hline
4\,0 \\
-3\,6 \\
\hline
4 \\
\end{array}
$$

(f)
$$
\begin{array}{r}
0.6\,6\,6\,6 \approx 0.667 \\
0.15\overline{)0.1} = 15\overline{)\,1\,0.0\,0\,0\,0} \\
-\,9\,0 \\
\hline
1\,0\,0 \\
-\,9\,0 \\
\hline
1\,0\,0 \\
-\,9\,0 \\
\hline
1\,0\,0 \\
-\,9\,0 \\
\hline
1\,0 \\
\end{array}
$$

(g)
$$
\begin{array}{r}
1\,9\,0.4 \approx 190 \\
0.021\overline{)4} = 21\overline{)\,4\,0\,0\,0.0} \\
-2\,1 \\
\hline
1\,9\,0 \\
-1\,8\,9 \\
\hline
1\,0\,0 \\
-\,8\,4 \\
\hline
1\,6 \\
\end{array}
$$

(h)
$$
\begin{array}{r}
1\,4.9 \approx 15 \\
2.4\overline{)35.9} = 24\overline{)\,3\,5\,9.0} \\
-2\,4 \\
\hline
1\,1\,9 \\
-\,9\,6 \\
\hline
2\,3\,0 \\
-2\,1\,6 \\
\hline
1\,4 \\
\end{array}
$$

$$0.1\ 2 \approx 0.1$$

(i) $16.8\overline{)2.1395}$ = $168\overline{)2\ 1.3\ 9\ 5}$

$$\begin{array}{r} -1\ 6\ 8 \\ \hline 4\ 5\ 9 \\ -3\ 3\ 6 \\ \hline 1\ 2\ 3 \end{array}$$

$$6.0\ 9 \approx 6.1$$

(j) $0.825\overline{)5.03}$ = $825\overline{)5\ 0\ 3\ 0.0\ 0}$

$$\begin{array}{r} -4\ 9\ 5\ 0 \\ \hline 8\ 0\ 0\ 0 \\ -7\ 4\ 2\ 5 \\ \hline 5\ 7\ 5 \end{array}$$

PROBLEM 9-12   Divide by 10: (a) $25.3 \div 10$ (b) $125.49 \div 10$ (c) $1.6 \div 10$
(d) $3.75 \div 10$ (e) $43 \div 10$ (f) $20 \div 10$ (g) $600 \div 10$ (h) $7000 \div 10$ (i) $0.6 \div 10$
(j) $0.0025 \div 10$

Solution: Recall that to divide by 10, you move the decimal point 1 decimal place to the left [see Examples 9-23 and 9-24]:

(a) $25.3 \div 10 = 2.53$

(b) $125.49 \div 10 = 12.549$

(c) $1.6 \div 10 = .16 = 0.16$

(d) $3.75 \div 10 = .375 = 0.375$

(e) $43 \div 10 = 43. \div 10 = 4.3$

(f) $20 \div 10 = 20. \div 10 = 2$ or $2\emptyset \div 1\emptyset = 2 \div 1 = 2$

(g) $6\emptyset\emptyset \div 1\emptyset = 60 \div 1 = 60$

(h) $700\emptyset \div 1\emptyset = 700 \div 1 = 700$

(i) $0.6 \div 10 = 0.06$

(j) $0.0025 \div 10 = 0.00025$

PROBLEM 9-13   Divide by 100: (a) $525.3 \div 100$ (b) $60035.1 \div 100$ (c) $23.4 \div 100$
(d) $1.5 \div 100$ (e) $0.3 \div 100$ (f) $0.05 \div 100$ (g) $2 \div 100$ (h) $30 \div 100$
(i) $900 \div 100$

Solution: Recall that to divide by 100, you move the decimal point 2 decimal places to the left [see Examples 9-25 and 9-26]:

(a) $525.3 \div 100 = 5.253$

(b) $60035.1 \div 100 = 600.351$

(c) $23.4 \div 100 = .234 = 0.234$

(d) $1.5 \div 100 = .015 = 0.015$

(e) $0.3 \div 100 = .003 = 0.003$

(f) $0.05 \div 100 = .0005 = 0.0005$

(g) $2 \div 100 = 2. \div 100 = .02 = 0.02$

(h) $30 \div 100 = 30. \div 100 = .30 = 0.3$

(i) $900 \div 100 = 900. \div 100 = 9$ or $9\emptyset\emptyset \div 1\emptyset\emptyset = 9 \div 1 = 9$

PROBLEM 9-14   Divide by 1000: (a) $1257.5 \div 1000$ (c) $382.4 \div 1000$ (c) $25.6 \div 1000$
(d) $1.85 \div 1000$ (e) $0.3 \div 1000$ (f) $0.025 \div 1000$ (g) $8 \div 1000$ (h) $900 \div 1000$
(i) $7000 \div 1000$ (j) $1,000,000 \div 1000$

Solution: Recall that to divide by 1000, you move the decimal point 3 decimal places to the left [see Examples 9-27 and 9-28]:

(a) $1257.5 \div 1000 = 1.2575$

(b) $382.4 \div 1000 = .3824 = 0.3824$

(c) $25.6 \div 1000 = .0256 = 0.0256$

(d) $1.85 \div 1000 = 0.00185$

(e) $0.3 \div 1000 = 0.0003$

(f) $0.025 \div 1000 = 0.000025$

(g) $8 \div 1000 = 0.008$

(h) $900 \div 1000 = 0.900 = 0.9$

(i) $7000 \div 1000 = 7$ or $700\emptyset \div 100\emptyset = 7 \div 1 = 7$

(j) $1,000,\emptyset\emptyset\emptyset \div 1\emptyset\emptyset\emptyset = 1000$

PROBLEM 9-15   Divide by powers of 10: (a) $2.8 \div 10$ (b) $2.8 \div 100$ (c) $2.8 \div 1000$
(d) $2.8 \div 10,000$ (e) $5857 \div 100,000$ (f) $258,973 \div 1,000,000$
(g) $373,582,961 \div 10,000,000$ (h) $80,000,000 \div 100,000,000$ (i) $24.99 \div 10$
(j) $359.86 \div 1000$

Solution: Recall that to divide by a power of 10, you move the decimal point 1 decimal place to the left for each zero in the power of 10 [see Example 9-29]:

(a) $2.8 \div 10 = 0.28$

(b) $2.8 \div 100 = 0.028$

(c) $2.8 \div 1000 = 0.0028$ **(d)** $2.8 \div 10,000 = 0.00028$

(e) $5857 \div 100,000 = 0.05857$ **(f)** $258,973 \div 1,000,000 = 0.258973$

(g) $373,582,961 \div 10,000,000 = 37.3582961$ **(h)** $80,000,000 \div 100,000,000 = 0.8$

(i) $24.99 \div 10 = 2.499$ **(j)** $359.86 \div 1000 = 0.35986$

PROBLEM 9-16 Find the average (mean) of each group of numbers. Round to the nearest tenth when the average is a repeating decimal: **(a)** 5, 2, 4, 9, 3, 7   **(b)** 18, 52, 34, 56   **(c)** 25, 36, 79 **(d)** 123, 142, 5, 83, 91, 125   **(e)** 0.3, 0.2, 0.9, 0.7   **(f)** 1.5, 1.8, 2.7, 2.8, 3.5, 1.6, 0.2, 3.1 **(g)** 15.82, 4.89, 5.83   **(h)** 25.82, 42.63, 92.87, 1.87, 5, 8, 9.23   **(i)** $\frac{1}{5}, \frac{3}{5}$ **(j)** $3\frac{1}{2}, 5\frac{1}{4}, 2\frac{7}{8}, \frac{3}{4}$

Solution: Recall that to find the average (mean) to two or more numbers, you must add all the numbers together and then divide the sum by the number of given numbers [see Example 9-30]:

$$\overset{\text{sum}}{} \qquad \text{number of given numbers}$$

(a) $5 + 2 + 4 + 9 + 3 + 7 = 30$ and $30 \div 6 = 5$
(b) $18 + 52 + 34 + 56 = 160$ and $160 \div 4 = 40$
(c) $25 + 36 + 79 = 140$ and $140 \div 3 = 46.666\cdots \approx 46.7$ (nearest tenth)
(d) $123 + 142 + 5 + 83 + 91 + 125 = 569$ and $569 \div 6 = 94.8333\cdots \approx 94.8$
(e) $0.3 + 0.2 + 0.9 + 0.7 = 2.1$ and $2.1 \div 4 = 0.525$
(f) $1.5 + 1.8 + 2.7 + 2.8 + 3.5 + 1.6 + 0.2 + 3.1 = 17.2$ and $17.2 \div 8 = 2.15$
(g) $15.82 + 4.89 + 5.83 = 26.54$ and $26.54 \div 3 = 8.84666\cdots \approx 8.8$
(h) $25.82 + 42.63 + 92.87 + 1.87 + 5 + 8 + 9.23 = 185.42$ and $185.42 \div 7 \approx 26.5$
(i) $\frac{1}{5} + \frac{3}{5} = \frac{4}{5}$ and $\frac{4}{5} \div 2 = \frac{4}{5} \div \frac{2}{1} = \frac{4}{5} \times \frac{1}{2} = \frac{2 \times \not{2}}{5} \times \frac{1}{\not{2}} = \frac{2}{5}$
(j) $3\frac{1}{2} + 5\frac{1}{4} + 2\frac{7}{8} + \frac{3}{4} = \frac{7}{2} + \frac{21}{4} + \frac{23}{8} + \frac{3}{4} = \frac{28}{8} + \frac{42}{8} + \frac{23}{8} + \frac{6}{8} = \frac{99}{8}$ and $\frac{99}{8} \div 4 = \frac{99}{8} \div \frac{4}{1} = \frac{99}{8} \div \frac{1}{4} = \frac{99}{32} = 3\frac{3}{32}$

PROBLEM 9-17 Find the average (mean) of each given number of addends using the given sum. Round to the nearest tenth when the average is a reapeating decimal: **(a)** 258, 4 addends **(b)** 192, 2 addends   **(c)** 354, 3 addends   **(d)** 7890, 6 addends   **(e)** 25.5, 5 addends **(f)** 135.2, 8 addends   **(g)** 51.25, 7 addends   **(h)** 3048.25, 9 addends   **(i)** $\frac{1}{2}$, 2 addends **(j)** $5\frac{1}{4}$, 3 addends

Solution: Recall that to find the average (mean) of a given number of addends, you divide the sum by the number of addends [see Example 9-31]:

$$\overset{\text{sum}}{\downarrow} \quad \overset{\text{number of addends}}{\downarrow}$$

(a) $258 \div 4 = 64.5$ **(b)** $192 \div 2 = 96$ **(c)** $354 \div 3 = 118$ **(d)** $7890 \div 6 = 1315$
(e) $25.5 \div 5 = 5.1$ **(f)** $135.2 \div 8 = 16.9$ **(g)** $51.25 \div 7 = 7.3$ **(h)** $3048.25 \div 9 = 338.7$
(i) $\frac{1}{2} \div 2 = \frac{1}{2} \div \frac{2}{1} = \frac{1}{2} \times \frac{1}{2} = \frac{1}{4}$ **(j)** $5\frac{1}{4} \div 3 = \frac{21}{4} \div \frac{3}{1} = \frac{21}{4} \times \frac{1}{3} = \frac{\not{3} \times 7}{4} \times \frac{1}{\not{3}} = \frac{7}{4} = 1\frac{3}{4}$

PROBLEM 9-18 Identify each decimal as either terminating or repeating: **(a)** 0.3   **(b)** 0.333 **(c)** $0.\overline{3}$ **(d)** $0.333\cdots$ **(e)** 5.1   **(f)** 5.16   **(g)** 5.1666   **(h)** $5.1666\cdots$ **(i)** $5.1\overline{6}$ **(j)** $63.\overline{63}$

Solution: Recall that a terminating decimal ends in a given place and a repeating decimal repeats one or more digits forever [see Example 9-34]:

(a) 0.3 is a terminating decimal because it ends in the tenths place.
(b) 0.333 is a terminating decimal because it ends in the thousandths place.
(c) $0.\overline{3}$ is a repeating decimal because it repeats 3s forever.
(d) $0.333\cdots$ is a repeating decimal because it repeats 3s forever.
(e) 5.1 is a terminating decimal because it ends in the tenths place.
(f) 5.16 is a terminating decimal because it ends in the hundredths place.
(g) 5.1666 is a terminating decimal because it ends in the ten thousandths place.

(h) 5.1666 ⋯ is a repeating decimal because it repeats 6s forever.
(i) 5.1$\overline{6}$ is a repeating decimal because it repeats 6s forever.
(j) 63.$\overline{63}$ is a repeating decimal because it repeats 63s forever.

PROBLEM 9-19   Rename each repeating decimal using bar notation in simplest form: **(a)** 0.333 ⋯
(b) 0.666 ⋯   **(c)** 33.333 ⋯   **(d)** 66.666 ⋯   **(e)** 0.1666 ⋯   **(f)** 0.8333 ⋯
(g) 0.181818 ⋯   **(h)** 0.090909 ⋯   **(i)** 0.142857142857142857 ⋯
(j) 714285.714285714285714285 ⋯

Solution: Recall that to write a repeating decimal using bar notation, you write each digit in the decimal part that repeats under the bar [see Example 9-30]:

(a) 0.333 ⋯ = 0.$\overline{3}$ **(b)** 0.666 ⋯ = 0.$\overline{6}$ **(c)** 33.333 ⋯ = 33.$\overline{3}$ **(d)** 66.666 ⋯ = 66.$\overline{6}$
(e) 0.1666 ⋯ = 0.1$\overline{6}$ **(f)** 0.8333 ⋯ = 0.8$\overline{3}$ **(g)** 0.181818 ⋯ = 0.$\overline{18}$ **(h)** 0.090909 ⋯ = 0.$\overline{09}$
(i) 0.142857142857142857 ⋯ = 0.$\overline{142857}$ **(j)** 714285.714285714285714285 ⋯ = 714285.$\overline{714285}$

PROBLEM 9-20   Rename each fraction as an equal terminating decimal: **(a)** $\frac{1}{2}$ **(b)** $\frac{1}{4}$ **(c)** $\frac{1}{5}$
(d) $\frac{4}{5}$ **(e)** $\frac{1}{8}$ **(f)** $\frac{5}{8}$ **(g)** $\frac{1}{10}$ **(h)** $\frac{7}{10}$ **(i)** $\frac{1}{16}$ **(j)** $\frac{5}{32}$

Solution: Recall that to rename a fraction as an equal terminating decimal, you divide the denominator into the numerator to get a terminating decimal quotient [see Example 9-27]:

(a)
$$\frac{0.5}{2)\,1.0} \leftarrow \tfrac{1}{2}$$
$$\frac{-1\,0}{0}$$

(b)
$$\frac{0.2\,5}{4)\,1.0\,0} \leftarrow \tfrac{1}{4}$$
$$-\ 8$$
$$\frac{}{2\,0}$$
$$-2\,0$$
$$\frac{}{0}$$

(c)
$$\frac{0.2}{5)\,1.0} \leftarrow \tfrac{1}{5}$$
$$\frac{-1\,0}{0}$$

(d)
$$\frac{0.8}{5)\,4.0} \leftarrow \tfrac{4}{5}$$
$$\frac{-4\,0}{0}$$

(e)
$$\frac{0.1\,2\,5}{8)\,1.0\,0\,0} \leftarrow \tfrac{1}{8}$$
$$-\ 8$$
$$\frac{}{2\,0}$$
$$-1\,6$$
$$\frac{}{4\,0}$$
$$-4\,0$$
$$\frac{}{0}$$

(f)
$$\frac{0.6\,2\,5}{8)\,5.0\,0\,0} \leftarrow \tfrac{5}{8}$$
$$-4\,8$$
$$\frac{}{2\,0}$$
$$-1\,6$$
$$\frac{}{4\,0}$$
$$-4\,0$$
$$\frac{}{0}$$

(g)
$$\frac{0.1}{10)\,1.0} \leftarrow \tfrac{1}{10}$$
$$\frac{-1\,0}{0}$$

(h)
$$\frac{0.7}{10)\,7.0} \leftarrow \tfrac{7}{10}$$
$$\frac{-7\,0}{0}$$

(i)
$$\frac{0.0\,6\,2\,5}{16)\,1.0\,0\,0\,0} \leftarrow \tfrac{1}{16}$$
$$-\ 9\,6$$
$$\frac{}{4\ 0}$$
$$-3\ 2$$
$$\frac{}{8\ 0}$$
$$-8\ 0$$
$$\frac{}{0}$$

(j)
$$\frac{0.1\,5\,6\,2\,5}{32)\,5.0\,0\,0\,0\,0} \leftarrow \tfrac{5}{32}$$
$$-3\,2$$
$$\frac{}{1\ 8\ 0}$$
$$-1\ 6\ 0$$
$$\frac{}{2\ 0\ 0}$$
$$-1\ 9\ 2$$
$$\frac{}{8\ 0}$$
$$-6\ 4$$
$$\frac{}{1\ 6\ 0}$$
$$-1\ 6\ 0$$
$$\frac{}{0}$$

PROBLEM 9-21   Rename each fraction as an equal repeating decimal in simplest bar notation:

(a) $\frac{1}{3}$ **(b)** $\frac{1}{6}$ **(c)** $\frac{1}{9}$ **(d)** $\frac{5}{9}$ **(e)** $\frac{1}{11}$ **(f)** $\frac{7}{11}$ **(g)** $\frac{1}{12}$ **(h)** $\frac{5}{12}$ **(i)** $\frac{1}{7}$ **(j)** $\frac{5}{7}$

Solution: Recall that to rename a fraction as an equal repeating decimal, you divide the denominator into the numerator to get a repeating decimal quotient [see Example 9-37]:

(a)
$$
\begin{array}{r}
0.3\ 3\ 3\ \cdots = 0.\overline{3} \longleftarrow \tfrac{1}{3} \\
3\,)\ \overline{1.0\ 0\ 0} \\
-\ 9 \\
\hline
1\ 0 \\
-\ 9 \\
\hline
1\ 0 \\
-\ 9 \\
\hline
1
\end{array}
$$

(b)
$$
\begin{array}{r}
0.1\ 6\ 6\ 6\ \cdots = 0.1\overline{6} \longleftarrow \tfrac{1}{6} \\
6\,)\ \overline{1.0\ 0\ 0\ 0} \\
-\ 6 \\
\hline
4\ 0 \\
-3\ 6 \\
\hline
4\ 0 \\
-3\ 6 \\
\hline
4\ 0 \\
-3\ 6 \\
\hline
4
\end{array}
$$

(c)
$$
\begin{array}{r}
0.1\ 1\ 1\ \cdots = 0.\overline{1} \longleftarrow \tfrac{1}{9} \\
9\,)\ \overline{1.0\ 0\ 0} \\
-\ 9 \\
\hline
1\ 0 \\
-\ 9 \\
\hline
1\ 0 \\
-\ 9 \\
\hline
1
\end{array}
$$

(d)
$$
\begin{array}{r}
0.5\ 5\ 5\ \cdots = 0.\overline{5} \longleftarrow \tfrac{5}{9} \\
9\,)\ \overline{5.0\ 0\ 0} \\
-4\ 5 \\
\hline
5\ 0 \\
-4\ 5 \\
\hline
5\ 0 \\
-4\ 5 \\
\hline
5
\end{array}
$$

(e)
$$
\begin{array}{r}
0.0\ 9\ 0\ 9\ 0\ 9\ \cdots = 0.\overline{09} \longleftarrow \tfrac{1}{11} \\
11\,)\ \overline{1.0\ 0\ 0\ 0\ 0\ 0} \\
-\ 9\ 9 \\
\hline
1\ 0\ 0 \\
-\ 9\ 9 \\
\hline
1\ 0\ 0 \\
-\ 9\ 9 \\
\hline
1
\end{array}
$$

(f)
$$
\begin{array}{r}
0.6\ 3\ 6\ 3\ 6\ 3\ \cdots = 0.\overline{63} \longleftarrow \tfrac{7}{11} \\
11\,)\ \overline{7.0\ 0\ 0\ 0\ 0\ 0} \\
-6\ 6 \\
\hline
4\ 0 \\
-3\ 3 \\
\hline
7\ 0 \\
-6\ 6 \\
\hline
4\ 0 \\
-3\ 3 \\
\hline
7\ 0 \\
-6\ 6 \\
\hline
4\ 0 \\
-3\ 3 \\
\hline
7
\end{array}
$$

(g)
$$
\begin{array}{r}
0.0\ 8\ 3\ 3\ 3\ \cdots = 0.08\overline{3} \longleftarrow \tfrac{1}{12} \\
12\,)\ \overline{1.0\ 0\ 0\ 0\ 0} \\
-\ 9\ 6 \\
\hline
4\ 0 \\
-3\ 6 \\
\hline
4\ 0 \\
-3\ 6 \\
\hline
4\ 0 \\
-3\ 6 \\
\hline
4
\end{array}
$$

(h)
$$
\begin{array}{r}
0.4\ 1\ 6\ 6\ 6\ \cdots = 0.41\overline{6} \longleftarrow \tfrac{5}{12} \\
12\,)\ \overline{5.0\ 0\ 0\ 0\ 0} \\
-4\ 8 \\
\hline
2\ 0 \\
-1\ 2 \\
\hline
8\ 0 \\
-7\ 2 \\
\hline
8\ 0 \\
-7\ 2 \\
\hline
8\ 0 \\
-7\ 2 \\
\hline
8
\end{array}
$$

(i)
$$
\begin{array}{r}
0.142857142857142857 \cdots = 0.\overline{142857} \longleftarrow \tfrac{1}{7} \\
7\,)\ \overline{1.000000000000000000}
\end{array}
$$

(j)
$$
\begin{array}{r}
0.714285714285714285 \cdots = 0.\overline{714285} \longleftarrow \tfrac{5}{7} \\
7\,)\ \overline{5.000000000000000000}
\end{array}
$$

PROBLEM 9-22   Decide what to do (add, subtract, multiply, and/or divide) and then solve the following word problems:

(a) A 6-pack of *Drink Me* soda costs $1.99. How many cents does each bottle costs, remembering that stores always round up?

(b) Each bottle of *Drink Me* soda costs 25¢ on sale. How much does a 6-pack of *Drink Me* cost on sale if there is an additional discount of 1 cent when a full 6-pack is purchased?

(c) A can of corn costs $1.05. A can of beans costs 49¢. How much more does the corn cost than the beans?

(d) A large loaf of bread costs $1.49. A small loaf of bread costs 99¢. How much will it cost for both a large and small loaf of bread?

(e) A case of noodles costs $19.08. Each box of noodles in a case costs $1.59. How many boxes are in a case?

(f) Brand X oil costs $29.88 per case. Brand Y oil costs $9.96 for the same size case. How many times more expensive is brand X than brand Y?

(g) The road mileage between two cities is 125.6 km. How many kilometers must be driven to make the trip 20 times?

(h) Sam gives the clerk $20 for a $6.88 purchase amount. How much change should Sam receive?

Solution: Recall that to solve addition word problems containing decimals, you find how much two or more amounts are all together. To solve subtraction word problems containing decimals, you find the difference between two amounts. To solve multiplication word problems with decimals, you find how much two or more equal amounts are all together. To solve division word problems with decimals, you find how many equal amounts there are in all [see Example 9-38]:

(a) divide and then round up to the next cent: $1.99 \div 6 = 0.331\overline{6} \approx 0.34$ or 34¢
(b) multiply and then subtract 1 cent: $6 \times 25 = 150$ and $150 - 1 = 149$ (cents) or $1.49
(c) subtract: $1.05 - 0.49 = 0.56$ (dollars) or 56 cents   (d) add: $1.49 + 0.99 = 2.48$ (dollars)
(e) divide: $19.08 \div 1.59 = 12$ (boxes)   (f) divide: $29.88 \div 9.96 = 3$ (times more)
(g) multiply: $20 \times 125.6 = 2512$ (km)   (h) subtract: $20 - 6.88 = 13.12$ (dollars)

Supplementary Exercises

PROBLEM 9-23   Multiply basic facts given vertical form:

Row (a)	3 ×9	0 ×3	2 ×8	7 ×2	6 ×0	1 ×5	5 ×3	1 ×6	1 ×9	5 ×0
Row (b)	9 ×1	7 ×7	9 ×3	0 ×5	7 ×6	1 ×1	5 ×9	8 ×8	4 ×8	8 ×9
Row (c)	3 ×2	4 ×6	1 ×3	9 ×7	0 ×0	9 ×5	3 ×1	5 ×8	8 ×0	7 ×5
Row (d)	6 ×2	1 ×2	7 ×3	3 ×0	6 ×9	0 ×1	6 ×4	3 ×7	0 ×8	5 ×5
Row (e)	4 ×2	1 ×7	5 ×7	5 ×1	8 ×2	4 ×3	6 ×5	3 ×5	9 ×9	0 ×7

Row (**f**)	7 ×1	3 ×6	3 ×8	1 ×8	6 ×1	0 ×4	7 ×9	5 ×4	2 ×3	2 ×6
Row (**g**)	2 ×9	2 ×1	7 ×8	8 ×4	9 ×2	2 ×0	4 ×4	8 ×6	3 ×3	6 ×7
Row (**h**)	8 ×5	2 ×5	1 ×4	5 ×6	0 ×9	5 ×2	9 ×6	4 ×0	1 ×0	7 ×4
Row (**i**)	4 ×9	9 ×0	9 ×4	4 ×5	8 ×7	4 ×7	6 ×8	4 ×1	6 ×3	8 ×1
Row (**j**)	2 ×2	6 ×6	8 ×3	2 ×7	7 ×0	3 ×4	0 ×6	9 ×8	2 ×4	0 ×2

PROBLEM 9-24 Multiply with decimals: (**a**) 8×0.3 (**b**) 1.2×5 (**c**) 0.2×0.3
(**d**) 1.5×2.4 (**e**) 3.25×0.7 (**f**) 8×0.125 (**g**) 5.67×8.02 (**h**) 0.6×0.375
(**i**) 0.03×0.04 (**j**) 0.0005×0.001

PROBLEM 9-25 Multiply by powers of 10: (**a**) 10×0.2 (**b**) 10×80 (**c**) 100×0.3
(**d**) 100×20 (**e**) 1000×0.005 (**f**) 1000×7000 (**g**) $10,000 \times 0.15$ (**h**) $10,000 \times 2300$
(**i**) $100,000 \times 0.0025$ (**j**) $1,000,000 \times 1000$

PROBLEM 9-26 Divide basic facts in division box form:

Row (**a**)	9⟌54	8⟌40	2⟌6	2⟌16	3⟌18	5⟌35	1⟌7	9⟌72	8⟌0	3⟌21
Row (**b**)	8⟌48	9⟌9	8⟌64	2⟌0	3⟌9	1⟌8	7⟌28	4⟌36	1⟌1	9⟌45
Row (**c**)	3⟌15	4⟌20	6⟌18	6⟌24	6⟌6	3⟌6	9⟌36	1⟌0	6⟌48	5⟌15
Row (**d**)	6⟌0	8⟌56	7⟌35	3⟌24	9⟌0	3⟌12	4⟌16	9⟌63	7⟌63	3⟌27
Row (**e**)	2⟌12	2⟌10	4⟌28	2⟌18	5⟌45	3⟌0	9⟌18	4⟌4	4⟌24	8⟌16
Row (**f**)	9⟌9	7⟌21	7⟌42	8⟌32	9⟌81	9⟌27	5⟌10	8⟌24	7⟌49	7⟌56
Row (**g**)	4⟌32	3⟌3	8⟌72	5⟌25	1⟌2	1⟌5	4⟌8	5⟌40	2⟌14	5⟌0
Row (**h**)	1⟌6	5⟌5	1⟌3	5⟌20	1⟌4	6⟌42	7⟌0	6⟌12	4⟌12	2⟌4
Row (**i**)	6⟌54	6⟌36	6⟌30	7⟌7	4⟌0	8⟌8	2⟌8	2⟌2	7⟌14	5⟌30

PROBLEM 9-27 Divide with decimals: (**a**) $1.5 \div 3$ (**b**) $1.25 \div 5$ (**c**) $0.012 \div 0.2$
(**d**) $0.492 \div 0.4$ (**e**) $1.2 \div 1.5$ (**f**) $10.419 \div 2.3$ (**g**) $17.55 \div 0.75$ (**h**) $0.0004 \div 0.08$
(**i**) $1.5625 \div 1.25$ (**j**) $0.000002 \div 0.001$

PROBLEM 9-28 Divide and round each quotient to the nearest whole, tenth, hundredth, and
thousandth: (**a**) $1.87 \div 3$ (**b**) $0.3 \div 7$ (**c**) $0.285 \div 9$ (**d**) $1 \div 64$ (**e**) $0.283 \div 0.3$
(**f**) $25 \div 0.6$ (**g**) $31 \div 1.5$ (**h**) $4.85 \div 1.2$ (**i**) $800 \div 0.07$ (**j**) $1 \div 1.05$

PROBLEM 9-29 Divide by powers of 10; (**a**) $0.2 \div 10$ (**b**) $5 \div 10$ (**c**) $1.3 \div 100$
(**d**) $60 \div 100$ (**e**) $25.8 \div 1000$ (**f**) $100 \div 1000$ (**g**) $2.5 \div 10,000$
(**h**) $1,000,000 \div 10,000$ (**i**) $0.1 \div 100,000$ (**j**) $1 \div 1,000,000$

PROBLEM 9-30   Find the average (mean) of: **(a)** 82, 96, 87, 93   **(b)** 1.25, 1.46, 1.26, 1.75, 0.82
(c) 0.2, 0.1, 0.9, 1.1   **(d)** $\frac{3}{4}, \frac{7}{8}$ **(e)** 486 if there are 2 addends
(f) 1.08 if there are 4 addends   **(g)** 250.87 if there are 3 addends
(h) 0.91 if there are 6 addends   **(i)** $7\frac{1}{2}$ if there are 3 addends   **(j)** $\frac{3}{4}$ if there are 6 addends

PROBLEM 9-31   Rename each repeating decimal using bar notation in simplest form: **(a)** $0.0333\cdots$
(b) $0.00666\cdots$ **(c)** $16.666\cdots$ **(d)** $83.333\cdots$ **(e)** $0.454545\cdots$ **(f)** $0.636363\cdots$
(g) $0.123123123\cdots$ **(h)** $1.010101\cdots$ **(i)** $0.285714285714285714\cdots$
(j) $857142.857142857142857142\cdots$

PROBLEM 9-32   Rename each fraction as an equal terminating or repeating decimal: **(a)** $\frac{3}{4}$
(b) $\frac{2}{5}$ **(c)** $\frac{3}{5}$ **(d)** $\frac{3}{8}$ **(e)** $\frac{7}{8}$ **(f)** $\frac{3}{10}$ **(g)** $\frac{9}{10}$ **(h)** $\frac{3}{16}$ **(i)** $\frac{11}{32}$ **(j)** $\frac{1}{64}$ **(k)** $\frac{2}{3}$
(l) $\frac{5}{6}$ **(m)** $\frac{2}{9}$ **(n)** $\frac{7}{9}$ **(o)** $\frac{2}{11}$ **(p)** $\frac{10}{11}$ **(q)** $\frac{7}{12}$ **(r)** $\frac{11}{12}$ **(s)** $\frac{2}{7}$ **(t)** $\frac{5}{7}$

PROBLEM 9-33   Solve each word problem containing decimals using addition, subtraction, multiplication, and/or division:

(a) Lyle received math scores for the semester of 96, 84, 98, 72, and 83. If an A is 96–100, and a B is 88–95, and a C is 77–87, then what grade will Lyle earn?

(b) What is the lowest score that Lyle, in problem **(a)**, can get on the next test for a B average if each test is worth 100 points?

(c) Melody pays $650.75 rent each month. How much rent does she pay during a year (12 months)?

(d) Yale owes $10,269 for a new car he just purchased. The amount owed is to be paid back over a 36-month period. What are the monthly payments?

(e) There are 365 days in a normal year. How many 7-day weeks are in a normal year?

(f) There are 366 days in a leap year. How many days are in an average month during a leap year?

(g) How many more hours are in an average month during a leap year than a normal year?

(h) If Meg works 40 hours per week at $12.25 per hour, how much will she earn during an average 365-day year?

Answers to Supplementary Exercises

(9-23)

Row **(a)**	27	0	16	14	0	5	15	6	9	0
Row **(b)**	9	49	27	0	42	1	45	64	32	72
Row **(c)**	6	24	3	63	0	45	3	40	0	35
Row **(d)**	12	2	21	0	54	0	24	21	0	25
Row **(e)**	8	7	35	5	16	12	30	15	81	0
Row **(f)**	7	18	24	8	6	0	63	20	6	12
Row **(g)**	18	2	56	32	18	0	16	48	9	42
Row **(h)**	40	10	4	30	0	10	54	0	0	28
Row **(i)**	36	0	36	20	56	28	48	4	18	8
Row **(j)**	4	36	24	14	0	12	0	72	8	0

(9-24) **(a)** 2.4   **(b)** 6   **(c)** 0.06   **(d)** 3.6   **(e)** 2.275   **(f)** 1   **(g)** 45.4734   **(h)** 0.225
(i) 0.0012   **(j)** 0.0000005

(9-25) **(a)** 2   **(b)** 800   **(c)** 30   **(d)** 2000   **(e)** 5   **(f)** 7,000,000   **(g)** 1500
(g) 23,000,000   **(i)** 250   **(j)** 1,000,000,000

(9-26)

Row (a)	6	5	3	8	6	7	7	8	0	7
Row (b)	6	1	8	0	3	8	4	9	1	5
Row (c)	5	5	3	4	1	2	4	0	8	3
Row (d)	0	7	5	8	0	4	4	7	9	9
Row (e)	6	5	7	9	9	0	2	1	6	2
Row (f)	1	3	6	4	9	3	2	3	7	8
Row (g)	8	1	9	5	2	5	2	8	7	0
Row (h)	6	1	3	4	4	7	0	2	3	2
Row (i)	9	6	5	1	0	1	4	1	2	6

(9-27)   **(a)** 0.5   **(b)** 0.25   **(c)** 0.06   **(d)** 1.23   **(e)** 0.8   **(f)** 4.53   **(g)** 23.4   **(h)** 0.005
(i) 1.25   **(j)** 0.002

(9-28)   **(a)** 1, 0.6; 0.62, 0.623   **(b)** 0, 0, 0.04, 0.043   **(c)** 0, 0, 0.03, 0.032   **(d)** 0, 0, 0.02, 0.016
(e) 1, 0.9, 0.94, 0.943   **(f)** 42, 41.7, 41.67, 41.667   **(g)** 21, 20.7, 20.67, 20.667
(h) 4, 4.0, 4.04, 4.042   **(i)** 11,429, 11,428.6, 11,428.57, 11,428.571   **(j)** 1, 1.0, 0.95, 0.952

(9-29)   **(a)** 0.02   **(b)** 0.5   **(c)** 0.013   **(d)** 0.6   **(e)** 0.0258   **(f)** 0.1   **(g)** 0.00025
(h) 100   **(i)** 0.000001   **(j)** 0.000001

(9-30)   **(a)** 89.5   **(b)** 1.308   **(c)** 0.575   **(d)** $\frac{13}{16}$ or 0.8125   **(e)** 243   **(f)** 0.27   **(g)** $83.62\overline{3}$
(h) $0.151\overline{6}$ **(i)** $2\frac{1}{2}$ or 2.5   **(j)** $\frac{1}{8}$ or 0.125

(9-31)   **(a)** $0.0\overline{3}$ **(b)** $0.00\overline{6}$ **(c)** $16.\overline{6}$ **(d)** $83.\overline{3}$ **(e)** $0.\overline{45}$ **(f)** $0.\overline{63}$ **(g)** $0.\overline{123}$
(h) $1.\overline{01}$ **(i)** $0.\overline{285714}$ **(j)** $857142.\overline{857142}$

(9-32)   **(a)** 0.75   **(b)** 0.4   **(c)** 0.6   **(d)** 0.375   **(e)** 0.875   **(f)** 0.3   **(g)** 0.9
(h) 0.1875   **(i)** 0.34375   **(j)** 0.015625   **(k)** $0.\overline{6}$ **(l)** $0.8\overline{3}$ **(m)** $0.\overline{2}$ **(n)** $0.\overline{7}$
(o) $0.\overline{18}$ **(p)** $0.\overline{90}$ **(q)** $0.58\overline{3}$ **(r)** $0.19\overline{6}$ **(s)** $0.\overline{285715}$ **(t)** $0.\overline{714285}$

(9-33)   **(a)** $86.\overline{6}$ = grade of C   **(b)** 92 points   **(c)** $7809   **(d)** $285.25
(e) $52.\overline{142857}$ 7-day weeks   **(f)** 30.5 days   **(g)** 2 hours   **(h)** $25,550

MIDTERM EXAMINATION

Chapters 1–9

Part 1: Skills and Concepts (88 questions)

1. The place-value name of the underlined digit in 718,054,693 is
   (a) hundred millions   (b) ten thousands   (c) zero   (d) hundred thousands
   (e) none of these

2. What is the value of 2 in 619,082,583?
   (a) thousands   (b) millions   (c) ten thousands   (d) hundreds
   (e) none of these

3. Six thousand twenty-five written as a whole number is
   (a) 625   (b) 6025   (c) 6250   (d) 60,025   (e) none of these

4. 705 written as a whole-number-word name is
   (a) seventy-five   (b) seven hundred and five   (c) seven hundred five
   (d) seven hundred fifty   (e) none of these

5. The correct symbol for ? in 17,843 ? 17,834 is
   (a) <   (b) >   (c) =   (d) all of these   (e) none of these

6. 546, 456, 465, 645, 654, 564 listed from smallest to largest is
   (a) 654, 645, 564, 546, 456, 465   (b) 456, 465, 564, 546, 645, 654
   (c) 465, 456, 546, 564, 645, 654   (d) 654, 645, 564, 546, 465, 456   (e) none of these

7. 6,375,418 rounded to the nearest ten thousand is
   (a) 6,400,000   (b) 6,370,000   (c) 6,380,000   (d) 6,375,000   (e) none of these

8. The sum of 83 + 585 + 109 is
   (a) 677   (b) 667   (c) 767   (d) 777   (e) none of these

9. 642 − 285 equals
   (a) 357   (b) 457   (c) 367   (d) 467   (e) none of these

10. 803 − 504 equals
    (a) 399   (b) 299   (c) 309   (d) 209   (e) none of these

11. The product of 4 × 368 is
    (a) 1272   (b) 1442   (c) 1242   (d) 1482   (e) none of these

12. The product of 25 × 846 is
    (a) 21,050   (b) 22,150   (c) 21,150   (d) 5859   (e) none of these

13. The product of 235 × 346 is
    (a) 91,310   (b) 89,310   (c) 81,210   (d) 81,310   (e) none of these

14. The product of 800 × 3000 is
    (a) 2400   (b) 240,000   (c) 24,000   (d) 2,400,000   (e) none of these

15. The product of 203 × 324 is
    (a) 65,772   (b) 7452   (c) 65,762   (d) 648,972   (e) none of these

16. The quotient of 58 ÷ 3 is
    (a) 19   (b) 18 R1   (c) 18 R2   (d) 19 R2   (e) none of these

17. The quotient of 108 ÷ 8 is
    (a) 13 R1   (b) 13 R2   (c) 13 R3   (d) 13 R4   (e) none of these

18. The quotient of 816 ÷ 4 is
    (a) 204   (b) 24   (c) 240   (d) 200 R16   (e) none of these

19. The quotient of 385 ÷ 20 is
    (a) 18 R5   (b) 19   (c) 14 R5   (d) 14   (e) none of these

20. The quotient of 542 ÷ 29 is
    (a) 19   (b) 18 R20   (c) 18   (d) 19 R20   (e) none of these

21. The quotient of 2654 ÷ 18 is
    (a) 137 R8   (b) 147   (c) 147 R8   (d) 14 R13   (e) none of these

22. The quotient of 8400 ÷ 200 is
    (a) 420   (b) 4200   (c) 42   (d) 42,000   (e) none of these

23. The whole-number factors of 18 are
    (a) 2, 3, 6, 9   (b) 1, 18   (c) 1, 2, 3, 6, 9   (d) 1, 2, 3, 6, 9, 18
    (e) none of these

24. 18 factored as a product of primes is
    (a) $2 \times 2 \times 3$ (b) 2×9 (c) $2 \times 3 \times 3$ (d) 3×6 (e) none of these

25. Which of the following fractions is not defined?
    (a) $\frac{0}{3}$ (b) $\frac{3}{0}$ (c) $\frac{3}{3}$ (d) $\frac{0}{1}$ (e) none of these

26. $\frac{12}{18}$ is equal to
    (a) $\frac{4}{6}$ (b) $\frac{3}{2}$ (c) $\frac{3}{4}$ (d) $\frac{6}{8}$ (e) none of these

27. 2 ÷ 5 renamed as a fraction is
    (a) $\frac{2}{5}$ (b) $\frac{5}{2}$ (c) $2\frac{1}{2}$ (d) $2\frac{1}{5}$ (e) none of these

28. $\frac{4}{3}$ renamed as a division problem is
    (a) 3 ÷ 4   (b) 4 ÷ 3   (c) $1\frac{1}{3}$ (d) $4\overline{\smash{\big)}3}$ (e) none of these

29. 4 renamed as an equal fraction is
    (a) $\frac{4}{1}$ (b) $\frac{12}{3}$ (c) $\frac{8}{2}$ (d) $\frac{16}{4}$ (e) all of these

30. $4\frac{2}{3}$ renamed as an equal fraction is
    (a) $\frac{2}{3}$ (b) $\frac{12}{3}$ (c) $\frac{24}{3}$ (d) $\frac{4}{1}$ (e) none of these

31. $\frac{15}{4}$ renamed as an equal mixed number is
    (a) $3\frac{1}{4}$ (b) $4\frac{1}{4}$ (c) $2\frac{3}{4}$ (d) $3\frac{3}{4}$ (e) none of these

32. Which of the following fractions is in lowest terms?
    (a) $\frac{4}{6}$ (b) $\frac{8}{18}$ (c) $\frac{15}{9}$ (d) $\frac{21}{14}$ (e) none of these

33. $\frac{18}{12}$ in simplest form is
    (a) $\frac{4}{6}$ (b) $\frac{1}{2}$ (c) $\frac{2}{3}$ (d) $\frac{8}{12}$ (e) none of these

34. $\frac{15}{9}$ in simplest form is
    (a) $\frac{15}{9}$ (b) $\frac{5}{3}$ (c) $1\frac{6}{9}$ (d) $1\frac{2}{3}$ (e) none of these

35. $2\frac{9}{3}$ in simplest form
    (a) $2\frac{3}{1}$ (b) 3   (c) $2\frac{9}{3}$ (d) $2 + \frac{9}{3}$ (e) none of these

36. The product of $\frac{2}{3} \times \frac{3}{4}$ in simplest form is
    (a) $\frac{1}{2}$ (b) $\frac{6}{16}$ (c) $\frac{2}{4}$ (d) $\frac{3}{6}$ (e) all of these

37. The product of $5 \times \frac{7}{10}$ in simplest form is
    (a) $\frac{35}{10}$ (b) $\frac{7}{50}$ (c) $5\frac{7}{10}$ (d) $\frac{57}{10}$ (e) none of these

38. The product of $2\frac{2}{3} \times 2\frac{1}{4}$ in simplest form is
    (a) $4\frac{1}{6}$ (b) $\frac{6}{1}$ (c) $\frac{72}{12}$ (d) 6   (e) none of these

39. The reciprocal of $\frac{8}{6}$ is
    (a) $\frac{8}{6}$ (b) $\frac{6}{8}$ (c) $\frac{4}{3}$ (d) $\frac{3}{4}$ (e) none of these

40. The quotient of $\frac{3}{4} \div \frac{8}{9}$ in simplest form is
    (a) $\frac{2}{3}$ (b) $\frac{3}{2}$ (c) $\frac{27}{32}$ (d) $\frac{32}{27}$ (e) none of these

41. The quotient of $\frac{5}{6} \div 10$ in simplest form is
    (a) $\frac{1}{12}$ (b) $\frac{25}{3}$ (c) $8\frac{1}{4}$ (d) $\frac{5}{60}$ (e) none of these

42. The quotient of $12 \div \frac{3}{4}$ in simplest form is
    (a) $\frac{9}{1}$ (b) 9   (c) $\frac{48}{3}$ (d) $\frac{16}{1}$ (e) none of these

43. The quotient of $2\frac{1}{2} \div 1\frac{1}{4}$ in simplest form is
    (a) $\frac{25}{8}$ (b) $3\frac{1}{8}$ (c) $\frac{4}{2}$ (d) 2   (e) none of these

44. Which of the following fraction pairs are like fractions?
    (a) $\frac{2}{3}$ and $\frac{3}{2}$ (b) $\frac{2}{4}$ and $\frac{4}{6}$ (c) $\frac{2}{3}$ and $\frac{2}{5}$ (d) $\frac{5}{8}$ and $\frac{2}{2}$ (e) none of these

45. The sum of $\frac{3}{4} + \frac{3}{4}$ in simplest form is
    (a) $\frac{9}{16}$ (b) $\frac{6}{4}$ (c) $\frac{3}{2}$ (d) $1\frac{1}{2}$ (e) none of these

46. The LCD for $\frac{5}{6}$ and $\frac{7}{8}$ is
    (a) 48   (b) 8   (c) 24   (d) 14   (e) none of these

47. $\frac{3}{4}$ built up as an equal fraction with a denominator of 20 is
    (a) $\frac{3}{20}$ (b) $\frac{7}{20}$ (c) $\frac{15}{20}$ (d) $\frac{16}{20}$ (e) none of these

48. $\frac{2}{3}, \frac{3}{4}$, and $\frac{5}{6}$ renamed as equivalent like fractions using the LCD are
    (a) $\frac{16}{24}, \frac{18}{24}, \frac{20}{24}$ (b) $\frac{8}{12}, \frac{9}{12}, \frac{10}{12}$ (c) $\frac{2}{12}, \frac{3}{12}, \frac{5}{12}$ (d) $\frac{2}{24}, \frac{3}{24}, \frac{5}{24}$
    (e) none of these

49. The sum of $\frac{2}{3} + \frac{3}{4}$ in simplest form is
    (a) $\frac{17}{12}$ (b) $\frac{5}{12}$ (c) $\frac{6}{12}$ (d) $\frac{1}{2}$ (e) none of these

50. The sum of $\frac{1}{2}$, and $\frac{3}{4}$ and $\frac{5}{8}$ in simplest form is
    (a) $1\frac{7}{8}$ (b) $\frac{15}{8}$ (c) $\frac{15}{64}$ (d) 1   (e) none of these

51. The sum of $3\frac{1}{2} + 4$ in simplest form is
    (a) $7\frac{1}{2}$ (b) $12\frac{1}{2}$ (c) $1\frac{1}{2}$ (d) 14   (e) none of these

52. The sum of $\frac{3}{4} + 2\frac{1}{6}$ in simplest form is
    (a) $2\frac{3}{12}$ (b) $2\frac{1}{4}$ (c) $\frac{35}{12}$ (d) $2\frac{11}{12}$ (e) none of these

53. The sum of $2\frac{1}{2} + 2\frac{1}{2} + 3$ in simplest form is
    (a) $12\frac{1}{4}$ (b) $3\frac{10}{2}$ (c) $3\frac{25}{4}$ (d) $7\frac{1}{2}$ (e) none of these

54. $\frac{5}{8} - \frac{3}{8}$ in simplest form equals

(a) $\frac{2}{8}$ (b) $\frac{8}{8}$ (c) $\frac{1}{4}$ (d) 1   (e) none of these

55. $\frac{7}{10} - \frac{1}{5}$ in simplest form equals

(a) $\frac{6}{5}$ (b) $\frac{1}{2}$ (c) $\frac{5}{10}$ (d) $\frac{9}{10}$ (e) none of these

56. $5\frac{1}{2} - 4$ in simplest form equals

(a) $1\frac{1}{2}$ (b) $9\frac{1}{2}$ (c) $20\frac{1}{2}$ (d) $4\frac{11}{2}$ (e) none of these

57. $2\frac{5}{8} - \frac{1}{3}$ in simplest form equals

(a) $2\frac{23}{24}$ (b) $\frac{21}{5}$ (c) $4\frac{1}{5}$ (d) $2\frac{1}{4}$ (e) none of these

58. $8\frac{1}{10} - 3\frac{1}{12}$ in simplest form equals

(a) $5\frac{1}{60}$ (b) $11\frac{11}{60}$ (c) $24\frac{1}{120}$ (d) $5\frac{1}{2}$ (e) none of these

59. $6 - 1\frac{3}{4}$ in simplest form equals

(a) $5\frac{3}{4}$ (b) $5\frac{1}{4}$ (c) $4\frac{1}{4}$ (d) $4\frac{3}{4}$ (e) none of these

60. $3\frac{1}{3} - \frac{1}{2}$ in simplest form equals

(a) $3\frac{1}{6}$ (b) $2\frac{5}{6}$ (c) $2\frac{11}{6}$ (d) $3\frac{5}{6}$ (e) none of these

61. Which digit is in the thousandths place in 1325.416908?

(a) 1   (b) 8   (c) 0   (d) 9   (e) none of these

62. The decimal-word name for 1.03 is

(a) thirteen   (b) 13 tenths   (c) 1 and 3 tenths   (d) 1 and 3 hundredths

(e) none of these

63. The decimal number for 25 and 8 thousandths is

(a) 25.8   (b) 258   (c) 25.08   (d) 25.008   (e) none of these

64. The correct symbol for ? in 365.24717 ? 365.24709 is

(a) <   (b) >   (c) =   (d) all of these   (e) none of these

65. The correct order from largest to smallest of 0.3, 0.03, 3, 3.03 is

(a) 0.03, 0.3, 3, 3.03   (b) 3, 3.03, 0.3, 0.03   (c) 3.03, 3, 0.03, 0.3   (d) 3, 3.03, 0.03, 0.3

(e) none of these

66. 325.963 rounded to the nearest tenth is

(a) 325.9   (b) 325   (c) 326.0   (d) 326   (e) none of these

67. 5 renamed as an equal decimal is

(a) 5.0   (b) 5.00   (c) 5.000   (d) 5.0000   (e) all of these

68. 245.00 renamed as an equal whole number is

(a) 2.45   (b) 24.5   (c) 245   (d) 245.0   (e) none of these

69. $\frac{9}{100}$ renamed as an equal decimal is

(a) 0.9   (b) 0.09   (c) 0.90   (d) 0.009   (e) none of these

70. 2.3 renamed as an equal decimal with 3 decimal places is

(a) 2.300   (b) 2.30   (c) 2.3   (d) 0.023   (e) none of these

71. The sum of 8.25 + 4.63 + 0.25 + 0.125 + 128 is

(a) 1566   (b) 141.255   (c) 141.155   (d) 140.255   (e) none of these

72. The sum of $14.89 + 99¢ is
   (a) $15.88   **(b)** $14.88   **(c)** $113.89   **(d)** $15.87   **(e)** none of these

73. 8.005 − 0.579 equals
   (a) 7.426   **(b)** 8.426   **(c)** 7.996   **(d)** 8.584   **(e)** none of these

74. 1.5 − 0.125 equals **(a)** 2.5   **(b)** 1.425   **(c)** 1.375   **(d)** 1.625   **(e)** none of these

75. 3.75 − 2 equals **(a)** 5.75   **(b)** 2.25   **(c)** 5.25   **(d)** 1.75   **(e)** none of these

76. $8 − 67¢ equals **(a)** $75   **(b)** 75¢   **(c)** $8.67   **(d)** $7.33   **(e)** none of these

77. The product of 5 × 0.2 is
   (a) 10   **(b)** 1   **(c)** 0.1   **(d)** 0.01   **(e)** none of these

78. The product of 3.4 × 1.02 is
   (a) 3.468   **(b)** 34.68   **(c)** 346.8   **(d)** 3468   **(e)** none of these

79. The product of 0.002 × 0.003 is
   (a) 0.06   **(b)** 0.006   **(c)** 0.0006   **(d)** 0.00006   **(e)** none of these

80. The product of 1000 × 0.05 is
   (a) 5   **(b)** 50   **(c)** 500   **(d)** 5000   **(e)** none of these

81. The quotient of 0.2 ÷ 5 is
   (a) 4   **(b)** 0.4   **(c)** 0.04   **(d)** 0.004   **(e)** none of these

82. The quotient of 25.2 ÷ 0.02 is
   (a) 126   **(b)** 12.6   **(c)** 1260   **(d)** 12,600   **(e)** none of these

83. The quotient of 0.001 ÷ 4 is
   (a) 0.25   **(b)** 0.025   **(c)** 0.0025   **(d)** 0.00025   **(e)** none of these

84. The quotient of 25,000 ÷ 1000 is
   (a) 25   **(b)** 250   **(c)** 2.5   **(d)** 25,000,000   **(e)** none of these

85. The average of 86, 87, 92, 53, 69 is
   (a) 77   **(b)** 774   **(c)** 7.74   **(d)** 77.4   **(e)** none of these

86. $\frac{5}{8}$ renamed as an equal terminating decimal is
   (a) 6.25   **(b)** 0.75   **(c)** 1.6   **(d)** 0.625   **(e)** none of these

87. $\frac{2}{3}$ renamed as an equal repeating decimal is
   (a) $0.\overline{6}$ **(b)** $66.\overline{6}$ **(c)** $0.\overline{3}$ **(d)** 1.5   **(e)** none of these

88. $\frac{3}{10}$ renamed as an equal decimal is
   (a) 3   **(b)** 0.3   **(c)** 0.03   **(d)** $3.\overline{3}$ **(e)** none of these

Part 2: Problem Solving (12 questions)

89. There are 24 cans in one box and 8 cans in another box. How many cans are there in all?

90. There are 24 cans in each box. There are 8 boxes in all. How many cans are there in all?

91. There are 24 cans in one box and 8 cans in another box. What is the difference between the number of cans in the two two boxes?

92. There are 24 cans in 8 boxes. If each box contains the same number of cans, how many cans are in each box?

93. A certain fence is 100 feet long. On Monday $\frac{3}{4}$ of the fence was painted. How many feet long was the painted section of the fence?

94. 40 feet of a certain fence was painted on Tuesday. This was $\frac{1}{5}$ of its total length. How long is the fence?

95. On Monday $\frac{1}{2}$ of a certain fence was painted. Tuesday another $\frac{1}{4}$ of the fence was painted. How much of the total fence was painted by the end of the day on Tuesday?

96. $\frac{1}{3}$ of a certain fence was painted on Monday. What fractional part of the fence remained to be painted on Tuesday?

97. Handkerchiefs are 3 for $1. How much will the store charge for 1 handkerchief?

98. Workshirts are $12.89 each. How much will 5 workshirts cost if the sales tax is $3.87?

99. Dress pants are $43.89 each and blouses are $25.99 each. How much more will 3 pairs of dress pants cost than 2 blouses, not including sales tax?

100. Jan earns $18.25 per hour. How much does Jan earn in a year if she works 40 hours per week, 52 weeks per year?

Midterm Examination Answers

Part 1

1. (d)	**2.** (a)	**3.** (b)	**4.** (c)	**5.** (b)
6. (e)	**7.** (c)	**8.** (d)	**9.** (a)	**10.** (b)
11. (e)	**12.** (c)	**13.** (d)	**14.** (d)	**15.** (a)
16. (e)	**17.** (d)	**18.** (a)	**19.** (e)	**20.** (b)
21. (e)	**22.** (c)	**23.** (d)	**24.** (c)	**25.** (b)
26. (a)	**27.** (a)	**28.** (b)	**29.** (e)	**30.** (e)
31. (d)	**32.** (e)	**33.** (e)	**34.** (d)	**35.** (e)
36. (a)	**37.** (e)	**38.** (d)	**39.** (b)	**40.** (c)
41. (a)	**42.** (e)	**43.** (d)	**44.** (e)	**45.** (d)
46. (c)	**47.** (c)	**48.** (b)	**49.** (e)	**50.** (a)
51. (a)	**52.** (d)	**53.** (e)	**54.** (c)	**55.** (b)
56. (a)	**57.** (e)	**58.** (a)	**59.** (c)	**60.** (b)
61. (e)	**62.** (d)	**63.** (d)	**64.** (b)	**65.** (e)
66. (c)	**67.** (e)	**68.** (c)	**69.** (b)	**70.** (a)
71. (b)	**72.** (a)	**73.** (a)	**74.** (c)	**75.** (d)
76. (d)	**77.** (b)	**78.** (a)	**79.** (e)	**80.** (b)
81. (c)	**82.** (c)	**83.** (d)	**84.** (a)	**85.** (d)
86. (d)	**87.** (a)	**88.** (b)		

Part 2

89. 32 cans	**90.** 192 cans	**91.** 16 cans	**92.** 3 cans
93. 75 feet	**94.** 200 feet	**95.** $\frac{3}{4}$ of the fence	**96.** $\frac{2}{3}$ of the fence
97. 34¢	**98.** $68.32	**99.** $79.69	**100.** $37,960 per year

10 PERCENT SKILLS

THIS CHAPTER IS ABOUT

- ☑ **Renaming Percents as Decimals and Fractions**
- ☑ **Renaming Decimals and Fractions as Percents**
- ☑ **Finding the Amount in a Percent Problem**
- ☑ **Finding the Base in a Percent Problem**
- ☑ **Find the Percent in a Percent Problem**
- ☑ **Finding Percent Increase and Decrease**

10-1. Renaming Percents as Decimals and Fractions

To rename part of a whole, you can use a number value, decimal, fraction, or **percent.**

EXAMPLE 10-1: Write part of a whole as a number value, decimal, fraction, and percent.

Solution:

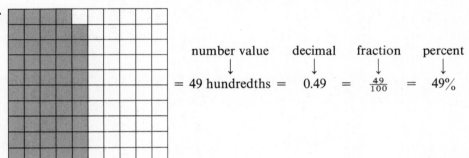

$$= 49 \text{ hundredths} = 0.49 = \frac{49}{100} = 49\%$$

number value   decimal   fraction   percent

The symbol % stands for the word *percent*. Read "49%" as "forty-nine percent." *Percent* can mean (a) hundredths, (b) out of one hundred, (c) per hundred, (d) $\times \frac{1}{100}$, or (e) $\div 100$.

EXAMPLE 10-2: Rename 50% using **(a)** hundredths **(b)** out of one hundred **(c)** per hundred **(d)** $\times \frac{1}{100}$ **(e)** $\div 100$.

Solution: **(a)** 50% = 50 hundredths   **(b)** 50% = 50 out of one hundred
(c) 50% = 50 per hundred   **(d)** 50% = 50 $\times \frac{1}{100}$ **(e)** 50% = 50 $\div$ 100

You can rename every percent as either a decimal or a fraction.

A. Rename percents as decimals.

To rename a percent as a decimal, you can use the following formula: % means "$\div 100$."

EXAMPLE 10-3: Rename 50% as a decimal using the following formula: % means "$\div 100$."

Solution: 50% = 50 $\div$ 100   *Think:* % means "$\div 100$."
            = 0.5   50 $\div$ 100 = 0.5

To rename a percent as a decimal, you can also use Shortcut 10-1.

Shortcut 10-1: To rename a percent as a decimal:
1. Move the decimal point two places to the **left** (divide by 100).
2. Eliminate the percent symbol (%).

EXAMPLE 10-4: Rename 125% as a decimal using Shortcut 10-1.

Solution: 125% = 1.25%   Move the decimal point two places to the left.

         = 1.25   Eliminate the percent symbol.

Check: Does 125% = 1.25 using the formula % means "÷ 100?" Yes: 125% = 125 ÷ 100 = 1.25

To rename a **fractional percent** (like $5\frac{3}{4}\%$) as a terminating decimal, you must first rename the fractional percent as a **decimal percent.**

EXAMPLE 10-5: Rename $5\frac{3}{4}\%$ as a decimal by first renaming as a decimal percent.

             fractional   decimal
              percent   percent

Solution: $5\frac{3}{4}\%$ = 5.75% $5\frac{3}{4}$ = 5.75

               = 0.0575%   Move the decimal point two places to the left.

               = 0.0575   Eliminate the percent symbol.

Check: Using the formula % means "÷ 100," $5\frac{3}{4}\%$ = 5.75 ÷ 100 = 0.0575.

B. Rename percents as fractions.

To rename a percent as a fraction, you can use the same formula you used to rename percents as decimals: % means "÷ 100."

EXAMPLE 10-6: Rename 50% as a fraction using the following formula: % means "÷ 100."

Solution: 50% = 50 ÷ 100   *Think:* % means "÷ 100."

$$= \frac{50}{100} \qquad\qquad 50 \div 100 = \frac{50}{100}$$

$$= \frac{1}{2} \qquad\qquad \frac{50}{100} = \frac{1 \times 50}{2 \times 50} = \frac{1}{2}$$

Note: 50%, 0.5, and $\frac{1}{2}$ all name the same value since $0.5 = \frac{5}{10} = \frac{1}{2}$.

EXAMPLE 10-7: Rename (a) $33\frac{1}{3}\%$ and (b) 62.5% as fractions using the following formula: % means "$\times \frac{1}{100}$."

Solution

(a) $33\frac{1}{3}$ = $33\frac{1}{3} \times \frac{1}{100}$ *Think:* % means "$\times \frac{1}{100}$."

$$= \frac{100}{3} \times \frac{1}{100} \qquad\qquad 33\frac{1}{3} = \frac{100}{3}$$

$$= \frac{1}{3} \qquad\qquad\qquad \frac{100}{3} \times \frac{1}{100} = \frac{1}{3}$$

(b) 62.5% = $62.5 \times \frac{1}{100}$ *Think:* % means "$\times \frac{1}{100}$."

$$= 62\frac{1}{2} \times \frac{1}{100} \qquad\qquad 62.5 = 62\frac{1}{2}$$

$$= \frac{125}{2} \times \frac{1}{100} \qquad\qquad 62\frac{1}{2} = \frac{125}{2}$$

$$= \frac{5 \times 25}{2} \times \frac{1}{4 \times 25} \qquad \text{Multiply fractions.}$$

$$= \frac{5}{2} \times \frac{1}{4} = \frac{5}{8}$$

10-2. Renaming Decimals and Fractions as Percents

You can rename every terminating decimal, repeating decimal, and fraction as a percent.

A. Rename decimals as percents.

Remember Shortcut 10-1 stated that to rename a percent as a decimal, you move the decimal point two places to the left, and then eliminate the percent symbol.

To rename a decimal as a percent, you do the opposite of each step in Shortcut 10-1. Shortcut 10-2 shows how to rename a decimal as a percent.

Shortcut 10-2: To rename a decimal as a percent:
1. Move the decimal point two places to the **right** (multiply by 100).
2. Add the percent symbol (%).

EXAMPLE 10-8: Rename 0.25 as a percent using Shortcut 10-2.

Solution: $0.25 = 025.\%$ Move the decimal point two places to the right.
                          Add the percent symbol (%).

         $= 25\%$ *Think:* 025. = 25

Check: Does $25\% = 0.25$ using the formula % means "$\div 100$?" Yes: $25\% = 25 \div 100 = 0.25$

B. Rename fractions as percents.

To rename a fraction as a percent, you can first rename as a decimal and then as a percent.

EXAMPLE 10-9: Rename $\frac{5}{8}$ as a percent by first renaming as a decimal.

Solution: $\frac{5}{8} = 0.625 \longleftarrow 8\overline{)5.000}^{\,0.625}$

         $= 62.5\% \longleftarrow 0.625 = 62.5\%$

Check: Does $62.5\% = \frac{5}{8}$ using the formula % means "$\div 100$?" Yes:

$$62.5\% = 62.5 \div 100 = 0.625 = \frac{625}{1000} = \frac{5 \times \cancel{125}}{8 \times \cancel{125}} = \frac{5}{8}$$

Note: To check $62.5\% = \frac{5}{8}$, you divide by 100 and eliminate the percent symbol (%): $62.5\% = 62.5 \div 100 = \frac{5}{8}$

You can also rename a fraction as a percent by using Shortcut 10-3.

Shortcut 10-3: To rename a fraction as a percent:
1. Multiply by 100.
2. Write the percent symbol (%).

EXAMPLE 10-10: Rename $\frac{2}{3}$ as a percent using Shortcut 10-3:

Solution: $\frac{2}{3} = (\frac{2}{3} \times 100)\%$ Multiply by 100. Write the percent symbol (%).

         $= \frac{200}{3}\%$ $\frac{2}{3} \times 100 = \frac{2}{3} \times \frac{100}{1} = \frac{200}{3}$

         $= 66\frac{2}{3}\%$ $3\overline{)200}^{\,66\frac{2}{3}}$

Note: When renaming a number like $\frac{2}{3}$ as a percent, it is easier to use Shortcut 10-3. It would be more difficult to rename $\frac{2}{3}$ as a percent by first renaming as a decimal because $\frac{2}{3}$ equals the repeating decimal of $0.\overline{6}$.

10-3. Finding the Amount in a Percent Problem

To solve percent problems, you can use the following **Percent Diagram.**

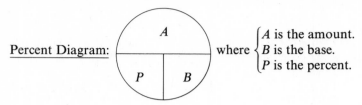

Percent Diagram: where $\begin{cases} A \text{ is the amount.} \\ B \text{ is the base.} \\ P \text{ is the percent.} \end{cases}$

When both the percent *P* and base *B* are known, you can use the Percent Diagram to write a formula for finding the amount *A* as follows:

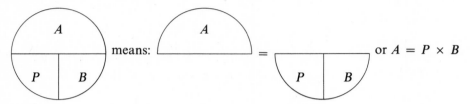

means: =   or $A = P \times B$

The formula $A = P \times B$ is called the **Amount Formula.** The Amount Formula states that to find the amount *A* when both the percent *P* and base *B* are known, you multiply the base by the percent.

Caution: To multiply by a percent, you must first rename the percent as either a terminating decimal or a fraction.

A. Find the amount by renaming the percent as a decimal.

When the percent can be renamed as a terminating decimal, you can multiply to find the amount.

EXAMPLE 10-11: Find the following amount by renaming the percent as a decimal: What number is 75% of 140?

Solution: $\overset{A}{\overbrace{\text{What number}}}$ is $\overset{P}{\widehat{75\%}}$ of $\overset{B}{\widehat{140}}$?   *Identify:* *P* means %.
 B comes after *of*.
 A is the remaining *number*.

means: $A = P \times B$ Write the Amount Formula.

$A = 75\% \times 140$ Substitute: $P = 75\%$ and $B = 140$

$A = 0.75 \times 140$ Rename the percent as a decimal.

$A = 105$ Multiply by a decimal.

Note: Because 75% can be renamed as either a terminating decimal (0.75) or a fraction ($\frac{3}{4}$), you can use either form to multiply.

B. Find the amount by renaming the percent as a fraction.

When the percent is equal to a repeating decimal, you must rename the percent as a fraction and then multiply to find the amount.

EXAMPLE 10-12: Find the amount by renaming the following percent as a fraction: $66\frac{2}{3}\%$ of 24 is what number?

Solution: $\overset{P}{\widehat{66\frac{2}{3}\%}}$ of $\overset{B}{\widehat{24}}$ is $\overset{A}{\overbrace{\text{what number}}}$?   *Identify:* *P* means %.
 B comes after *of*.
 A is the remaining *number*.

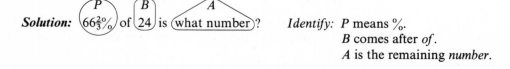

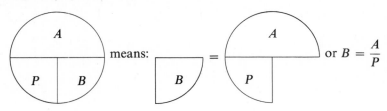

means: $A = P \times B$ Write the Amount Formula.

$A = 66\frac{2}{3}\% \times 24$ Substitute: $P = 66\frac{2}{3}\%$ and $B = 24$

$A = \frac{2}{3} \times 24$ Rename the percent as a fraction.

$A = 16$ Multiply by a fraction.

Note: Because $66\frac{2}{3}\%$ is equal to the repeating decimal $0.\overline{6}$, you must rename $66\frac{2}{3}\%$ as the fraction $\frac{2}{3}$ in order to multiply.

10-4. Finding the Base in a Percent Problem

When both the percent P and the amount A are known, you can use the Percent Diagram to write a formula for finding the base B as follows:

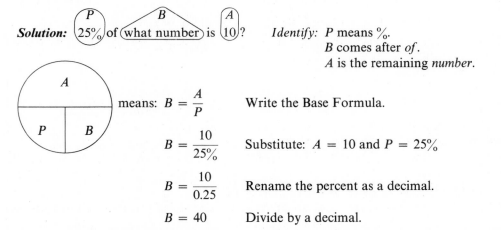

The formula $B = \frac{A}{P}$ is called the **Base Formula.** The Base Formula states that to find the base B when both the percent P and amount A are known, you divide the amount by the percent.

     To divide by a percent, you must first rename the percent as either a terminating decimal or a fraction.

A. Find the base by renaming the percent as a decimal.

When the percent can be renamed as a terminating decimal, you can divide to find the amount.

EXAMPLE 10-13: Find the base by renaming the following percent as a decimal: 25% of what number is 10?

Solution: $\overset{P}{\boxed{25\%}}$ of $\overset{B}{\boxed{\text{what number}}}$ is $\overset{A}{\boxed{10}}$?   *Identify:* P means %.

                                                  B comes after *of*.

                                                  A is the remaining *number*.

means: $B = \frac{A}{P}$ Write the Base Formula.

$B = \frac{10}{25\%}$ Substitute: $A = 10$ and $P = 25\%$

$B = \frac{10}{0.25}$ Rename the percent as a decimal.

$B = 40$ Divide by a decimal.

Check: Does the original percent times the proposed base equal the original amount?
Yes: $25\% \times 40 = 0.25 \times 40 = 10$

Note: Because 25% can be renamed as either a terminating decimal (0.25) or a fraction ($\frac{1}{4}$), you can use either form to divide.

B. Find the base by renaming the percent as a fraction.

When the percent is equal to a repeating decimal, you must rename the percent as a fraction and then divide to find the base.

EXAMPLE 10-14: Find the base by renaming the following percent as a fraction: $5\frac{1}{2}$ is $16\frac{2}{3}\%$ of what number?

Solution: $\overset{A}{5\frac{1}{2}}$ is $\overset{P}{16\frac{2}{3}\%}$ of $\overset{B}{\text{what number}}$?   *Identify:* P means %
B comes after *of*.
A is the remaining *number*.

means: $B = \dfrac{A}{P}$ Write the Base Formula.

$B = \dfrac{5\frac{1}{2}}{16\frac{2}{3}\%}$ Substitute: $A = 5\frac{1}{2}$ and $P = 16\frac{2}{3}\%$

$B = \dfrac{5\frac{1}{2}}{\frac{1}{6}}$ Rename the percent as a fraction.

$B = 5\frac{1}{2} \div \frac{1}{6}$ Divide by a fraction.

$B = \frac{11}{2} \times \frac{6}{1}$

$B = 33$

Check: Does the original percent times the proposed base equal the original amount?
Yes: $16\frac{2}{3}\% \times 33 = \frac{1}{6} \times 33 = \frac{11}{2} = 5\frac{1}{2}$

Note: Because $16\frac{2}{3}\%$ is equal to the repeating decimal $0.1\overline{6}$, you must rename $16\frac{2}{3}\%$ as the fraction $\frac{1}{6}$ in order to divide.

10-5. Finding the Percent in a Percent Problem

When both the amount A and base B are known, you can use the Percent Diagram to write a formula for finding the percent P as follows:

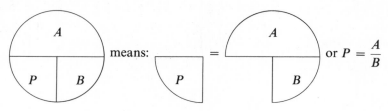

The formula $P = \dfrac{A}{B}$ is called the **Percent Formula.** The Percent Formula states that to find the percent P when both the amount A and base B are known, divide the amount by the base.

After dividing the amount by the base, you must rename the decimal or fractional quotient as a percent.

A. Find the percent by dividing to get a decimal quotient.

EXAMPLE 10-15: Find the following percent by dividing to get a decimal quotient: What percent of 16 is 5?

Solution: $\overset{P}{\text{What percent}}$ of $\overset{B}{16}$ is $\overset{A}{5}$?   *Identify:* P means *percent*.
B comes after *of*.
A is the remaining *number*.

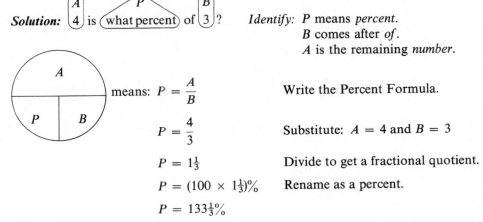

means: $P = \dfrac{A}{B}$ Write the Percent Formula.

$P = \dfrac{5}{16}$ Substitute: $A = 5$ and $B = 16$

$P = 0.3125$ Divide to get a decimal quotient.

$P = 31.25\%$ Rename as a percent.

Check: Does the proposed percent times the original base equal the original amount?
Yes: $31.25\% \times 16 = 0.3125 \times 16 = 5$

Note: Because 0.3125 is a terminating decimal, you can rename 0.3125 as either a decimal percent (31.25%) or a fractional percent ($31\frac{1}{4}\%$).

B. Find the percent by dividing to get a fractional quotient.

EXAMPLE 10-16: Find the following percent by dividing to get a fractional quotient: 4 is what percent of 3?

Solution: $\boxed{\overset{A}{4}}$ is (what percent) of $\boxed{\overset{B}{3}}$?   *Identify:* P means *percent*.
B comes after *of*.
A is the remaining *number*.

means: $P = \dfrac{A}{B}$ Write the Percent Formula.

$P = \dfrac{4}{3}$ Substitute: $A = 4$ and $B = 3$

$P = 1\frac{1}{3}$ Divide to get a fractional quotient.

$P = (100 \times 1\frac{1}{3})\%$ Rename as a percent.

$P = 133\frac{1}{3}\%$

Check: Does the proposed percent times the original base equal the original amount?
Yes: $133\frac{1}{3}\% \times 3 = \frac{400}{3}\% \times 3 = \frac{4}{3} \times 3 = 4$

Note: $1\frac{1}{3}$ is equal to the repeating decimal $1.\overline{3}$. You should rename $1\frac{1}{3}$ as the fractional percent $133\frac{1}{3}\%$ because percents are not usually written in repeating decimal form, such as $133.\overline{3}\%$.

10-6. Finding Percent Increase and Decrease

When the **original amount** of something increases to form a **new amount,** the difference between the two amounts is called the **amount of increase.** The percent found by dividing the amount of increase by the original amount is called the **percent increase.**

Similarly, when the original amount of something decreases, the difference between the two amounts is called the **amount of decrease.** The percent found by dividing the amount of decrease by the original amount is called the **percent decrease.**

A. Find the percent increase.

Recall that to find the percent P in a percent problem, you must know the base B and the amount A.

In percent increase problems.
1. The given original amount (the smaller number) is used as the base B.
2. The amount of increase (the difference between the original amount and the new amount) is used as the amount A.

EXAMPLE 10-17: What is the percent increase from 2 to 3?

Solution: The given original amount is 2. ←————— base B
The amount of increase from 2 to 3 is 1. ←— amount A

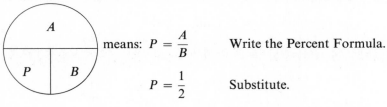

means: $P = \dfrac{A}{B}$ Write the Percent Formula.

$P = \dfrac{1}{2}$ Substitute.

$P = 50\%$ Rename as a percent.

Note: The percent increase from 2 to 3 is 50%.

B. Find the percent decrease.

In percent decrease problems:
1. The given original amount (the larger number) is used as the base B.
2. The amount of decrease (the difference between the original amount and the new amount) is used as the amount A.

EXAMPLE 10-18: What is the percent decrease from 3 to 2?

Solution: The given original amount is 3. ←————— base B
The amount of decrease from 3 to 2 is 1. ←— amount A

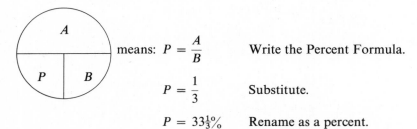

means: $P = \dfrac{A}{B}$ Write the Percent Formula.

$P = \dfrac{1}{3}$ Substitute.

$P = 33\frac{1}{3}\%$ Rename as a percent.

Note: The percent decrease from 3 to 2 is $33\frac{1}{3}\%$, while the percent increase from 2 to 3 is 50%.

To help you understand percent increase and decrease better, think of a \$3 necktie that has been marked down $33\frac{1}{3}\%$ to \$2 and then marked back up 50% to \$3, the original price.

RAISE YOUR GRADES
Can you ... ?

☑ write part of a whole as a number value, decimal, fraction, and percent
☑ rename a percent using hundredths, out of one hundred, per hundred, $\times \frac{1}{100}$, and $\div 100$
☑ rename a percent as a terminating decimal or a fraction when it does not equal a repeating decimal
☑ rename a percent that equals a repeating decimal as a fraction
☑ rename a terminating decimal as a percent
☑ rename a fraction as a percent
☑ rename a repeating decimal as a percent
☑ find the amount A when both the percent P and base B are known
☑ find the base B when both the percent P and amount A are known
☑ find the percent P when both the base B and amount A are known

☑ find the percent increase when both the original amount and new amount are known
☑ find the percent decrease when both the original amount and new amount are known

SUMMARY

1. Percent means **(a)** hundredths, **(b)** out of one hundred, **(c)** per hundred, **(d)** $\times \frac{1}{100}$, or **(e)** $\div 100$.

2. To rename a percent as a decimal:
   (a) Move the decimal point two places to the **left** (divide by 100).
   (b) Eliminate the percent symbol (%).

3. To rename a percent as a fraction:
   (a) Multiply by $\frac{1}{100}$ ($\times \frac{1}{100}$).
   (b) Eliminate the percent symbol (%).

4. To rename a decimal as a percent:
   (a) Move the decimal point two places to the **right** (multiply by 100).
   (b) Add the percent symbol (%).

5. To rename a fraction as a percent:
   (a) Multiply by 100 ($\times 100$).
   (b) Add the percent symbol (%).

6. The Percent Diagram:

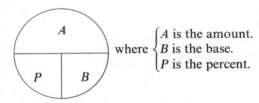

where $\begin{cases} A \text{ is the amount.} \\ B \text{ is the base.} \\ P \text{ is the percent.} \end{cases}$

7. The Amount Formula: $A = P \times B$

8. The Base Formula: $B = \dfrac{A}{P}$

9. The Percent Formula: $P = \dfrac{A}{B}$

10. To multiply or divide by a percent, you must first rename the percent as either a terminating decimal or a fraction.

11. When the percent equals a repeating decimal, you must rename the percent as a fraction in order to multiply or divide.

12. In percent problems, you identify the
   (a) percent P by locating the number that has the percent symbol (%) or the word *percent*.
   (b) base B by locating the number that comes after the word *of*.
   (c) amount A by locating the remaining *number* after the percent P and the base B have been identified.

13. In percent increase problems
   (a) the given original amount is used as the base B.
   (b) the amount of increase (the difference between the original amount and the new amount) is used as the amount A.

14. In percent decrease problems:
   (a) The original given amount is used as the base B.
   (b) The amount of decrease (the difference between the original amount and the new amount) is used as the amount A.

SOLVED PROBLEMS

PROBLEM 10-1   Rename each percent using both "$\times \frac{1}{100}$" and "$\div 100$:" **(a)** 25%   **(b)** 75%
(c) 10%   **(d)** 1%   **(e)** 90%   **(f)** 100%   **(g)** 150%   **(h)** 5%   **(i)** $\frac{1}{2}$%   **(j)** 0.25%
(k) $133\frac{1}{3}$%   **(l)** 0.01%   **(m)** 1000%   **(n)** 12.5%   **(o)** $66\frac{2}{3}$%

Solution: Recall that % means "hundredths," "out of one hundred," "per hundred," "$\times \frac{1}{100}$," or
"$\div 100$" [see Example 10-2]: **(a)** $25 \times \frac{1}{100}, 25 \div 100$ **(b)** $75 \times \frac{1}{100}, 75 \div \frac{1}{100}$
(c) $10 \times \frac{1}{100}, 10 \div 100$ **(d)** $1 \times \frac{1}{100}, 1 \div 100$ **(e)** $90 \times \frac{1}{100}, 90 \div 100$
(f) $100 \times \frac{1}{100}, 100 \div 100$ **(g)** $150 \times \frac{1}{100}, 150 \div 100$ **(h)** $5 \times \frac{1}{100}, 5 \div 100$
(i) $\frac{1}{2} \times \frac{1}{100}, \frac{1}{2} \div 100$ **(j)** $0.25 \times \frac{1}{100}, 0.25 \div 100$ **(k)** $133\frac{1}{3} \times \frac{1}{100}, 133\frac{1}{3} \div 100$
(l) $0.01 \times \frac{1}{100}, 0.01 \div 100$ **(m)** $1000 \times \frac{1}{100}, 1000 \div 100$ **(n)** $12.5 \times \frac{1}{100}, 12.5 \div 100$
(o) $66\frac{2}{3} \times \frac{1}{100}, 66\frac{2}{3} \div 100$

PROBLEM 10-2   Rename each percent as a decimal: **(a)** 37.5%   **(b)** 12.5%   **(c)** 7.5%
(d) 6.25%   **(e)** 0.5%   **(f)** 0.125%   **(g)** 10%   **(h)** 5%   **(i)** 150%   **(j)** 100%   **(k)** $\frac{1}{4}$%
(l) $\frac{7}{10}$%   **(m)** $4\frac{3}{4}$%   **(n)** $12\frac{5}{8}$%   **(o)** $215\frac{1}{2}$%

Solution: To rename a percent as a decimal, you move the decimal point two places to the left
and then eliminate the percent symbol [see Example 10-4]:

(a) $37.5\% = 0.375$ **(b)** $12.5\% = 0.125$ **(c)** $7.5\% = 0.075$ **(d)** $6.25\% = 0.0625$

(e) $0.5\% = 0.005$ **(f)** $0.125\% = 0.00125$ **(g)** $10\% = 0.10 = 0.1$ **(h)** $5\% = 0.05$

(i) $150\% = 1.50 = 1.5$ **(j)** $100\% = 1.00 = 1$ **(k)** $\frac{1}{4}\% = 0.25\% = 0.0025$

(l) $\frac{7}{10}\% = 0.7\% = 0.007$ **(m)** $4\frac{3}{4}\% = 4.75\% = 0.0475$ **(n)** $12\frac{5}{8}\% = 12.625\% = 0.12625$

(o) $215\frac{1}{2}\% = 215.5\% = 2.155$

PROBLEM 10-3   Rename each percent as a fraction: **(a)** 37.5%   **(b)** 12.5%   **(c)** 0.25%
(d) 0.5%   **(e)** 0.1%   **(f)** 0.01%   **(g)** 50%   **(h)** 25%   **(i)** 175%   **(j)** 100%   **(k)** $\frac{2}{3}$%
(l) $\frac{5}{6}$%   **(m)** $\frac{5}{12}$%   **(n)** $16\frac{2}{3}$%   **(o)** $83\frac{1}{3}$%

Solution: To rename a percent as a fraction, you multiply by $\frac{1}{100}$ and then eliminate the percent
symbol [see Example 10-7]:

(a) $37.5\% = 37.5 \times \frac{1}{100} = 37\frac{1}{2} \times \frac{1}{100} = \frac{75}{2} \times \frac{1}{100} = \frac{3}{2} \times \frac{1}{4} = \frac{3}{8}$
(b) $12.5\% = 12.5 \times \frac{1}{100} = 12\frac{1}{2} \times \frac{1}{100} = \frac{25}{2} \times \frac{1}{100} = \frac{1}{2} \times \frac{1}{4} = \frac{1}{8}$
(c) $0.25\% = 0.25 \times \frac{1}{100} = \frac{1}{4} \times \frac{1}{100} = \frac{1}{400}$
(d) $0.5\% = 0.5 \times \frac{1}{100} = \frac{1}{2} \times \frac{1}{100} = \frac{1}{200}$
(e) $0.1\% = 0.1 \times \frac{1}{100} = \frac{1}{10} \times \frac{1}{100} = \frac{1}{1000}$
(f) $0.01\% = 0.01 \times \frac{1}{100} = \frac{1}{100} \times \frac{1}{100} = \frac{1}{10,000}$
(g) $50\% = 50 \times \frac{1}{100} = \frac{50}{1} \times \frac{1}{100} = \frac{1}{1} \times \frac{1}{2} = \frac{1}{2}$
(h) $25\% = 25 \times \frac{1}{100} = \frac{25}{1} \times \frac{1}{100} = \frac{1}{1} \times \frac{1}{4} = \frac{1}{4}$
(i) $175\% = 175 \times \frac{1}{100} = \frac{175}{1} \times \frac{1}{100} = \frac{7}{1} \times \frac{1}{4} = \frac{7}{4}$ or $1\frac{3}{4}$
(j) $100\% = 100 \times \frac{1}{100} = \frac{100}{1} \times \frac{1}{100} = \frac{1}{1} = 1$
(k) $\frac{2}{3}\% = \frac{2}{3} \times \frac{1}{100} = \frac{1}{3} \times \frac{1}{50} = \frac{1}{150}$
(l) $\frac{5}{6}\% = \frac{5}{6} \times \frac{1}{100} = \frac{1}{6} \times \frac{1}{20} = \frac{1}{120}$
(m) $\frac{5}{12}\% = \frac{5}{12} \times \frac{1}{100} = \frac{1}{12} \times \frac{1}{20} = \frac{1}{240}$
(n) $16\frac{2}{3}\% = 16\frac{2}{3} \times \frac{1}{100} = \frac{50}{3} \times \frac{1}{100} = \frac{1}{3} \times \frac{1}{2} = \frac{1}{6}$
(o) $83\frac{1}{3}\% = 83\frac{1}{3} \times \frac{1}{100} = \frac{250}{3} \times \frac{1}{100} = \frac{5}{3} \times \frac{1}{2} = \frac{5}{6}$

PROBLEM 10-4   Rename each decimal as a percent: **(a)** 0.375   **(b)** 0.125   **(c)** 0.75
(d) 0.25   **(e)** 0.4   **(f)** 0.2   **(g)** 0.01   **(h)** 0.05   **(i)** 1   **(j)** 2.5   **(k)** 0.6   **(l)** 0.625
(m) 1.075   **(n)** 2.875   **(o)** 0.9

Solution: To rename a decimal as a percent, you move the decimal point two places to the right and then write the percent symbol [see Example 10-8]:

(a) $0.375 = 037.5\% = 37.5\%$ **(b)** $0.125 = 012.5\% = 12.5\%$ **(c)** $0.75 = 075.\% = 75\%$

(d) $0.25 = 025.\% = 25\%$ **(e)** $0.4 = 040.\% = 40\%$ **(f)** $0.2 = 020.\% = 20\%$

(g) $0.01 = 001.\% = 1\%$ **(h)** $0.05 = 005.\% = 5\%$ **(i)** $1 = 1. = 100.\% = 100\%$

(j) $2.5 = 250.\% = 250\%$ **(k)** $0.6 = 060.\% = 60\%$ **(l)** $0.625 = 062.5\% = 62.5\%$

(m) $1.075 = 107.5\%$ **(n)** $2.875 = 287.5\%$ **(o)** $0.9 = 090.\% = 90\%$

PROBLEM 10-5   Rename each fraction or mixed number as a percent: **(a)** $\frac{1}{2}$ **(b)** $\frac{1}{4}$ **(c)** $\frac{3}{4}$
(d) $\frac{1}{3}$ **(e)** $\frac{2}{3}$ **(f)** $\frac{1}{5}$ **(g)** $\frac{1}{8}$ **(h)** $\frac{5}{12}$ **(i)** $\frac{9}{10}$ **(j)** $\frac{1}{50}$ **(k)** $\frac{1}{100}$ **(l)** $\frac{1}{300}$ **(m)** $\frac{1}{500}$
(n) $1\frac{3}{8}$ **(o)** $2\frac{3}{5}$

Solution: To rename a fraction or mixed number as a percent, you multiply by 100 and then draw the percent symbol [see Example 10-10]:

(a) $\frac{1}{2} = (100 \times \frac{1}{2})\% = 50\%$ **(b)** $\frac{1}{4} = (100 \times \frac{1}{4})\% = 25\%$ **(c)** $\frac{3}{4} = (100 \times \frac{3}{4})\% = 75\%$
(d) $\frac{1}{3} = (100 \times \frac{1}{3})\% = 33\frac{1}{3}\%$ **(e)** $\frac{2}{3} = (100 \times \frac{2}{3})\% = 66\frac{2}{3}\%$ **(f)** $\frac{1}{5} = (100 \times \frac{1}{5})\% = 20\%$
(g) $\frac{1}{8} = (100 \times \frac{1}{8})\% = 12\frac{1}{2}\%$ **(h)** $\frac{5}{12} = (100 \times \frac{5}{12})\% = 41\frac{2}{3}\%$ **(i)** $\frac{9}{10} = (100 \times \frac{9}{10})\% = 90\%$
(j) $\frac{1}{50} = (100 \times \frac{1}{50})\% = 2\%$ **(k)** $\frac{1}{100} = (100 \times \frac{1}{100})\% = 1\%$ **(l)** $\frac{1}{300} = (100 \times \frac{1}{300})\% = \frac{1}{3}\%$
(m) $\frac{1}{500} = (100 \times \frac{1}{500})\% = \frac{1}{5}\%$ **(n)** $1\frac{3}{8}\% = (100 \times 1\frac{3}{8})\% = 137\frac{1}{2}\%$
(o) $2\frac{3}{5} = (100 \times 2\frac{3}{5})\% = 260\%$

PROBLEM 10-6   Find the following amounts by renaming the percent as a decimal.

(a)   What number is 20% of 55?   **(b)**   40% of 25 is what number?
(c)   What number is 25% of 16?   **(d)**   5% of 20 is what number?
(e)   What number is $\frac{1}{2}$% of 1600?   **(f)**   200% of 4.5 is what number?
(g)   What number is 30% of 50?   **(h)**   10% of 400 is what number?
(i)   What number is 15% of 20?   **(j)**   1% of 200 is what number?
(k)   What number is $\frac{1}{4}$% of 2400?   **(l)**   175% of 4.5 is what number?
(m)   What number is 60% of 3?   **(n)**   2% of 65 is what number?
(o)   What number is 100% of 5.75?

Solution: To find a decimal amount, you write the Amount Formula, substitute for *P* and *B*, rename the percent as a decimal, and then multiply [see Example 10-11]:

(a) $A = P \times B = 20\% \times 55 = 0.2 \times 55 = 11$
(b) $A = P \times B = 40\% \times 25 = 0.4 \times 25 = 10$
(c) $A = P \times B = 25\% \times 16 = 0.25 \times 16 = 4$
(d) $A = P \times B = 5\% \times 20 = 0.05 \times 20 = 1$
(e) $A = P \times B = \frac{1}{2}\% \times 1600 = 0.5\% \times 1600 = 0.005 \times 1600 = 8$
(f) $A = P \times B = 200\% \times 4.5 = 2 \times 4.5 = 9$
(g) $A = P \times B = 30\% \times 50 = 0.3 \times 50 = 15$
(h) $A = P \times B = 10\% \times 400 = 0.1 \times 400 = 40$
(i) $A = P \times B = 15\% \times 20 = 0.15 \times 20 = 3$
(j) $A = P \times B = 1\% \times 200 = 0.01 \times 200 = 2$
(k) $A = P \times B = \frac{1}{4}\% \times 2400 = 0.25\% \times 2400 = 0.0025 \times 2400 = 6$
(l) $A = P \times B = 175\% \times 4.5 = 1.75 \times 4.5 = 7.875$
(m) $A = P \times B = 60\% \times 3 = 0.6 \times 3 = 1.8$
(n) $A = P \times B = 2\% \times 65 = 0.02 \times 65 = 1.3$
(o) $A = P \times B = 100\% \times 5.75 = 1 \times 5.75 = 5.75$

PROBLEM 10-7   Find the following amounts by renaming the percent as a fraction:

(a)   What number is $66\frac{2}{3}\%$ of 90?   **(b)** $8\frac{1}{3}\%$ of 36 is what number?

(c)   What number is 25% of $5\frac{3}{5}$?   **(d)**   75% of 4.8 is what number?

(e)   What number is $\frac{5}{6}\%$ of 24?   **(f)** $291\frac{2}{3}\%$ of 4 is what number?

(g)   What number is $33\frac{1}{3}\%$ of 45?   **(h)** $16\frac{2}{3}\%$ of 24 is what number?

(i)   What number is $41\frac{2}{3}\%$ of $2\frac{1}{10}$?   **(j)**   50% of 0.8 is what number?

(k)   What number is $\frac{1}{6}\%$ of 100?   **(l)** $183\frac{1}{3}\%$ of 12 is what number?

(m)   What number is $\frac{2}{3}\%$ of 1500?   **(n)** $\frac{5}{6}\%$ of 60 is what number?

(o)   What number is $58\frac{1}{3}\%$ of 30?

Solution: To find a fractional amount, you write the Amount Formula, substitute for *P* and *B*, rename the percent as a fraction, and then multiply [see Example 10-12]:

(a) $A = P \times B = 66\frac{2}{3}\% \times 90 = \frac{2}{3} \times 90 = 60$

(b) $A = P \times B = 8\frac{1}{3}\% \times 36 = \frac{1}{12} \times 36 = 3$

(c) $A = P \times B = 25\% \times 5\frac{3}{5} = \frac{1}{4} \times \frac{28}{5} = \frac{7}{5}$ or $1\frac{2}{5}$

(d) $A = P \times B = 75\% \times 4.8 = \frac{3}{4} \times 4\frac{4}{5} = \frac{3}{4} \times \frac{24}{5} = \frac{18}{5}$ or $3\frac{3}{5}$

(e) $A = P \times B = \frac{5}{6}\% \times 24 = \frac{1}{120} \times 24 = \frac{1}{5}$

(f) $A = P \times B = 291\frac{2}{3}\% \times 4 = 2\frac{11}{12} \times 4 = 11\frac{2}{3}$

(g) $A = P \times B = 33\frac{1}{3}\% \times 45 = \frac{1}{3} \times 45 = 15$

(h) $A = P \times B = 16\frac{2}{3}\% \times 24 = \frac{1}{6} \times 24 = 4$

(i) $A = P \times B = 41\frac{2}{3}\% \times 2\frac{1}{10} = \frac{5}{12} \times \frac{21}{10} = \frac{7}{8}$

(j) $A = P \times B = 50\% \times 0.8 = \frac{1}{2} \times \frac{8}{10} = \frac{4}{10} = \frac{2}{5}$

(k) $A = P \times B = \frac{1}{6}\% \times 100 = \frac{1}{600} \times \frac{100}{1} = \frac{1}{6}$

(l) $A = P \times B = 183\frac{1}{3}\% \times 12 = 1\frac{5}{6} \times 12 = \frac{11}{6} \times \frac{12}{1} = 22$

(m) $A = P \times B = \frac{2}{3}\% \times 1500 = \frac{1}{150} \times \frac{1500}{1} = 10$

(n) $A = P \times B = \frac{5}{6}\% \times 60 = \frac{1}{120} \times \frac{60}{1} = \frac{1}{2}$

(o) $A = P \times B = 58\frac{1}{3}\% \times 30 = \frac{7}{12} \times \frac{30}{1} = 17\frac{1}{2}$

PROBLEM 10-8   Find the base by renaming the percent as a decimal:

(a)   80% of what number is 5?   **(b)**   20 is 50% of what number?   **(c)**   65% of what number is 39?
(d)   2.25 is 9% of what number?   **(e)** $\frac{1}{2}\%$ of what number is 7?   **(f)**   35 is 20% of what number?
(g)   45% of what number is 9?   **(h)**   0.24 is 3% of what number?   **(i)** $\frac{3}{4}\%$ of what number is 6?
(j)   18 is 150% of what number?   **(k)**   10% of what number is 10?   **(l)**   2.5 is 125% of what number?
(m)   60% of what number is 1.5?   **(n)**   1.25 is 0.25% of what number?
(o)   95% of what number is 22.8?

Solution: To find the base in decimal form, you write the Base Formula, substitute for *A* and *P*, rename the percent as a decimal, and then divide [see Example 10-13]:

(a) $B = \dfrac{A}{P} = \dfrac{5}{80\%} = \dfrac{5}{0.8} = 6.25$ **(b)** $B = \dfrac{A}{P} = \dfrac{20}{50\%} = \dfrac{20}{0.5} = 40$ **(c)** $B = \dfrac{A}{P} = \dfrac{39}{65\%} = \dfrac{39}{0.65} = 60$

(d) $B = \dfrac{A}{P} = \dfrac{2.25}{9\%} = \dfrac{2.25}{0.09} = 25$ **(e)** $B = \dfrac{A}{P} = \dfrac{7}{\frac{1}{2}\%} = \dfrac{7}{0.5\%} = \dfrac{7}{0.005} = 1400$

(f) $B = \dfrac{A}{P} = \dfrac{35}{20\%} = \dfrac{35}{0.2} = 175$ **(g)** $B = \dfrac{A}{P} = \dfrac{9}{45\%} = \dfrac{9}{0.45} = 20$ **(h)** $B = \dfrac{A}{P} = \dfrac{0.24}{3\%} = \dfrac{0.24}{0.03} = 8$

(i) $B = \dfrac{A}{P} = \dfrac{6}{\frac{3}{4}\%} = \dfrac{6}{0.75\%} = \dfrac{6}{0.0075} = 800$ **(j)** $B = \dfrac{A}{P} = \dfrac{18}{150\%} = \dfrac{18}{1.5} = 12$

(k) $B = \dfrac{A}{P} = \dfrac{10}{10\%} = \dfrac{10}{0.1} = 100$ **(l)** $B = \dfrac{A}{P} = \dfrac{2.5}{125\%} = \dfrac{2.5}{1.25} = 2$

(m) $B = \dfrac{A}{P} = \dfrac{1.5}{60\%} = \dfrac{1.5}{0.6} = 2.5$ **(n)** $B = \dfrac{A}{P} = \dfrac{1.25}{0.25\%} = \dfrac{1.25}{0.0025} = 500$

(o) $B = \dfrac{A}{P} = \dfrac{22.8}{95\%} = \dfrac{22.8}{0.95} = 24$

PROBLEM 10-9 Find each base by renaming the percent as a fraction:

(a) $83\frac{1}{3}\%$ of what number is 25? **(b)** 5 is $8\frac{1}{3}\%$ of what number? **(c)** $91\frac{2}{3}\%$ of what number is $1\frac{5}{6}$?

(d) 0.4 is 50% of what number? **(e)** $\frac{5}{6}\%$ of what number is $\frac{1}{100}$? **(f)** 70 is $16\frac{2}{3}\%$ of what number?

(g) 75% of what number is $\frac{1}{2}$? **(h)** 2.6 is 25% of what number? **(i)** $\frac{2}{3}\%$ of what number is $\frac{1}{10}$?

(j) $25\frac{1}{5}$ is 70% of what number? **(k)** $66\frac{2}{3}\%$ of what number is 5? **(l)** $\frac{1}{8}$ is $16\frac{2}{3}\%$ of what number?

(m) 60% of what number is 1.5? **(n)** $2\frac{3}{4}$ is $33\frac{1}{3}\%$ of what number? **(o)** $41\frac{2}{3}\%$ of what number is 250?

Solution: To find the base in fraction form, you write the Base Formula, substitute for A and P, rename the percent as a fraction, and then divide [see Example 10-14]:

(a) $B = \dfrac{A}{P} = \dfrac{25}{83\frac{1}{3}\%} = \dfrac{25}{\frac{5}{6}} = 25 \div \frac{5}{6} = \frac{25}{1} \times \frac{6}{5} = 30$

(b) $B = \dfrac{A}{P} = \dfrac{5}{8\frac{1}{3}\%} = \dfrac{5}{\frac{1}{12}} = 5 \div \frac{1}{12} = \frac{5}{1} \times \frac{12}{1} = 60$

(c) $B = \dfrac{A}{P} = \dfrac{1\frac{5}{6}}{91\frac{2}{3}\%} = \dfrac{1\frac{5}{6}}{\frac{11}{12}} = 1\frac{5}{6} \div \frac{11}{12} = \frac{11}{12} \times \frac{12}{11} = 1$

(d) $B = \dfrac{A}{P} = \dfrac{0.4}{50\%} = \dfrac{0.4}{\frac{1}{2}} = 0.4 \div \frac{1}{2} = \frac{4}{10} \times \frac{2}{1} = \frac{4}{5}$

(e) $B = \dfrac{A}{P} = \dfrac{\frac{1}{100}}{\frac{5}{6}\%} = \dfrac{\frac{1}{100}}{\frac{1}{120}} = \frac{1}{100} \div \frac{1}{120} = \frac{1}{100} \times \frac{120}{1} = \frac{6}{5} \text{ or } 1\frac{1}{5}$

(f) $B = \dfrac{A}{P} = \dfrac{70}{16\frac{2}{3}\%} = \dfrac{70}{\frac{1}{6}} = 70 \div \frac{1}{6} = \frac{70}{1} \times \frac{6}{1} = 420$

(g) $B = \dfrac{A}{P} = \dfrac{\frac{1}{2}}{75\%} = \dfrac{\frac{1}{2}}{\frac{3}{4}} = \frac{1}{2} \div \frac{3}{4} = \frac{1}{2} \times \frac{4}{3} = \frac{2}{3}$

(h) $B = \dfrac{A}{P} = \dfrac{2.6}{25\%} = \dfrac{2.6}{\frac{1}{4}} = 2\frac{3}{5} \div \frac{1}{4} = \frac{13}{5} \times \frac{4}{1} = \frac{52}{5} \text{ or } 10\frac{2}{5}$

(i) $B = \dfrac{A}{P} = \dfrac{\frac{1}{10}}{\frac{2}{3}\%} = \dfrac{\frac{1}{10}}{\frac{1}{150}} = \frac{1}{10} \div \frac{1}{150} = \frac{1}{10} \times \frac{150}{1} = 15$

(j) $B = \dfrac{A}{P} = \dfrac{25\frac{1}{5}}{70\%} = \dfrac{25\frac{1}{5}}{\frac{7}{10}} = 25\frac{1}{5} \div \frac{7}{10} = \frac{126}{5} \times \frac{10}{7} = 36$

(k) $B = \dfrac{A}{P} = \dfrac{5}{66\frac{2}{3}\%} = \dfrac{5}{\frac{2}{3}} = 5 \div \frac{2}{3} = \frac{5}{1} \times \frac{3}{2} = \frac{15}{2} \text{ or } 7\frac{1}{2}$

(l) $B = \dfrac{A}{P} = \dfrac{\frac{1}{8}}{16\frac{2}{3}\%} = \dfrac{\frac{1}{8}}{\frac{1}{6}} = \frac{1}{8} \div \frac{1}{6} = \frac{1}{8} \times \frac{6}{1} = \frac{3}{4}$

(m) $B = \dfrac{A}{P} = \dfrac{1.5}{60\%} = \dfrac{1.5}{\frac{3}{5}} = 1\frac{1}{2} \div \frac{3}{5} = \frac{3}{2} \times \frac{5}{3} = \frac{5}{2} \text{ or } 2\frac{1}{2}$

(n) $B = \dfrac{A}{P} = \dfrac{2\frac{3}{4}}{33\frac{1}{3}\%} = \dfrac{2\frac{3}{4}}{\frac{1}{3}} = 2\frac{3}{4} \div \frac{1}{3} = \frac{11}{4} \times \frac{3}{1} = \frac{33}{4} \text{ or } 8\frac{1}{4}$

(o) $B = \dfrac{A}{P} = \dfrac{250}{41\frac{2}{3}\%} = \dfrac{250}{\frac{5}{12}} = 250 \div \frac{5}{12} = \frac{250}{1} \times \frac{12}{5} = 600$

PROBLEM 10-10   Find each percent by dividing to get a decimal quotient:

(a) What percent of 125 is 25?   (b) 7 is what percent of 10?   (c) What percent of 200 is 130?
(d) 110 is what percent of 160?   (e) What percent of 84 is 0.63?   (f) 3 is what percent of 5?
(g) What percent of 40 is 18?   (h) 27 is what percent of 48?   (i) What percent of 24 is 0.12?
(j) 5 is what percent of 8?   (k) What percent of 75 is 30?   (l) 1 is what percent of 2?
(m) What percent of 8 is 3?   (n) 2 is what percent of 1?   (o) What percent of 5 is 7.5?

Solution: To find a percent in decimal quotient form, you write the Percent Formula, substitute for A and B, divide to get a decimal quotient, and then rename as a percent [see Example 10-15]:

(a) $P = \dfrac{A}{B} = \dfrac{25}{125} = 0.2 = 20\%$
(b) $P = \dfrac{A}{B} = \dfrac{7}{10} = 0.7 = 70\%$

(c) $P = \dfrac{A}{B} = \dfrac{130}{200} = 0.65 = 65\%$
(d) $P = \dfrac{A}{B} = \dfrac{110}{160} = 0.6875 = 68.75\%$

(e) $P = \dfrac{A}{B} = \dfrac{0.63}{84} = 0.0075 = 0.75\%$
(f) $P = \dfrac{A}{B} = \dfrac{3}{5} = 0.6 = 60\%$

(g) $P = \dfrac{A}{B} = \dfrac{18}{40} = 0.45 = 45\%$
(h) $P = \dfrac{A}{B} = \dfrac{27}{48} = 0.5625 = 56.25\%$

(i) $P = \dfrac{A}{B} = \dfrac{0.12}{24} = 0.005 = 0.5\%$
(j) $P = \dfrac{A}{B} = \dfrac{5}{8} = 0.625 = 62.5\%$

(k) $P = \dfrac{A}{B} = \dfrac{30}{75} = 0.4 = 40\%$
(l) $P = \dfrac{A}{B} = \dfrac{1}{2} = 0.5 = 50\%$

(m) $P = \dfrac{A}{B} = \dfrac{3}{8} = 0.375 = 37.5\%$
(n) $P = \dfrac{A}{B} = \dfrac{2}{1} = 2 = 200\%$

(o) $P = \dfrac{A}{B} = \dfrac{7.5}{5} = 1.5 = 150\%$

PROBLEM 10-11   Find each percent by dividing to get a fractional quotient:

(a) What percent of $\frac{3}{5}$ is $\frac{1}{5}$?   (b) $\frac{1}{6}$ is what percent of $\frac{2}{3}$?
(c) What percent of $7\frac{1}{2}$ is 3?   (d) $\frac{1}{4}$ is what percent of $1\frac{1}{2}$?
(e) What percent of 120 is 10?   (f) $\frac{1}{5}$ is what percent of $\frac{3}{10}$?
(g) What percent of $5\frac{1}{3}$ is 4?   (h) $\frac{2}{3}$ is what percent of $3\frac{1}{3}$?
(i) What percent of 12 is 10?   (j) $\frac{1}{2}$ is what percent of $\frac{5}{8}$?
(k) What percent of $\frac{3}{4}$ is $\frac{3}{8}$?   (l) 11 is what percent of 12?
(m) What percent of $5\frac{1}{3}$ is $\frac{1}{3}$?   (n) $\frac{3}{4}$ is what percent of $\frac{1}{2}$?
(o) What percent of 3 is 12?

Solution: To find a percent in fractional quotient form, you write the Percent Formula, substitute for A and B, divide to get a fractional quotient, and then rename as a percent [see Example 10-16]:

(a) $P = \dfrac{A}{B} = \dfrac{\frac{1}{5}}{\frac{3}{5}} = \frac{1}{5} \div \frac{3}{5} = \frac{1}{5} \times \frac{5}{3} = \frac{1}{3} = 33\frac{1}{3}\%$
(b) $P = \dfrac{A}{B} = \dfrac{\frac{1}{6}}{\frac{2}{3}} = \frac{1}{6} \div \frac{2}{3} = \frac{1}{6} \times \frac{3}{2} = \frac{1}{4} = 25\%$

(c) $P = \dfrac{A}{B} = \dfrac{3}{7\frac{1}{2}} = 3 \div \frac{15}{2} = \frac{3}{1} \times \frac{2}{15} = \frac{2}{5} = 40\%$
(d) $P = \dfrac{A}{B} = \dfrac{\frac{1}{4}}{1\frac{1}{2}} = \frac{1}{4} \div \frac{3}{2} = \frac{1}{4} \times \frac{2}{3} = \frac{1}{6} = 16\frac{2}{3}\%$

(e) $P = \dfrac{A}{B} = \dfrac{10}{120} = \frac{1}{12} = 8\frac{1}{3}\%$
(f) $P = \dfrac{A}{B} = \dfrac{\frac{1}{5}}{\frac{3}{10}} = \frac{1}{5} \div \frac{3}{10} = \frac{1}{5} \times \frac{10}{3} = \frac{2}{3} = 66\frac{2}{3}\%$

(g) $P = \dfrac{A}{B} = \dfrac{4}{5\frac{1}{3}} = 4 \div \frac{16}{3} = \frac{4}{1} \times \frac{3}{16} = \frac{3}{4} = 75\%$
(h) $P = \dfrac{A}{B} = \dfrac{\frac{2}{3}}{3\frac{1}{3}} = \frac{2}{3} \div \frac{10}{3} = \frac{2}{3} \times \frac{3}{10} = \frac{1}{5} = 20\%$

(i) $P = \dfrac{A}{B} = \dfrac{10}{12} = \dfrac{5}{6} = 83\frac{1}{3}\%$

(j) $P = \dfrac{A}{B} = \dfrac{\frac{1}{2}}{\frac{5}{8}} = \frac{1}{2} \div \frac{5}{8} = \frac{1}{2} \times \frac{8}{5} = \frac{4}{5} = 80\%$

(k) $P = \dfrac{A}{B} = \dfrac{\frac{3}{8}}{\frac{3}{4}} = \frac{3}{8} \div \frac{3}{4} = \frac{3}{8} \times \frac{4}{3} = \frac{1}{2} = 50\%$

(l) $P = \dfrac{A}{B} = \dfrac{11}{12} = 91\frac{2}{3}\%$

(m) $P = \dfrac{A}{B} = \dfrac{\frac{1}{3}}{5\frac{1}{3}} = \frac{1}{3} \div \frac{16}{3} = \frac{1}{3} \times \frac{3}{16} = \frac{1}{16} = 6\frac{1}{4}\%$

(n) $P = \dfrac{A}{B} = \dfrac{\frac{3}{4}}{\frac{1}{2}} = \frac{3}{4} \div \frac{1}{2} = \frac{3}{4} \times \frac{2}{1} = \frac{3}{2} = 1\frac{1}{2} = 150\%$

(o) $P = \dfrac{A}{B} = \dfrac{12}{3} = 4 = 400\%$

PROBLEM 10-12   Find each percent increase from: **(a)** 50 to 75   **(b)** 8.24 to 14.42
(c) $18\frac{3}{4}$ to $31\frac{1}{4}$ **(d)** 80 to 112   **(e)** 5.1 to 9.18   **(f)** $21\frac{1}{3}$ to $29\frac{1}{3}$ **(g)** 60 to 75
(h) 3.87 to 5.16   **(i)** $25\frac{1}{2}$ to $30\frac{3}{5}$ **(j)** 115 to 184   **(k)** 28.4 to 31.95   **(l)** $10\frac{2}{3}$ to $17\frac{1}{3}$
(m) 100 to 107   **(n)** 50 to 57   **(o)** 10 to 17

Solution: To find a percent increase, you identify the given original amount (the smaller number) as the base B, the amount of increase (the difference between the two given amounts) as the amount A, and then find the percent P using the Percent Formula [see Example 10-17]:

(a) $B = 50,\ A = 75 - 50 = 25,\ P = \dfrac{A}{B} = \dfrac{25}{50} = \dfrac{1}{2} = 50\%$

(b) $B = 8.24,\ A = 14.42 - 8.24 = 6.18,\ P = \dfrac{A}{B} = \dfrac{6.18}{8.24} = 0.75 = 75\%$

(c) $B = 18\frac{3}{4},\ A = 31\frac{1}{4} - 18\frac{3}{4} = 12\frac{1}{2},\ P = \dfrac{A}{B} = \dfrac{12\frac{1}{2}}{18\frac{3}{4}} = \frac{25}{2} \div \frac{75}{4} = \frac{25}{2} \times \frac{4}{75} = \frac{2}{3} = 66\frac{2}{3}\%$

(d) $B = 80,\ A = 112 - 80 = 32,\ P = \dfrac{A}{B} = \dfrac{32}{80} = \dfrac{2}{5} = 40\%$

(e) $B = 5.1,\ A = 9.18 - 5.1 = 4.08,\ P = \dfrac{A}{B} = \dfrac{4.08}{5.1} = 0.8 = 80\%$

(f) $B = 21\frac{1}{3},\ A = 29\frac{1}{3} - 21\frac{1}{3} = 8,\ P = \dfrac{A}{B} = \dfrac{8}{21\frac{1}{3}} = \frac{8}{1} \div \frac{64}{3} = \frac{8}{1} \times \frac{3}{64} = \frac{3}{8} = 37\frac{1}{2}\%$

(g) $B = 60,\ A = 75 - 60 = 15,\ P = \dfrac{A}{B} = \dfrac{15}{60} = \dfrac{1}{4} = 25\%$

(h) $B = 3.87,\ A = 5.16 - 3.87 = 1.29,\ P = \dfrac{A}{B} = \dfrac{1.29}{3.87} = 0.\overline{3} = \frac{1}{3} = 33\frac{1}{3}\%$

(i) $B = 25\frac{1}{2},\ A = 30\frac{3}{5} - 25\frac{1}{2} = 5\frac{1}{10},\ P = \dfrac{A}{B} = \dfrac{5\frac{1}{10}}{25\frac{1}{2}} = \frac{51}{10} \div \frac{51}{2} = \frac{51}{10} \times \frac{2}{51} = \frac{1}{5} = 20\%$

(j) $B = 115,\ A = 184 - 115 = 69,\ P = \dfrac{A}{B} = \dfrac{69}{115} = 0.6 = 60\%$

(k) $B = 28.4,\ A = 31.95 - 28.4 = 3.55,\ P = \dfrac{A}{B} = \dfrac{3.55}{28.4} = 0.125 = 12.5\%$

(l) $B = 10\frac{2}{3},\ A = 17\frac{1}{3} - 10\frac{2}{3} = 6\frac{2}{3},\ P = \dfrac{A}{B} = \dfrac{6\frac{2}{3}}{10\frac{2}{3}} = \frac{20}{3} \div \frac{32}{3} = \frac{20}{3} \times \frac{3}{32} = \frac{5}{8} = 62\frac{1}{2}\%$

(m) $B = 100,\ A = 107 - 100 = 7,\ P = \dfrac{A}{B} = \dfrac{7}{100} = 7\%$

(n) $B = 50, A = 57 - 50 = 7, P = \dfrac{A}{B} = \dfrac{7}{50} = 0.14 = 14\%$

(o) $B = 10, A = 17 - 10 = 7, P = \dfrac{A}{B} = \dfrac{7}{10} = 70\%$

PROBLEM 10-13   Find each percent decrease from: **(a)** 120 to 42   **(b)** 36.96 to 21.56
(c) $28\frac{4}{5}$ to $2\frac{2}{5}$ **(d)** 500 to 480   **(e)** 35.6 to 33.82   **(f)** $52\frac{1}{2}$ to $34\frac{1}{8}$ **(g)** 80 to 44
(h) 30.36 to 27.83   **(i)** $22\frac{1}{2}$ to $9\frac{3}{8}$ **(j)** 200 to 194   **(k)** 4.5 to 4.14   **(l)** $11\frac{1}{9}$ to $9\frac{4}{9}$
(m) 100 to 99   **(n)** 50 to 47   **(o)** 10 to 9

Solution: To find percent decrease, you identify the given original amount (the larger number) as the base *B*, the amount of decrease (the difference between the two given amounts) as the amount *A*, and then find the percent *P* using the Percent Formula [see Example 10-18]:

(a) $B = 120, A = 120 - 42 = 78, P = \dfrac{A}{B} = \dfrac{78}{120} = 0.65 = 65\%$

(b) $B = 36.96, A = 36.96 - 21.56 = 15.4, P = \dfrac{A}{B} = \dfrac{15.4}{36.96} = 0.41\overline{6} = 41\frac{2}{3}\%$

(c) $B = 28\frac{4}{5}, A = 28\frac{4}{5} - 2\frac{2}{5} = 26\frac{2}{5}, P = \dfrac{A}{B} = \dfrac{26\frac{2}{5}}{28\frac{4}{5}} = \frac{132}{5} \div \frac{144}{5} = \frac{132}{5} \times \frac{5}{144} = \frac{11}{12} = 91\frac{2}{3}\%$

(d) $B = 500, A = 500 - 480 = 20, P = \dfrac{A}{B} = \dfrac{20}{500} = \frac{1}{25} = 4\%$

(e) $B = 35.6, A = 35.6 - 33.82 = 1.78, P = \dfrac{A}{B} = \dfrac{1.78}{35.6} = 0.05 = 5\%$

(f) $B = 52\frac{1}{2}, A = 52\frac{1}{2} - 34\frac{1}{8} = 18\frac{3}{8}, P = \dfrac{A}{B} = \dfrac{18\frac{3}{8}}{52\frac{1}{2}} = \frac{147}{8} \div \frac{105}{2} = \frac{147}{8} \times \frac{2}{105} = \frac{7}{20} = 35\%$

(g) $B = 80, A = 80 - 44 = 36, P = \dfrac{A}{B} = \dfrac{36}{80} = \frac{9}{20} = 45\%$

(h) $B = 30.36, A = 30.36 - 27.83 = 2.53, P = \dfrac{A}{B} = \dfrac{2.53}{30.36} = 0.08\overline{3} = 8\frac{1}{3}\%$

(i) $B = 22\frac{1}{2}, A = 22\frac{1}{2} - 9\frac{3}{8} = 13\frac{1}{8}, P = \dfrac{A}{B} = \dfrac{13\frac{1}{8}}{22\frac{1}{2}} = \frac{105}{8} \div \frac{45}{2} = \frac{105}{8} \times \frac{2}{45} = \frac{7}{12} = 58\frac{1}{3}\%$

(j) $B = 200, A = 200 - 194 = 6, P = \dfrac{A}{B} = \dfrac{6}{200} = \frac{3}{100} = 3\%$

(k) $B = 4.5, A = 4.5 - 4.14 = 0.36, P = \dfrac{A}{B} = \dfrac{0.36}{4.5} = 0.08 = 8\%$

(l) $B = 11\frac{1}{9}, A = 11\frac{1}{9} - 9\frac{4}{9} = 1\frac{6}{9} = 1\frac{2}{3}, P = \dfrac{A}{B} = \dfrac{1\frac{2}{3}}{11\frac{1}{9}} = \frac{5}{3} \div \frac{100}{9} = \frac{5}{3} \times \frac{9}{100} = \frac{3}{20} = 15\%$

(m) $B = 100, A = 100 - 99 = 1, P = \dfrac{A}{B} = \dfrac{1}{100} = 1\%$

(n) $B = 50, A = 50 - 47 = 3, P = \dfrac{A}{B} = \dfrac{3}{50} = 6\%$

(o) $B = 10, A = 10 - 9 = 1, P = \dfrac{A}{B} = \dfrac{1}{10} = 10\%$

Supplementary Exercises

PROBLEM 10-14   Rename each percent using both "$\times \frac{1}{100}$" and "$\div 100$:" **(a)** 30%   **(b)** 5%
(c) 90%   **(d)** $33\frac{1}{3}$%   **(e)** $166\frac{2}{3}$%   **(f)** 150%   **(g)** 37.5%   **(h)** 87.5%   **(i)** $16\frac{2}{3}$%
(j) $83\frac{1}{3}$%   **(k)** 62.5%   **(l)** 20%   **(m)** 80%   **(n)** $\frac{1}{3}$%   **(o)** $\frac{3}{4}$%   **(p)** $6\frac{1}{4}$%   **(q)** $\frac{2}{3}$%
(r) $8\frac{1}{2}$%   **(s)** 175%   **(t)** 225%   **(u)** 0.5%   **(v)** 0.125%

PROBLEM 10-15   Rename each percent as a decimal: **(a)** 25%   **(b)** 50%   **(c)** 75%   **(d)** 90%
(e) 10%   **(f)** 20%   **(g)** 12.5%   **(h)** 37.5%   **(i)** 30%   **(j)** 40%   **(k)** $62\frac{1}{2}$%   **(l)** 87.5%
(m) 60%   **(n)** 80%   **(o)** $6\frac{1}{4}$%   **(p)** 18.75%   **(q)** 100%   **(r)** 125%   **(s)** 31.25%
(t) $43\frac{3}{4}$%   **(u)** 250%   **(v)** 1000%

PROBLEM 10-16   Rename each percent as a fraction: **(a)** $16\frac{2}{3}$%   **(b)** $83\frac{1}{3}$%   **(c)** $33\frac{1}{3}$%
(d) $66\frac{2}{3}$%   **(e)** $8\frac{1}{3}$%   **(f)** $41\frac{2}{3}$%   **(g)** $58\frac{1}{3}$%   **(h)** $91\frac{2}{3}$%   **(i)** 1%   **(j)** 25%   **(k)** 50%
(l) 75%   **(m)** 10%   **(n)** 20%   **(o)** 30%   **(p)** 40%   **(q)** 60%   **(r)** 70%   **(s)** 80%
(t) 90%   **(u)** 5%   **(v)** 4%

PROBLEM 10-17   Rename each decimal as a percent: **(a)** 0.1   **(b)** 0.2   **(c)** 0.625   **(d)** 0.875
(e) 0.5   **(f)** 0.6   **(g)** 0.3125   **(h)** 0.4375   **(i)** 0.9   **(j)** 0.25   **(k)** 0.75   **(l)** 0.05
(m) 0.125   **(n)** 0.375   **(o)** 0.3   **(p)** 0.4   **(q)** 0.0625   **(r)** 0.1875   **(s)** 0.7   **(t)** 0.8
(u) 0.5625   **(v)** 0.6875

PROBLEM 10-18   Rename each fraction as a percent: **(a)** $\frac{1}{2}$ **(b)** $\frac{1}{4}$ **(c)** $\frac{1}{5}$ **(d)** $\frac{1}{8}$
(e) $\frac{1}{10}$ **(f)** $\frac{1}{12}$ **(g)** $\frac{1}{16}$ **(h)** $\frac{3}{4}$ **(i)** $\frac{2}{5}$ **(j)** $\frac{3}{8}$ **(k)** $\frac{3}{10}$ **(l)** $\frac{5}{12}$ **(m)** $\frac{3}{16}$ **(n)** $\frac{3}{5}$
(o) $\frac{5}{8}$ **(p)** $\frac{7}{10}$ **(q)** $\frac{7}{12}$ **(r)** $\frac{5}{16}$ **(s)** $\frac{4}{5}$ **(t)** $\frac{7}{8}$ **(u)** $\frac{9}{10}$ **(v)** $\frac{7}{16}$

PROBLEM 10-19   Find each amount using the Percent Diagram to write the Amount Formula:

(a)   What number is 50% of 30?
(b) $33\frac{1}{3}$% of 60 is what number?
(c)   What number is $66\frac{2}{3}$% of 15?
(d)   25% of 100 is what number?
(e)   What number is 75% of 84?
(f)   20% of 56 is what number?
(g)   What number is 40% of 102?
(h)   60% of 1 is what number?
(i)   What number is 80% of 2.5?
(j) $16\frac{2}{3}$% of 3.6 is what number?
(k)   What number is $83\frac{1}{3}$% of 6?
(l)   12.5% of 0.1 is what number?
(m)   What number is $37\frac{1}{2}$% of 16.50?
(n)   62.5% of 800 is what number?
(o)   What number is $87\frac{1}{2}$% of 160?
(p)   10% of $\frac{1}{2}$ is what number?
(q)   What number is 20% of $3\frac{1}{4}$?
(r) $8\frac{1}{3}$% of 12 is what number?
(s)   What number is $41\frac{2}{3}$% of 4.8?
(t)   5% of 5 is what number?
(u)   What number is 6.25% of 32?
(v) $18\frac{3}{4}$% of $4\frac{1}{2}$ is what number?

PROBLEM 10-20   Find each base using the Percent Diagram to write the Base Formula:

(a)   50% of what number is 8?
(b)   28 is 70% of what number?
(c) $33\frac{1}{3}$% of what number is 100?
(d)   22.5 is 90% of what number?
(e) $66\frac{2}{3}$% of what number is 2.1?
(f)   2.1 is $58\frac{1}{3}$% of what number?
(g)   25% of what number is 5.2?
(h)   8.8 is $91\frac{2}{3}$% of what number?
(i)   75% of what number is $4\frac{4}{5}$?
(j) $2\frac{1}{2}$ is 31.25% of what number?
(k)   20% of what number is $5\frac{2}{3}$?
(l) $1\frac{3}{4}$ is 43.75% of what number?
(m)   40% of what number is 90?
(n)   55 is $68\frac{3}{4}$% of what number?
(o)   60% of what number is 82?
(p)   42.25 is $81\frac{1}{4}$% of what number?

(q) 80% of what number is 80?

(s) $16\frac{2}{3}\%$ of what number is $\frac{1}{6}$?

(u) $83\frac{1}{3}\%$ of what number is $\frac{5}{8}$?

(r) 0.6 is 93.75% of what number?

(t) 9.2 is 4% of what number?

(v) $8\frac{3}{4}$ is 2% of what number?

PROBLEM 10-21   Find each percent using the Percent Diagram to write the Percent Formula:

(a) What percent of 2 is 1?

(c) What percent of 30 is 10?

(e) What percent of 4.5 is 3?

(g) What percent of 1 is $\frac{1}{4}$?

(i) What percent of 4 is 3?

(k) What percent of 50 is 10?

(m) What percent of 32.5 is 13?

(o) What percent of $\frac{1}{2}$ is $\frac{3}{10}$?

(q) What percent of 39.25 is 31.4?

(s) What percent of 1 is 6?

(u) What percent of $\frac{5}{6}$ is 1?

(b) 1 is what percent of 8?

(d) 30 is what percent of 80?

(f) 1.25 is what percent of 2?

(h) $\frac{7}{16}$ is what percent of $\frac{1}{2}$?

(j) 1 is what percent of 10?

(l) 30 is what percent of 100?

(n) 8.75 is what percent of 12.5?

(p) $\frac{9}{10}$ is what percent of 1?

(r) 1000 is what percent of 12,000?

(t) 12 is what percent of 5?

(v) $1\frac{1}{6}$ is what percent of 2?

PROBLEM 10-22   Find the amount, base, or percent using the Percent Diagram to write the correct formula:

(a) What number is $16\frac{2}{3}\%$ of 210?

(c) $41\frac{2}{3}\%$ of what number is 144?

(e) What percent of 16 is 1?

(g) $18\frac{3}{4}\%$ of 20 is what number?

(i) 16 is what percent of 5?

(k) $8\frac{1}{4}$ is 56.25% of what number?

(m) What number is 150% of 6?

(o) 68.75% of what number is $5\frac{1}{2}$?

(q) What percent of 32 is 26?

(s) $93\frac{3}{4}\%$ of 18 is what number?

(u) 8.04 is $6\frac{1}{4}\%$ of what number?

(b) $83\frac{1}{3}\%$ of 21 is what number?

(d) 33 is $91\frac{2}{3}\%$ of what number?

(f) 20 is what percent of 1?

(h) 4% of what number is 9.2?

(j) What number is $43\frac{3}{4}\%$ of 100?

(l) What percent of 3 is 8?

(n) $56\frac{3}{4}\%$ of 25 is what number?

(p) 11.5 is $5\frac{3}{4}\%$ of what number?

(r) 2 is what percent of 50?

(t) 1% of what number is $1\frac{1}{2}$?

(v) What number is 100% of 8?

PROBLEM 10-23   Find each percent increase from: **(a)** 8 to 13   **(b)** 1 to 2   **(c)** 20 to 30
(d) 45 to 60   **(e)** 4 to 5   **(f)** 50 to 60   **(g)** 72 to 84   **(h)** 8 to 9   **(i)** 100 to 110
(j) 1.2 to 1.3   **(k)** 16 to 17   **(l)** 20 to 21   **(m)** 25 to 26   **(n)** 50 to 51   **(o)** 1 to 1.01
(p) 40 to 46   **(q)** $\frac{4}{5}$ to $1\frac{2}{5}$ **(r)** 1 to $1\frac{2}{3}$ **(s)** $\frac{5}{8}$ to $\frac{7}{8}$ **(t)** 8 to 11   **(u)** 760 to 1425
(v) 2000 to 2375

PROBLEM 10-24   Find each percent decrease from: **(a)** 2 to 1   **(b)** $\frac{3}{4}$ to $\frac{1}{4}$ **(c)** 40 to 10
(d) 1.25 to 0.75   **(e)** 5 to 1   **(f)** $\frac{3}{4}$ to $\frac{1}{8}$ **(g)** 80 to 50   **(h)** 1 to 0.125   **(i)** 12 to 0
(j) 100 to 70   **(k)** $\frac{1}{2}$ to $\frac{1}{20}$ **(l)** 6 to 2.5   **(m)** 120 to 70   **(n)** 4 to $3\frac{2}{3}$ **(o)** 160 to 150
(p) 4.5 to 3.75   **(q)** $\frac{5}{8}$ to $\frac{1}{4}$ **(r)** 2 to 1.9   **(s)** $1\frac{2}{3}$ to $1\frac{1}{3}$ **(t)** 4 to 3   **(u)** 250 to 240
(v) 0.09 to 0.045

Answers to Supplementary Exercises

(10-14)   **(a)** $30 \times \frac{1}{100}, 30 \div 100$

(d) $33\frac{1}{3} \times \frac{1}{100}, 33\frac{1}{3} \div 100$

(g) $37.5 \times \frac{1}{100}, 37.5 \div 100$

(b) $5 \times \frac{1}{100}, 5 \div 100$

(e) $166\frac{2}{3} \times \frac{1}{100}, 166\frac{2}{3} \div 100$

(h) $87.5 \times \frac{1}{100}, 87.5 \div 100$

(c) $90 \times \frac{1}{100}, 90 \div 100$

(f) $150 \times \frac{1}{100}, 150 \div 100$

(i) $16\frac{2}{3} \times \frac{1}{100}, 16\frac{2}{3} \div 100$

(j) $83\frac{1}{3} \times \frac{1}{100}, 83\frac{1}{3} \div 100$ **(k)** $62.5 \times \frac{1}{100}, 62.5 \div 100$ **(l)** $20 \times \frac{1}{100}, 20 \div 100$

(m) $80 \times \frac{1}{100}, 80 \div 100$ **(n)** $\frac{1}{3} \times \frac{1}{100}, \frac{1}{3} \div 100$ **(o)** $\frac{3}{4} \times \frac{1}{100}, \frac{3}{4} \div 100$

(p) $6\frac{1}{4} \times \frac{1}{100}, 6\frac{1}{4} \div 100$ **(q)** $\frac{2}{3} \times \frac{1}{100}, \frac{2}{3} \div 100$ **(r)** $8\frac{1}{2} \times \frac{1}{100}, 8\frac{1}{2} \div 100$

(s) $175 \times \frac{1}{100}, 175 \div 100$ **(t)** $225 \times \frac{1}{100}, 225 \div 100$ **(u)** $0.5 \times \frac{1}{100}, 0.5 \div 100$

(v) $0.125 \times \frac{1}{100}, 0.125 \div 100$

(10-15) **(a)** 0.25   **(b)** 0.5   **(c)** 0.75   **(d)** 0.9   **(e)** 0.1

(f) 0.2   **(g)** 0.125   **(h)** 0.375   **(i)** 0.3   **(j)** 0.4

(k) 0.625   **(l)** 0.875   **(m)** 0.6   **(n)** 0.8   **(o)** 0.0625

(p) 0.1875   **(q)** 1   **(r)** 1.25   **(s)** 0.3125   **(t)** 0.4375

(u) 2.5   **(v)** 10

(10-16) **(a)** $\frac{1}{6}$ **(b)** $\frac{5}{6}$ **(c)** $\frac{1}{3}$ **(d)** $\frac{2}{3}$ **(e)** $\frac{1}{12}$

(f) $\frac{5}{12}$ **(g)** $\frac{7}{12}$ **(h)** $\frac{11}{12}$ **(i)** $\frac{1}{100}$ **(j)** $\frac{1}{4}$

(k) $\frac{1}{2}$ **(l)** $\frac{3}{4}$ **(m)** $\frac{1}{10}$ **(n)** $\frac{1}{5}$ **(o)** $\frac{3}{10}$

(p) $\frac{2}{5}$ **(q)** $\frac{3}{5}$ **(r)** $\frac{7}{10}$ **(s)** $\frac{4}{5}$ **(t)** $\frac{9}{10}$

(u) $\frac{1}{20}$ **(v)** $\frac{1}{25}$

(10-17) **(a)** 10%   **(b)** 20%   **(c)** 62.5% or $62\frac{1}{2}$%   **(d)** 87.5% or $87\frac{1}{2}$%

(e) 50%   **(f)** 60%   **(g)** 31.25% or $31\frac{1}{4}$%   **(h)** 43.75% or $43\frac{3}{4}$%

(i) 90%   **(j)** 25%   **(k)** 75%   **(l)** 5%

(m) 12.5% or $12\frac{1}{2}$%   **(n)** 37.5% or $37\frac{1}{2}$%   **(o)** 30%   **(p)** 40%

(q) 6.25% or $6\frac{1}{4}$%   **(r)** 18.75% or $18\frac{3}{4}$%   **(s)** 70%   **(t)** 80%

(u) 56.25% or $56\frac{1}{4}$%   **(v)** 68.75% or $68\frac{3}{4}$%

(10-18) **(a)** 50%   **(b)** 25%   **(c)** 20%   **(d)** $12\frac{1}{2}$% or 12.5%

(e) 10%   **(f)** $8\frac{1}{3}$%   **(g)** $6\frac{1}{4}$% or 6.25%   **(h)** 75%

(i) 40%   **(j)** $37\frac{1}{2}$% or 37.5%   **(k)** 30%   **(l)** $41\frac{2}{3}$%

(m) $18\frac{3}{4}$% or 18.75%   **(n)** 60%   **(o)** $62\frac{1}{2}$% or 62.5%   **(p)** 70%

(q) $58\frac{1}{3}$%   **(r)** $31\frac{1}{4}$% or 31.25%   **(s)** 80%   **(t)** $87\frac{1}{2}$% or 87.5%

(u) 90%   **(v)** $43\frac{3}{4}$% or 43.75%

(10-19) **(a)** 15   **(b)** 20   **(c)** 10   **(d)** 25   **(e)** 63

(f) 11.2   **(g)** 40.8   **(h)** 0.6   **(i)** 2   **(j)** $\frac{3}{5}$ or 0.6

(k) 5   **(l)** 0.0125   **(m)** $6\frac{3}{16}$ or 6.1875   **(n)** 500   **(o)** 140

(p) $\frac{1}{20}$ or 0.05   **(q)** $\frac{13}{20}$ or 0.65   **(r)** 1   **(s)** 2   **(t)** 0.25

(u) 2   **(v)** $\frac{27}{32}$ or 0.84375

(10-20) **(a)** 16   **(b)** 40   **(c)** 300   **(d)** 25   **(e)** 3.15 or $3\frac{3}{20}$

(f) 3.6   **(g)** 20.8   **(h)** 9.6 or $9\frac{3}{5}$ **(i)** $6\frac{2}{5}$ or 6.4   **(j)** 8

(k) $28\frac{1}{3}$ **(l)** 4   **(m)** 225   **(n)** 80   **(o)** $136\frac{2}{3}$

(p) 52   **(q)** 100   **(r)** 0.64   **(s)** 1   **(t)** 230

(u) $\frac{3}{4}$ **(v)** $437\frac{1}{2}$

(10-21) **(a)** 50%   **(b)** 12.5% or $12\frac{1}{2}$%   **(c)** $33\frac{1}{3}$%   **(d)** 37.5% or $37\frac{1}{2}$%

(e) $66\frac{2}{3}$%   **(f)** 62.5% or $62\frac{1}{2}$%   **(g)** 25%   **(h)** 87.5% or $87\frac{1}{2}$%

(i) 75%   **(j)** 10%   **(k)** 20%   **(l)** 30%

(m) 40%   **(n)** 70%   **(o)** 60%   **(p)** 90%

(q) 80%   **(r)** $8\frac{1}{3}$%   **(s)** 600%   **(t)** 240%

(u) 120%   **(v)** $58\frac{1}{3}$%

(10-22) (a) 35 (b) 17.5 (c) $345\frac{3}{5}$ or 345.6 (d) 36

(e) 6.25% or $6\frac{1}{4}$% (f) 5% (g) 3.75 or $3\frac{3}{4}$ (h) 230

(i) 320% (j) 43.75 (k) $15\frac{1}{9}$ (l) $266\frac{2}{3}$%

(m) 9 (n) 14.1875 or $14\frac{3}{16}$ (o) 8 (p) 200

(q) 81.25% or $81\frac{1}{4}$% (r) 4% (s) 16.875 or $16\frac{7}{8}$ (t) 150

(u) 128.64 (v) 8

(10-23) (a) 62.5% or $62\frac{1}{2}$% (b) 100% (c) 50% (d) $33\frac{1}{3}$%

(e) 25% (f) 20% (g) $16\frac{2}{3}$% (h) 12.5% or $12\frac{1}{2}$%

(i) 10% (j) $8\frac{1}{3}$% (k) 6.25% or $6\frac{1}{4}$% (l) 5%

(m) 4% (n) 2% (o) 1% (p) 15%

(q) 75% (r) $66\frac{2}{3}$% (s) 40% (t) 37.5% or $37\frac{1}{2}$%

(u) 87.5% or $87\frac{1}{2}$% (v) 18.75% or $18\frac{3}{4}$%

(10-24) (a) 50% (b) $66\frac{2}{3}$% (c) 75% (d) 40% (e) 80%

(f) $83\frac{1}{3}$% (g) 37.5% or $37\frac{1}{2}$% (h) 87.5% or $87\frac{1}{2}$% (i) 100% (j) 30%

(k) 90% (l) $58\frac{1}{3}$% (m) $41\frac{2}{3}$% (n) $8\frac{1}{3}$% (o) 6.25% or $6\frac{1}{4}$%

(p) $16\frac{2}{3}$% (q) 60% (r) 5% (s) 20% (t) 25%

(u) 4% (v) 50%

11 PERCENT APPLICATIONS

THIS CHAPTER IS ABOUT

☑ **Solving Problems Involving Sales Tax**
☑ **Solving Problems Involving Discounts**
☑ **Solving Problems Involving Commissions**
☑ **Solving Problems Involving Simple Interest**
☑ **Solving Problems Involving Compound Interest**

11-1. Solving Problems Involving Sales Tax

The price marked on an item in a retail store is called the **purchase price.** On most items sold in retail stores, you must also pay **sales tax.** The **total price** for an item is the purchase price plus any sales tax that is applicable.

A. Find the total price.

To find the total price for an item, you add the purchase price and applicable sales tax (if any) together.

Total Price Formula: Purchase Price + Sales Tax = Total Price

EXAMPLE 11-1: A certain coat has a purchase price of $89.99 with a sales tax of $5.40. What is the total price of the coat?

$$\overset{\text{purchase price}}{\$89.99} + \overset{\text{sales tax}}{\$5.40} = \overset{\text{total price}}{\$95.39} \qquad \text{Add.}$$

Solution:

B. Find the sales tax.

To find the sales tax on an item, you multiply the purchase price by the **sales-tax rate.**

Sales Tax Formula: Sales-Tax Rate × Purchase Price = Sales Tax

Caution: When the sales-tax rate is given as a percent, you must first rename the sales-tax rate as either a terminating decimal or a fraction in order to multiply.

EXAMPLE 11-2: The sales-tax rate in Atlanta, Georgia, is 4%. What is the sales tax on a dress that has a purchase price of $45 in Atlanta, Georgia?

Solution:
$$\overset{\substack{\text{sales-tax}\\\text{rate}}}{4\%} \times \overset{\substack{\text{purchase}\\\text{price}}}{\$45} = 0.04 \times \$45 \qquad \text{Rename the percent [see Example 10-4].}$$
$$= \$1.80 \longleftarrow \text{sales tax} \qquad \text{Multiply.}$$

Note: The total price for the dress in Example 11-2 is

$$\$45 + \$1.80 = \$46.80 \longleftarrow \text{total price [See Example 11-1.]}$$

C. Find the sales tax rate.

To find the sales-tax rate for an item, you can divide the sales tax by the purchase price.

Sales Tax Rate Formula: Sales Tax ÷ Purchase Price = Sales-Tax Rate

Caution: After you divide the sales tax by the purchase price, you must rename the quotient as a percent to represent the sales-tax rate in standard form.

EXAMPLE 11-3: The sales tax on a $120 suit is $6. What is the sales-tax rate?

$$\begin{array}{ll}
\text{sales} \quad \text{purchase} \\
\text{tax} \quad\;\; \text{price}
\end{array}$$

Solution: $\overset{\frown}{\$6} \div \overset{\frown}{\$120}$ = 0.05   Divide.

                = 005.%   Rename as a percent [see Example 10-8].

                = 5% ⟵ sales-tax rate

D. Find the purchase price.

To find the purchase price when both the sales tax and sales-tax rate are known, you divide the sales tax by the sales-tax rate.

Purchase Price Formula: Sales Tax ÷ Sales-Tax Rate = Purchase Price

Caution: When the sales-tax rate is given as a percent, you must first rename the sales-tax rate as either a terminating decimal or a fraction in order to divide.

EXAMPLE 11-4: The sales tax on a certain book is $1 when the sales-tax rate is 4%. What is the purchase price of the book?

Solution

$$\begin{array}{ll}
\text{sales-tax} \\
\text{sales tax} \quad\;\;\; \text{rate}
\end{array}$$

   $\overset{\frown}{\$1}$ ÷ $\overset{\frown}{4\%}$ = $1 ÷ 0.04   Rename the percent [see Example 10-4].

             = $25 ⟵ purchase price   Divide.

Note: The total price for the book in Example 11-4 is:

           $25 + $1 = $26 ⟵ total price [See Example 11-1.]

11-2. Solving Problems Involving Discounts

The price at which an item normally sells is called the **regular price.** When the regular price of an item is reduced to a lower price, the new lower price is called the **sale price.** The dollar amount by which the regular price is reduced to get the sale price is called the **discount.**

A. Find the sale price.

To find the sale price for an item, you subtract the discount from the regular price.

Sale Price Formula: Regular Price − Discount = Sale Price

EXAMPLE 11-5: A certain radio has a regular price of $49.99. What is the sale price if the radio is discounted $25?

$$\text{regular price} \quad\;\; \text{discount} \quad\;\; \text{sale price}$$

Solution: $\overset{\frown}{\$49.99}$ − $\overset{\frown}{\$25}$ = $\overset{\frown}{\$24.99}$ Subtract.

B. **Find the discount.**
To find the discount on a given sale item, you multiply the regular price by the **discount rate** or **rate of discount.**

Discount Formula: Discount Rate × Regular Price = Discount

Caution: When the discount rate is given as a percent, you must rename the discount rate as either a terminating decimal or a fraction in order to multiply.

EXAMPLE 11-6: A $19.99 calculator is marked down 25%. How much is the discount?

discount rate   regular price

Solution: 25%   × $19.99   = 0.25 × $19.99 Rename the percent [see Example 10-4].

= $4.9975   Multiply.

≈ $5.00 Round to the nearest cent [see Example 1-16].

= $5 ⟵ discount

Note: The sale price for the calculator in Example 11-6 is:

$19.99 − $5 = $14.99 ⟵ sale price [See Example 11-5.]

C. **Find the discount rate.**
To find the discount rate for an item, you divide the discount by the regular price.

Discount Rate Formula: Discount ÷ Regular Price = Discount Rate

Caution: After you divide the discount by the regular price, you must rename the quotient as a percent to represent the discount rate in standard form.

EXAMPLE 11-7: The discount is $17 on a watch that costs $85. What is the discount rate?

discount   regular price

Solution: $17   ÷ $85   = 0.2   Divide.

= 020.%   Rename as a percent [see Example 10-8].

= 20% ⟵ discount rate

D. **Find the regular price.**
To find the regular price when both the discount and discount rate are known, you divide the discount by the discount rate.

Regular Price Formula: Discount ÷ Discount Rate = Regular Price

EXAMPLE 11-8: The discount on a certain pair of shoes is $32 when the discount rate is 40%. What is the regular price of the shoes?

discount   discount rate

Solution: $32   ÷   40%   = $32 ÷ 0.40 Rename the percent [see Example 10-4].

= $80 ⟵ regular price   Divide

Note: The sale price for the pair of shoes in Example 11-8 is:

$80 − $32 = $48 ⟵ sale price [See Example 11-5.]

11-3. Solving Problems Involving Commissions

The amount paid for working a given number of hours is called **salary.** The amount paid for selling a given dollar amount is called **commission.** The **total pay** is the salary plus any commission that is applicable.

A. Find the total pay.

To find the total pay for salary plus commission work, you add the salary and the applicable commission (if any) together.

Total Pay Formula: Salary + Commission = Total Pay

EXAMPLE 11-9: A used car salesperson earns $200 salary per week. This week the salesperson earned $450 in commissions. What is the salesperson's total pay this week?

$$\overbrace{\$200}^{\text{salary}} + \overbrace{\$450}^{\text{commission}} = \overbrace{\$650}^{\text{total pay}} \qquad \text{Add.}$$

Solution:

B. Find the commission.

To find the commission on sales, you multiply the sales by the commission rate.

Commission Formula: Commission Rate × Sales = Commission

Caution: When the commission rate is given as a percent, you must rename the commission rate as either a terminating decimal or a fraction in order to multiply.

EXAMPLE 11-10: What is the commission on sales of $500 if the commission rate is 15%?

Solution:
$$\overbrace{15\%}^{\text{commission rate}} \times \overbrace{\$500}^{\text{sales}} = 0.15 \times \$500 \quad \text{Rename the percent [see Example 10-4].}$$
$$= \$75 \longleftarrow \text{commission} \qquad \text{Multiply.}$$

C. Find the commission rate.

To find the commission rate, you divide the commission by the sales.

Commission Rate Formula: Commission ÷ Sales = Commission Rate

Caution: After you divide the commission by the sales, you must rename the quotient as a percent to represent the commission rate in standard form.

EXAMPLE 11-11: A magazine salesperson earned $250 in commission based on $1000 in sales. What is the commission rate?

Solution:
$$\overbrace{\$250}^{\text{commission}} \div \overbrace{\$1000}^{\text{sales}} = 0.25 \qquad \text{Divide.}$$
$$= 025.\% \qquad \text{Rename as a percent [see Example 10-8].}$$
$$= 25\% \longleftarrow \text{commission rate}$$

D. Find the sales.

To find the sales when both the commission and commission rate are known, you divide the commission by the commission rate.

Sales Formula: Commission ÷ Commission Rate = Sales

Caution: When the commission rate is given as a percent, you must rename the commission rate as either a terminating decimal or a fraction in order to divide.

EXAMPLE 11-12: The commission on a certain amount of book sales is $1500 when the commission rate is 5%. What is the amount of sales?

$$\overbrace{\$1500}^{\text{commission}} \div \overbrace{5\%}^{\text{commission rate}}$$

Solution: $\quad \$1500 \div 5\% \quad = \$1500 \div 0.05$ Rename the percent [see Example 10-4].

$$= \$30,000 \longleftarrow \text{ sales} \quad \text{Divide.}$$

11-4. Solving Problems Involving Simple Interest

A. Understand interest rates and time periods.

Money that is paid for the use of money is called **interest.** The money on which interest is paid is called the **principal.** The rate that interest is charged as a percent of the principal over a period of time is called the **interest rate.** The period of time that interest is paid on the principal is called the **time period.** The three most popular time periods for interest rates are the day, the month, and the year.

EXAMPLE 11-13: Explain the meaning of **(a)** 15% per year   **(b)** 1%/month   **(c)** $\dfrac{0.04\%}{\text{day}}$

Solution

(a) 15% per year means the interest is paid on the principal at a rate of 15% each year.

(b) 1%/month or 1% per month means the interest is paid on the principal at a rate of 1% each month.

(c) $\dfrac{0.04\%}{\text{day}}$ or 0.04% per day means the interest is paid on the principal at a rate of 0.04% each day.

Interest rates can be written in several different ways.

EXAMPLE 11-14: Write the following interest rate in three different ways: 12% per year

Solution: $\quad \overbrace{12\% \text{ per year}}^{\textbf{word form}} = 12\%/\text{year} \longleftarrow$ **slash form**

$$= \frac{12\%}{\text{year}} \longleftarrow \textbf{fraction form}$$

When computing interest in the text, the following **business time units** will be used:

$$1 \textbf{ business year} = 360 \text{ days} \qquad 1 \textbf{ business month} = 30 \text{ days}$$

Note: For the remainder of this chapter, all years and months are assumed to be business years and business months.

When computing interest, it is often necessary to rename a given time period in terms of a different time unit.

EXAMPLE 11-15: Rename 8 months as **(a)** days   **(b)** years.

Solution

(a) To rename a time period (like 8 months) in terms of a shorter time unit (like days), you multiply.

$$8 \text{ months} = 8 \times 30 \text{ days} = 240 \text{ days} \qquad \textit{Think:} \text{ 1 month} = 30 \text{ days}$$

(b) To rename a time period (like 8 months) in terms of a longer time unit (like years), you divide.

$$8 \text{ months} = \frac{8}{12} \text{ year} = \frac{2}{3} \text{ year} \qquad \textit{Think:} \text{ 12 months} = 1 \text{ year}$$

B. Find the simple interest.

Interest that is paid on principal only is called **simple interest.** To find the simple interest, you multiply the principal times the interest rate times the time period.

Simple Interest Formula: Simple Interest = Principal × Interest Rate × Time Period

$$\text{or } I = P \times r \times t$$

Caution: Before you can evaluate the simple interest formula
(a) both the interest rate r and the time period t must be expressed with a common time unit
(b) the interest rate r must be renamed as either a terminating decimal or a fraction

EXAMPLE 11-16: What is the simple interest paid on $500 at 8% per year simple interest for 90 days?

Solution: $I = P \times r \times t$ Write the simple interest formula.

$= \$500 \times \dfrac{8\%}{\text{year}} \times 90 \text{ days}$ Substitute.

$= \$500 \times \dfrac{8\%}{\text{year}} \times \dfrac{90}{360} \text{ year}$ Rename to get a common time unit [see Example 11-15].

$= \$500 \times \dfrac{8\%}{\text{year}} \times \dfrac{90}{360} \text{ year}$ Eliminate the common time unit.

$= \$500 \times 8\% \times \dfrac{1}{4}$ Simplify the fraction [see Example 4-24].

$= 500 \times 0.08 \times \dfrac{1}{4}$ Rename the percent [see Example 10-4].

$= \$40 \times \dfrac{1}{4}$ Multiply.

$= \$10 \longleftarrow$ simple interest on $500 at 8% for 90 days

Note: If you borrow $500 at 8% per year simple interest for 90 days, you will need to repay the $500 principal plus an additional $10 in interest by the end of the 90-day time period.

C. Find the total amount.

To find the **total amount,** you add the principal and the interest together.

Total Amount Formula: Total Amount = Principal + Interest

$$\text{or } A = P + I$$

Caution: Before you can evaluate the total amount formula, you must find the interest.

EXAMPLE 11-17: What is the total amount to be repaid for a loan of $2000 at $1\frac{1}{2}$% per month simple interest for 540 days?

Solution: $I = P \times r \times t$ Find the simple interest.

$= \$2000 \times \dfrac{1\frac{1}{2}\%}{\text{month}} \times 540 \text{ days}$ Substitute.

$= \$2000 \times \dfrac{1\frac{1}{2}\%}{\text{month}} \times 18 \text{ months}$ Rename to get a common time unit [see Example 11-15].

$= \$2000 \times 1\frac{1}{2}\% \times 18$ Eliminate the common time unit.

$= \$2000 \times 0.015 \times 18$ Rename the percent: $1\frac{1}{2}\% = 1.5\%$

$= \$30 \times 18$ Multiply.

$= \$540 \longleftarrow$ simple interest

$$A = P + I \qquad\qquad \text{Find the total amount to be repaid.}$$

$$= \$2000 + \$540$$

$$= \$2540 \ \longleftarrow \ \text{total amount to be repaid for a loan of \$2000}$$
$$\text{at } 1\tfrac{1}{2}\% \text{ per month simple interest for 540 days}$$

11-5. Solving Problems Involving Compound Interest

A. Understand compound interest rates and compound time periods.

Interest paid on principal plus previously earned interest is called **compound interest.** Savings accounts usually earn compound interest. The time period between interest payments is called the **compounding period** or **compound time period.** The five most popular time periods for **compound interest rates** are **compounded annually, compounded semi-annually, compounded quarterly, compounded monthly,** and **compounded daily.**

EXAMPLE 11-18: Explain the meaning of (**a**) compounded annually   (**b**) compounded semi-annually   (**c**) compounded quarterly   (**d**) compounded monthly   (**e**) compounded daily

Solution
(**a**) "Compounded annually" means the interest is paid 1 time each year at the end of the 12-month time period ($t = 1$ year).
(**b**) "Compounded semi-annually" means the interest is paid 2 times each year at the end of each 6-month time period ($t = \frac{1}{2}$ year).
(**c**) "Compounded quarterly" means the interest is paid 4 times each year at the end of each 3-month time period ($t = \frac{1}{4}$ year).
(**d**) "Compounded monthly" means the interest is paid 12 times each year at the end of each 1-month time period ($t = \frac{1}{12}$ year).
(**e**) "Compounded daily" means the interest is paid 360 times each year at the end of each business day ($t = \frac{1}{360}$ year).

B. Find the compound interest and total amount for 2 compounding periods.

To find the compound interest and total amount for 2 compounding periods, you
1. Find the simple interest for the first compound time period using $I = P \times r \times t$ [see Example 11-16].
2. Find the total amount at the end of the first compounding period using $A = P + I$ [see Example 11-17].
3. Find the compound interest for the second compounding period using the total amount from Step 2 as a **new principal** P in $I = P \times r \times t$.
4. Find the total amount at the end of the second compounding period using the total amount from Step 2 as the new principal P and the compound interest from Step 3 as I in $A = P + I$.

EXAMPLE 11-19: What is the total amount in savings at the end of 1 year if you start with $100 and earn 5% per year compounded semi-annually?

Solution: Compounded semi-annually means the compound time period is 6 months or $\dfrac{1}{2}$ year $\left(t = \dfrac{1}{2} \text{ year} \right)$.

In the given time period of 1 year there are **2** compounding periods of $\dfrac{1}{2}$ year each $\left(1 = \mathbf{2} \times \dfrac{1}{2} \right)$.

$I = P \times r \times t$ Find the simple interest for the first of the 2 compounding periods.

$= \$100 \times \dfrac{5\%}{\text{year}} \times \dfrac{1}{2} \text{ year}$ Substitute.

$= \$100 \times 5\% \times \dfrac{1}{2}$ Eliminate the common time unit.

$= \$100 \times 0.05 \times 0.5$ Rename percents and fractions as decimals.

$= \$5 \times 0.5$ Multiply.

$= \$2.50$ ⟵ simple interest for the first of the 2 compounding periods

$A = P + I$ Find the total amount at the end of the first of the 2 compounding periods.

$= \$100 + \$2.50$ Substitute.

$= \$102.50$ ⟵ total amount (new principal P) at the end of the first of the 2 compounding periods

$I = P \times r \times t$ Find the compound interest for the second of the 2 compounding periods.

$= \$102.50 \times \dfrac{5\%}{\text{year}} \times \dfrac{1}{2} \text{ year}$ Substitute the new principal $102.50 for P.

$= \$102.50 \times 5\% \times \dfrac{1}{2}$

$= \$102.50 \times 0.05 \times 0.5$

$= \$5.125 \times 0.5$

$= \$2.5625$

$\approx \$2.56$ ⟵ compound interest for the last of the 2 compounding periods (rounded to the nearest cent)

$A = P + I$ Find the total amount at the end of the second of the 2 compounding periods.

$\approx \$102.50 + \$2.56$

$= \$105.06$ ⟵ total amount at the end of the given time period (1 year)

Note 1: In Example 11-19, the total amount in savings at the end of 1 year if you start with $100 and earn 5% per year compounded semi-annually is $105.06.

Note 2: In Example 11-19, the total interest earned at the end of 1 year is the sum of the simple interest and the compound interest: $2.50 + $2.56 = $5.06

C. Find the compound interest and total amount for more than 2 compounding periods.

If the given time period in Example 11-19 was $1\frac{1}{2}$ years instead of 1 year, then you would need to find the interest and total amount once more (3 times in all) using $105.06 as the new principal P in $I = P \times r \times t$

EXAMPLE 11-20: What is the total amount in savings at the end of $1\frac{1}{2}$ years if you start with $100 and earn 5% per year compounded semi-annually?

Solution: In the given time period of $1\frac{1}{2}$ years there are 3 compounding periods of $\frac{1}{2}$ year each ($t = \frac{1}{2}$ year).

$$I = P \times r \times t$$

Find the total amount at the end of the second of the 3 compounding periods [see Example 11-19].

$$= \$100 \times \frac{5\%}{\text{year}} \times \frac{1}{2} \text{ year}$$

$$= \$2.50 \longleftarrow \text{ simple interest } I$$

$$A = P + I$$

$$= \$100 + \$2.50$$

$$= \$102.50 \longleftarrow \text{ new principal } P$$

$$I = P \times r \times t$$

$$= \$102.50 \times \frac{5\%}{\text{year}} \times \frac{1}{2} \text{ year}$$

$$\approx \$2.56 \longleftarrow \text{ compound interest } I$$

$$A = P + I$$

$$\approx \$102.50 + \$2.56$$

$$= \$105.06 \longleftarrow \text{ total amount (new principal } P) \text{ at the end} \\ \text{of the second of the 3 compounding periods}$$

$$I = P \times r \times t$$

Find the compound interest for the last of the 3 compounding periods.

$$= \$105.06 \times \frac{5\%}{\text{year}} \times \frac{1}{2} \text{ year}$$

Substitute using the new principal \$105.06 for P.

$$= \$105.06 \times 5\% \times \frac{1}{2}$$

$$= \$105.06 \times 0.05 \times 0.5$$

$$= \$5.253 \times 0.5$$

$$= \$2.6265$$

$$\approx 2.63 \longleftarrow \text{ compound interest for the last of the 3 compounding periods to the nearest cent}$$

$$A = P + I \qquad \text{Find the total amount at the end of the last of the 3 compounding periods.}$$

$$\approx \$105.06 + \$2.63$$

$$= \$107.69 \longleftarrow \text{ total amount at the end of the given time period } (1\tfrac{1}{2} \text{ years})$$

Note 1: In Example 11-20, the total amount in savings at the end of $1\frac{1}{2}$ years if you start with \$100 and earn 5% per year compounded semi-annually is \$107.69.

Note 2: In Example 11-20, the total interest earned over the $1\frac{1}{2}$ year time period was \$7.69: \$2.50 + \$2.56 + \$2.63 = \$7.69.

Note 3: If the given time period in Example 11-20 was 2 years instead of $1\frac{1}{2}$ years, then you would need to find the interest and the total amount once more (4 times in all) using \$107.69 as the new principal in $I = P \times r \times t$, and so on.

RAISE YOUR GRADES

Can you . . . ?

☑ find the total price given the purchase price and sales tax
☑ find the sales tax given the purchase price and sales tax rate
☑ find the sales tax rate given the sales tax and the purchase price
☑ find the purchase price given the sales tax and sales tax rate

☑ find the sale price given the regular price and discount
☑ find the discount given the regular price and discount rate
☑ find the discount rate given the regular price and discount
☑ find the regular price given the discount and discount rate
☑ find the total pay given the salary and commission
☑ find the commission given the sales and commission rate
☑ find the commission rate given the sales and commission
☑ find the sales given the commission and commission rate
☑ name and explain the three most popular time periods for interest rates
☑ write a simple interest rate in three different forms
☑ write the number of days in a business year and a business month
☑ rename time periods as days, months, and years
☑ find the simple interest given the principal, the interest rate, and the time period
☑ find the total amount given the principal and interest
☑ explain the meaning of compounded (**a**) annually (**b**) semi-annually (**c**) quarterly (**d**) daily
☑ find the compound interest and total amount for 2 or more compounding periods given the principal, interest rate, and time period

SUMMARY

1. To find the total price for an item, you add the purchase price and applicable sales tax (if any) together.
2. To find the sales tax on an item, you multiply the purchase price by the sales-tax rate.
3. When necessary, always round the sales tax to the nearest cent.
4. To find the sales-tax rate for an item, you divide the sales tax by the purchase price.
5. To find the purchase price when both the sales tax and sales-tax rate are known, you divide the sales tax by the sales-tax rate.
6. To find the sale price for an item, you subtract the discount from the regular price.
7. To find the discount on an item, you multiply the regular price by the discount rate.
8. When necessary, always round the discount to the nearest cent.
9. To find the discount rate for an item, you divide the discount by the regular price.
10. To find the regular price when both the discount and discount rate are known, you divide the discount by the discount rate.
11. To find the total pay for salary plus commission work, you add the salary and the applicable commission (if any) together.
12. To find the commission on sales, you multiply the sales by the commission rate.
13. When necessary, always round the commission to the nearest cent.
14. To find the commission rate, you divide the commission by the sales.
15. To find the sales when both the commission and commission rate are known, you divide the commission by the commission rate.
16. One business year equals 360 days and one business month equals thirty days.
17. To rename a time period in terms of a shorter time period, you multiply. To rename a period in terms of a longer time period, you divide.
18. To find the simple interest, you multiply the principal times the interest rate times the time period: $I = P \times r \times t$
19. When necessary, always round the interest to the nearest cent.
20. To find the total amount, you add the principal and interest together: $A = P + I$
21. "Compounded annually" means the interest is paid 1 time each year at the end of the 12-month time period ($t = 1$ year).
22. "Compounded semi-annually" means the interest is paid 2 times each year at the end of each 6-month time period ($t = \frac{1}{2}$ year).
23. "Compounded quarterly" means the interest is paid 4 times each year at the end of each 3-month time period ($t = \frac{1}{4}$ year).
24. "Compounded monthly" means the interest is paid 12 times each year at the end of each 1-month time period ($t = \frac{1}{12}$ year).

25. "Compounded daily" means the interest is paid 360 times each year at the end of each business day ($t = \frac{1}{360}$ year).

26. To find the compound interest and total amount for 2 compounding periods:

    (a) Find the simple interest for the first compounding period using $I = P \times r \times t$.

    (b) Find the total amount at the end of the first compounding period using $A = P + I$.

    (c) Find the compound interest for the second compounding period using the total amount from Step **(b)** as the new principal P in $I = P \times r \times t$.

    (d) Find the total amount at the end of the second compounding period using the total amount from Step **(b)** as the new principal P and the compound interest from Step **(c)** as I in $A = P + I$.

27. To find the total amount for more than 2 compounding periods, you repeat Steps **(c)** and **(d)** in number 27 until the interest and total amount have been computed for each compounding period in the given time period.

SOLVED PROBLEMS

PROBLEM 11-1   Find the sales tax and total price given the purchase price and sales-tax rate. Round the sales tax to the nearest cent when necessary.

	Purchase Price	Sales-Tax Rate	Sales Tax	Total Price
(a)	$45	5% (Chicago, IL)	?	?
(b)	$89	6% (Denver, CO)	?	?
(c)	$25.50	7% (Buffalo, NY)	?	?
(d)	$18.39	8% (New York, NY)	?	?
(e)	$50	$4\frac{1}{2}$% (Cincinnati, OH)	?	?
(f)	$39	$6\frac{1}{2}$% (San Francisco, CA)	?	?
(g)	$15.80	$4\frac{1}{8}$% (St. Louis, MO)	?	?
(h)	$62.99	$3\frac{5}{8}$% (Kansas City, MO)	?	?

Solution: Recall that to find the sales tax and total price on an item, you first multiply the purchase price by the sales-tax rate to find the sales tax and then you add the purchase price and sales tax together to find the total price [see Examples 11-1 and 11-2]:

(a) 5% × $45 = 0.05 × $45 = $2.25 and $45 + $2.25 = $47.25

(b) 6% × $89 = 0.06 × $89 = $5.34 and $89 + $5.34 = $94.34

(c) 7% × $25.50 = 0.07 × $25.50 = $1.785 ≈ $1.79 and $25.50 + $1.79 = $27.29

(d) 8% × $18.39 = 0.08 × $18.39 = $1.4712 ≈ $1.47 and $18.39 + $1.47 = $19.86

(e) $4\frac{1}{2}$% × $50 = 0.045 × $50 = $2.25 and $50 + $2.25 = $52.25

(f) $6\frac{1}{2}$% × $39 = 0.065 × $39 = $2.535 ≈ $2.54 and $39 + $2.54 = $41.54

(g) $4\frac{1}{8}$% × $15.80 = 0.04125 × $15.80 = $0.65175 ≈ $0.65 and $15.80 + $0.65 = $16.45

(h) $3\frac{5}{8}$% × $62.99 = 0.03625 × $62.99 = $2.2833875 ≈ $2.28 and $62.99 + $2.28 = $65.27

PROBLEM 11-2   Find the sales-tax rate and total price given the purchase price and sales tax:

	Purchase Price	Sales-Tax Rate	Sales Tax	Total Price
(a)	$80	?	$2.40	?
(b)	$25	?	$1.25	?
(c)	$14.50	?	$0.87	?
(d)	$33	?	$2.64	?
(e)	$10	?	$0.25	?
(f)	$42	?	$1.47	?
(g)	$80	?	$3.80	?
(h)	$60	?	$3.75	?

Solution: Recall that to find the sales-tax rate for a given item, you divide the sales tax by the purchase price. To find the total price, you add the purchase price and the sales tax together [see Example 11-3]:

(a) $2.40 ÷ $80 = 0.03 = 3% and $80 + $2.40 = $82.40
(b) $1.25 ÷ $25 = 0.05 = 5% and $25 + $1.25 = $26.25
(c) $0.87 ÷ $14.50 = 0.06 = 6% and $14.50 + $0.87 = $15.37
(d) $2.64 ÷ $33 = 0.08 = 8% and $33 + $2.64 = $35.64
(e) $0.25 ÷ $10 = 0.025 = 2.5% and $10 + $0.25 = $10.25
(f) $1.47 ÷ $42 = 0.035 = 3.5% and $42 + $1.47 = $43.47
(g) $3.80 ÷ $80 = 0.0475 = 4.75% and $80 + $3.80 = $83.80
(h) $3.75 ÷ $60 = 0.0625 = 6.25% and $60 + $3.75 = $63.75

PROBLEM 11-3 Find the purchase price and total price given the sales tax and sales-tax rate:

	Purchase Price	Sales-Tax Rate	Sales Tax	Total Price
(a)	?	3%	$0.15	?
(b)	?	4%	$1.20	?
(c)	?	5%	$1.75	?
(d)	?	6%	$2.79	?
(e)	?	$7\frac{1}{2}$%	$6.90	?
(f)	?	$2\frac{1}{4}$%	$0.45	?
(g)	?	$3\frac{1}{8}$%	$1.25	?
(h)	?	$5\frac{7}{8}$%	$4.70	?

Solution: Recall that to find the purchase price, you divide the sales tax by the sales-tax rate [see Example 11-4]:

(a) $0.15 ÷ 3% = $0.15 ÷ 0.03 = $5 and $5 + $0.15 = $5.15
(b) $1.20 ÷ 4% = $1.20 ÷ 0.04 = $30 and $30 + $1.20 = $31.20
(c) $1.75 ÷ 5% = $1.75 ÷ 0.05 = $35 and $35 + $1.75 = $36.75
(d) $2.79 ÷ 6% = $2.79 ÷ 0.06 = $46.50 and $46.50 + $2.79 = $49.29
(e) $6.90 ÷ $7\frac{1}{2}$% = $6.90 ÷ 0.075 = $92 and $92 + $6.90 = $98.90
(f) $0.45 ÷ $2\frac{1}{4}$% = $0.45 ÷ 0.0225 = $20 and $20 + $0.45 = $20.45
(g) $1.25 ÷ $3\frac{1}{8}$% = $1.25 ÷ 0.03125 = $40 and $40 + $1.25 = $41.25
(h) $4.70 ÷ $5\frac{7}{8}$% = $4.70 ÷ 0.05875 = $80 and $80 + $4.70 = $84.70

PROBLEM 11-4 Find the discount and sale price given the regular price and the discount rate. (Round the discount to the nearest cent when necessary):

	Regular Price	Discount Rate	Discount	Sale Price
(a)	$25	$\frac{1}{2}$	?	?
(b)	$10	$\frac{1}{3}$	?	?
(c)	$32	30%	?	?
(d)	$15	40%	?	?
(e)	$59.30	15%	?	?
(f)	$75.85	25%	?	?

Solution: Recall that to find the discount and sale price on a given item, you first multiply the regular price by the discount rate to find the discount and then you subtract the discount from the regular price to find the sale price [see Examples 11-5 and 11-6]:

(a) $\frac{1}{2}$ × $25 = $\frac{$25}{2}$ = $12.50 and $25 − $12.50 = $12.50
(b) $\frac{1}{3}$ × $10 = $\frac{$10}{3}$ = $3.333 ≈ $3.33 and $10 − $3.33 = $6.67
(c) 30% × $32 = 0.3 × $32 = $9.60 and $32 − $9.60 = $22.40

(d) $40\% \times \$15 = 0.4 \times \$15 = \$6$ and $\$15 - \$6 = \$9$

(e) $15\% \times \$59.30 = 0.15 \times \$59.30 = \$8.895 \approx \$8.90$ and $\$59.30 - \$8.90 = \$50.40$

(f) $25\% \times \$75.85 = 0.25 \times \$75.85 = \$18.9625 \approx \$18.96$ and $\$75.85 - \$18.96 = \$56.89$

PROBLEM 11-5 Find the discount rate and sale price given the regular price and the discount:

	Regular Price	Discount Rate	Discount	Sale Price
(a)	$ 80	?	$20	?
(b)	$ 30	?	$20	?
(c)	$ 52	?	$10.40	?
(d)	$ 49.98	?	$24.99	?
(e)	$ 5.40	?	$ 0.27	?
(f)	$129	?	$96.75	?

Solution: Recall that to find the discount rate for a given item, you divide the discount by the regular price [see Example 11-7]:

(a) $\$20 \div \$80 = 0.25 = 25\%$ or $\frac{1}{4}$ and $\$80 - \$20 = \$60$

(b) $\$20 \div \$30 = 0.66\overline{6} = 66\frac{2}{3}\%$ or $\frac{2}{3}$ and $\$30 - \$20 = \$10$

(c) $\$10.40 \div \$52 = 0.2 = 20\%$ or $\frac{1}{5}$ and $\$52 - \$10.40 = \$41.60$

(d) $\$24.99 \div \$49.98 = 0.5 = 50\%$ or $\frac{1}{2}$ and $\$49.98 - \$24.99 = \$24.99$

(e) $\$0.27 \div \$5.40 = 0.05 = 5\%$ or $\frac{1}{20}$ and $\$5.40 - \$0.27 = \$5.13$

(f) $\$96.75 \div \$129 = 0.75 = 75\%$ or $\frac{3}{4}$ and $\$129 - \$96.75 = \$32.25$

PROBLEM 11-6 Find the regular price and sale price given the discount and discount rate:

	Regular Price	Discount Rate	Discount	Sale Price
(a)	?	$\frac{3}{4}$	$ 6	?
(b)	?	$\frac{2}{3}$	$ 24	?
(c)	?	20%	$ 15	?
(d)	?	25%	$105	?
(e)	?	50%	$ 24.98	?
(f)	?	80%	$ 43.88	?

Solution: Recall that to find the regular price, you can divide the discount by the discount rate [see Example 11-8]:

(a) $\$6 \div \frac{3}{4} = \$6 \times \frac{4}{3} = \$8$ and $\$8 - \$6 = \$2$

(b) $\$24 \div \frac{2}{3} = \$24 \times \frac{3}{2} = \$36$ and $\$36 - \$24 = \$12$

(c) $\$15 \div 20\% = \$15 \div 0.2 = \$75$ and $\$75 - \$15 = \$60$

(d) $\$105 \div 25\% = \$105 \div 0.25 = \$420$ and $\$420 - \$105 = \$315$

(e) $\$24.98 \div 50\% = \$24.98 \div 0.5 = \$49.96$ and $\$49.96 - \$24.98 = \$24.98$

(f) $\$43.88 \div 80\% = \$43.88 \div 0.8 = \$54.85$ and $\$54.85 - \$43.88 = \$10.97$

PROBLEM 11-7 Find the commission and total pay given the sales, commission rate, and salary. (Round the commission to the nearest cent when necessary):

	Sales	Commission Rate	Commission	Salary	Total Pay
(a)	$100	5%	?	$ 50	?
(b)	$500	10%	?	$100	?
(c)	$250	20%	?	$ 75	?
(d)	$725	25%	?	$200	?
(e)	$ 85.98	$12\frac{1}{2}\%$	?	$ 25	?
(f)	$112.53	$6\frac{1}{4}\%$	?	$ 52	?

Solution: Recall that to find the commission and total pay, you first multiply the sales by the commission rate to find the commission and then you add the salary to the commission to find the total pay [see Examples 11-9 and 11-10]:

	Commission Rate	Sales			Commission		Salary		Total Pay
(a)	5%	× $100	= 0.05	× $100	= $ 5		and $ 50+$ 5		= $ 55
(b)	10%	× $500	= 0.1	× $500	= $ 50		and $100+$ 50		= $150
(c)	20%	× $250	= 0.2	× $250	= $ 50		and $ 75+$ 50		= $125
(d)	25%	× $725	= 0.25	× $725	= $181.25		and $200+$181.25		= $381.25
(e)	$12\frac{1}{2}$%	× $ 85.98	= 0.125	× $ 85.98	= $ 10.7475	$\approx$$10.75	and $ 25+$ 10.75		= $ 35.75
(f)	$6\frac{1}{4}$%	× $112.53	= 0.0625	× $112.53	= $ 7.033125	$\approx$$7.03	and $ 52+$ 7.03		= $ 59.03

PROBLEM 11-8 Find the commission rate and total pay given the sales, commission, and salary:

	Sales	Commission Rate	Commission	Salary	Total Pay
(a)	$200	?	$ 21	$ 25	?
(b)	$300	?	$ 36.75	$100	?
(c)	$850	?	$127.50	$250	?
(d)	$725	?	$130.50	$175	?
(e)	$475.30	?	$142.59	$200	?
(f)	$615.24	?	$307.62	$125	?

Solution: Recall that to find the commission rate, you divide the commission by the sales [see Example 11-11]:

	Commission		Sales		Commission Rate		Salary		Commission		Total Pay
(a)	$ 21	÷	$200	= 0.105	= 10.5%	or $10\frac{1}{2}$% and	$ 25	+	$ 21		= $ 46
(b)	$ 36.75	÷	$300	= 0.1225	= 12.25%	or $12\frac{1}{4}$% and	$100	+	$ 36.75		= $136.75
(c)	$127.50	÷	$850	= 0.15	= 15%	and	$250	+	$127.50		= $377.50
(d)	$130.50	÷	$725	= 0.18	= 18%	and	$175	+	$130.50		= $305.50
(e)	$142.59	÷	$475.30	= 0.3	= 30%	and	$200	+	$142.59		= $342.59
(f)	$307.62	÷	$615.24	= 0.5	= 50%	and	$125	+	$307.62		= $432.62

PROBLEM 11-9 Find the sales and total pay given the commission rate, commission, and salary:

	Sales	Commission Rate	Commission	Salary	Total Pay
(a)	?	3%	$ 25.50	$100	?
(b)	?	8%	$ 10	$200	?
(c)	?	15%	$ 54.75	$ 75	?
(d)	?	25%	$120.56	$125	?
(e)	?	30%	$ 37.68	$ 50	?
(f)	?	40%	$302.50	$200	?

Solution: Recall that to find the sales, you can divide the commission by the commission rate [see Example 11-12]:

	Commission		Commission Rate		Sales	Salary		Commission		Total Pay
(a)	$ 25.50	÷	3%	= $ 25.50 ÷ 0.03	= $850	and $100	+	$ 25.50		= $125.50
(b)	$ 10	÷	8%	= $ 10 ÷ 0.08	= $125	and $200	+	$ 10		= $210
(c)	$ 54.75	÷	15%	= $ 54.75 ÷ 0.15	= $365	and $ 75	+	$ 54.75		= $129.75
(d)	$120.56	÷	25%	= $120.56 ÷ 0.25	= $482.24	and $125	+	$120.56		= $245.56
(e)	$ 37.68	÷	30%	= $ 37.68 ÷ 0.3	= $125.60	and $ 50	+	$ 37.68		= $ 87.68
(f)	$302.50	÷	40%	= $302.50 ÷ 0.4	= $756.25	and $200	+	$302.50		= $502.50

PROBLEM 11-10   Rename each time period in terms of the given shorter time unit:

(a) 3 years = ? months   (b) $2\frac{1}{2}$ years = ? months   (c) 2 years = ? days
(d) $1\frac{1}{2}$ years = ? days   (e) 6 months = ? days   (f) $5\frac{1}{2}$ months = ? days

Solution: Recall that to rename a time period in terms of a shorter time unit, you multiply [see Example 11-15a]:

(a) 3 years = 3 × 12 months = 36 months   (b) $2\frac{1}{2}$ years = $2\frac{1}{2}$ × 12 months = 30 months
(c) 2 years = 2 × 360 days = 720 days   (d) $1\frac{1}{2}$ years = $1\frac{1}{2}$ × 360 days = 540 days
(e) 6 months = 6 × 30 days = 180 days   (f) $5\frac{1}{2}$ months = $5\frac{1}{2}$ × 30 days = 165 days

PROBLEM 11-11   Rename each time period in terms of the given larger time unit:

(a) 48 months = ? years   (b) 42 months = ? years   (c) 270 days = ? months
(d) 135 days = ? months   (e) 1440 days = ? years   (f) 900 days = ? years

Solution: Recall that to rename a time period in terms of a longer time unit, you divide [see Example 11-15b]:

(a) 48 months = $\frac{48}{12}$ years = 4 years   (b) 42 months = $\frac{42}{12}$ years = $3\frac{1}{2}$ years
(c) 270 days = $\frac{270}{30}$ months = 9 months   (d) 135 days = $\frac{135}{30}$ months = $4\frac{1}{2}$ months
(e) 1440 days = $\frac{1440}{360}$ years = 4 years   (f) 900 days = $\frac{900}{360}$ years = $2\frac{1}{2}$ years

PROBLEM 11-12   Find the simple interest and total amount:

	Principal (P)	Interest Rate (r)	Time Period (t)	Simple Interest	Total Amount
(a)	$100	10% per year	5 years	?	?
(b)	$250	$6\frac{1}{2}$% per year	18 months	?	?
(c)	$425	1% per month	9 months	?	?
(d)	$ 80.50	$1\frac{1}{2}$% per month	90 days	?	?
(e)	$525.10	$\frac{1}{2}$% per day	30 days	?	?
(f)	$135.68	0.45% per day	3 months	?	?

Solution: Recall that to find the simple interest and total amount, you first multiply the principal, interest rate, and time period together to find the simple interest and then add the principal to the simple interest to find the total amount [See Examples 11-16 and 11-17]:

(a) $I = P \times r \times t$

$= \$100 \times \dfrac{10\%}{\text{year}} \times 5 \text{ years}$

$= \$100 \times 0.1 \times 5$

$= \$10 \times 5$

$= \$50 \longleftarrow$ simple interest

$A = P + I$

$= \$100 + \$50$

$= \$150 \longleftarrow$ total amount

(b) $I = P \times r \times t$

$= \$250 \times \dfrac{6\frac{1}{2}\%}{\text{year}} \times 18 \text{ months}$

$= \$250 \times \dfrac{0.065}{\text{year}} \times 1\frac{1}{2} \text{ years}$

$= \$16.25 \times 1.5$

$= \$24.375$

$\approx \$24.38 \longleftarrow$ simple interest

$A = P + I$

$\approx \$250 + \$24.38$

$= \$274.38 \longleftarrow$ total amount

(c) $I = P \times r \times t$

$\qquad = \$425 \times \dfrac{1\%}{\text{month}} \times 9 \text{ \sout{months}}$

$\qquad = \$425 \times 0.01 \times 9$

$\qquad = \$4.25 \times 9$

$\qquad = \$38.25 \longleftarrow \text{ simple interest}$

$A = P + I$

$\qquad = \$425 + \$38.25$

$\qquad = \$463.25 \longleftarrow \text{ total amount}$

(d) $I = P \times r \times t$

$\qquad = \$80.50 \times \dfrac{1\frac{1}{2}\%}{\text{month}} \times 90 \text{ days}$

$\qquad = \$80.50 \times \dfrac{0.015}{\text{\sout{month}}} \times 3 \text{ \sout{month}}$

$\qquad = \$1.2075 \times 3$

$\qquad = \$3.6225$

$\qquad \approx \$3.62 \longleftarrow \text{ simple interest}$

$A = P + I$

$\qquad \approx \$80.50 + \$3.62$

$\qquad = \$84.12 \longleftarrow \text{ total amount}$

(e) $I = P \times r \times t$

$\qquad = \$525.10 \times \dfrac{\frac{1}{2}\%}{\text{day}} \times 30 \text{ \sout{days}}$

$\qquad = \$525.10 \times 0.005 \times 30$

$\qquad = \$2.6255 \times 30$

$\qquad = \$78.765$

$\qquad \approx \$78.77$

$A = P + I \longleftarrow \text{ simple interest}$

$\qquad \approx \$525.10 + \$78.77$

$\qquad = \$603.87 \longleftarrow \text{ total amount}$

(f) $I = P \times r \times t$

$\qquad = \$135.68 \times \dfrac{0.45\%}{\text{day}} \times 3 \text{ months}$

$\qquad = \$135.68 \times \dfrac{0.0045}{\text{day}} \times 90 \text{ \sout{days}}$

$\qquad = \$0.61056 \times 90$

$\qquad = \$54.9504$

$\qquad \approx \$54.95 \longleftarrow \text{ simple interest}$

$A = P + I$

$\qquad \approx \$135.68 + \$54.95$

$\qquad = \$190.63 \longleftarrow \text{ total amount}$

PROBLEM 11-13   Find the compound interest and total amount for 2 compounding periods:

	Principal (P)	Interest Rate (r)	Time Period (t)	Compound Interest	Total Amount
(a)	$ 100	5% compounded annually	2 years	?	?
(b)	$ 50	6% compounded semi-annually	1 year	?	?
(c)	$ 200	$6\frac{1}{2}\%$ compounded quarterly	6 months	?	?
(d)	$1000	$8\frac{1}{4}\%$ compounded monthly	2 months	?	?

Solution: Recall that to find the compound interest and total amount for 2 compounding periods, you:

1. Find the simple interest for the first compounding period using: $I = P \times r \times t$
2. Find the total amount at the end of the first compounding period using: $A = P + I$
3. Find the compound interest for the second compounding period using the total amount from Step 2 as the new principal P in: $I = P \times r \times t$
4. Find the total amount at the end of the second compounding period using the total amount from Step 2 as the new principal P and the compound interest from Step 3 as I in: $A = P + I$ [See Example 11-19.]

(a) $I = P \times r \times t$

$\quad = \$100 \times \dfrac{5\%}{\text{year}} \times 1 \text{ year}$

$\quad = \$100 \times 0.05 \times 1$

$\quad = \$5 \longleftarrow$ simple interest

$A = P + I$

$\quad = \$100 + \$5$

$\quad = \$105 \longleftarrow$ new principal

$I = P \times r \times t$

$\quad = \$105 \times \dfrac{5\%}{\text{year}} \times 1 \text{ year}$

$\quad = \$105 \times 0.05 \times 1$

$\quad = \$5.25 \longleftarrow$ compound interest

$A = P + I$

$\quad = \$105 + \$5.25$

$\quad = \$110.25 \longleftarrow$ total amount

(b) $I = P \times r \times t$

$\quad = \$50 \times \dfrac{6\%}{\text{year}} \times \dfrac{1}{2} \text{ year}$

$\quad = \$50 \times 0.06 \times 0.5$

$\quad = \$1.50 \longleftarrow$ simple interest

$A = P + I$

$\quad = \$50 + \$1.50$

$\quad = \$51.50 \longleftarrow$ new principal

$I = P \times r \times t$

$\quad = \$51.50 \times \dfrac{6\%}{\text{year}} \times \dfrac{1}{2} \text{ year}$

$\quad = \$51.50 \times 0.06 \times 0.5$

$\quad = \$1.545$

$\quad \approx \$1.55 \longleftarrow$ compound interest

$A = P + I$

$\quad \approx \$51.50 + \$1.55$

$\quad = \$53.05 \longleftarrow$ total amount

(c) $I = P \times r \times t$

$\quad = \$200 \times \dfrac{6\frac{1}{2}\%}{\text{year}} \times \dfrac{1}{4} \text{ year}$

$\quad = \$200 \times 0.065 \times 0.25$

$\quad = \$3.25 \longleftarrow$ simple interest

$A = P + I$

$\quad = \$200 + \$3.25$

$\quad = \$203.25 \longleftarrow$ new principal

$I = P \times r \times t$

$\quad = \$203.25 \times \dfrac{6\frac{1}{2}\%}{\text{year}} \times \dfrac{1}{4} \text{ year}$

$\quad = \$203.25 \times 0.065 \times 0.25$

$\quad = \$3.3028125$

$\quad \approx \$3.30 \longleftarrow$ compound interest

$A = P + I$

$\quad \approx \$203.25 + \$3.30$

$\quad = \$206.55 \longleftarrow$ total amount

(d) $I = P \times r \times t$

$\quad = \$1000 \times \dfrac{8\frac{1}{4}\%}{\text{year}} \times \dfrac{1}{12} \text{ year}$

$\quad = \$1000 \times 0.0825 \times \dfrac{1}{12}$

$\quad = \$6.875$

$\quad \approx \$6.88 \longleftarrow$ simple interest

$A = P + I$

$\quad = \$1000 + 6.88$

$\quad = \$1006.88 \longleftarrow$ new principal

$I = P \times r \times t$

$\quad = \$1006.88 \times \dfrac{8\frac{1}{4}\%}{\text{year}} \times \dfrac{1}{12} \text{ year}$

$\quad = \$1006.88 \times 0.0825 \times \dfrac{1}{12}$

$\quad = \$6.9223$

$\quad \approx \$6.92 \longleftarrow$ compound interest

$A = P + I$

$\quad \approx \$1006.88 + \$6.92$

$\quad = \$1013.80 \longleftarrow$ total amount

PROBLEM 11-14   Find the total amount for more than 2 compounding periods:

	Principal P	Interest Rate r	Time Period t	Total Interest	Total Amount
(a)	\$100	10% compounded annually	3 years	?	?
(b)	\$500	$8\frac{1}{2}$% compounded daily	4 days	?	?

Solution: Recall that to find the total amount for more than 2 compounding periods, you find the interest and total amount for each compounding period in the given time period [see Example 11-20]:

(a) $I = P \times r \times t$

$\quad = \$100 \times \dfrac{10\%}{\text{year}} \times 1 \text{ year}$

$\quad = \$100 \times 0.1 \times 1$

$\quad = \$10 \longleftarrow$ simple interest

$A = P + I$

$\quad = \$100 + \$10$

$\quad = \$110 \longleftarrow$ new principal

$I = P \times r \times t$

$\quad = \$110 \times \dfrac{10\%}{\text{year}} \times 1 \text{ year}$

$\quad = \$110 \times 0.1 \times 1$

$\quad = \$11 \longleftarrow$ compound interest

$A = P + I$

$\quad = \$110 + \$11$

$\quad = \$121 \longleftarrow$ new principal

$I = P \times r \times t$

$\quad = \$121 \times \dfrac{10\%}{\text{year}} \times 1 \text{ year}$

$\quad = \$121 \times 0.1 \times 1$

$\quad = \$12.10 \longleftarrow$ compound interest

$A = P + I$

$\quad = \$121 + \$12.10$

$\quad = \$133.10 \longleftarrow$ total amount

(b) $I = P \times r \times t$

$\quad = \$500 \times \dfrac{8\frac{1}{2}\%}{\text{year}} \times \dfrac{1}{360} \text{ year}$

$\quad = \$500 \times 0.085 \times \dfrac{1}{360}$

$\quad = \$0.1180\overline{5}$

$\quad \approx \$0.12 \longleftarrow$ simple interest

$A = P + I$

$\quad \approx \$500 + \$0.12$

$\quad = \$500.12 \longleftarrow$ new principal

$I = P \times r \times t$

$\quad \approx \$500.12 \times \dfrac{8\frac{1}{2}\%}{\text{year}} \times \dfrac{1}{360} \text{ year}$

$\quad = \$500.12 \times 0.085 \times \dfrac{1}{360}$

$\quad = \$0.1180838\overline{8}$

$\quad \approx \$0.12 \longleftarrow$ compound interest

$A = P + I$

$\quad \approx \$500.12 + \$0.12$

$\quad = \$500.24 \longleftarrow$ new principal

$I = P \times r \times t$

$\quad \approx \$500.24 \times \dfrac{8\frac{1}{2}\%}{\text{year}} \times \dfrac{1}{360} \text{ year}$

$\quad = \$500.24 \times 0.085 \times \dfrac{1}{360}$

$\quad = \$0.118111\overline{2}$

$\quad \approx \$0.12 \longleftarrow$ compound interest

$A = P + I$

$\quad \approx \$500.24 + \$0.12$

$\quad = \$500.36 \longleftarrow$ new principal

$$I = P \times r \times t$$

$$\approx \$500.36 \times \frac{8\frac{1}{2}\%}{\text{year}} \times \frac{1}{360} \text{ year}$$

$$= \$500.36 \times 0.085 \times \frac{1}{360}$$

$$= \$0.1181405\overline{5}$$

$$\approx \$0.12 \longleftarrow \text{ compound interest}$$

$$A = P + I$$

$$\approx \$500.36 + \$0.12$$

$$= \$500.48 \longleftarrow \text{ total amount}$$

Supplementary Exercises

PROBLEM 11-15   Find the missing total price and sales tax, sales-tax rate, or purchase price:

	Purchase Price	Sales-Tax Rate	Sales Tax	Total Price
(a)	$ 59	4%	?	?
(b)	$145	5%	?	?
(c)	$349.99	$6\frac{1}{2}\%$	?	?
(d)	$ 20	?	$0.85	?
(e)	$ 65.50	?	$5.24	?
(f)	$208	?	$6.24	?
(g)	?	7%	$3.50	?
(h)	?	8%	$2.08	?
(i)	?	$4\frac{1}{4}\%$	$2.21	?

PROBLEM 11-16   Find the missing sale price and discount, or discount rate, or regular price:

	Regular Price	Discount Rate	Discount	Sale Price
(a)	$ 85	20%	?	?
(b)	$106	10%	?	?
(c)	$ 7.99	$\frac{1}{4}$ off	?	?
(d)	$ 50	?	$ 15	?
(e)	$249	?	$ 83	?
(f)	$110	?	$ 22	?
(g)	?	15%	$105	?
(h)	?	$12\frac{1}{2}\%$	$ 25	?
(i)	?	$\frac{2}{3}$ off	$ 8.50	?

PROBLEM 11-17   Find the missing total pay and sales, or commission rate, or commission:

	Sales	Commission Rate	Commission	Salary	Total Pay
(a)	$ 100	2%	?	$ 75	?
(b)	$ 50	3%	?	$ 10	?
(c)	$ 200	5%	?	$ 100	?
(d)	$ 125	?	$ 10	$ 125	?
(e)	$ 500	?	$ 50	$ 200	?
(f)	$1500	?	$225	$1000	?

	Sales	Commission Rate	Commission	Salary	Total Pay
(g)	?	6%	$ 3.60	$ 30	?
(h)	?	20%	$ 50	$ 200	?
(i)	?	25%	$200	$ 500	?

PROBLEM 11-18   Rename each time period in terms of each given time unit:

	Years (yr)	Quarterly (q)	Months (mo)	Days (d)
(a)	5	?	?	?
(b)	$4\frac{1}{2}$	?	?	?
(c)	4	?	?	?
(d)	$3\frac{1}{4}$	?	?	?
(e)	3	?	?	?
(f)	$2\frac{3}{4}$	?	?	?
(g)	2	?	?	?
(h)	$1\frac{1}{2}$	?	?	?
(i)	1	?	?	?
(j)	$\frac{1}{2}$	?	?	?
(k)	?	6	?	?
(l)	?	10	?	?
(m)	?	3	?	?
(n)	?	?	48	?
(o)	?	?	30	?
(p)	?	?	9	?
(q)	?	?	?	360
(r)	?	?	?	990
(s)	?	?	?	1530
(t)	?	?	?	90

PROBLEM 11-19   Find the simple interest and total amount for each simple interest problem:

	Principal P	Interest Rate r	Time Period t	Simple Interest	Total Amount
(a)	$ 100	10% per year	5 years	?	?
(b)	$ 200	$1\frac{1}{2}$% per month	$2\frac{1}{2}$ years	?	?
(c)	$ 350	0.04% per day	$\frac{1}{4}$ year	?	?
(d)	$1000	5% per year	12 quarters	?	?
(e)	$3000	$1\frac{1}{4}$% per month	2 quarters	?	?
(f)	$4500	0.05% per day	1 quarter	?	?
(g)	$ 300	6% per year	48 months	?	?
(h)	$2000	$1\frac{1}{3}$% per month	30 months	?	?
(i)	$ 250	0.0475% per day	3 months	?	?
(j)	$1500	12% per year	720 days	?	?
(k)	$ 500	$1\frac{2}{3}$% per month	120 days	?	?
(l)	$5000	0.0525% per day	180 days	?	?

PROBLEM 11-20   Find the total interest and total amount for each compound interest problem:

	Principal P	Interest Rate r	Time Period t	Total Interest	Total Amount
(a)	$ 100	5% compounded annually	2 years	?	?
(b)	$ 200	6% compounded semi-annually	1 year	?	?
(c)	$1000	8% compounded quarterly	$\frac{3}{4}$ year	?	?
(d)	$2000	10% compounded annually	8 quarters	?	?
(e)	$ 300	4% compounded semi-annually	4 quarters	?	?
(f)	$3000	12% compounded quarterly	3 quarters	?	?
(g)	$ 125	7% compounded annually	24 months	?	?

	Principal P	Interest Rate r	Time Period t	Total Interest	Total Amount
(h)	$ 975.50	9% compounded semi-annually	12 months	?	?
(i)	$ 250	$5\frac{1}{2}\%$ compounded monthly	4 months	?	?
(j)	$2125.95	$6\frac{1}{4}\%$ compounded semi-annually	360 days	?	?
(k)	$ 526.49	$5\frac{3}{4}\%$ compounded quarterly	180 days	?	?
(l)	$5125	$4\frac{1}{8}\%$ compounded daily	4 days	?	?

Answers to Supplementary Exercises

(11-15)   **(a)** $2.36, $61.36   **(b)** $7.25, $152.25   **(c)** $22.75, $372.74   **(d)** 4.25% or $4\frac{1}{4}\%$, $20.85
(e) 8%, $70.74   **(f)** 3%, $214.24   **(g)** $50, $53.50   **(h)** $26, $28.08   **(i)** $52, $54.21

(11-16)   **(a)** $17, $68   **(b)** $10.60, $95.40   **(c)** $2, $5.99   **(d)** 30%, $35   **(e)** $33\frac{1}{3}\%$ or $\frac{1}{3}$, $166
(f) 20%, $88   **(g)** $700, $595   **(h)** $200, $175   **(i)** $12.75, $4.25

(11-17)   **(a)** $2, $77   **(b)** $1.50, $11.50   **(c)** $10, $110   **(d)** 8%, $135   **(e)** 10%, $250
(f) 15%, $1225   **(g)** $60, $33.60   **(h)** $250, $250   **(i)** $800, $700

(11-18)   **(a)** 20 q, 60 mo, 1800 d   **(b)** 18 q, 54 mo, 1620 d   **(c)** 16 q, 48 mo, 1440 d
(d) 13 q, 39 mo, 1170 d   **(e)** 12 q, 36 mo, 1080 d   **(f)** 11 q, 33 mo, 990 d
(g) 8 q, 24 mo, 720 d   **(h)** 6 q, 18 mo, 540 d   **(i)** 4 q, 12 mo, 360 d
(j) 2 q, 6 mo, 180 d   **(k)** $1\frac{1}{2}$ yr, 18 mo, 540 d   **(l)** $2\frac{1}{2}$ yr, 30 mo, 900 d
(m) $\frac{3}{4}$ yr, 9 mo, 270 d   **(n)** 4 yr, 16 q, 1440 d   **(o)** $2\frac{1}{2}$ yr, 10 q, 900 d
(p) $\frac{3}{4}$ yr, 3 q, 270 d   **(q)** 1 yr, 4 q, 12 mo   **(r)** $2\frac{3}{4}$ yr, 11 q, 33 mo
(s) $4\frac{1}{4}$ yr, 17 q, 51 mo   **(t)** $\frac{1}{4}$ yr, 1 q, 3 mo

(11-19)   **(a)** $50, $150   **(b)** $90, $290   **(c)** $12.60, $362.60   **(d)** $150, $1150
(e) $225, $3225   **(f)** $202.50, $4702.50   **(g)** $72, $372   **(h)** $800, $2800
(i) $10.69, $260.69   **(j)** $360, $1860   **(k)** $33.33, $533.33   **(l)** $472.50, $5472.50

(11-20)   **(a)** $10.25, $110.25   **(b)** $12.18, $212.18   **(c)** $61.21, $1061.21   **(d)** $420, $2420
(e) $12.12, $312.12   **(f)** $278.18, $3278.18   **(g)** $18.11, $143.11   **(h)** $89.77, $1065.27
(i) $4.61, $254.61   **(j)** $134.95, $2260.90   **(k)** $15.25, $541.74   **(l)** $2.36, $5127.36

12 MEASUREMENT

THIS CHAPTER IS ABOUT

☑ **Identifying U.S. Customary Measures**
☑ **Renaming U.S. Customary Measures**
☑ **Identifying Metric Measures**
☑ **Renaming Metric Measures**
☑ **Computing with Measures**

12-1. Identifying U.S. Customary Measures

A. Introduction to U.S. Customary measures.

Every **measure** has two parts, an **amount** and a **unit of measure.**

EXAMPLE 12-1: Identify the amount and unit of measure in 5 feet.

Solution: In 5 feet, 5 is the amount and
feet is the unit of measure.

There are two different measurement systems used in the world today: the **United States (U.S.) Customary measurement system** and the **metric measurement system.** At the present time, the United States is the only industrial nation in the world that has not completely converted to the metric system of measures.

The basic U.S. Customary measures are **length, capacity** (liquid volume), **weight, area,** and **volume** (space volume).

B. Identify U.S. Customary length measures.

The common units of measure for U.S. Customary length measures are the **inch** (in.), **foot** (ft), **yard** (yd), and **mile** (mi).

Note: The only common U.S. Customary unit of measure that ends in a period is the inch (in.).

EXAMPLE 12-2: Describe the distance represented by **(a)** 1 inch   **(b)** 1 foot   **(c)** 1 yard
(d) 1 mile.

Solution
(a) For the average person, the distance between the first and second knuckle on the index finger is about 1 inch (in.).
(b) For the average person, the distance between the elbow and the wrist is about 1 foot (1 ft).
(c) The width of an average house door is about 1 yard (1 yd).
(d) The length of 10 average city blocks is about 1 mile (1 mi).

The previous descriptions of 1 inch, 1 foot, 1 yard, and 1 mile can help you to identify the correct U.S. Customary length measure.

EXAMPLE 12-3: Which of the following is the approximate height of an average house door? **(a)** 7 in. **(b)** 7 ft **(c)** 7 yd **(d)** 7 mi

Solution
(a) Wrong. 7 in. (or 7 inches) is about the length of a new pencil.
(b) Correct! 7 ft (or 7 feet) is about the height of an average door.
(c) Wrong. 7 yd (or 7 yards) is about the height of an average house.
(d) Wrong. 7 mi (or 7 miles) is about the height that a jet airliner cruises above land.

C. Identify U.S. Customary capacity measures.

The common units of measure for U.S. Customary capacity measures are the **teaspoon** (tsp), **tablespoon** (tbsp), **fluid ounce** (fl oz), **cup** (c), **pint** (pt), **quart** (qt), and **gallon** (gal).

EXAMPLE 12-4: Describe the amount of liquid represented by **(a)** 1 teaspoon **(b)** 1 tablespoon **(c)** 1 fluid ounce **(d)** 1 cup **(e)** 1 pint **(f)** 1 quart **(g)** 1 gallon.

Solution
(a) A household teaspoon measures 1 teaspoon (1 tsp) of liquid.
(b) A household tablespoon measures 1 tablespoon (1 tbsp) of liquid.
(c) Two household tablespoons contain about 1 fluid ounce (1 fl oz) of liquid.
(d) An average coffee cup contains about 1 cup (1 c) of liquid.
(e) A large glass of milk contains about 1 pint (1 pt) of liquid.
(f) An average carton of milk contains about 1 quart (1 qt) of liquid.
(g) An average can of house paint contains about 1 gallon (1 gal) of liquid.

The previous descriptions of 1 teaspoon, 1 tablespoon, 1 fluid ounce, 1 cup, 1 pint, 1 quart, and 1 gallon can help you to identify the correct U.S. Customary capacity measure.

EXAMPLE 12-5: Which of the following is the approximate amount of liquid that an average bathtub might contain? **(a)** 40 tsp **(b)** 40 c **(c)** 40 gal

Solution
(a) Wrong. 40 tsp (or 40 teaspoons) is about the amount of liquid that an average teacup might contain.
(b) Wrong. 40 c (or 40 cups) is about the amount of liquid that an average bathroom sink might contain.
(c) Correct! 40 gal (or 40 gallons) is about the amount of liquid that an average bathtub might contain.

D. Identify U.S. Customary weight measures.

The common units of measure for U.S. Customary weight measures are the **ounce** (oz), **pound** (lb), and **ton** (T).

Note: In the U.S. Customary system, there is a **long ton** measure and a **short ton** measure. In this text, the use of *ton* will always refer to the short ton measure.

EXAMPLE 12-6: Describe the weight represented by **(a)** 1 ounce **(b)** 1 pound **(c)** 1 ton.

Solution
(a) Nine U.S. pennies weigh about 1 ounce (1 oz).
(b) An average can of soda pop weighs about 1 pound (1 lb).
(c) A small compact car weighs about 1 ton (1 T).

The previous descriptions of 1 ounce, 1 pound, and 1 ton can help you to identify the correct U.S. Customary weight measure.

EXAMPLE 12-7: Which of the following is the approximate weight of an average cat? **(a)** 5 oz **(b)** 5 lb **(c)** 5 T

Solution
(a) Wrong. 5 oz (or 5 ounces) is about the weight of a hamburger patty.
(b) Correct! 5 lb (or 5 pounds) is about the weight of an average cat.
(c) Wrong. 5 T (or 5 tons) is about the weight of an African elephant.

E. **Identify U.S. Customary area measures.**

The common units of measure for U.S. Customary area measures are the **square inch** (sq in. or in.^{2}), **square foot** (sq ft or ft^2), **square yard** (sq yd or yd^2), and **acre** (A).

EXAMPLE 12-8: Describe the area represented by (a) 1 square inch   (b) 1 square foot
(c) 1 square yard   (d) 1 acre.

Solution
(a) The surface of an average U.S. postage stamp is about 1 square inch (1 in.^{2}).
(b) The surface of a 33⅓ record album is about 1 square foot (1 ft^2).
(c) The top (or bottom) half of an average house door surface is about 1 square yard (1 yd^2).
(d) The combined area of two ice hockey rinks is a little smaller than 1 acre (1 A).

The previous descriptions of 1 square inch, 1 square foot, 1 square yard, and 1 acre can help you to identify the correct U.S. Customary area measure.

EXAMPLE 12-9: Which of the following is the approximate area of an average kitchen floor?
(a) 72 in.^{2} (b) 72 ft^2 (c) 72 yd^2 (d) 72 A

Solution
(a) Wrong. 72 in.^{2} (or 72 square inches) is about one half the surface of a 33⅓ record album.
(b) Correct! 72 ft^2 (or 72 square feet) is about the area of an average kitchen floor.
(c) Wrong. 72 yd^2 (or 72 square yards) is larger than the surface of a U.S. basketball court.
(d) Wrong. 72 A (or 72 acres) is about the area of a small farm.

F. **Identify U.S. Customary volume measures.**

The common units of measure for U.S. Customary volume measures are the **cubic inch** (cu in. or in.^{3}), **cubic foot** (cu ft or ft^3), and **cubic yard** (cu yd or yd^3).

EXAMPLE 12-10: Describe the volume represented by (a) 1 cubic inch   (b) 1 cubic foot
(c) 1 cubic yard.

Solution
(a) The space occupied by a small pillbox measuring 1 inch on each side is 1 cubic inch (1 in.^{3}).
(b) The space occupied by a small portable TV is about 1 cubic foot (1 ft^3).
(c) The space occupied by a standard floor model TV is 1 cubic yard (1 yd^3).

The previous description of 1 cubic inch, 1 cubic foot, and 1 cubic yard can help you to identify the correct U.S. Customary volume measure.

EXAMPLE 12-11: Which of the following is the approximate space occupied by a standard four-drawer file cabinet? (a) 14 in.^{3} (b) 14 ft^3 (c) 14 yd^3

Solution
(a) Wrong. 14 in.^{3} (or 14 cubic inches) is about the amount of space occupied by a standard soup can.
(b) Correct! 14 ft^3 (or 14 cubic feet) is about the amount of space occupied by a standard four-drawer file cabinet.
(c) Wrong. 14 yd^3 (or 14 cubic yards) is about the amount of space inside a small kitchen.

12-2. Renaming U.S. Customary Measures

A. **Introduction to U.S. Customary unit conversions and unit fractions.**

A **number sentence** relating two different units of measure in which one of the amounts is 1 is called a **unit conversion**.

EXAMPLE 12-12: Which of the following number sentences are unit conversions? **(a)** 1 ft = 12 in. **(b)** 2 c = 1 pt   **(c)** $1\frac{1}{2}$ lb = 24 oz

Solution
(a) The number sentence 1 ft = 12 in. is a unit conversion because it relates two diffferent units of measure (feet and inches) and one of the amounts is 1 (1 ft).
(b) The number sentence 2 c = 1 pt is a unit conversion because it relates two different units of measure (cups and pints) and one of the amounts is 1 (1 pt).
(c) The number sentence $1\frac{1}{2}$ lb = 24 oz is not a unit conversion even though it relates two different units of measure (pounds and ounces) because neither one of the amounts is 1 ($1\frac{1}{2}$ lb and 24 oz).

The following are the common **U.S. Customary unit conversions:**

Length
12 in. = 1 ft
3 ft = 1 yd
5280 ft = 1 mi

Capacity	
3 tsp = 1 tbsp	2 c = 1 pt
2 tbsp = 1 fl oz	2 pt = 1 qt
8 fl oz = 1 c	4 qt = 1 gal

Weight	Area	Volume
16 oz = 1 lb	144 in.^{2} = 1 ft^2	1728 in.^{3} = 1 ft^3
2000 lb = 1 T	9 ft^2 = 1 yd^2	27 ft^3 = 1 yd^3

A **unit fraction** is formed when one measure in a unit conversion is used for the numerator and the other measure is used for the denominator.

EXAMPLE 12-13: Write two different unit fractions using the following unit conversion: 1 yd = 3 ft

Solution: $\dfrac{1\text{ yd}}{3\text{ ft}}$

$\dfrac{3\text{ ft}}{1\text{ yd}}$ $\Big\}$ the two unit fractions associated with 1 yd = 3 ft

Note: Every unit conversion has exactly two different unit fractions associated with it.

Since the numerator and denominator of a unit fraction are equal, every unit fraction is equal to 1.

EXAMPLE 12-14: Write two different number sentences relating the number 1 and the unit fractions associated with the following unit conversion: 1 yd = 3 ft

Solution: $1 = \dfrac{1\text{ yd}}{3\text{ ft}}$

$1 = \dfrac{3\text{ ft}}{1\text{ yd}}$ $\Big\}$ the two number sentences relating 1 and the unit fractions associated with 1 yd = 3 ft

Note: Number sentences like those in Example 12-14 are used to rename a given measure in terms of a smaller or larger unit of measure.

B. Rename a U.S. Customary measure using a unit conversion.
To rename a given measure in terms of a different unit of measure when there is a unit conversion relating the two units of measure, you multiply the given measure by the correct unit fraction.

EXAMPLE 12-15: Rename 42 in. as feet.

Solution: Rename inches as feet using the following unit conversion: 12 in. = 1 ft

42 in. = 42 in. × 1   Multiply by 1.

Replace 1 with the correct unit fraction.

$$= 42 \text{ in.} \times \frac{1 \text{ ft}}{12 \text{ in.}} \longleftarrow 12 \text{ in.} = 1 \text{ ft means } 1 = \frac{1 \text{ ft}}{12 \text{ in.}}$$

$$= \frac{42 \text{ in.}}{1} \times \frac{1 \text{ ft}}{12 \text{ in.}}$$ Eliminate the common unit of measure. (See the following *Note* and *Caution*.)

$$= \frac{42}{12} \text{ ft}$$ Multiply fractions.

$$= \frac{7}{2} \text{ ft}$$ Simplify.

$$= 3\tfrac{1}{2} \text{ ft}$$

Note: In Example 12-15, the unit fraction $\dfrac{1 \text{ ft}}{12 \text{ in.}}$ was used instead of $\dfrac{12 \text{ in.}}{1 \text{ ft}}$ so that the common unit of measure (inches) could be eliminated in the next step.

Caution: To eliminate a common unit of measure, one of the common units must be in the numerator and the other common unit must be in the denominator.

EXAMPLE 12-16: Can the common unit of measure be eliminated in either of the following?

(a) $30 \text{ ft} \times \dfrac{3 \text{ ft}}{1 \text{ yd}}$ (b) $30 \text{ ft} \times \dfrac{1 \text{ yd}}{3 \text{ ft}}$

Solution

(a) The common unit of measure (feet) cannot be eliminated in $\dfrac{30 \text{ ft}}{1} \times \dfrac{3 \text{ ft}}{1 \text{ yd}}$ because neither denominator contains feet as a unit of measure.

(b) The common unit of measure (feet) can be eliminated in $\dfrac{30 \text{ ft}}{1} \times \dfrac{1 \text{ yd}}{3 \text{ ft}}$ because feet appears in both a numerator and a denominator:

$$\frac{30 \text{ ft}}{1} \times \frac{1 \text{ yd}}{3 \text{ ft}} = \frac{30}{3} \text{ yd} = 10 \text{ yd}$$

C. Rename U.S. Customary measures when more than one unit conversion is necessary.

To rename a measure when more than one unit conversion is necessary, you first list each necessary unit conversion in the order needed.

EXAMPLE 12-17: Rename $\tfrac{3}{4}$ gal as cups.

Solution: Rename gallons as cups using the following unit conversions:

gallons to quarts	quarts to pints	pints to cups
1 gal = 4 qt	1 qt = 2 pt	1 pt = 2 c

$$\frac{3}{4} \text{ gal} = \frac{3}{4} \text{ gal} \times \; 1 \; \times \; 1 \; \times \; 1$$ Multiply by 1 for each necessary unit conversion.

$$= \frac{3 \text{ gal}}{4} \times \frac{4 \text{ qt}}{1 \text{ gal}} \times \frac{2 \text{ pt}}{1 \text{ qt}} \times \frac{2 \text{ c}}{1 \text{ pt}}$$ Substitute the correct unit fractions for 1.

$$= \frac{3 \text{ gal}}{4} \times \frac{4 \text{ qt}}{1 \text{ gal}} \times \frac{2 \text{ pt}}{1 \text{ qt}} \times \frac{2 \text{ c}}{1 \text{ pt}}$$ Eliminate all common units of measure. [See the following *Note*.]

$$= \frac{3 \times 4 \times 2 \times 2}{4} \, c \qquad \text{Simplify.}$$

$$= \frac{48}{4} \, c$$

$$= 12 \, c \longleftarrow \frac{3}{4} \text{ gallon contains 12 cups}$$

Note: In Example 12-17, one unit fraction is substituted for each necessary unit conversion in one step so that the common units of measure can all be eliminated in the next step.

D. Rename U.S. Customary mixed measures.

When two or more different units of measure are used to write one single measure, that measure is called a **mixed measure**.

EXAMPLE 12-18: Rename the following units of measure as a mixed measure: 1 T + 800 lb

Solution:

$$1 \text{ T} + 800 \text{ lb} = \overbrace{1 \text{ T } 800 \text{ lb}}^{\text{mixed measure}} \qquad \textit{Think:} \text{ 1 T 800 lb means 1 T } + 800 \text{ lb}$$

To rename a mixed measure, you first write the mixed measure as a sum.

EXAMPLE 12-19: Rename 2 T 800 lb as pounds.

Solution: Rename tons as pounds using the unit conversion 2000 lb = 1 T.

2 T 800 lb = 2 T + 800 lb	Rename as a sum.
$= (2 \text{ T} \times 1) + 800 \text{ lb}$	Multiply 2 T by 1 for the necessary unit conversion.
$= \left(\frac{2 \text{ T}}{1} \times \frac{2000 \text{ lb}}{1 \text{ T}} \right) + 800 \text{ lb}$	Substitute the correct unit fraction. $2000 \text{ lb} = 1 \text{ T means } 1 = \frac{2000 \text{ lb}}{1 \text{ T}}$
$= (2 \times 2000) \text{ lb} + 800 \text{ lb}$	Eliminate the common unit of measure.
$= 4000 \text{ lb} + 800 \text{ lb}$	Multiply.
$= 4800 \text{ lb}$	Add pounds: 4000 + 800 = 4800

Note: 2 T 800 lb = 4800 lb or 2.4 T

The following table will help you identify metric units and their abbreviations.

Metric Units of Measure

Place Values	Thousands	Hundreds	Tens	Ones	Tenths	Hundredth	Thousandths
Digit Values	1000	100	10	1	0.1	0.01	0.001
U.S. Money Values	$1000 bill	$100 bill	$10 bill	dollar	dime	cent	mill
Metric Length Units	kilometer km	hectometer hm	dekameter dam	meter m	decimeter dm	centimeter cm	millimeter mm
Metric Capacity Units	kiloliter kL	hectoliter hL	dekaliter daL	liter L	deciliter dL	centiliter cL	milliliter mL
Metric Mass Units	kilogram kg	hectogram hg	dekagram dag	gram g	decigram dg	centigram cg	milligram mg

To rename U.S. Customary measures as metric measures, you can use the unit conversions in Appendix Table 1.

12-3. Identifying Metric Measures

A. Introduction to metric measures.

The metric measurement system has been used in Europe for about 200 years. Presently, the United States is slowly trying to convert to the metric system. If you take the time and effort to learn the metric system now, you will have a well-planned head start when metric conversion becomes mandatory in the United States.

The basic metric measures are **length, capacity** (liquid volume), **mass*** (weight), **area,** and **volume** (space volume).

* There is a scientific difference between mass and weight. But in everyday situations and problems you can think of mass and weight as being the same thing.

The basic metric units of measure for length, capacity, and mass (weight) are **meter** (m), **liter** (L), and **gram** (g), respectively.

Note: The symbol ℓ (a cursive "el") is sometimes used to represent the word "liter" instead of the symbol L. In this text, however, the symbol L will always be used for "liter."

To name other metric units of measure, you attach the correct prefix to the correct basic unit.

EXAMPLE 12-20: Identify the meaning and symbol of the following prefixes: **(a)** kilo- **(b)** hecto- **(c)** deka- **(d)** deci- **(e)** centi- **(f)** milli-

Solution:

	Prefix	Symbol	Meaning
(a)	kilo-	k	1000 (one thousand)
(b)	hecto-	h	100 (one hundred)
(c)	deka-	da	10 (ten)
(d)	deci-	d	$\frac{1}{10}$ or 0.1 (one tenth)
(e)	centi-	c	$\frac{1}{100}$ or 0.01 (one hundredth)
(f)	milli-	m	$\frac{1}{1000}$ or 0.001 (one thousandth)

Note: In the previous chart, each metric prefix is 10 times greater than the prefix directly below it.

B. Identify metric length measures.

The common units of measure for metric length measures are the **millimeter** (mm), **centimeter** (cm), **meter** (m), and **kilometer** (km).

EXAMPLE 12-21: Describe the distance represented by **(a)** 1 millimeter **(b)** 1 centimeter **(c)** 1 meter **(d)** 1 kilometer.

Solution
(a) The thickness of a U.S. dime is about 1 millimeter (1 mm).
(b) For the average person, the thickness at the end of the little finger is about 1 centimeter (1 cm).
(c) The width of an average house door is about 1 meter (1 m).
(d) The length of 6 average city blocks is about 1 kilometer (1 km).

The previous descriptions of 1 millimeter, 1 centimeter, 1 meter, and 1 kilometer can help you to identify the correct metric length measure.

EXAMPLE 12-22: Which of the following is the approximate length of a standard paper clip?
(a) 3 mm **(b)** 3 cm **(c)** 3 m **(d)** 3 km

Solution
(a) Wrong. 3 mm (or 3 millimeters) is about the thickness of a standard paper clip.
(b) Correct! 3 cm (or 3 centimeters) is about the length of a standard paper clip.
(c) Wrong. 3 m (or 3 meters) is about the length of a compact car.
(d) Wrong. 3 km (or 3 kilometers) is about the height of a large mountain.

C. **Identify metric capacity measures.**

The common units of measure for metric capacity measures are the **milliliter** (mL), **liter** (L), and **kiloliter** (kL).

EXAMPLE 12-23: Describe the amount of liquid represented by (**a**) 1 milliliter   (**b**) 1 liter (**c**) 1 kiloliter.

Solution
(**a**) It takes about 15 drops of water to make 1 milliliter (1 mL) of liquid.
(**b**) An average carton of milk (quart-size) contains about 1 liter (1 L) of liquid.
(**c**) A shallow children's wading pool contains about 1 kiloliter (1 kL) of liquid.

The previous descriptions of 1 milliliter, 1 liter, and 1 kiloliter can help you to identify the correct metric capacity measure.

EXAMPLE 12-24: Which of the following is the approximate amount of liquid that an average water bucket might contain? (**a**) 5 mL   (**b**) 5 L   (**c**) 5 kL

Solution
(**a**) Wrong. 5 mL (or 5 milliliters) is about the amount of liquid that an average household teaspoon might contain.
(**b**) Correct! 5 L (or 5 liters) is about the amount of liquid that an average water bucket might hold.
(**c**) Wrong. 5 kL (or 5 kiloliters) is about the amount of liquid that a small water tank might contain.

D. **Identify metric mass (weight) measures.**

The common units of measure for metric mass (weight) measures are the **milligram** (mg), **gram** (g), **kilogram** (kg), and **tonne** (t).

Note: A tonne is sometimes referred to as a **metric ton.** In this text, the term and abbreviation tonne (t) will always be used.

EXAMPLE 12-25: Describe the mass (weight) represented by (**a**) 1 milligram   (**b**) 1 gram (**c**) 1 kilogram   (**d**) 1 tonne.

Solution
(**a**) A long human hair has a mass (weight) of about 1 milligram (1 mg).
(**b**) Two standard paper clips have a mass (weight) of about 1 gram (1 g).
(**c**) A pair of men's shoes have a mass (weight) of about 1 kilogram (1 kg).
(**d**) A large horse has a mass (weight) of about 1 tonne (1 t).

The previous descriptions of 1 milligram, 1 gram, 1 kilogram, and 1 tonne can help you to identify the correct metric mass (weight) measure.

EXAMPLE 12-26: Which of the following is the approximate weight of a U.S. nickel (5¢ piece)?
(**a**) 5 mg   (**b**) 5 g   (**c**) 5 kg   (**d**) 5 t
Solution
(**a**) Wrong. 5 mg (or 5 milligrams) is about the mass (weight) of a tiny bird feather.
(**b**) Correct! 5 g (or 5 grams) is about the mass (weight) of a U.S. nickel.
(**c**) Wrong. 5 kg (or 5 kilograms) is about the mass (weight) of an average bowling ball.
(**d**) Wrong. 5 t (or 5 tonnes) is about the mass (weight) of an African elephant.

E. **Identify metric area measures.**

The common units of measure for metric area measures are the **square centimeter** (cm^2), **square meter** (m^2), and **hectare** (ha).

EXAMPLE 12-27: Describe the area represented by (**a**) 1 square centimeter   (**b**) 1 square meter (**c**) 1 hectare.

Solution

(a) Any one of the six square surfaces of a standard sugar cube measures about 1 square centimeter (1 cm^2).

(b) The top (or bottom) half of a standard house door measures about 1 square meter (1 m^2).

(c) The combined area of two U.S. football fields is a little larger than 1 hectare (1 ha).

The previous descriptions of 1 square centimeter, 1 square meter, and 1 hectare can help you to identify the correct metric area measure.

EXAMPLE 12-28: Which of the following is the approximate area of a standard garage door?
(a) 8 cm^2 (b) 8 m^2 (c) 8 ha

Solution

(a) Wrong. 8 cm^2 (or 8 square centimeters) is about the area of a postage stamp.

(b) Correct! 8 m^2 (or 8 square meters) is about the area of a standard garage door.

(c) Wrong. 8 ha (or 8 hectares) is about the area of a city park.

F. Identify metric volume measures.

The common units of measure for metric volume measures are the **cubic centimeter** (cm^3 or cc) and **cubic meter** (m^3).

EXAMPLE 12-29: Describe the volume represented by (a) 1 cubic centimeter (b) 1 cubic meter.

Solution

(a) The space occupied by a standard sugar cube is about 1 cubic centimeter (1 cm^3 or cc).

(b) The space occupied by a standard kitchen stove is about 1 cubic meter (1 m^3).

The previous descriptions of 1 cubic centimeter and 1 cubic meter can help you to identify the correct metric volume measure.

EXAMPLE 12-30: Which of the following is the approximate space enclosed by an average living-room? (a) 25 cm^3 (b) 25 m^3

Solution

(a) Wrong. 25 cm^3 (or 25 cubic centimeters) is about the space occupied by a standard finger-ring box.

(b) Correct! 25 m^3 (or 25 cubic meters) is about the space enclosed by an average living room.

12-4. Renaming Metric Measures

A. Introduction to metric unit conversion and unit fractions.

You should recall from Section 12-2 that a number sentence relating two different units of measure in which one of the amounts is 1 is called a unit conversion.

The following are the common **metric unit conversions:**

Length	Capacity	Mass (Weight)
10 mm = 1 cm	1000 mL = 1 L	1000 mg = 1 g
100 cm = 1 m	1000 L = 1 kL	1000 g = 1 kg
1000 m = 1 km		1000 kg = 1 t

Area	Volume
100 mm^2 = 1 cm^2	10,000 mm^3 = 1 cm^3
10,000 cm^2 = 1 m^2	100,000,000 cm^3 = 1 m^3

You should also recall that a unit fraction is formed when one measure in a unit conversion is used for the numerator and the other measure in the unit conversion is used for the denominator. Since the numerator and denominator are equal, every unit fraction is equal to 1. Every unit conversion has two different unit fractions associated with it.

B. **Rename a metric measure using a unit conversion.**

Recall from Example 12-15 that to rename a given measure as a different unit of measure when there is a unit conversion relating the two units of measure, you multiply the given measure by the correct unit fraction.

EXAMPLE 12-31: Rename 2.5 m as centimeters.

Solution: Rename meters as centimeters using the following unit conversion: 100 cm = 1 m

$$2.5 \text{ m} = 2.5 \text{ m} \times 1 \qquad \text{Multiply by 1.}$$

Replace 1 with the correct unit fraction.

$$= 2.5 \text{ m} \times \frac{100 \text{ cm}}{1 \text{ m}} \longleftarrow 100 \text{ cm} = 1 \text{ m means } 1 = \frac{100 \text{ cm}}{1 \text{ m}}$$

$$= \frac{2.5 \text{ m}}{1} \times \frac{100 \text{ cm}}{1 \text{ m}} \qquad \text{Eliminate the common unit of measure.}$$

$$= \left(\frac{2.5 \times 100}{1}\right) \text{cm} \qquad \text{Multiply fractions.}$$

$$= 250 \text{ cm} \qquad \text{Simplify.}$$

Note: In Example 12-31, the unit fraction $\frac{100 \text{ cm}}{1 \text{ m}}$ was used instead of $\frac{1 \text{ m}}{100 \text{ cm}}$ so that the common unit of measure (meters) could be eliminated in the next step.

C. **Rename metric measures when more than one unit conversion is necessary.**

To rename a measure when more than one unit conversion is necessary, you first list each necessary unit conversion in the order needed.

EXAMPLE 12-32: Rename 185,000 mL as kiloliters.

Solution: Rename milliliters as kiloliters using the following unit conversions:

milliliters to liters	liters to kiloliters
1000 mL = 1 L	1000 L = 1 kL

$$185{,}000 \text{ mL} = 185{,}000 \text{ mL} \times 1 \times 1 \qquad \begin{array}{l}\text{Multiply by 1 for each necessary unit}\\ \text{conversion.}\end{array}$$

$$= \frac{185{,}000 \text{ mL}}{1} \times \frac{1 \text{ L}}{1000 \text{ mL}} \times \frac{1 \text{ kL}}{1000 \text{ L}} \qquad \text{Substitute the correct unit fractions for 1.}$$

$$= \frac{185{,}000 \text{ mL}}{1} \times \frac{1 \text{ L}}{1000 \text{ mL}} \times \frac{1 \text{ kL}}{1000 \text{ L}} \qquad \text{Eliminate all common units of measure.}$$

$$= \frac{185{,}000}{1000 \times 1000} \text{ kL} \qquad \text{Simplify.}$$

$$= \frac{185{,}000}{1{,}000{,}000} \text{ kL}$$

$$= \frac{185}{1000} \text{ kL}$$

$$= 0.185 \text{ kL} \longleftarrow 185{,}000 \text{ mL is contained in 0.185 kiloliters}$$

Note: In Example 12-32, one unit fraction is substituted for each necessary unit conversion in one step so that the common units of measure can all be eliminated in the next step.

To rename metric measures as U.S. Customary measures, you can use the unit conversions in Appendix Table 1.

12-5. Computing with Measures

Measures that have the same unit of measure are called **like measures.** Measures that have different units of measure are called **unlike measures.**

EXAMPLE 12-33: Which of the following are like measures? **(a)** 5 m, 5 cm   **(b)** 6 ft, 2 ft

Solution
(a) 5 m and 5 cm are unlike measures because the units of measure (m and cm) are different.
(b) 6 ft and 2 ft are like measures because the unit of measure (ft) is the same.

A. Add measures.

To **add like measures,** you add the amounts and then write the same unit of measure on the sum.

EXAMPLE 12-34: Add: 35.8 g + 26.75 g

Solution: 35.8 g + 26.75 g = (35.8 + 26.75) g   Add the amounts.

$\qquad\qquad\qquad\qquad$ = 62.55 g   Write the same unit of measure.

To **add mixed measures,** you first line up corresponding like measures in columns, then add the like measures, and then simplify the sum if possible.

EXAMPLE 12-35: Add: 3 yd 2 ft 8 in. + 5 yd 9 in.

Solution:

yards	feet	inches	
3 yd	2 ft	8 in.	Line up the measures in columns.
+5 yd		9 in.	
8 yd	2 ft	17 in.	Add like measures.

8 yd   2 ft   17 in. = 8 yd + 2 ft + 17 in.   Simplify.

$\qquad\qquad\qquad$ = 8 yd + 2 ft + 12 in. + 5 in.

$\qquad\qquad\qquad$ = 8 yd + 2 ft + 1 ft + 5 in.   *Think:* 12 in. = 1 ft

$\qquad\qquad\qquad$ = 8 yd + 3 ft + 5 in.

$\qquad\qquad\qquad$ = 8 yd + 1 yd + 5 in.   *Think:* 3 ft = 1 yd

$\qquad\qquad\qquad$ = 9 yd + 5 in.

$\qquad\qquad\qquad$ = 9 yd 5 in. ⟵ simplest form

Note: 3 yd 2 ft 8 in. + 5 yd 9 in. = 9 yd 5 in., or 329 in.

B. Subtract measures.

To **subtract like measures,** you subtract the amounts and then write the same unit of measure on the difference.

EXAMPLE 12-36: Subtract: 500 m − 150 m

Solution: 500 m − 150 m = (500 − 150) m   Subtract the amounts.

$\qquad\qquad\qquad\qquad$ = 350 m   Write the same unit of measure.

To **subtract mixed measures,** you first line up corresponding like measures in columns, and then subtract like measures, renaming when necessary.

EXAMPLE 12-37: Subtract: 5 gal 4 qt − 6 qt 1 pt

Solution: gallons   quarts   pints

Step 1:   5 gal   4 qt

−   6 qt   1 pt

Line up like measures.

Step 2:

4 qts

5 gal   3 qt   2 pt

−   6 qt   1 pt

1 pt

Rename quarts as pints to get more pints.

4 qt = 3 qt + 1 qt

= 3 qt + 2 pt

= 3 qt 2 pt

Subtract pints.

Step 3:   5 gal   3 qt

4 gal   7 qt   2 pt

−   6 qt   1 pt

1 qt   1 pt

Rename gallons as quarts to get more quarts.

5 gal 3 qt = 4 gal + 1 gal + 3 qt

= 4 gal + 4 qt + 3 qt

= 4 gal 7 qt

Subtract quarts.

Step 4:   4 gal   7 qt   2 pt

−   6 qt   1 pt

4 gal   1 qt   1 pt ⟵ simplest form

Subtract gallons.

Note: 5 gal 4 qt − 6 qt 1 pt = 4 gal 1 qt 1 pt

C. **Multiply measures.**

To **multiply a measure by a number,** you multiply the amount by the number and then write the same unit of measure on the product.

EXAMPLE 12-38: Multiply: 5×8 L

Solution: 5×8 L $= (5 \times 8)$ L   Multiply the amount by the number.

= 40 L   Write the same unit of measure on the product.

To **multiply a mixed measure by a number,** you multiply each individual measure within the mixed measure by the number.

EXAMPLE 12-39: Multiply: 6×4 lb 8 oz

Solution: 4 lb   8 oz   Write in vertical form.

×   6

24 lb   48 oz   Multiply: 6×4 lb $= 24$ lb

6×8 oz $= 48$ oz

24 lb   48 oz = 24 lb + 48 oz   Simplify.

= 24 lb + 3 lb ⟵ 16 oz = 1 lb means 48 oz = 3 lb

= 27 lb ⟵ simplest form

Note: 6×4 lb 8 oz = 27 lb, or 432 oz

To **multiply two or more like measures,** you multiply the amounts and multiply the units of measure separately.

Multiplication Rule for Two Like Length Measures
When two like length measures are multiplied, the product will always be an associated area measure (for example, in.^{2}, ft^2, m^2, cm^2, etc.).

EXAMPLE 12-40: Multiply the following like length measures: **(a)** 5 in. × 4 in.   **(b)** 2 ft × 1 ft
(c) 6 yd × 3 yd   **(d)** 1 mm × 8 mm   **(e)** 4 cm × 8 cm   **(f)** 1 m × 1 m

Solution: two like length measures   associated area measure

$$
\begin{aligned}
\textbf{(a)} \ & \overbrace{5 \text{ in.} \times 4 \text{ in.}} && = (5 \times 4)(\text{in.} \times \text{in.}) && = \overbrace{20 \text{ in.}^2} \\
\textbf{(b)} \ & 2 \text{ ft} \times 1 \text{ ft} && = (2 \times 1)(\text{ft} \times \text{ft}) && = 2 \text{ ft}^2 \\
\textbf{(c)} \ & 6 \text{ yd} \times 3 \text{ yd} && = (6 \times 3)(\text{yd} \times \text{yd}) && = 18 \text{ yd}^2 \\
\textbf{(d)} \ & 1 \text{ mm} \times 8 \text{ mm} && = (1 \times 8)(\text{mm} \times \text{mm}) && = 8 \text{ mm}^2 \\
\textbf{(e)} \ & 4 \text{ cm} \times 8 \text{ cm} && = (4 \times 8)(\text{cm} \times \text{cm}) && = 32 \text{ cm}^2 \\
\textbf{(f)} \ & 1 \text{ m} \times 1 \text{ m} && = (1 \times 1)(\text{m} \times \text{m}) && = 1 \text{ m}^2
\end{aligned}
$$

Multiplication Rule for Three Like Length Measures
When three like length measures are multiplied, the product will always be an associated volume measure (for example, in.^{3}, ft^3, m^3, cm^3, etc.).

EXAMPLE 12-41: Multiply the following like length measures: **(a)** 1 in. × 4 in. × 1 in.
(b) 2 ft × 1 ft × 5 ft   **(c)** 3 yd × 2 yd × 4 yd   **(d)** 1 cm × 1 cm × 1 cm
(e) 5 m × 4 m × 2 m

Solution: three like length measures   associated volume measure

$$
\begin{aligned}
\textbf{(a)} \ & \overbrace{1 \text{ in.} \times 4 \text{ in.} \times 1 \text{ in.}} && = (1 \times 4 \times 1)(\text{in.} \times \text{in.} \times \text{in.}) && = \overbrace{4 \text{ in.}^3} \\
\textbf{(b)} \ & 2 \text{ ft} \times 1 \text{ ft} \times 5 \text{ ft} && = (2 \times 1 \times 5)(\text{ft} \times \text{ft} \times \text{ft}) && = 10 \text{ ft}^3 \\
\textbf{(c)} \ & 3 \text{ yd} \times 2 \text{ yd} \times 4 \text{ yd} && = (3 \times 2 \times 4)(\text{yd} \times \text{yd} \times \text{yd}) && = 24 \text{ yd}^3 \\
\textbf{(d)} \ & 1 \text{ cm} \times 1 \text{ cm} \times 1 \text{ cm} && = (1 \times 1 \times 1)(\text{cm} \times \text{cm} \times \text{cm}) && = 1 \text{ cm}^3 \\
\textbf{(e)} \ & 5 \text{ m} \times 4 \text{ m} \times 2 \text{ m} && = (5 \times 4 \times 2)(\text{m} \times \text{m} \times \text{m}) && = 40 \text{ m}^3
\end{aligned}
$$

When two unlike measures are multiplied, you multiply the amounts and the units of measure separately. For example, in a **work problem,** feet (ft) and pounds (lb) are multiplied to get **foot-pounds** (ft-lb).

EXAMPLE 12-42: Multiply two unlike measures: 6 ft × 5 lb

Solution: 6 ft × 5 lb = (6 × 5)(ft × lb)   Multiply the amounts and
                                                              units of measure separately.

$$= 30 \text{ ft} \times \text{lb or } 30 \text{ ft-lb} \longleftarrow \text{ simplest form}$$

Note: 30 ft-lb or 30 ft × lb are both read as "thirty foot-pounds."

D. Divide measures.

To **divide a measure by a number,** you divide the amount by the number and then write the same unit of measure on the quotient.

EXAMPLE 12-43: Divide: 6 km ÷ 2

Solution: 6 km ÷ 2 = (6 ÷ 2) km   Divide the amount by the number.

$$= 3 \text{ km}$$ Write the same unit of measure on the quotient.

To **divide a mixed measure by a number,** you divide each individual measure within the mixed measure by the number while being careful to leave space in the dividend for any missing units of measure that may be needed.

EXAMPLE 12-44: Divide: 9 yd 3 in. ÷ 2

Solution: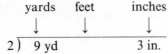

Write in division box form while leaving space for the missing units of measure:
9 yd 3 in. = 9 yd 0 ft 3 in.

$$
\begin{array}{r}
4 \text{ yd} \\
2\overline{)\,9 \text{ yd} \qquad\qquad 3 \text{ in.}} \\
-8 \text{ yd} \longleftarrow \qquad 2 \times 4 \text{ yd} \\
\overline{\rlap{/}{1 \text{ yd}}} \quad 3 \text{ ft}
\end{array}
$$

Divide yards.

Rename yards as feet.

$$
\begin{array}{r}
4 \text{ yd} \quad 1 \text{ ft} \quad\; 7 \text{ in.} \\
2\overline{)\,9 \text{ yd} \qquad\quad\; 3 \text{ in.}} \\
-8 \text{ yd} \\
\overline{\rlap{/}{1 \text{ yd}}} \quad 3 \text{ ft} \\
-2 \text{ ft} \\
\overline{\rlap{/}{1 \text{ ft}}} \longrightarrow 12 \text{ in.} \\
15 \text{ in.} \\
-14 \text{ in.} \longleftarrow 2 \times 7 \text{ in.} \\
\overline{1 \text{ in.}}
\end{array}
$$

Divide feet.

Rename feet as inches.
Divide inches.

Note: 9 yd 3 in. ÷ 2 = 4 yd 1 ft 7 in. R 1 in.

To **divide two like measures,** you divide the amounts and divide the units of measure separately.

Division Rule for Two Like Measures
When two like measures are divided, the quotient will always be a number, with *no* unit of measure.

EXAMPLE 12-45: Divide the following like measures: **(a)** 1 in. ÷ 1 in.   **(b)** 5 lb ÷ 2 lb
(c) 18 pt ÷ 10 pt   **(d)** 1 mm ÷ 2 mm   **(e)** 10 kg ÷ 3 kg   **(f)** 100 L ÷ 1000 L

Solution: two like measures   number

(a) $\overbrace{1 \text{ in.} \div 1 \text{ in.}} = (1 \div 1)\dfrac{\cancel{\text{in.}}}{\cancel{\text{in.}}} = \overbrace{1}$

(b) $5 \text{ lb} \div 2 \text{ lb} = (5 \div 2)\dfrac{\cancel{\text{lb}}}{\cancel{\text{lb}}} = \dfrac{5}{2} \text{ or } 2\dfrac{1}{2} \text{ or } 2.5$

(c) $18 \text{ pt} \div 10 \text{ pt} = (18 \div 10)\dfrac{\cancel{\text{pt}}}{\cancel{\text{pt}}} = \dfrac{9}{5} \text{ or } 1\dfrac{4}{5} \text{ or } 1.8$

(d) $1 \text{ mm} \div 2 \text{ mm} = (1 \div 2)\dfrac{\cancel{\text{mm}}}{\cancel{\text{mm}}} = \dfrac{1}{2} \text{ or } 0.5$

(e) $10 \text{ kg} \div 3 \text{ kg} = (10 \div 3)\dfrac{\cancel{\text{kg}}}{\cancel{\text{kg}}} = \dfrac{10}{3} \text{ or } 3\dfrac{1}{3} \text{ or } 3.\overline{3}$

(f) $100 \text{ L} \div 1000 \text{ L} = (100 \div 1000)\dfrac{\cancel{\text{L}}}{\cancel{\text{L}}} = \dfrac{1}{10} \text{ or } 0.1$

To **divide two unlike measures,** you divide the amounts and the units of measure separately.

Division Rule for Two Unlike Measures
When two unlike measures are divided, the quotient is called a **rate**.

EXAMPLE 12-46: Divide the following unlike measures: 20 lb ÷ 5 ft

Solution: $20 \text{ lb} \div 5 \text{ ft} = (20 \div 5)\dfrac{\text{lb}}{\text{ft}}$ Divide the amounts and units of measure separately.

$$= 4\dfrac{\text{lb}}{\text{ft}} \longleftarrow \text{ rate}$$

Note: The rate $4\frac{\text{lb}}{\text{ft}}$ can also be written as 4 lb/ft, or 4 lb per ft, and all three forms are read as "four pounds per foot."

Caution: The unlike measures in a rate cannot be eliminated.

RAISE YOUR GRADES
Can you . . . ?

☑ identify the amount and unit of measure for a given measure

☑ state the names of the two different measurement systems that are used in the world

☑ list the basic U.S. Customary measures

☑ list the basic metric measures

☑ identify the common units of measure for U.S. Customary **(a)** length measures
  (b) capacity measures   **(c)** weight measures   **(d)** area measures
  (e) volume measures

☑ state the basic metric unit of measure for **(a)** length   **(b)** capacity   **(c)** mass

☑ identify the common units of measure for metric **(a)** length measures
  (b) capacity measures   **(c)** mass measures   **(d)** area measures
  (e) volume measures

☑ describe in words **(a)** 1 inch   **(b)** 1 foot   **(c)** 1 yard   **(d)** 1 mile
  (e) 1 teaspoon   **(f)** 1 tablespoon   **(g)** 1 fluid ounce   **(h)** 1 cup   **(i)** 1 pint
  (j) 1 quart   **(k)** 1 gallon   **(l)** 1 ounce   **(m)** 1 pound   **(n)** 1 ton
  (o) 1 square inch   **(p)** 1 square foot   **(q)** 1 square yard   **(r)** 1 acre
  (s) 1 cubic inch   **(t)** 1 cubic foot   **(u)** 1 cubic yard

☑ describe in words **(a)** 1 millimeter   **(b)** 1 centimeter   **(c)** 1 meter   **(d)** 1 kilometer
  (e) 1 milliliter   **(f)** 1 liter   **(g)** 1 kiloliter   **(h)** 1 milligram   **(i)** 1 gram
  (j) 1 kilogram   **(k)** 1 tonne   **(l)** 1 square centimeter   **(m)** 1 square meter
  (n) 1 hectare   **(o)** 1 cubic centimeter   **(p)** 1 cubic meter

☑ identify the correct U.S. Customary measure of an object from a list of different
  (a) length measures   **(b)** capacity measures   **(c)** weight measures
  (d) area measures   **(e)** volume measures

☑ identify the correct metric measure of an object from a list of different **(a)** length measures
  (b) capacity measures   **(c)** mass measures   **(d)** area measures
  (e) volume measures

☑ identify a unit conversion

☑ write two different unit fractions associated with a given unit conversion

☑ write the common U.S. Customary unit conversions for **(a)** length   **(b)** capacity
  (c) weight   **(d)** area   **(e)** volume

☑ write the common metric unit conversions for **(a)** length   **(b)** capacity   **(c)** mass
  (d) area   **(e)** volume

☑ Rename a measure
  (a) using a unit conversion
  (b) when more than one unit conversion is necessary
  (c) when the measure is a mixed measure

☑ identify like measures and unlike measures

☑ add like measures

☑ add mixed measures

☑ subtract like measures

☑ subtract mixed measures

☑ multiply a mixed measure by a number

☑ multiply two like measures to get an associated area measure
☑ multiply three like measures to get an associated volume measure
☑ multiply two unlike measures
☑ divide a measure by a number
☑ divide a mixed measure by a number
☑ divide like measures to get a number
☑ divide unlike measures to get a rate

SUMMARY

1. Every measure has two parts, an amount and a unit of measure.
2. There are two different measurement systems used in the world today: the U.S. Customary measurement system and the metric measurement system.
3. The basic U.S. Customary measures are length, capacity (liquid volume), weight, area, and volume (space volume).
4. The following table summarizes the common U.S. Customary units of measure and their abbreviations:

U.S. Customary Units of Measure

Length Units

inch(es)	foot (feet)	yard(s)	mile(s)
in.	ft	yd	mi

Capacity Units

teaspoon(s)	tablespoon(s)	fluid ounce(s)	cup(s)	pint(s)	quart(s)	gallon(s)
tsp	tbsp	fl oz	c	pt	qt	gal

Weight Units

ounce(s)	pound(s)	ton(s)
oz	lb	T

Area Units

square inch(es)	square foot (feet)	square yards(s)	acre(s)
sq in.	sq ft or ft^2	sq yd or yd^2	A

Volume Units

cubic inch(es)	cubic foot (feet)	cubic yards(s)
cu. in or $in.^3$	cu ft or ft^3	cu yd or yd^3

5. The basic metric measures are length, capacity, mass (weight), area, and volume.
6. The basic metric unit of measure for length, capacity, and mass are meter, liter, and gram, respectively.
7. To name other metric units of measure, you attach one of the following prefixes to meter (m), liter (L), or gram (g).

Prefix	Symbol	Meaning
kilo-	k	1000 (one thousand)
hecto-	h	100 (one hundred)
deka-	da	10 (ten)
deci-	d	$\frac{1}{10}$ or 0.1 (one tenth)
centi-	c	$\frac{1}{100}$ or 0.01 (one hundredth)
milli-	m	$\frac{1}{1000}$ or 0.001 (one thousandth)

8. The following table can help you understand and remember the basic metric units of measure and their abbreviations.

Metric Units of Measure

Place Values	Thousands	Hundreds	Tens	Ones	Tenths	Hundredth	Thousandths
Digit Values	1000	100	10	1	0.1	0.01	0.001
U.S. Money Values	$1000 bill	$100 bill	$10 bill	dollar	dime	cent	mill
Metric Length Units	kilometer km	hectometer hm	dekameter dam	meter m	decimeter dm	centimeter cm	millimeter mm
Metric Capacity Units	kiloliter kL	hectoliter hL	dekaliter daL	liter L	deciliter dL	centiliter cL	milliliter mL
Metric Mass Units	kilogram kg	hectogram hg	dekagram dag	gram g	decigram dg	centigram cg	milligram mg

9. A number sentence relating two different units of measure in which one of the amounts is 1 is called a unit conversion.

10. The common U.S. Customary unit conversions are:

Length	Capacity	
12 in. = 1 ft	3 tsp = 1 tbsp	2 c = 1 pt
3 ft = 1 yd	2 tbsp = 1 fl oz	2 pt = 1 qt
5280 ft = 1 mi	8 fl oz = 1 cup	4 qt = 1 gal

Weight	Area	Volume
16 oz = 1 lb	$144 \text{ in.}^2 = 1 \text{ ft}^2$	$1728 \text{ in.}^3 = 1 \text{ ft}^3$
2000 lb = 1 T	$9 \text{ ft}^2 = 1 \text{ yd}^2$	$27 \text{ ft}^3 = 1 \text{ yd}^3$

11. The common metric unit conversions are:

Length	Capacity	Mass (weight)
10 mm = 1 cm	1000 mL = 1 L	1000 mg = 1 g
100 cm = 1 m	1000 L = 1 kL	1000 g = 1 kg
1000 m = 1 km		1000 kg = 1 t

Area	Volume
$100 \text{ mm}^2 = 1 \text{ cm}^2$	$10,000 \text{ mm}^3 = 1 \text{ cm}^3$
$10,000 \text{ cm}^2 = 1 \text{ m}^2$	$100,000,000 \text{ cm}^3 = 1 \text{ m}^3$

12. A unit fraction is formed when one measure in a unit conversion is used for the numerator and the other measure is used for the denominator.

13. Every unit conversion has exactly two different unit fractions associated with it.

14. To rename a given measure as a different unit of measure when there is one or more unit conversions relating the two units of measure, you multiply the given measure by the correct unit fraction(s).

15. When two or more different units of measure are used to write one single measure, that measure is called a mixed measure.

16. To rename a mixed measure, you first write the mixed measure as a sum.

17. Measures that have the same units of measure are called like measures.

18. Measures that have different units of measure are called unlike measures.

19. To add like measures, you add the amounts and then write the same unit of measure on the sum.

20. To add mixed measures, you first line up corresponding like measures in columns, then add the like measures, and then simplify the sum if possible.

21. To subtract like measures, you subtract the amounts and then write the same unit of measure on the difference.

22. To subtract mixed measures, you first line up corresponding like measures in columns, and then subtract like measures, renaming when necessary.
23. To multiply a measure by a number, you multiply the amount by the number and then write the same unit of measure on the product.
24. To multiply a mixed measure by a number, you multiply each individual measure within the mixed measure by the number.
25. To multiply two or more like measures, you multiply the amounts and multiply the units of measure separately.
26. When two like length measures are multiplied, the product will always be an associated area measure.
27. When three like length measures are multiplied, the product will always be an associated volume measure.
28. When two unlike measures are multiplied, you multiply the amounts and the units of measure separately.
29. To divide a measure by a number, you divide the amount by the number and then write the same unit of measure on the quotient.
30. To divide a mixed measure by a number, you divide each individual measure within the mixed measure while being careful to leave space in the dividend for any missing units of measure that may be needed.
31. To divide two like measures, you divide the amounts and divide the units of measure separately.
32. When two like measures are divided, the quotient will always be a number.
33. When two unlike measures are divided, the quotient will usually be a rate.

SOLVED PROBLEMS

PROBLEM 12-1 Match each given statement on the left with the most appropriate U.S. Customary measure on the right:

Length

(a)	The length of a garage	3 in.
(b)	The length of a used pencil	3 ft
(c)	The height of a kitchen chair	3 yd
(d)	The distance between two towns	3 mi

Capacity

(e)	The capacity of a standard can of wall paint	1 tsp
(f)	The capacity of a standard can of car oil	1 c
(g)	The capacity of an individual soup serving	1 qt
(h)	The capacity of 80 drops of water	1 gal

Weight

(i)	The weight of a standard bag of sugar	5 oz
(j)	The weight of an African elephant	5 lb
(k)	The weight of a hamburger patty	5 T

Area

(l)	The area of a standard room carpet	90 in.^{2}
(m)	The floor area of a small house	90 ft^2
(n)	The area of a farm	90 yd^2
(o)	The area of the front cover of this text	90 A

Volume

(p)	The space in a standard house room	800 in.^{3}
(q)	The space in a large house	800 ft^3
(r)	The space taken up by a portable radio	800 yd^3

Solution: Recall that to help identify U.S. Customary measures, you can use your intuitive idea of 1 foot, 1 quart, 1 pound, and so forth [see Examples 12-2 through 12-11]:

(a) The length of a garage might be 3 yards.
(b) The length of a used pencil might be 3 inches.
(c) The height of a kitchen chair might be 3 feet.
(d) The distance between towns might be 3 miles.
(e) The capacity of a standard can of wall paint is 1 gallon.
(f) The capacity of a standard can of car oil is 1 quart.
(g) The capacity of an individual soup serving is about 1 cup.
(h) The capacity of 80 drops of water is 1 teaspoon.
(i) The weight of a standard bag of sugar is 5 pounds.
(j) The weight of an African elephant might be 5 tons.
(k) The weight of a hamburger patty might be 5 ounces.
(l) The area of a standard room carpet might be 90 square feet.
(m) The floor area of a small house might be 90 square yards.
(n) The area of a farm might be 90 acres.
(o) The area of the front cover of this text is about 90 square inches.
(p) The space in a standard house room might be 800 cubic feet.
(q) The space in a large house might be 800 cubic yards.
(r) The space taken up by a portable radio might be 800 cubic inches.

PROBLEM 12-2   Rename each U.S. Customary measure using the correct unit conversion:

(a) 2 ft = ? in.   **(b)** 9 ft = ? yd   **(c)** 2 mi = ? ft   **(d)** 2 pt = ? c   **(e)** 5 pt = ? qt
(f) 2 gal = ? qt   **(g)** 2 lb = ? oz   **(h)** 500 lb = ? T   **(i)** $3\frac{1}{2}$ T = ? lb   **(j)** 68 oz = ? lb
(k) 2 ft^2 = ? in.^{2} **(l)** 27 ft^2 = ? yd^2 **(m)** 2 ft^3 = ? in.^{3} **(n)** 63 ft^3 = ? yd^3

Solution: Recall that to rename a given measure in terms of a different unit of measure when there is a unit conversion relating the two units of measure, you multiply the given measure by the correct unit fractions [see Example 12-15]:

(a) $2 \text{ ft} = 2 \text{ ft} \times 1 = \dfrac{2 \text{ ft}}{1} \times \dfrac{12 \text{ in.}}{1 \text{ ft}} = (2 \times 12) \text{ in.} = 24 \text{ in.}$

(b) $9 \text{ ft} = 9 \text{ ft} \times 1 = \dfrac{9 \text{ ft}}{1} \times \dfrac{1 \text{ yd}}{3 \text{ ft}} = \dfrac{9}{3} \text{ yd} = 3 \text{ yd}$

(c) $2 \text{ mi} = 2 \text{ mi} \times 1 = \dfrac{2 \text{ mi}}{1} \times \dfrac{5280 \text{ ft}}{1 \text{ mi}} = (2 \times 5280) \text{ ft} = 10{,}560 \text{ ft}$

(d) $2 \text{ pt} = 2 \text{ pt} \times 1 = \dfrac{2 \text{ pt}}{1} \times \dfrac{2 \text{ c}}{1 \text{ pt}} = (2 \times 2) \text{ c} = 4 \text{ c}$

(e) $5 \text{ pt} = 5 \text{ pt} \times 1 = \dfrac{5 \text{ pt}}{1} \times \dfrac{1 \text{ qt}}{2 \text{ pt}} = \dfrac{5}{2} \text{ qt} = 2\dfrac{1}{2} \text{ qt}$

(f) $2 \text{ gal} = 2 \text{ gal} \times 1 = \dfrac{2 \text{ gal}}{1} \times \dfrac{4 \text{ qt}}{1 \text{ gal}} = (2 \times 4) \text{ qt} = 8 \text{ qt}$

(g) $2 \text{ lb} = 2 \text{ lb} \times 1 = \dfrac{2 \text{ lb}}{1} \times \dfrac{16 \text{ oz}}{1 \text{ lb}} = (2 \times 16) \text{ oz} = 32 \text{ oz}$

(h) $500 \text{ lb} = 500 \text{ lb} \times 1 = \dfrac{500 \text{ lb}}{1} \times \dfrac{1 \text{ T}}{2000 \text{ lb}} = \dfrac{500}{2000} \text{ T} = \dfrac{1}{4} \text{ T}$

(i) $3\dfrac{1}{2} \text{ T} = 3\dfrac{1}{2} \text{ T} \times 1 = \dfrac{3\frac{1}{2} \text{ T}}{1} \times \dfrac{2000 \text{ lb}}{1 \text{ T}} = \left(3\dfrac{1}{2} \times 2000\right) \text{ lb} = 7000 \text{ lb}$

(j) $68 \text{ oz} = 68 \text{ oz} \times 1 = \dfrac{68 \text{ oz}}{1} \times \dfrac{1 \text{ lb}}{16 \text{ oz}} = \dfrac{68}{16} \text{ lb} = \dfrac{17}{4} \text{ lb} = 4\dfrac{1}{4} \text{ lb}$

(k) $2 \text{ ft}^2 = 2 \text{ ft}^2 \times 1 = \dfrac{2 \text{ ft}^2}{1} \times \dfrac{144 \text{ in.}^2}{1 \text{ ft}^2} = (2 \times 144) \text{ in.}^2 = 288 \text{ in.}^2$

(l) $27 \text{ ft}^2 = 27 \text{ ft}^2 \times 1 = \dfrac{27 \text{ ft}^2}{1} \times \dfrac{1 \text{ yd}^2}{9 \text{ ft}^2} = \dfrac{27}{9} \text{ yd}^2 = 3 \text{ yd}^2$

(m) $2 \text{ ft}^3 = 2 \text{ ft}^3 \times 1 = \dfrac{2 \text{ ft}^3}{1} \times \dfrac{1728 \text{ in.}^3}{1 \text{ ft}^3} = (2 \times 1728) \text{ in.}^3 = 3456 \text{ in.}^3$

(n) $63 \text{ ft}^3 = 63 \text{ ft}^3 \times 1 = \dfrac{63 \text{ ft}^3}{1} \times \dfrac{1 \text{ yd}^3}{27 \text{ ft}^3} = \dfrac{63}{27} \text{ yd}^3 = \dfrac{7}{3} \text{ yd}^3 = 2\dfrac{1}{3} \text{ yd}^3$

PROBLEM 12-3 Rename each U.S. Customary measure when more than one unit conversion is necessary:

(a) 1 mi = ? yd.   **(b)** 60 in. = ? yd   **(c)** 1 gal = ? c   **(d)** 160 fl oz = ? qt   **(e)** 1 T = ? oz
(f) 48,000 oz = ? T   **(g)** 1 yd^2 = ? in.^{2} **(h)** 72 in.^{2} = ? yd^2 **(i)** 1 yd^3 = ? in.^{3}
(j) 8640 in.^{3} = ? yd^3

Solution: Recall that to rename a given measure when more than one unit conversion is necessary, you first list each necessary unit conversion in the order needed and then multiply by all the correct unit fractions [see Example 12-17]:

(a) 1 mi = 5280 ft and 3 ft = 1 yd are the necessary unit conversions:

$$1 \text{ mi} = \frac{1 \text{ mi}}{1} \times \frac{5280 \text{ ft}}{1 \text{ mi}} \times \frac{1 \text{ yd}}{3 \text{ ft}} = \frac{5280}{3} \text{ yd} = 1760 \text{ yd}$$

(b) 12 in. = 1 ft and 3 ft = 1 yd are the necessary unit conversions:

$$60 \text{ in.} = \frac{60 \text{ in.}}{1} \times \frac{1 \text{ ft}}{12 \text{ in.}} \times \frac{1 \text{ yd}}{3 \text{ ft}} = \frac{60}{12 \times 3} \text{ yd} = \frac{5}{3} \text{ yd} = 1\frac{2}{3} \text{ yd}$$

(c) 1 gal = 4 qt, 1 qt = 2 pt, and 1 pt = 2 c are the necessary unit conversions:

$$1 \text{ gal} = \frac{1 \text{ gal}}{1} \times \frac{4 \text{ qt}}{1 \text{ gal}} \times \frac{2 \text{ pt}}{1 \text{ qt}} \times \frac{2 \text{ c}}{1 \text{ pt}} = (4 \times 2 \times 2) \text{ c} = 16 \text{ c}$$

(d) 8 fl oz = 1 c, 2 c = 1 pt, and 2 pt = 1 qt are the necessary unit conversions:

$$160 \text{ fl oz} = \frac{160 \text{ fl oz}}{1} \times \frac{1 \text{ c}}{8 \text{ fl oz}} \times \frac{1 \text{ pt}}{2 \text{ c}} \times \frac{1 \text{ qt}}{2 \text{ pt}} = \frac{160}{8 \times 2 \times 2} \text{ qt} = \frac{5}{1} \text{ qt} = 5 \text{ qt}$$

(e) 1 T = 2000 lb and 1 lb = 16 oz are the necessary unit conversions:

$$1 \text{ T} = \frac{1 \text{ T}}{1} \times \frac{2000 \text{ lb}}{1 \text{ T}} \times \frac{16 \text{ oz}}{1 \text{ lb}} = (2000 \times 16) \text{ oz} = 32,000 \text{ oz}$$

(f) 16 oz = 1 lb and 2000 lb = 1 T are the necessary unit conversions:

$$48,000 \text{ oz} = \frac{48,000 \text{ oz}}{1} \times \frac{1 \text{ lb}}{16 \text{ oz}} \times \frac{1 \text{ T}}{2000 \text{ lb}} = \frac{48,000}{16 \times 2000} \text{ T} = \frac{3}{2} \text{ T} = 1\frac{1}{2} \text{ T}$$

(g) 1 yd^2 = 9 ft^2 and 1 ft^2 = 144 in.^{2} are the necessary unit conversions:

$$1 \text{ yd}^2 = \frac{1 \text{ yd}^2}{1} \times \frac{9 \text{ ft}^2}{1 \text{ yd}^2} \times \frac{144 \text{ in.}^2}{1 \text{ ft}^2} = (9 \times 144) \text{ in.}^2 = 1296 \text{ in.}^2$$

(h) 144 in.^{2} = 1 ft^2 and 9 ft^2 = 1 yd^2 are the necessary unit conversions:

$$72 \text{ in.}^2 = \frac{72 \text{ in.}^2}{1} \times \frac{1 \text{ ft}^2}{144 \text{ in.}^2} \times \frac{1 \text{ yd}^2}{9 \text{ ft}^2} = \frac{72}{144 \times 9} \text{ yd}^2 = \frac{1}{18} \text{ yd}^2$$

(i) 1 yd^3 = 27 ft^3 and 1 ft^3 = 1728 in.^{3} are the necessary unit conversions:

$$1 \text{ yd}^3 = \frac{1 \text{ yd}^3}{1} \times \frac{27 \text{ ft}^3}{1 \text{ yd}^3} \times \frac{1728 \text{ in.}^3}{1 \text{ ft}^3} = (27 \times 1728) \text{ in.}^3 = 46,656 \text{ in.}^3$$

(j) 1728 in.^{3} = 1 ft^3 and 27 ft^3 = 1 yd^3 are the necessary unit conversions:

$$8640 \text{ in.}^3 = \frac{8640 \text{ in.}^3}{1} \times \frac{1 \text{ ft}^3}{1728 \text{ in.}^3} \times \frac{1 \text{ yd}^3}{27 \text{ ft}^3} = \frac{8640}{1728 \times 27} \text{ yd}^3 = \frac{5}{27} \text{ yd}^3$$

PROBLEM 12-4   Rename each U.S. Customary mixed measure: **(a)** 3 yd 2 ft = ? ft
(b) 3 yd 2 ft = ? yd   **(c)** 5 qt 1 pt = ? pt   **(d)** 5 qt 1 pt = ? qt   **(e)** 2 T 1500 lb = ? lb
(f) 2 T 1500 lb = ? T   **(g)** 1 yd^2 3 ft^2 = ? ft^2 **(h)** 1 yd^2 3 ft^2 = ? yd^2
(i) 2 yd^3 18 ft^3 = ? ft^3 **(j)** 2 yd^3 18 ft^3 = ? yd^3

Solution: Recall that to rename a mixed measure, you first write the mixed measure as a sum [see Example 12-19]:

(a) $3 \text{ yd } 2 \text{ ft} = 3 \text{ yd} + 2 \text{ ft} = \frac{3 \text{ yd}}{1} \times \frac{3 \text{ ft}}{1 \text{ yd}} + 2 \text{ ft} = 9 \text{ ft} + 2 \text{ ft} = 11 \text{ ft}$

(b) $3 \text{ yd } 2 \text{ ft} = 3 \text{ yd} + 2 \text{ ft} = 3 \text{ yd} + \frac{2 \text{ ft}}{1} \times \frac{1 \text{ yd}}{3 \text{ ft}} = 3 \text{ yd} + \frac{2}{3} \text{ yd} = 3\frac{2}{3} \text{ yd}$

(c) $5 \text{ qt } 1 \text{ pt} = 5 \text{ qt} + 1 \text{ pt} = \frac{5 \text{ qt}}{1} \times \frac{2 \text{ pt}}{1 \text{ qt}} + 1 \text{ pt} = 10 \text{ pt} + 1 \text{ pt} = 11 \text{ pt}$

(d) $5 \text{ qt } 1 \text{ pt} = 5 \text{ qt} + 1 \text{ pt} = 5 \text{ qt} + \frac{1 \text{ pt}}{1} \times \frac{1 \text{ qt}}{2 \text{ pt}} = 5 \text{ qt} + \frac{1}{2} \text{ qt} = 5\frac{1}{2} \text{ qt}$

(e) $2 \text{ T } 1500 \text{ lb} = 2 \text{ T} + 1500 \text{ lb} = \frac{2 \text{ T}}{1} \times \frac{2000 \text{ lb}}{1 \text{ T}} + 1500 \text{ lb} = 4000 \text{ lb} + 1500 \text{ lb} = 5500 \text{ lb}$

(f) $2 \text{ T } 1500 \text{ lb} = 2 \text{ T} + 1500 \text{ lb} = 2 \text{ T} + \frac{1500 \text{ lb}}{1} \times \frac{1 \text{ T}}{2000 \text{ lb}} = 2 \text{ T} + \frac{3}{4} \text{ T} = 2\frac{3}{4} \text{ T}$

(g) $1 \text{ yd}^2 3 \text{ ft}^2 = 1 \text{ yd}^2 + 3 \text{ ft}^2 = \frac{1 \text{ yd}^2}{1} \times \frac{9 \text{ ft}^2}{1 \text{ yd}^2} + 3 \text{ ft}^2 = 9 \text{ ft}^2 + 3 \text{ ft}^2 = 12 \text{ ft}^2$

(h) $1 \text{ yd}^2 3 \text{ ft}^2 = 1 \text{ yd}^2 + 3 \text{ ft}^2 = 1 \text{ yd}^2 + \frac{3 \text{ ft}^2}{1} \times \frac{1 \text{ yd}^2}{9 \text{ ft}^2} = 1 \text{ yd}^2 + \frac{1}{3} \text{ yd}^2 = 1\frac{1}{3} \text{ yd}^2$

(i) $2 \text{ yd}^3 18 \text{ ft}^3 = 2 \text{ yd}^3 + 18 \text{ ft}^3 = \frac{2 \text{ yd}^3}{1} \times \frac{27 \text{ ft}^3}{1 \text{ yd}^3} + 18 \text{ ft}^3 = 54 \text{ ft}^3 + 18 \text{ ft}^3 = 72 \text{ ft}^3$

(j) $2 \text{ yd}^3 18 \text{ ft}^3 = 2 \text{ yd}^3 + 18 \text{ ft}^3 = 2 \text{ yd}^3 + \frac{18 \text{ ft}^3}{1} \times \frac{1 \text{ yd}^3}{27 \text{ ft}^3} = 2 \text{ yd}^3 + \frac{2}{3} \text{ yd}^3 = 2\frac{2}{3} \text{ yd}^3$

PROBLEM 12-5   Match each given item on the left with the most appropriate metric measure on the right:

Length

(a) The height of a room	3 mm
(b) The width of a paper clip	3 cm
(c) The distance from home to work	3 m
(d) The length of an eraser	3 km

Capacity

(e) The capacity of a large water bottle	20 mL
(f) The capacity of a large table spoon	20 L
(g) The capacity of a city water tank	20 kL

Mass

(h) The mass of a high school girl	50 mg
(i) The mass of an alarm clock	50 g
(j) The mass of a train engine	50 kg
(k) The mass of a large bird feather	50 t

Area

(l)	The surface area of a double house door	5 cm²
(m)	The surface area of a city block	5 m²
(n)	The surface area of a U.S. nickel	5 ha

Volume

(o)	The space taken up by a hand-held calculator	20 cm³
(p)	The space inside a swimming pool	20 m³

Solution: Recall that to help identify metric measures, you can use your intuitive idea of 1 meter, 1 liter, 1 gram, and so forth [see Examples 12-21 through 12-30]:

(a) The height of a room might be 3 m. **(b)** The width of a paper clip might be 3 mm.
(c) The distance from home to work might be 3 km. **(d)** The length of an eraser might be 3 cm.
(e) The capacity of a large water bottle might be 20 L.
(f) The capacity of a large table spoon might be 20 mL.
(g) The capacity of a city water tank might be 20 kL.
(h) The mass of a high school girl might be 50 kg. **(i)** The mass of an alarm clock might be 50 g.
(j) The mass of a train engine might be 50 t. **(k)** The mass of a large bird feather might be 50 mg.
(l) The surface area of a double house door might be 5 m².
(m) The surface area of one city block might be 5 ha.
(n) The surface area of a U.S. nickel is about 5 cm².
(o) The space taken up by a hand-held calculator might be 20 cm³.
(p) The space inside a swimming pool might be 20 m³.

PROBLEM 12-6 Rename each metric measure using the correct unit conversions:

(a) 2 cm = ? mm **(b)** 300 cm = ? m **(c)** 2.5 km = ? m **(d)** 500 mL = ? L
(e) 2 kL = ? L **(f)** 3 g = ? mg **(g)** 250 g = ? kg **(h)** 0.25 t = ? kg **(i)** 2 cm² = ? mm²
(j) 1000 cm² = ? m² **(k)** 1.5 cm³ = ? mm³ **(l)** 500,000 cm³ = ? m³

Solution: Recall that to rename a given measure in terms of a different unit of measure when there is a unit conversion relating the two units of measure, you multiply the given measure by the correct unit fraction [see Example 12-31]:

(a) $2 \text{ cm} = 2 \text{ cm} \times 1 = \dfrac{2 \text{ cm}}{1} \times \dfrac{10 \text{ mm}}{1 \text{ cm}} = (2 \times 10) \text{ mm} = 20 \text{ mm}$

(b) $300 \text{ cm} = 300 \text{ cm} \times 1 = \dfrac{300 \text{ cm}}{1} \times \dfrac{1 \text{ m}}{100 \text{ cm}} = \dfrac{300}{100} \text{ m} = 3 \text{ m}$

(c) $2.5 \text{ km} = 2.5 \text{ km} \times 1 = \dfrac{2.5 \text{ km}}{1} \times \dfrac{1000 \text{ m}}{1 \text{ km}} = (2.5 \times 1000) \text{ m} = 2500 \text{ m}$

(d) $500 \text{ mL} = 500 \text{ mL} \times 1 = \dfrac{500 \text{ mL}}{1} \times \dfrac{1 \text{ L}}{1000 \text{ mL}} = \dfrac{500}{1000} \text{ L} = \dfrac{1}{2} \text{ L or } 0.5 \text{ L}$

(e) $2 \text{ kL} = 2 \text{ kL} \times 1 = \dfrac{2 \text{ kL}}{1} \times \dfrac{1000 \text{ L}}{1 \text{ kL}} = (2 \times 1000) \text{ L} = 2000 \text{ L}$

(f) $3 \text{ g} = 3 \text{ g} \times 1 = \dfrac{3 \text{ g}}{1} \times \dfrac{1000 \text{ mg}}{1 \text{ g}} = (3 \times 1000) \text{ mg} = 3000 \text{ mg}$

(g) $250 \text{ g} = 250 \text{ g} \times 1 = \dfrac{250 \text{ g}}{1} \times \dfrac{1 \text{ kg}}{1000 \text{ g}} = \dfrac{250}{1000} \text{ kg} = \dfrac{1}{4} \text{ kg or } 0.25 \text{ kg}$

(h) $0.25 \text{ } t = 0.25 \text{ } t \times 1 = \dfrac{0.25 \text{ } t}{1} \times \dfrac{1000 \text{ kg}}{1 \text{ } t} = (0.25 \times 1000) \text{ kg} = 250 \text{ kg}$

(i) $2 \text{ cm}^2 = 2 \text{ cm}^2 \times 1 = \dfrac{2 \text{ cm}^2}{1} \times \dfrac{100 \text{ mm}^2}{1 \text{ cm}^2} = (2 \times 100) \text{ mm}^2 = 200 \text{ mm}^2$

(j) $1000 \text{ cm}^2 = 1000 \text{ cm}^2 \times 1 = \dfrac{1000 \text{ cm}^2}{1} \times \dfrac{1 \text{ m}^2}{10,000 \text{ cm}^2} = \dfrac{1000}{10,000} \text{ m}^2 = \dfrac{1}{10} \text{ m}^2 \text{ or } 0.1 \text{ m}^2$

(k) $1.5 \text{ cm}^3 = 1.5 \text{ cm}^3 \times 1 = \dfrac{1.5 \text{ cm}^3}{1} \times \dfrac{10,000 \text{ mm}^3}{1 \text{ cm}^3} = (1.5 \times 10,000) \text{ mm}^3 = 15,000 \text{ mm}^3$

(l) $500,000 \text{ cm}^3 = 500,000 \text{ cm}^3 \times 1 = \dfrac{500,000 \text{ cm}^3}{1} \times \dfrac{1 \text{ m}^3}{100,000,000 \text{ cm}^3} = \dfrac{500,000}{100,000,000} \text{ m}^3 = 0.005 \text{ m}^3$

PROBLEM 12-7   Rename each metric measure when more than one unit conversion is necessary:

(a) $2 \text{ m} = ? \text{ mm}$ **(b)** $750,000 \text{ mm} = ? \text{ km}$ **(c)** $3 \text{ kL} = ? \text{ mL}$ **(d)** $100,000 \text{ mL} = ? \text{ kL}$
(e) $2.5 \text{ kg} = ? \text{ mg}$ **(f)** $500,000 \text{ g} = ? \text{ t}$ **(g)** $1 \text{ m}^2 = ? \text{ mm}^2$ **(h)** $10,000,000 \text{ mm}^3 = ? \text{ m}^3$

Solution: Recall that to rename a measure when more than one unit conversion is necessary, you first list each necessary unit conversion in the order needed and then multiply by all the correct unit fractions [see Example 12-24]:

(a) $1 \text{ m} = 100 \text{ cm}$ and $1 \text{ cm} = 10 \text{ mm}$ are the necessary unit conversions:

$$2 \text{ m} = \dfrac{2 \text{ m}}{1} \times \dfrac{100 \text{ cm}}{1 \text{ m}} \times \dfrac{10 \text{ mm}}{1 \text{ cm}} = (2 \times 100 \times 10) \text{ mm} = 2000 \text{ mm}$$

(b) $10 \text{ mm} = 1 \text{ cm}$, $100 \text{ cm} = 1 \text{ m}$, and $1000 \text{ m} = 1 \text{ km}$ are the necessary unit conversions:

$$750,000 \text{ mm} = \dfrac{750,000 \text{ mm}}{1} \times \dfrac{1 \text{ cm}}{10 \text{ mm}} \times \dfrac{1 \text{ m}}{100 \text{ cm}} \times \dfrac{1 \text{ km}}{1000 \text{ m}} = \dfrac{750,000}{1,000,000} \text{ km} = 0.75 \text{ km or } \dfrac{3}{4} \text{ km}$$

(c) $1 \text{ kL} = 1000 \text{ L}$ and $1 \text{ L} = 1000 \text{ mL}$ are the necessary unit conversions:

$$3 \text{ kL} = \dfrac{3 \text{ kL}}{1} \times \dfrac{1000 \text{ L}}{1 \text{ kL}} \times \dfrac{1000 \text{ mL}}{1 \text{ L}} = (3 \times 1000 \times 1000) \text{ mL} = 3,000,000 \text{ mL}$$

(d) $1000 \text{ mL} = 1 \text{ L}$ and $1000 \text{ L} = 1 \text{ kL}$ are the necessary unit conversions:

$$100,000 \text{ mL} = \dfrac{100,000 \text{ mL}}{1} \times \dfrac{1 \text{ L}}{1000 \text{ mL}} \times \dfrac{1 \text{ kL}}{1000 \text{ L}} = \dfrac{100,000}{1000 \times 1000} \text{ kL} = \dfrac{1}{10} \text{ kL} = 0.1 \text{ kL}$$

(e) $1 \text{ kg} = 1000 \text{ g}$ and $1 \text{ g} = 1000 \text{ mg}$ are the necessary unit conversions:

$$2.5 \text{ kg} = \dfrac{2.5 \text{ kg}}{1} \times \dfrac{1000 \text{ g}}{1 \text{ kg}} \times \dfrac{1000 \text{ mg}}{1 \text{ g}} = (2.5 \times 1000 \times 1000) \text{ mg} = 2,500,000 \text{ mg}$$

(f) $1000 \text{ g} = 1 \text{ kg}$ and $1000 \text{ kg} = 1 \text{ t}$ are the necessary unit conversions:

$$500,000 \text{ g} = \dfrac{500,000 \text{ g}}{1} \times \dfrac{1 \text{ kg}}{1000 \text{ g}} \times \dfrac{1 \text{ t}}{1000 \text{ kg}} = \dfrac{500,000}{1000 \times 1000} \text{ t} = \dfrac{1}{2} \text{ t} = 0.5 \text{ t}$$

(g) $1 \text{ m}^2 = 10,000 \text{ cm}^2$ and $1 \text{ cm}^2 = 100 \text{ mm}^2$ are the necessary unit conversions:

$$1 \text{ m}^2 = \dfrac{1 \text{ m}^2}{1} \times \dfrac{10,000 \text{ cm}^2}{1 \text{ m}^2} \times \dfrac{100 \text{ mm}^2}{1 \text{ cm}^2} = (10,000 \times 100) \text{ mm}^2 = 1,000,000 \text{ mm}^2$$

(h) $10,000 \text{ mm}^3 = 1 \text{ cm}^3$ and $100,000,000 \text{ cm}^3 = 1 \text{ m}^3$ are the necessary unit conversions:

$$10,000,000 \text{ mm}^3 = \dfrac{10,000,000 \text{ mm}^3}{1} \times \dfrac{1 \text{ cm}^3}{10,000 \text{ mm}^3} \times \dfrac{1 \text{ m}^3}{100,000,000 \text{ cm}^3} =$$

$$\dfrac{10,000,000}{10,000 \times 100,000,000} \text{ m}^3 = \dfrac{1}{100,000} \text{ m}^3 = 0.00001 \text{ m}^3$$

PROBLEM 12-8   Add like measures: **(a)** $5 \text{ m} + 8 \text{ m}$ **(b)** $2\frac{1}{2} \text{ ft} + 3 \text{ ft}$ **(c)** $0.3 \text{ g} + 1.8 \text{ g}$
(d) $\frac{1}{2} \text{ c} + \frac{3}{4} \text{ c}$

Solution: Recall that to add like measures, you add the amounts and then write the same unit of measure on the sum:

(a) $5 \text{ m} + 8 \text{ m} = (5 + 8) \text{ m} = 13 \text{ m}$ **(b)** $2\frac{1}{2} \text{ ft} + 3 \text{ ft} = (2\frac{1}{2} + 3) \text{ ft} = 5\frac{1}{2} \text{ ft}$

(c) $0.3 \text{ g} + 1.8 \text{ g} = (0.3 + 1.8) \text{ g} = 2.1 \text{ g}$ **(d)** $\frac{1}{2} \text{ c} + \frac{3}{4} \text{ c} = (\frac{1}{2} + \frac{3}{4}) \text{ c} = \frac{5}{4} \text{ c} = 1\frac{1}{4} \text{ c}$

PROBLEM 12-9   Add mixed measures: **(a)** 5 yd 2 ft + 6 yd 1 ft   **(b)** 2 ft $8\frac{1}{2}$ in. + 4 yd $7\frac{1}{2}$ in.
(c) 8 lb 10 oz + 6 lb 11 oz   **(d)** 3 gal 2 qt 1 pt + 3 qt 1 pt

Solution: Recall that to add mixed measures, you first line up corresponding like measures in columns, then add like measures, and then simplify the sum if possible [see Example 12-35]:

(a)
$$
\begin{array}{rl}
5 \text{ yd} \quad 2 \text{ ft} & \\
+ \ 6 \text{ yd} \quad 1 \text{ ft} & \\
\hline
11 \text{ yd} \quad 3 \text{ ft} & = 11 \text{ yd} + 3 \text{ ft} \\
& = 11 \text{ yd} + 1 \text{ yd} \\
& = 12 \text{ yd}
\end{array}
$$

(b)
$$
\begin{array}{rl}
2 \text{ ft} \quad 8\frac{1}{2} \text{ in.} & \\
+ 4 \text{ yd} \qquad 7\frac{1}{2} \text{ in.} & \\
\hline
4 \text{ yd} \quad 2 \text{ ft} \quad 16 \text{ in.} & = 4 \text{ yd} + 2 \text{ ft} + 16 \text{ in.} \\
& = 4 \text{ yd} + 2 \text{ ft} + 12 \text{ in.} + 4 \text{ in.} \\
& = 4 \text{ yd} + 2 \text{ ft} + 1 \text{ ft} + 4 \text{ in.} \\
& = 4 \text{ yd} + 3 \text{ ft} + 4 \text{ in.} \\
& = 4 \text{ yd} + 1 \text{ yd} + 4 \text{ in.} \\
& = 5 \text{ yd } 4 \text{ in.}
\end{array}
$$

(c)
$$
\begin{array}{rl}
8 \text{ lb} \quad 10 \text{ oz} & \\
+ \ 6 \text{ lb} \quad 11 \text{ oz} & \\
\hline
14 \text{ lb} \quad 21 \text{ oz} & = 14 \text{ lb} + 21 \text{ oz} \\
& = 14 \text{ lb} + 16 \text{ oz} + 5 \text{ oz} \\
& = 14 \text{ lb} + 1 \text{ lb} + 5 \text{ oz} \\
& = 15 \text{ lb } 5 \text{ oz}
\end{array}
$$

(d)
$$
\begin{array}{rl}
3 \text{ gal} \quad 2 \text{ qt} \quad 1 \text{ pt} & \\
+ \qquad\quad 3 \text{ qt} \quad 1 \text{ pt} & \\
\hline
3 \text{ gal} \quad 5 \text{ qt} \quad 2 \text{ pt} & = 3 \text{ gal} + 5 \text{ qt} + 2 \text{ pt} \\
& = 3 \text{ gal} + 4 \text{ qt} + 1 \text{ qt} + 1 \text{ qt} \\
& = 3 \text{ gal} + 1 \text{ gal} + 2 \text{ qt} \\
& = 4 \text{ gal } 2 \text{ qt}
\end{array}
$$

PROBLEM 12-10   Subtract like measures: **(a)** 3 L $-$ 1.5 L   **(b)** 6 lb $- 4\frac{1}{2}$ lb
(c) 2.8 km $-$ 0.75 km   **(d)** $\frac{3}{4}$ in. $- \frac{5}{8}$ in.

Solution: Recall that to subtract like measures, you subtract the amounts and then write the same unit of measure on the difference [see Example 12-36]:

(a) $3 \text{ L} - 1.5 \text{ L} = (3 - 1.5) \text{ L} = 1.5 \text{ L}$ **(b)** $6 \text{ lb} - 4\frac{1}{2} \text{ lb} = (6 - 4\frac{1}{2}) \text{ lb} = 1\frac{1}{2} \text{ lb}$

(c) $2.8 \text{ km} - 0.75 \text{ km} = (2.8 - 0.75) \text{ km} = 2.05 \text{ km}$ **(d)** $\frac{3}{4} \text{ in.} - \frac{5}{8} \text{ in.} = (\frac{3}{4} - \frac{5}{8}) \text{ in.} = \frac{1}{8} \text{ in.}$

PROBLEM 12-11   Subtract mixed measures: **(a)** 8 lb 9 oz $-$ 3 lb $7\frac{1}{2}$ oz   **(b)** 2 yd $-$ 2 ft 6 in.
(c) 5 gal 1 pt $-$ 3 qt 1 c   **(d)** $4\frac{1}{2}$ T 500 lb $-$ 1800 lb

Solution: Recall that to subtract mixed measures, you first line up corresponding like measures in columns, then rename when necessary, and then subtract like measures [see Example 12-37]:

(a)
$$
\begin{array}{r}
8 \text{ lb} \quad 9 \text{ oz} \\
- 3 \text{ lb} \quad 7\frac{1}{2} \text{ oz} \\
\hline
5 \text{ lb} \quad 1\frac{1}{2} \text{ oz}
\end{array}
$$

(b)
$$
\begin{array}{rl}
 & \overset{\displaystyle 2 \text{ yd}}{\overbrace{\quad\quad}} \\
2 \text{ yd} & = 1 \text{ yd} \quad 3 \text{ ft} \\
- \quad 2 \text{ ft} \quad 6 \text{ in.} & = \qquad\quad 2 \text{ ft} \quad 6 \text{ in.}
\end{array}
$$

$$
\begin{array}{rl}
 & \overset{\displaystyle 3 \text{ ft}}{\overbrace{\quad\quad}} \\
= 1 \text{ yd} & 2 \text{ ft} \quad 12 \text{ in.} \\
 & 2 \text{ ft} \quad\ \ 6 \text{ in.} \\
\hline
1 \text{ yd} & 0 \text{ ft} \quad\ \ 6 \text{ in.} = 1 \text{ yd } 6 \text{ in.}
\end{array}
$$

(c)
$$
\begin{array}{rl}
 & \qquad \overset{\displaystyle 5 \text{ gal}}{\overbrace{\quad\quad\quad}} \qquad \overset{\displaystyle 1 \text{ pt}}{\overbrace{\quad\quad}} \\
5 \text{ gal} \qquad\quad 1 \text{ pt} & = 4 \text{ gal} \quad 4 \text{ qt} \quad 0 \text{ pt} \quad 2 \text{ c} \\
- \qquad\quad 3 \text{ qt} \quad 1 \text{ c} & = \qquad\quad\ \ 3 \text{ qt} \qquad\quad\ 1 \text{ c} \\
\hline
 & \quad\ 4 \text{ gal} \quad 1 \text{ qt} \quad 0 \text{ pt} \quad 1 \text{ c} = 4 \text{ gal } 1 \text{ qt } 1 \text{c}
\end{array}
$$

$$4\tfrac{1}{2}\,\text{T}\ 500\,\text{lb}$$

(d) $4\tfrac{1}{2}\,\text{T}\quad 500\,\text{lb} = \overbrace{3\tfrac{1}{2}\,\text{T}\quad 2500\,\text{lb}} \longleftarrow 4\tfrac{1}{2} + 500\,\text{lb} = 3\tfrac{1}{2}\,\text{T} + \quad 1\,\text{T}\quad + 500\,\text{lb}$

$\underline{-\qquad\quad 1800\,\text{lb}} = \underline{\qquad\quad 1800\,\text{lb}} \qquad\qquad\qquad = 3\tfrac{1}{2}\,\text{T} + 2000\,\text{lb} + 500\,\text{lb}$

$\qquad\qquad\qquad\qquad\quad 3\tfrac{1}{2}\,\text{T}\quad 700\,\text{lb}$

PROBLEM 12-12   Multiply a measure by a number: **(a)** $3 \times 5\,\text{cm}$ **(b)** $5.6\,\text{mL} \times 0.25$
(c) $8 \times 4\tfrac{1}{2}\,\text{ft}$ **(d)** $\tfrac{3}{4}\,\text{lb} \times \tfrac{2}{3}$

Solution: Recall that to multiply a measure by a number, you multiply the amounts and then write the same unit of measure on the product [see Example 12-38]:

(a) $3 \times 5\,\text{cm} = (3 \times 5)\,\text{cm} = 15\,\text{cm}$ **(b)** $5.6\,\text{mL} \times 0.25 = (5.6 \times 0.25)\,\text{mL} = 1.4\,\text{mL}$
(c) $8 \times 4\tfrac{1}{2}\,\text{ft} = (8 \times 4\tfrac{1}{2})\,\text{ft} = 36\,\text{ft}$ **(d)** $\tfrac{3}{4}\,\text{lb} \times \tfrac{2}{3} = (\tfrac{3}{4} \times \tfrac{2}{3})\,\text{lb} = \tfrac{1}{2}\,\text{lb}$

PROBLEM 12-13   Multiply a mixed measure by a number: **(a)** $2 \times 5\,\text{mi}\ 30\,\text{yd}$
(b) $5\,\text{T}\ 400\,\text{lb} \times 5$ **(c)** $\tfrac{1}{2} \times 2\,\text{gal}\ 3\,\text{qt}\ 1\,\text{pt}$ **(d)** $5\,\text{yd}\ 6\,\text{ft}\ 10\,\text{in.} \times 2\tfrac{1}{2}$

Solution: To multiply a mixed measure by a number, you multiply each individual measure within the mixed measure by the number [see Example 12-39]:

(a) $\begin{array}{r} 5\,\text{mi}\quad 30\,\text{yd} \\ \times \qquad\quad 2 \\ \hline 10\,\text{mi}\quad 60\,\text{yd} \end{array}$ **(b)** $\begin{array}{r} 5\,\text{T}\quad 400\,\text{lb} \\ \times \qquad\quad 5 \\ \hline 25\,\text{T}\quad 2000\,\text{lb} \end{array} = 25\,\text{T} + 2000\,\text{lb}$

$\qquad\qquad\qquad\qquad\qquad\qquad\qquad\qquad = 25\,\text{T} + 1\,\text{T}$

$\qquad\qquad\qquad\qquad\qquad\qquad\qquad\qquad = 26\,\text{T}$

(c) $\begin{array}{r} 2\,\text{gal}\quad 3\,\text{qt}\quad 1\,\text{pt} \\ \times \qquad\qquad\ \ \tfrac{1}{2} \\ \hline 1\,\text{gal}\quad 1\tfrac{1}{2}\,\text{qt}\quad \tfrac{1}{2}\,\text{pt} \end{array}$ **(d)** $\begin{array}{r} 5\,\text{yd}\quad 6\,\text{ft}\quad 10\,\text{in.} \\ \times \qquad\qquad\quad 2\tfrac{1}{2} \\ \hline 12\tfrac{1}{2}\,\text{yd}\quad 15\,\text{ft}\quad 25\,\text{in.} \end{array} = 12\tfrac{1}{2}\,\text{yd} + 15\,\text{ft} + 25\,\text{in.}$

$\qquad\qquad\qquad\qquad\qquad\qquad\qquad\qquad = 12\tfrac{1}{2}\,\text{yd} + \overbrace{5\,\text{yd}} + \overbrace{2\,\text{ft} + 1\,\text{in.}}$

$\qquad\qquad\qquad\qquad\qquad\qquad\qquad\qquad = 17\tfrac{1}{2}\,\text{yd}\ 2\,\text{ft}\ 1\,\text{in.}$

PROBLEM 12-14   Multiply two like length measures: **(a)** $3\,\text{in.} \times 5\,\text{in.}$ **(b)** $1\tfrac{1}{2}\,\text{ft} \times 4\,\text{ft}$
(c) $\tfrac{1}{2}\,\text{yd} \times \tfrac{3}{4}\,\text{yd}$ **(d)** $2\,\text{mm} \times 3\,\text{mm}$ **(e)** $5\,\text{cm} \times 0.5\,\text{cm}$ **(f)** $2.3\,\text{m} \times 0.8\,\text{m}$

Solution: Recall that when two like length measures are multiplied, the product will always be the associated area measure [see Example 12-40]:

(a) $3\,\text{in.} \times 5\,\text{in.} = (3 \times 5)\,\text{in.}^2 = 15\,\text{in.}^2$ **(b)** $1\tfrac{1}{2}\,\text{ft} \times 4\,\text{ft} = (1\tfrac{1}{2} \times 4)\,\text{ft}^2 = 6\,\text{ft}^2$
(c) $\tfrac{1}{2}\,\text{yd} \times \tfrac{3}{4}\,\text{yd} = (\tfrac{1}{2} \times \tfrac{3}{4})\,\text{yd}^2 = \tfrac{3}{8}\,\text{yd}^2$ **(d)** $2\,\text{mm} \times 3\,\text{mm} = (2 \times 3)\,\text{mm}^2 = 6\,\text{mm}^2$
(e) $5\,\text{cm} \times 0.5\,\text{cm} = (5 \times 0.5)\,\text{cm}^2 = 2.5\,\text{cm}^2$ **(f)** $2.3\,\text{m} \times 0.8\,\text{m} = (2.3 \times 0.8)\,\text{m}^2 = 1.84\,\text{m}^2$

PROBLEM 12-15   Multiply three like length measures: **(a)** $2\,\text{in.} \times 3\,\text{in.} \times 5\,\text{in.}$
(b) $\tfrac{1}{2}\,\text{yd} \times 3\,\text{yd} \times 2\tfrac{1}{2}\,\text{yd}$ **(c)** $\tfrac{1}{2}\,\text{ft} \times \tfrac{3}{4}\,\text{ft} \times \tfrac{5}{8}\,\text{ft}$ **(d)** $5\,\text{cm} \times 2\,\text{cm} \times 4\,\text{cm}$
(e) $0.5\,\text{m} \times 2.5\,\text{m} \times 1.25\,\text{m}$

Solution: Recall that when three like length measures are multiplied, the product will always be an associated volume measure [see Example 12-41]:

(a) $2\,\text{in.} \times 3\,\text{in.} \times 5\,\text{in.} = (2 \times 3 \times 5)\,\text{in.}^3 = 30\,\text{in.}^3$
(b) $\tfrac{1}{2}\,\text{yd} \times 3\,\text{yd} \times 2\tfrac{1}{2}\,\text{yd} = (\tfrac{1}{2} \times 3 \times 2\tfrac{1}{2})\,\text{yd}^3 = \tfrac{15}{4}\,\text{yd}^3 = 3\tfrac{3}{4}\,\text{yd}^3$
(c) $\tfrac{1}{2}\,\text{ft} \times \tfrac{3}{4}\,\text{ft} \times \tfrac{5}{8}\,\text{ft} = (\tfrac{1}{2} \times \tfrac{3}{4} \times \tfrac{5}{8})\,\text{ft}^3 = \tfrac{15}{64}\,\text{ft}^3$
(d) $5\,\text{cm} \times 2\,\text{cm} \times 4\,\text{cm} = (5 \times 2 \times 4)\,\text{cm}^3 = 40\,\text{cm}^3$
(e) $0.5\,\text{m} \times 2.5\,\text{m} \times 1.25\,\text{m} = (0.5 \times 2.5 \times 1.25)\,\text{m}^3 = 1.5625\,\text{m}^3$

PROBLEM 12-16   Divide a measure by a number: **(a)** $8\,\text{ft} \div 2$ **(b)** $7\tfrac{1}{2}\,\text{lb} \div 3$
(c) $18\,\text{m} \div 1.5$ **(d)** $0.75\,\text{L} \div 0.25$

Solution: Recall that to divide a measure by a number, you divide the amounts and then write the same unit of measure on the quotient [see Example 12-43]:

(a) $8 \text{ ft} \div 2 = (8 \div 2) \text{ ft} = 4 \text{ ft}$

(b) $7\frac{1}{2} \text{ lb} \div 3 = (7\frac{1}{2} \div 3) \text{ lb} = 2\frac{1}{2} \text{ lb}$

(c) $18 \text{ m} \div 1.5 = (18 \div 1.5) \text{ m} = 12 \text{ m}$

(d) $0.75 \text{ L} \div 0.25 = (0.75 \div 0.25) \text{ L} = 3 \text{ L}$

PROBLEM 12-16   Divide a mixed measure by a number: **(a)** 16 lb 8 oz ÷ 4   **(b)** 3 T 500 lb ÷ 5
(c) 8 yd 5 in. ÷ 3   **(d)** 9 gal 3 qt 1 c ÷ 2

Solution: Recall that to divide a mixed measure by a number, you divide each individual measure within the mixed measure by the number while being careful to leave space in the dividend for any missing units of measure that may be needed [see Example 12-44]:

$$
\begin{array}{r}
4 \text{ lb } 2 \text{ oz} \\
\textbf{(a)} \ 4\overline{)\ 16 \text{ lb } 8 \text{ oz}} \\
-16 \text{ lb} \\
\hline
\varnothing \quad 8 \text{ oz} \\
-8 \text{ oz} \\
\hline
0
\end{array}
$$

$$
\begin{array}{r}
1300 \text{ lb} \\
\textbf{(b)} \ 5\overline{)\ 3 \text{ T } \ 500 \text{ lb}} \\
\hookrightarrow 6000 \text{ lb} \\
\hline
6500 \text{ lb} \\
-6500 \text{ lb} \\
\hline
0
\end{array}
$$

$$
\begin{array}{r}
2 \text{ yd } 2 \text{ ft } 1 \text{ in. R } 2 \text{ in.} \\
\textbf{(c)} \ 3\overline{)\ 8 \text{ yd} \qquad 5 \text{ in.}} \\
-6 \text{ yd} \\
\hline
2 \text{ yd } 6 \text{ ft} \\
-6 \text{ ft} \\
\hline
0 \quad 5 \text{ in.} \\
-3 \text{ in.} \\
\hline
2 \text{ in.}
\end{array}
$$

$$
\begin{array}{r}
4 \text{ gal } 3 \text{ qt } 1 \text{ pt } \quad \text{R } 1 \text{ c} \\
\textbf{(d)} \ 2\overline{)\ 9 \text{ gal } 3 \text{ qt} \qquad 1 \text{ c}} \\
-8 \text{ gal} \\
\hline
1 \text{ gal } 4 \text{ qt} \\
7 \text{ qt} \\
-6 \text{ qt} \\
\hline
1 \text{ qt } 2 \text{ pt} \\
-2 \text{ pt} \\
\hline
0
\end{array}
$$

PROBLEM 12-18   Divide like measures: **(a)** 8 ft ÷ 2 ft   **(b)** 32 oz ÷ $1\frac{1}{2}$ oz   **(c)** $\frac{3}{4}$ c ÷ $\frac{1}{2}$ c
(d) 10 m ÷ 5 m   **(e)** 6.5 g ÷ 4 g   **(f)** 0.25 mL ÷ 0.4 mL

Solution: Recall that when two like measures are divided, the quotient will always be a number (no unit of measure) [see Example 12-45]:

(a) $8 \text{ ft} \div 2 \text{ ft} = (8 \div 2)\dfrac{\cancel{ft}}{\cancel{ft}} = 4$

(b) $32 \text{ oz} \div 1\frac{1}{2} \text{ oz} = (32 \div 1\frac{1}{2})\dfrac{\cancel{oz}}{\cancel{oz}} = 21\frac{1}{3}$

(c) $\frac{3}{4} \text{ c} \div \frac{1}{2} \text{ c} = (\frac{3}{4} \div \frac{1}{2})\dfrac{\cancel{c}}{\cancel{c}} = 1\frac{1}{2}$

(d) $10 \text{ m} \div 5 \text{ m} = (10 \div 5)\dfrac{\cancel{m}}{\cancel{m}} = 2$

(e) $6.5 \text{ g} \div 4 \text{ g} = (6.5 \div 4)\dfrac{\cancel{g}}{\cancel{g}} = 1.625$

(f) $0.25 \text{ mL} \div 0.4 \text{ mL} = (0.25 \div 0.4)\dfrac{\cancel{mL}}{\cancel{mL}} = 0.625$

PROBLEM 12-19   Divide unlike measures: **(a)** 8 gal ÷ 2 ft   **(b)** $6\frac{1}{2}$ oz ÷ $2\frac{1}{2}$ in.
(c) 12 kg ÷ 0.5 m   **(d)** 1.25 g ÷ 2 mL

Solution: Recall that when unlike measures are divided, the quotient will always be a rate [see Example 12-44]:

(a) $8 \text{ gal} \div 2 \text{ ft} = (8 \div 2)\dfrac{\text{gal}}{\text{ft}} = 4\dfrac{\text{gal}}{\text{ft}}$

(b) $6\frac{1}{2} \text{ oz} \div 2\frac{1}{2} \text{ in.} = (6\frac{1}{2} \div 2\frac{1}{2})\dfrac{\text{oz}}{\text{in.}} = 2.6\dfrac{\text{oz}}{\text{in.}}$

(c) $12 \text{ kg} \div 0.5 \text{ m} = (12 \div 0.5)\dfrac{\text{kg}}{\text{m}} = 24\dfrac{\text{kg}}{\text{m}}$

(d) $1.25 \text{ g} \div 2 \text{ mL} = (1.25 \div 2)\dfrac{\text{g}}{\text{mL}} = 0.625\dfrac{\text{g}}{\text{mL}}$

Supplementary Exercises

PROBLEM 12-20   Match each given statement on the left with the most appropriate U.S. Customary measure on the right:

(a) The weight of a portable typewriter 2 ft^3

(b) The length of a jumbo paper clip 7 ft

(c) The capacity of a small carton of cream 2 ft^2

(d) The weight of a golf ball 1 c

(e) The height of a basketball player 2 oz

(f) The capacity of a small milk glass 2 in.

(g) The top area of a coffee table 8 lb

(h) The capacity of a fish tank 1 pt

PROBLEM 12-21   Rename U.S. Customary measures: (a) 30 in. = ? ft   (b) $3\frac{1}{2}$ gal = ? pt
(c) 2 T 800 lb = ? lb   (d) 8 yd = ? ft   (e) $1\frac{1}{2}$ mi = ? yd   (f) 2 yd $1\frac{1}{2}$ ft = ? yd
(g) 3 pt = ? c   (h) $2\frac{1}{2}$ qt = ? c   (i) $\frac{1}{2}$ gal 1 qt = ? qt   (j) 24 oz = ? lb
(k) 3 lb 6 oz = ? lb   (l) 36 in. = ? ft   (m) 2 yd = ? ft   (n) 6600 ft = ? mi
(o) 1980 yd = ? mi   (p) 60 in. = ? yd   (q) 5 yd 2 ft = ? ft   (r) 7 c = ? pt   (s) 9 pt = ? qt
(t) 11 qt = ? gal   (u) 20 c = ? qt   (v) 3 qt 1 pt = ? qt   (w) 48 oz = ? lb
(x) 6000 lb = ? T   (y) 48,000 oz = ? T   (z) 3 T 1500 lb = ? T

PROBLEM 12-22   Match each given statement on the left with the most appropriate metric measure on the right:

(a) The length of a standard paper clip 5 kL

(b) The capacity of a car's gas tank 2 m^3

(c) The mass of a thumbtack 400 mg

(d) The distance from Los Angeles to New York 75 L

(e) The capacity of a farm water tank 2 m^2

(f) The mass of a 10-speed bicycle 3 cm

(g) The area of a standard house door 4500 km

(h) The capacity of a bathtub 13 kg

PROBLEM 12-23   Rename metric measures: (a) 325 cm = ? m   (b) 0.5 kL = ? mL
(c) 750 mg = ? g   (d) 250 mm = ? cm   (e) 2.5 m = ? cm   (f) 1.5 L = ? mL
(g) 3250 mL = ? L   (h) 2.5 kg = ? g   (i) 0.2 kg = ? mg   (j) 800 mm = ? m
(k) 900 cm = ? m   (l) 2000 m = ? km   (m) 1 m = ? mm   (n) 7000 mL = ? L
(o) 8500 mL = ? L   (p) 5250 L = ? kL   (q) 1 kL = ? mL   (r) 900 mg = ? g
(s) 215 g = ? kg   (t) 4000 kg = ? t   (u) 1 kg = ? mg   (v) 0.3 g = ? mg
(w) 0.8 m = ? cm   (x) 0.05 km = ? cm   (y) 0.005 t = ? g   (z) 1500 mm = ? m

PROBLEM 12-24   Add measures: (a) 6 m + 2 m   (b) 5 yd 2 ft 10 in. + 3 yd 8 in.
(c) 3 qt 1 pt 1 c + 1 pt 1 c   (d) 46 t + 25 t   (e) 8 mm + 7 mm   (f) 11 L + 35 L
(g) 1.5 t + 2.6 t   (h) 0.6 km + 0.85 km   (i) $5\frac{1}{2}$ mi + $6\frac{1}{4}$ mi   (j) $1\frac{1}{3}$ c + $3\frac{3}{4}$ c
(k) 5 yd 1 ft + 8 yd 1 ft   (l) 3 ft $6\frac{1}{2}$ in. + 2 ft $5\frac{1}{4}$ in.   (m) 8 lb 10 oz + 3 lb 9 oz
(n) 4 yd $7\frac{1}{4}$ in. + $9\frac{3}{8}$ in.   (o) 4 gal 3 qt 1 pt + 2 qt 1 pt   (p) 2 T 11 oz + 1999 lb 5 oz

PROBLEM 12-25   Subtract measures: (a) 9 mL − 5 mL   (b) 72 m − 15 m
(c) 2.5 g − 0.8 g   (d) 8.5 kL − 3.9 kL   (e) $3\frac{1}{3}$ T − $2\frac{1}{2}$ T   (f) $5\frac{1}{8}$ pt − $1\frac{1}{4}$ pt
(g) 8 lb 10 oz − 4 lb 7 oz   (h) 3 gal $2\frac{1}{2}$ qt − 1 gal $1\frac{3}{4}$ qt   (i) 6 yd 1 ft − 2 yd 2 ft
(j) 5 mi 25 yd − 750 yd   (k) 5 yd 2 ft − 5 ft 6 in.   (l) 5 gal 1 pt − 3 qt 1 c   (m) 6 m − 2 m
(n) 4 gal 2 qt − 3 qt 1 pt   (o) 5 yd 6 in. − 2 ft 8 in.   (p) 82 L − 17 L

PROBLEM 12-26   Multiply with measures: (a) 7 × 8 mg   (b) 85 kg × 4   (c) 0.6 × 1.5 m
(d) 2.8 L × 1.5   (e) 3 × 6 mi 5 yd   (f) 5 T 200 lb × 8   (g) $2\frac{1}{2}$ × 6 gal 3 qt 1 pt

(h) 6 yd 5 ft 10 in. × $3\frac{1}{2}$ **(i)** 7 in. × 8 in.   **(j)** $2\frac{1}{2}$ ft × 6 ft   **(k)** $\frac{1}{2}$ yd × $3\frac{1}{2}$ yd
(l) $\frac{3}{4}$ mi × $\frac{2}{3}$ mi   **(m)** 5 mm × 9 mm   **(n)** 0.5 cm × 8 cm   **(o)** 1.5 m × 3.2 m
(p) 0.1 km × 0.1 km   **(q)** 2 in. × 5 in. × 8 in.   **(r)** $2\frac{1}{2}$ ft × 2 ft × $\frac{1}{4}$ ft
(s) $\frac{1}{2}$ yd × $\frac{1}{2}$ yd × $\frac{1}{2}$ yd   **(t)** 0.5 mm × 2 mm × 1 mm   **(u)** 2 cm × 1.25 cm × 0.25 cm
(v) 0.1 m × 0.1 m × 0.1 m   **(w)** 4 × 8 lb 5 oz   **(x)** 6 × 2 m   **(y)** 1 km × 1 km
(z) 1 km × 1 km × 1 km

PROBLEM 12-27   Divide measures: **(a)** 6 m ÷ 2   **(b)** 7 yd 5 in. ÷ 4
(c) 8 gal 1 pt ÷ 3   **(d)** 2.4 m ÷ 1.5   **(e)** 15 g ÷ 3   **(f)** 81 t ÷ 6   **(g)** 930 m ÷ 0.6
(h) 8.52 mm ÷ 0.04   **(i)** $\frac{1}{4}$ ft ÷ $\frac{1}{2}$ **(j)** $\frac{5}{6}$ gal ÷ $\frac{2}{3}$ **(k)** $2\frac{2}{3}$ mi ÷ $1\frac{1}{3}$ **(l)** $1\frac{7}{8}$ in. ÷ $2\frac{1}{4}$
(m) 180 lb 15 oz ÷ 5   **(n)** 21 gal 3 qt ÷ 3   **(o)** 7 T 14 oz ÷ $3\frac{1}{2}$ **(p)** 25 yd 11 in. ÷ $2\frac{1}{2}$
(q) 12 in. ÷ 3 in.   **(r)** 20 ft ÷ $\frac{1}{2}$ ft   **(s)** $10\frac{1}{2}$ yd ÷ 3 yd   **(t)** $\frac{1}{2}$ mi ÷ $\frac{3}{4}$ mi
(u) 0.5 m ÷ 1.5 m   **(v)** 10.8 cm ÷ 2 cm   **(w)** 18 lb ÷ 2 ft   **(x)** 36 g ÷ 0.5 cm
(y) 100 L ÷ 20 m   **(z)** 1000 mm ÷ 100 mL

Answers to Supplementary Exercises

(12-20)   **(a)** 8 lb   **(b)** 2 in.   **(c)** 1 pt   **(d)** 2 oz   **(e)** 7 ft   **(f)** 1 c   **(g)** 2 ft^2 **(h)** 2 ft^3

(12-21)   **(a)** $2\frac{1}{2}$ ft   **(b)** 28 pt   **(c)** 4800 lb   **(d)** 24 ft   **(e)** 2640 yd   **(f)** $2\frac{1}{2}$ yd   **(g)** 6 c
(h) 10 c   **(i)** 3 qt   **(j)** $1\frac{1}{2}$ lb   **(k)** $3\frac{3}{8}$ **(l)** 3 ft   **(m)** 6 ft   **(n)** $1\frac{1}{4}$ mi   **(o)** $1\frac{1}{8}$ mi
(p) $1\frac{2}{3}$ yd   **(q)** 17 ft   **(r)** $3\frac{1}{2}$ pt   **(s)** $4\frac{1}{2}$ qt   **(t)** $2\frac{3}{4}$ gal   **(u)** 5 qt   **(v)** $3\frac{1}{2}$ qt
(w) 3 lb   **(x)** 3 T   **(y)** $1\frac{1}{2}$ T   **(z)** $3\frac{3}{4}$ T

(12-22)   **(a)** 3 cm   **(b)** 75 L   **(c)** 400 mg   **(d)** 4500 km   **(e)** 5 kL   **(f)** 13 kg
(g) 2 m^2 **(h)** 2 m^3

(12-23)   **(a)** 3.25 m   **(b)** 500,000 mL   **(c)** 0.75 g   **(d)** 25 cm   **(e)** 250 cm   **(f)** 1500 mL
(g) 3.25 L   **(h)** 2500 g   **(i)** 200,000 mg   **(j)** 0.8 m   **(k)** 9 m   **(l)** 2 km   **(m)** 1000 mm
(n) 7 L   **(o)** 8.5 L   **(p)** 5.25 kL   **(q)** 1,000,000 mL   **(r)** 0.9 g   **(s)** 0.215 kg   **(t)** 4 t
(u) 1,000,000 mg   **(v)** 300 mg   **(w)** 80 cm   **(x)** 5000 cm   **(y)** 5000 g   **(z)** 1.5 m

(12-24)   **(a)** 8 m   **(b)** 9 yd 6 in.   **(c)** 4 qt 1 pt   **(d)** 71 t   **(e)** 15 mm   **(f)** 46 L
(g) 4.1 t   **(h)** 1.45 km   **(i)** $11\frac{3}{4}$ mi   **(j)** $5\frac{1}{12}$ c   **(k)** 13 yd 2 ft   **(l)** 5 ft $11\frac{3}{4}$ in.
(m) 12 lb 3 oz   **(n)** 4 yd 1 ft $4\frac{5}{8}$ in.   **(o)** 5 gal 2 qt   **(p)** 3 T

(12-25)   **(a)** 4 mL   **(b)** 57 m   **(c)** 1.7 g   **(d)** 4.6 kL   **(e)** $\frac{5}{6}$ T   **(f)** $3\frac{7}{8}$ pt   **(g)** 4 lb 3 oz
(h) 2 gal $\frac{3}{4}$ qt   **(i)** 3 yd 2 ft   **(j)** 4 mi 1035 yd   **(k)** 3 yd 2 ft 6 in.   **(l)** 4 gal 1 qt 1 c
(m) 4 m   **(n)** 3 gal 2 qt 1 pt   **(o)** 4 yd 10 in.   **(p)** 65 L

(12-26)   **(a)** 56 mg   **(b)** 340 kg   **(c)** 0.9 m   **(d)** 4.2 L   **(e)** 18 mi 15 yd   **(f)** 40 T 1600 lb
(g) 17 gal $\frac{1}{2}$ qt $\frac{1}{2}$ pt   **(h)** 27 yd 2 ft 5 in.   **(i)** 56 in.^{2} **(j)** 15 ft^2 **(k)** $1\frac{3}{4}$ yd^2 **(l)** $\frac{1}{2}$ mi^2
(m) 45 mm^2 **(n)** 4 cm^2 **(o)** 4.8 m^2 **(p)** 0.01 km^2 **(q)** 80 in.^{3} **(r)** $1\frac{1}{4}$ ft^3 **(s)** $\frac{1}{8}$ yd^3
(t) 1 mm^3 **(u)** 0.625 cm^3 **(v)** 0.001 m^3 **(w)** 33 lb 4 oz   **(x)** 12 m   **(y)** 1 km^2
(z) 1 km^3

(12-27)   **(a)** 3 m   **(b)** 1 yd 2 ft 4 in. R 1 in.   **(c)** 3 pt   **(d)** 1.6 m   **(e)** 5 g   **(f)** 13.5 t
(g) 1550 m   **(h)** 213 mm   **(i)** $\frac{1}{2}$ ft   **(j)** $1\frac{1}{4}$ gal   **(k)** 2 mi   **(l)** $\frac{5}{6}$ in.   **(m)** 36 lb 3 oz
(n) 7 gal 1 qt   **(o)** 2 T 4 oz   **(p)** 10 yd 4 in. R 1 in.   **(q)** 4   **(r)** 40   **(s)** $3\frac{1}{2}$ **(t)** $\frac{2}{3}$
(u) $\frac{1}{3}$ or $0.\overline{3}$ **(v)** 5.4   **(w)** $9\,\dfrac{\text{lb}}{\text{ft}}$ **(x)** $72\,\dfrac{\text{g}}{\text{cm}}$ **(y)** $5\,\dfrac{\text{L}}{\text{m}}$ **(z)** $10\,\dfrac{\text{mm}}{\text{mL}}$

13 RATIO, RATE, AND PROPORTION

13-1. Finding the Ratio

A. Find the ratio of like measures.

Recall from Example 12-33 in the last chapter that measures with the same unit of measure are called like measures. You should recall from Example 12-45 in the last chapter that when two like measures are divided, the quotient will always be a number (with no unit of measure).

When two like measures are divided and the number quotient is written in fraction form, the fraction is called a **ratio.** To find the **ratio of two like measures in lowest terms,** you use the first like measure as the numerator of the fraction and the second like measure as the denominator of the fraction. Then you divide the like measures and rename the fraction quotient in lowest terms.

EXAMPLE 13-1: Find the ratio of 20 feet to 8 feet in lowest terms.

Solution: The ratio of 20 feet to 8 feet $= \dfrac{20 \text{ feet}}{8 \text{ feet}}$ ⟵ first given measure
⟵ second given measure

$= \dfrac{20 \text{ feet}}{8 \text{ feet}}$ Eliminate the common unit of measure [see Example 12-45].

$= \dfrac{5}{2}$ Simplify $\dfrac{20}{8}$ [see Example 4-24].

Ratios are commonly written in three different forms: **fraction form, word form,** and **colon form.**

EXAMPLE 13-2: Write the following ratio in fraction form, word form, and colon form:
20 feet to 8 feet

Solution: The ratio of 20 feet to 8 feet $= \frac{5}{2}$ ⟵ fraction form

$= 5$ to 2 ⟵ word form

$= 5:2$ ⟵ colon form

Note: All three forms of the ratio ($\frac{5}{2}$, 5 to 2, and 5:2) are read as "five to two."

EXAMPLE 13-3: What is the meaning of the following sentence? The ratio of 20 feet to 8 feet is $\frac{5}{2}$, or 5 to 2, or 5:2.

Solution: This sentence means there are 5 feet in the first measure (20 feet) for each 2 feet in the second measure (8 feet).

Caution: The ratio of 20 feet to 8 feet is not the same as the ratio of 8 feet to 20 feet.

EXAMPLE 13-4: Find the ratio of 8 feet to 20 feet in lowest terms.

Solution: The ratio of 8 feet to 20 feet $= \dfrac{8 \text{ feet}}{20 \text{ feet}}$ ⟵ first given measure
⟵ second given measure

$$= \dfrac{8 \text{ feet}}{20 \text{ feet}} \qquad \text{Eliminate the common unit of measure.}$$

$$= \dfrac{2}{5} \qquad \text{Simplify } \dfrac{8}{20}.$$

Note: The ratio of 20 feet to 8 feet is $\frac{5}{2}$, or 5 to 2, or 5:2.
The ratio of 8 feet to 20 feet is $\frac{2}{5}$, or 2 to 5, or 2:5.

B. Find the ratio of unlike measures that belong to the same measurement family.

Measures can be separated into **measurement families.**

EXAMPLE 13-5: Name the common measurement families.

Solution: The common measurement families are
(a) length measures;
(b) capacity measures;
(c) weight (mass) measures;
(d) area measures;
(e) volume measures;
(f) time measures;
(g) money measures.

To rename unlike U.S. Customary family measures, you can use the unit conversions on page 237.
To rename unlike metric family measures, you can use the unit conversions on page 242.
To rename U.S. Customary measures as metric measures (or vice versa), you can use the unit conversions in Appendix Table 1 on page 402.

To rename unlike time and money measures, you can use the following unit conversions.

Time			Money		
1 century	=	10 decades	1 dollar	=	100 cents
1 decade	=	10 years	1 half-dollar	=	50 cents
1 year	=	12 months	1 quarter	=	25 cents
1 normal year	=	365 days	1 dime	=	10 cents
1 leap year	=	366 days	1 nickel	=	5 cents
1 business year	=	360 days			
1 week	=	7 days			
1 day	=	24 hours			
1 hour	=	60 minutes			
1 minute	=	60 seconds			

To rename time or money measures, you multiply by the correct unit fraction.

EXAMPLE 13-6: Rename $2\frac{1}{2}$ hours as minutes.

Solution: The necessary unit conversion is 1 hour = 60 minutes.

$$2\frac{1}{2} \text{ hours} = 2\frac{1}{2} \text{ hours} \times 1 \qquad\qquad \text{Multiply by 1.}$$

Substitute the correct unit conversion.

$$= \frac{2\frac{1}{2} \text{ hours}}{1} \times \frac{60 \text{ minutes}}{1 \text{ hour}} \longleftarrow 1 \text{ hour} = 60 \text{ minutes means } 1 = \frac{60 \text{ minutes}}{1 \text{ hour}}$$

$$= \frac{2\frac{1}{2} \text{ hours}}{1} \times \frac{60 \text{ minutes}}{1 \text{ hour}} \qquad \text{Eliminate the common unit of measure.}$$

$$= \left(2\frac{1}{2} \times 60\right) \text{ minutes} \qquad \text{Multiply.}$$

$$= 150 \text{ minutes}$$

Note: $2\frac{1}{2}$ hours can be renamed as 150 minutes.

To find the **ratio of unlike measures that belong to the same measurement family,** you first rename the unlike measures as like measures and then find the ratio of the like measures.

EXAMPLE 13-7: Find the ratio of $2\frac{1}{2}$ hours to 30 minutes.

Solution: The necessary unit conversion is 1 hour = 60 minutes.

$$2\frac{1}{2} \text{ hours} = 150 \text{ minutes} \qquad\qquad \text{Rename to get like measures.}$$

$$\frac{2\frac{1}{2} \text{ hours}}{30 \text{ minutes}} = \frac{150 \text{ minutes}}{30 \text{ minutes}} \longleftarrow \begin{array}{l} \text{The ratio of } 2\frac{1}{2} \text{ hours to 30 minutes} \\ = \text{the ratio of 250 minutes to 30 minutes.} \end{array}$$

$$= \frac{150 \text{ minutes}}{30 \text{ minutes}} \qquad\qquad \text{Eliminate the common unit of measure.}$$

$$= \frac{5}{1} \qquad\qquad\qquad \text{Simplify } \frac{150}{30}.$$

Note: The ratio of $2\frac{1}{2}$ hours to 30 minutes is $\frac{5}{1}$, or 5 to 1, or 5:1. This means that there are 5 minutes in the first given measure ($2\frac{1}{2}$ hours) for each 1 minute in the second given measure (30 minutes).

13-2. Finding the Rate

A. Find the rate of unlike measures that cannot be renamed as like measures.

Sometimes unlike measures cannot be renamed as like measures because they come from different measurement families and there is no unit conversion that relates them.

EXAMPLE 13-8: Can 300 miles and 15 gallons be renamed as like measures?

Solution: No. 300 miles and 15 gallons cannot be renamed as like measures because they come from different measurement families. 300 miles is a length measure and 15 gallons is a capacity measure, and there is no unit conversion that relates length measures and capacity measures.

The quotient of two unlike measures that cannot be renamed as like measures is called a **rate.** To find **the rate in lowest terms** of two unlike measures that cannot be renamed as like measures, you use the first measure for the numerator of a fraction and the second measure for the denominator. Then you divide the unlike measures and rename the numerical part of the fractional quotient in lowest terms. You should recall from Example 12-46 that the unlike measures in a rate cannot be eliminated.

EXAMPLE 13-9: Find the rate of 300 miles to 8 gallons in lowest terms.

Solution: The rate of 300 miles to 8 gallons $= \dfrac{300 \text{ miles}}{8 \text{ gallons}}$ ⟵ first given measure
⟵ second given measure

$= \dfrac{300}{8} \dfrac{\text{miles}}{\text{gallon}}$ Divide unlike measures [see Example 12-46].

$= \dfrac{75}{2} \dfrac{\text{miles}}{\text{gallon}}$ Simplify $\dfrac{300}{8}$.

Rates are commonly written in three different forms: **fraction form, word form,** and **slash form.**

EXAMPLE 13-10: Write the following rate in fraction form, word form, and slash form: 300 miles to 8 gallons.

Solutions: The rate of 300 miles to 8 gallons $= \dfrac{75}{2} \dfrac{\text{miles}}{\text{gallon}}$ ⟵———— fraction form

$= \dfrac{75}{2} \text{ miles per gallon}$ ⟵ word form

$= \dfrac{75}{2} \text{ miles/gallon}$ ⟵——— slash form

Note 1: All three forms of the rate $\left(\dfrac{75}{2} \dfrac{\text{miles}}{\text{gallon}}, \dfrac{75}{2} \text{ miles per gallon, and } \dfrac{75}{2} \text{ miles/gallon} \right)$ are read as "seventy-five-halves miles per gallon."

Note 2: In the rate $\dfrac{75}{2} \dfrac{\text{miles}}{\text{gallon}}$, $\dfrac{75}{2}$ is called the **amount** and $\dfrac{\text{miles}}{\text{gallon}}$ is called the **fractional unit of measure.** In a fractional unit of measure $\left(\dfrac{\text{miles}}{\text{gallon}} \right)$, the numerator is always plural (miles) and the denominator is always singular (gallon).

EXAMPLE 13-11: What is the meaning of the following sentence? The rate of 300 miles to 8 gallons is $\dfrac{75}{2} \dfrac{\text{miles}}{\text{gallon}}$, or $\dfrac{75}{2}$ miles per gallon, or $\dfrac{75}{2}$ miles/gallon.

Solution: This sentence means there are 75 miles in the first measure (300 miles) for each 2 gallons in the second measure (8 gallons).

Note: A car traveling 300 miles on 8 gallons of gasoline would average 75 miles for each 2 gallons of gasoline used, or $\dfrac{75}{2} \dfrac{\text{miles}}{\text{gallon}}$.

Caution: The rate of 300 miles to 8 gallons is not the same as the rate of 8 gallons to 300 miles.

EXAMPLE 13-12: Find the rate of 8 gallons to 300 miles in lowest terms.

Solution: The rate of 8 gallons to 300 miles $= \dfrac{8 \text{ gallons}}{300 \text{ miles}}$ ⟵ first given measure
⟵ second given measure

$= \dfrac{8}{300} \dfrac{\text{gallons}}{\text{mile}}$ ⟵ plural
⟵ singular

$= \dfrac{2}{75} \dfrac{\text{gallons}}{\text{mile}}$ Simplify $\dfrac{8}{300}$.

Note 1: The rate of 8 gallons to three hundred miles means that there are 2 gallons in the first measure (8 gallons) for each 75 miles in the second measure (300 miles).

Note 2: A car using 8 gallons of gasoline to travel 300 miles would average 2 gallons of gasoline used for each 75 miles traveled, or $\dfrac{2}{75}\dfrac{\text{gallons}}{\text{mile}}$.

Note 3: The rate of 8 gallons to 300 miles is $\dfrac{2}{75}\dfrac{\text{gallons}}{\text{mile}}$, or $\dfrac{2}{75}$ gallons per mile, or $\dfrac{2}{75}$ gallons/mile.

The rate of 300 miles to 8 gallons is $\dfrac{75}{2}\dfrac{\text{miles}}{\text{gallon}}$, or $\dfrac{75}{2}$ miles per gallon, or $\dfrac{75}{2}$ miles/gallon.

B. Find the unit rate.

A rate that has an amount that is a whole number, mixed number, or decimal is called a **unit rate.** Every rate can be renamed as a unit rate by renaming the fractional amount as a whole number, mixed number, or decimal.

EXAMPLE 13-13: Write $\dfrac{75}{2}\dfrac{\text{miles}}{\text{gallon}}$ as a unit rate.

Solution: $\dfrac{75}{2} = 75 \div 2$ Rename to get a whole number, mixed number, or decimal quotient.

$$= 37\frac{1}{2} \text{ or } 37.5$$

$$\frac{75}{2}\frac{\text{miles}}{\text{gallon}} = 37\frac{1}{2}\frac{\text{miles}}{\text{gallon}} \text{ or } 37.5\frac{\text{miles}}{\text{gallon}} \longleftarrow \text{ unit rate}$$

Note 1: The unit rate is $37\dfrac{1}{2}\dfrac{\text{miles}}{\text{gallon}}$ or $\dfrac{37\frac{1}{2}}{1}\dfrac{\text{miles}}{\text{gallon}}$, which means that there are $37\dfrac{1}{2}$ miles in the first measure (300 miles) for each 1 gallon in the second measure (8 gallons).

Note 2: A car traveling 300 miles on 8 gallons of gasoline would average $37\frac{1}{2}$ miles for each gallon of gasoline used, or $37\frac{1}{2}$ miles per gallon (mpg).

Note 3: Whenever you are asked to find the rate in a problem, your answer should always be the unit rate (unless otherwise stated).

13-3. Writing Proportions

A. Determine if two given ratios are proportional.

Two given ratios that are equal are called **proportional.** A **proportion** is formed by placing an **equality symbol** ($=$) is between two proportional ratios.

Note: Two ratios are proportional if the ratios (fractions) are equal.

To determine if two given ratios are proportional, you can compare the two fractions using cross multiplication. If the cross products of the fractions are equal, then the ratios are proportional. If the cross products of the fractions are not equal, then the ratios are not proportional.

EXAMPLE 13-14: Are the following ratios proportional? $\dfrac{20}{8}$ and $\dfrac{5}{2}$

Solution: To determine whether the fractions $\dfrac{20}{8}$ and $\dfrac{5}{2}$ are proportional, you can cross multiply [see Example 4-15].

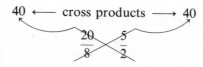

Since the cross products of $\frac{20}{8}$ and $\frac{5}{2}$ are equal ($40 = 40$), $\frac{20}{8}$ and $\frac{5}{2}$ are proportional ratios.

B. Write a proportion given two proportional ratios.

To write a proportion given two ratios that are proportional (equal), you just place an equality symbol ($=$) between them.

EXAMPLE 13-15: Write a proportion using the equal ratios $\frac{20}{8}$ and $\frac{5}{2}$.

Solution: The proportion for the equal ratios is $\frac{20}{8} = \frac{5}{2}$.

C. Identify like rates that are directly proportional.

Two rates that have identical fractional units of measure are called **like rates**. Two rates that have different fractional units of measure are called **unlike rates**.

EXAMPLE 13-16: Which of the following pairs of rates are like rates? (a) $\frac{8 \text{ people}}{2 \text{ pound}}$ and $\frac{12 \text{ people}}{3 \text{ pound}}$

(b) $\frac{72 \text{ beats}}{1 \text{ minute}}$ and $\frac{18 \text{ beats}}{15 \text{ second}}$

Solution

(a) $\frac{8 \text{ people}}{2 \text{ pound}}$ and $\frac{12 \text{ people}}{3 \text{ pound}}$ are like rates because they have identical fractional units of measure.

(b) $\frac{72 \text{ beats}}{1 \text{ minute}}$ and $\frac{18 \text{ beats}}{15 \text{ second}}$ are unlike rates because they have different fractional units of measure.

Two like rates are **directly proportional** if the amount of one like rate is proportional (equal) to the amount of the other like rate.

EXAMPLE 13-17: Show that the following two like rates are directly proportional:
$\frac{8 \text{ people}}{2 \text{ pound}}$ and $\frac{12 \text{ people}}{3 \text{ pound}}$

1. *Identify:*   To feed (8 people it takes 2 pounds)   Circle the two like rates.
   of hamburger. At that rate , to
   feed (12 people it will take 3 pounds)
   of hamburger.

2. *Understand:*   The amounts of like rates are $\frac{8}{2}$ and $\frac{12}{3}$, respectively.

3. *Decide:*   The amounts $\frac{8}{2}$ and $\frac{12}{3}$ are proportional because, when you cross-multiply,
   $3 \times 8 = 24$ and $2 \times 12 = 24$.

4. *Interpret:*   The two like rates $\frac{8}{2}$ and $\frac{12}{3}$ are directly proportional means that $\frac{8 \text{ people}}{2 \text{ pound}}$ is proportional (equal) to $\frac{12 \text{ people}}{3 \text{ pound}}$.

D. Write a direct proportion given two like rates that are directly proportional.

When two like rates are directly proportional, you can always write a **direct proportion** by using the two like rates to form two ratios and then equating the ratios, as shown in the following example.

EXAMPLE 13-18: Write a direct proportion given two like rates that are directly proportional.

1. *Identify:*   The normal human heart rate is   Circle the two like rates.
(72 beats in 60 seconds). At that
rate, the normal human heart
makes (18 beats in 15 seconds).

2. *Understand:*   The two like rates $\dfrac{72 \text{ beats}}{60 \text{ second}}$ and $\dfrac{18 \text{ beats}}{16 \text{ second}}$ are directly proportional because

$\dfrac{72}{60}$ and $\dfrac{18}{15}$ are proportional, since their cross products are equal:

$(15 \times 72 = 1080$ and $60 \times 18 = 1080)$.

3. *Decide:*   When two like rates are directly proportional, you can always write a direct
proportion as follows:

4. *Form ratios:* $\dfrac{72 \text{ beats}}{18 \text{ beats}}$ ⟵ 72 beats in 60 seconds ⟶ $\dfrac{60 \text{ seconds}}{15 \text{ seconds}}$
 ⟵ 18 beats in 15 seconds ⟶

direct proportion

5. *Write the proportion:*   beats ratio ⟶ $\dfrac{72}{18} = \dfrac{60}{15}$ seconds ratio

A proportion can always be written in four different ways, using either both original ratios, or using
both reciprocals of the original ratios. Example 13-19 shows how to form these proportions.

Example 13-19: Write the direct proportion $\dfrac{72}{18} = \dfrac{60}{15}$ in four different ways.

Solution

direct proportions

(a) beats ratio ⟶ $\dfrac{72}{18} = \dfrac{60}{15}$ ⟵ seconds ratio

(b) reciprocal of beats ratio ⟶ $\dfrac{18}{72} = \dfrac{15}{60}$ ⟵ reciprocal of seconds ratio

(c) seconds ratio ⟶ $\dfrac{60}{15} = \dfrac{72}{18}$ ⟵ beats ratio

(d) reciprocal of seconds ratio ⟶ $\dfrac{15}{60} = \dfrac{18}{72}$ ⟵ reciprocal of beats ratio

Note: To write a direct proportion, you use either both original ratios or both reciprocals of the
original ratios.

E. Identify like rates that are indirectly proportional.

Two like rates are **indirectly proportional** if the amount of one like rate is proportional (equal) to the
amount of the other like rate after interchanging both numerators or both denominators.

EXAMPLE 13-20: Show that the following two like rates are indirectly proportional:

$$\frac{30 \text{ inches}}{300 \text{ rpm}} \text{ and } \frac{15 \text{ inches}}{600 \text{ rpm}}$$

1. *Identify:*   A driving pulley with a diameter of
(30 inches turns at 300 revolutions per minute (rpm)).
The driven pulley with a
diameter of (15 inches turns at 600 rpm).

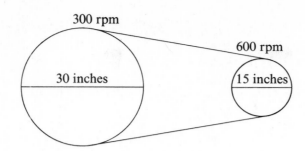

2. *Understand:*   The amounts of the two like rates are $\dfrac{30 \text{ inches}}{300 \text{ rpm}}$ and $\dfrac{15 \text{ inches}}{600 \text{ rpm}}$. To make the like rates indirectly proportional, the numerators can be interchanged:

interchange numerators

$$\dfrac{15}{300} \qquad \dfrac{30}{600}$$

3. *Decide:*   The amounts with the numerators interchanged are $\dfrac{15}{300}$ and $\dfrac{30}{600}$ and are proportional because $600 \times 15 = 300 \times 30 = 9000$.

4. *Interpret:*   Since $\dfrac{15}{300}$ and $\dfrac{30}{600}$ are proportional, the two like rates $\dfrac{30 \text{ inches}}{300 \text{ rpm}}$ and $\dfrac{15 \text{ inches}}{600 \text{ rpm}}$ are indirectly proportional.

Caution: Two like rates that are indirectly proportional are never directly proportional too. Likewise, two like rates that are directly proportional are never indirectly proportional too.

EXAMPLE 13-21: Show that the like rates from Example 13-20 are not directly proportional.

Solution: The like rates from Example 13-20 are $\dfrac{30 \text{ inches}}{300 \text{ rpm}}$ and $\dfrac{15 \text{ inches}}{600 \text{ rpm}}$ and are not directly proportional because $\dfrac{30}{300}$ and $\dfrac{15}{600}$ are not proportional, since their cross products are not equal $(600 \times 30 = 18{,}000$ and $300 \times 15 = 4500)$.

F. **Write an indirect proportion given two like rates that are indirectly proportional.**

When two like rates are indirectly proportional, you can always write an **indirect proportion** by using the two like rates to form two ratios and then equating one of the ratios with the reciprocal of the other ratio. This is illustrated in Example 13-22.

EXAMPLE 13-22: Write an indirect proportion given two like rates that are indirectly proportional.

1. *Identify:*   A driving gear turns at (100 rpm with 48 teeth).
The driven gear turns at (75 rpm with 64 teeth).

2. *Understand:*   The two like rates $\frac{100 \text{ rpm}}{48 \text{ teeth}}$ and $\frac{75 \text{ rpm}}{64 \text{ teeth}}$ are indirectly proportional because $\frac{75}{48}$

and $\frac{100}{64}$ are proportional ($64 \times 75 = 48 \times 100 = 4800$).

3. *Decide:*   When two like rates are indirectly proportional, you can always write an indirect proportion as follows:

4. *Form ratios:*

$$\frac{100 \text{ rpm}}{75 \text{ rpm}} \longleftarrow \begin{array}{l} \text{100 rpm with 48 teeth} \\ \text{75 rpm with 64 teeth} \end{array} \longrightarrow \frac{48 \text{ teeth}}{64 \text{ teeth}}$$

5. *Write the proportion:*   rpm ratio $\longrightarrow \overset{\text{indirect proportion}}{\frac{100}{75} = \frac{64}{48}} \longleftarrow$ reciprocal of teeth ratio

6. *Check:*

$\frac{100}{75}$ and $\frac{64}{48}$ are proportional because $48 \times 100 = 75 \times 64 = 4800$

Note: To write an indirect proportion, one of the ratios used to form the proportion must be a reciprocal ratio.

Recall: A proportion (direct or indirect) can always be written in four different ways, using various combinations of both original ratios and their reciprocals. Example 13-23 shows the four ways in which an indirect proportion can be written.

EXAMPLE 13-23: Write the indirect proportion $\frac{100}{75} = \frac{64}{48}$ in four different ways.

Solution

$$\overset{\text{indirect proportions}}{}$$

(a) rpm ratio $\longrightarrow \frac{100}{75} = \frac{64}{48} \longleftarrow$ reciprocal of teeth ratio

(b) reciprocal of rpm ratio $\longrightarrow \frac{75}{100} = \frac{48}{64} \longleftarrow$ teeth ratio

(c) teeth ratio $\longrightarrow \frac{48}{64} = \frac{75}{100} \longleftarrow$ reciprocal of rpm ratio

(d) reciprocal of teeth ratio $\longrightarrow \frac{64}{48} = \frac{100}{75} \longleftarrow$ rpm ratio

Caution: An indirect proportion always has exactly one reciprocal ratio.

13-4. Solving Proportions

A. Identify the terms of a proportion.

Each of the two numerators and two denominators that form a proportion is called a **term of the proportion.** Every proportion has exactly four terms.

EXAMPLE 13-24: What are the terms of the following proportion? $\frac{20}{8} = \frac{5}{2}$

Solution: The four terms of the proportion $\dfrac{20}{8} = \dfrac{5}{2}$ are 20, 5, 8, and 2.

B. Solve a proportion when one of the terms is unknown.

The proportion $\dfrac{n}{8} = \dfrac{5}{2}$ has one **unknown term** (*n*) and three **known terms** (5, 8, and 2). To **solve a proportion with one unknown term,** you must find a number that will replace the unknown term and thus form a proportion. This number is called the **solution of the proportion.**

To solve a proportion containing one unknown term, you can use the cross products to write a number sentence and then solve the number sentence.

EXAMPLE 13-25: Solve the following proportion: $\dfrac{n}{8} = \dfrac{5}{2}$

Solution: 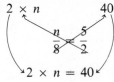   Find the cross products.

$\qquad\qquad 2 \times n = 40$ *Think:* In a proportion, the cross products are always equal.

$\qquad\qquad\quad n = 40 \div 2$ *Think:* To find an unknown factor, you divide the product by the known factor.

$\qquad\qquad\quad n = 20 \longleftarrow$ proposed solution

Check: $\qquad \dfrac{n}{8} = \dfrac{5}{2} \longleftarrow$ original proportion

$\qquad\qquad \dfrac{20}{8} \overset{?}{=} \dfrac{5}{2}$ Substitute the proposed solution $n = 20$.

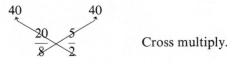   Cross multiply.

$n = 20$ is the correct solution for $\dfrac{n}{8} = \dfrac{5}{2}$ because $2 \times 20 = 8 \times 5 = 40$.

Note: The method shown in Example 13-25 will work when the unknown term is in the numerator or in the denominator.

EXAMPLE 13-26: Solve the following proportion: $\dfrac{20}{n} = \dfrac{5}{2}$

Solution: 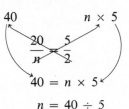   Find the cross products.

$\qquad\qquad 40 = n \times 5$ Equate the cross products.

$\qquad\qquad\quad n = 40 \div 5$ Divide the product by the known factor.

$\qquad\qquad\quad n = 8 \longleftarrow$ solution

Check: $\dfrac{20}{8} = \dfrac{5}{2}$ Substitute the solution in the original proportion.

Note: If you cross multiply, you can see that the solution is correct ($2 \times 20 = 8 \times 5 = 40$).

13-5. Solving Problems Using Proportions

A proportion can sometimes be used to solve a word problem. A word problem that can be solved using a proportion is called a **proportion problem.** To **solve a proportion problem,** you write and solve a proportion.

You can use a proportion to solve a word problem in which you are given a rate or ratio and a fixed amount for one of the units in that rate or ratio, and then asked to find the proportional amount of the other unit of measure in the rate or ratio.

A. Solve a proportion problem when a direct proportion is required.

When the two units of measure in a given rate (ratio) act in the same way, you solve the proportion problem using a **direct proportion.** Any two units of measure in a rate can be said to act in the same way if, as one unit of measure increases, the other unit of measure increases too. Or, if one unit of measure decreases, the other unit of measure decreases too.

Recall: You write a direct proportion by using two rates to form two ratios and then equating the ratios.

EXAMPLE 13-27: Solve this direct proportion problem:

1. *Identify:*   A certain car can travel (120 miles on 10 gallons) of gas. At that rate, how far can the car travel on (25 gallons) of gas?   Circle the given rate and the fixed amount of one of the measures involved in the rate.

2. *Understand:*   The two units of measure in the given rate (miles and gallons) act in the same way. That is, the number of miles increases as the number of gallons increases. (Or if the number of miles decreased the number of gallons would decrease.)

3. *Decide:*   To solve a proportion problem when the two units of measure in the given rate act in the same way, you write and solve a direct proportion.

4. *Form ratios:*

$$\overbrace{\hspace{3cm}}^{\text{ratios}}$$

$$\frac{120 \text{ miles}}{n \text{ miles}} \longleftarrow \frac{120 \text{ miles on 10 gallons}}{n \text{ miles on 25 gallons}} \longrightarrow \frac{10 \text{ gallons}}{25 \text{ gallons}}$$

5. *Write the proportion:*   miles ratio $\longrightarrow \dfrac{120}{n} \overbrace{=}^{\text{direct proportion}} \dfrac{10}{25} \longleftarrow$ gallons ratio

6. *Solve as before*

$$25 \times 120 \overset{\displaystyle\frac{120}{n} \times \frac{10}{25}}{\longleftarrow\qquad\longrightarrow} n \times 10$$ Find the cross products.

$$3000 = n \times 10$$ Equate the cross products.

$$n = 3000 \div 10$$ Divide the product by the known factor.

$$n = 300 \longleftarrow \text{ proposed solution}$$

7. *Interpret:* $n = 300$ means that at the given rate 25 gallons of gas will allow the car to travel 300 miles.

8. *Check:*   Are the two like rates $\dfrac{120}{10} \dfrac{\text{miles}}{\text{gallons}}$ and $\dfrac{300}{25} \dfrac{\text{miles}}{\text{gallons}}$ directly proportional?

Yes: $\dfrac{120}{10}$ and $\dfrac{300}{25}$ are directly proportional because

$25 \times 120 = 10 \times 300 = 3000$.

Recall: A proportion can always be written in four different ways.

EXAMPLE 13-28: Write the four direct proportions that can be used to solve the proportion problem just presented in Example 13-27.

Solution

direct proportions

(a) miles ratio ⟶ $\dfrac{120}{n} = \dfrac{10}{25}$ ⟵ gallons ratio

(b) reciprocal of miles ratio ⟶ $\dfrac{n}{120} = \dfrac{25}{10}$ ⟵ reciprocal of gallons ratio

(c) reciprocal of gallons ratio ⟶ $\dfrac{25}{10} = \dfrac{n}{120}$ ⟵ reciprocal of miles ratio

(d) gallons ratio ⟶ $\dfrac{10}{25} = \dfrac{120}{n}$ ⟵ miles ratio

Caution: To write a direct proportion, you either can use both the original ratios from the proportion problem or both of the reciprocals of the original ratios from the proportion problem. *Never* use one original ratio and one reciprocal ratio to write a direct proportion.

B. Solve a proportion problem when an indirect proportion is required.

When the two units of measure in the given rate (ratio) act in opposite ways, you solve the proportion problem using an **indirect proportion.** Any two units of measure can be said to act in opposite ways if as one unit of measure increases, the other unit of measure decreases. Or if one unit of measure decreases, the other unit of measure increases.

Recall: You write an indirect proportion by using two like rates to form two ratios and then equating one of the ratios with the reciprocal of the other ratio.

EXAMPLE 13-29: Solve this indirect proportion problem:

1. *Identify:* A certain job can be completed by ⟨5 women in 3 hours⟩. At that rate, how long would it take ⟨2 women⟩ to do the same job?

Circle the given rate and the fixed amount of one of the measures involved in the rate.

2. *Understand:* The two units of measure in the given rate (women and hours) act in opposite ways. That is, the number of women decreases as the number of hours increases. (Or if the number of women increased the number of hours would decrease.)

3. *Decide:* To solve a proportion problem when the two units of measure in the given rate act in opposite ways, you write and solve an indirect proportion.

⟵———————ratios———————⟶

4. *Form ratios:* $\dfrac{5 \text{ women}}{2 \text{ women}}$ ⟵ 5 women take 3 hours ⟶ $\dfrac{3 \text{ hours}}{n \text{ hours}}$
                      ⟵ 2 women take *n* hours ⟶

indirect proportion

5. *Write the proportion:* women ratio ⟶ $\dfrac{5}{2} = \dfrac{n}{3}$ ⟵ reciprocal of hours ratio

6. *Solve as before:*

$$15 \qquad n \times 2 \quad \text{Find the cross products.}$$

$$\dfrac{5}{2} \diagdown \dfrac{n}{3}$$

$$15 = n \times 2 \qquad \text{Equate the cross products.}$$

$$n = 15 \div 2 \qquad \text{Divide the product by the known factor.}$$

$$n = 7\dfrac{1}{2} \quad \longleftarrow \text{ solution}$$

7. *Interpret:* $n = 7\frac{1}{2}$ means that to complete the given job with 2 women would take $7\frac{1}{2}$ hours.

Check: Are the two like rates $\dfrac{5 \text{ women}}{3 \text{ hours}}$ and $\dfrac{2 \text{ women}}{7\frac{1}{2} \text{ hours}}$ indirectly proportional?

    Yes: After interchanging the numerators, you can see that $\dfrac{2}{3}$ and $\dfrac{5}{7\frac{1}{2}}$ are proportional because $7\frac{1}{2} \times 2 = 3 \times 5 = 15$.

Does it take 2 women longer to complete the job than it takes 5 women to complete the job? *Yes:* It takes 2 women $7\frac{1}{2}$ hours to complete the job, while it takes 5 women 3 hours to complete the job. The fact that it takes *less* women *more* time to do the job shows the rates to be indirectly proportional: as one unit of measure decreases, the other unit of measure increases. The indirect proportion can also be stated another way: it takes *more* women *less* time to do the job—so as one unit of measure increases, the other unit decreases.

Recall: A proportion can always be written in four different ways.

EXAMPLE 13-30: Write the four different indirect proportions that can be used to solve the proportion problem in Example 13-29.

Solution

indirect proportions

(a) women ratio $\longrightarrow$ $\overbrace{\dfrac{5}{2} = \dfrac{n}{3}}$ $\longleftarrow$ reciprocal of hour ratio

(b) reciprocal of women ratio $\longrightarrow$ $\dfrac{2}{5} = \dfrac{3}{n}$ $\longleftarrow$ hour ratio

(c) hour ratio $\longrightarrow$ $\dfrac{3}{n} = \dfrac{2}{5}$ $\longleftarrow$ reciprocal of women ratio

(d) reciprocal of hour ratio $\longrightarrow$ $\dfrac{n}{3} = \dfrac{5}{2}$ $\longleftarrow$ women ratio

Caution: An indirect proportion always uses exactly one reciprocal of one of the original ratios.

Caution: A direct proportion cannot be used to solve a proportion problem that requires an indirect proportion. Likewise, an indirect proportion cannot be used to solve a direct proportion.

EXAMPLE 13-31: Show that the direct proportion $\dfrac{5}{2} = \dfrac{3}{n}$ cannot be used to solve the proportion problem in Example 13-29.

1. *Solve as before:*   5n   6   Find the cross products.

                $5n = 6$ Equate the cross products.

                $n = \dfrac{6}{5}$ Divide the product by the known factor.

                $n = 1\dfrac{1}{5}$ $\longleftarrow$ proposed solution

2. *Interpret:* $n = 1\dfrac{1}{5}$ would mean that it takes 2 women less time to complete the given job in Example 13-29 than it did for 5 women. This doesn't seem possible. It should take fewer women more time, not less time.

3. *Conclusion:* The direct proportion $\dfrac{5}{2} = \dfrac{3}{n}$ cannot be used to solve the indirect proportion problem in Example 13-29.

RAISE YOUR GRADES

Can you . . . ?

☑ write the ratio of two like measures
☑ find the ratio of two like measures in lowest terms
☑ write a ratio in fraction form, word form, and colon form
☑ name the common measurement families
☑ find the ratio of unlike measures that belong to the same measurement family
☑ find the rate of two unlike measures that cannot be renamed as like measures in lowest terms
☑ write a rate in fraction form, word form, and slash form
☑ find the unit rate
☑ determine if two ratios are proportional (equal)
☑ write a proportion given two proportional ratios
☑ identify like rates
☑ identify two like rates that are directly proportional
☑ write a direct proportion for two like rates that are directly proportional
☑ write a direct proportion in four different ways
☑ identify two like rates that are indirectly proportional
☑ write an indirect proportion for two like rates that are indirectly proportional
☑ write an indirect proportion in four different ways
☑ identify the terms of a proportion
☑ solve a proportion containing one unknown term
☑ identify when a proportion can be used to solve a word problem
☑ identify when a direct proportion must be used to solve a proportion problem
☑ solve a proportion problem using a direct proportion
☑ identify when an indirect proportion must be used to solve a proportion problem
☑ solve a proportion problem using an indirect proportion

SUMMARY

1. To find the ratio of two like measures in lowest terms, you use the first like measure as the numerator of the fraction and the second like measure as the denominator of the fraction. Then you divide the like measures, and rename the fraction quotient in lowest terms.

2. The three different forms of a ratio are the fraction form $\left(\frac{5}{2}\right)$, the word form (5 to 2), and the colon form (5:2).

3. The common measurement families are length measures, capacity measures, weight (mass) measures, area measures, volume measures, time measures, and money measures.

4. To find the ratio of two unlike measures that belong to the same measurement family, you first use the correct unit conversion to rename the two unlike measures as like measures.

5. To rename unlike U.S. Customary family measures, you can use the unit conversions on page 237.

6. To rename unlike metric family measures, you can use the unit conversions that begin on page 242.

7. To rename unlike time and money measures, you can use the unit conversions on page 263.

8. To rename U.S. Customary measures as metric measures (or vice versa), you can use the unit conversions in Appendix Table 1 on page 402.

9. The quotient of two unlike measures that cannot be renamed as like measures is called a rate.

10. To find the rate in lowest terms of two unlike measures that cannot be renamed as like measures, you use the first measure for the numerator of a fraction and the second measure for the denominator. Then you divide the unlike measures and rename the numerical part of the fractional quotient in lowest terms.

11. The three different forms of a rate are the fraction form $\left(\frac{75}{2}\frac{\text{miles}}{\text{gallon}}\right)$, the word form $\left(\frac{75}{2}\text{ miles per gallon}\right)$, and the slash form $\left(\frac{75}{2}\text{ miles/gallon}\right)$.

12. In a rate's fractional unit of measure, the numerator is always plural and the denominator is always singular.

13. Every rate can be renamed as a unit rate by renaming the fractional amount as a whole number, mixed number, or decimal.

14. To determine if two given ratios are proportional, you can compare the two fractions using cross multiplication. If the fractions cross products of the fractions are equal, then the ratios are proportional. If the cross products of the fractions are not equal, then the ratios are not proportional.

15. To write a proportion given two proportional (equal) ratios, you place an equality symbol ($=$) between them.

16. Two rates that have different fractional units of measure are called unlike rates.

17. Two rates that have identical fractional units of measures are called like rates.

18. Two like rates are directly proportional if the amount of one like rate is proportional (equal) to the amount of the other like rate.

19. When two like rates are directly proportional, you can write a direct proportion by using the two like rates to form two ratios and then equating the two ratios.

20. Two like rates are indirectly proportional if the amount of one like rate is proportional (equal) to the amount of the other like rate after interchanging both numerators or both denominators.

21. When two like rates are indirectly proportional, you can write an indirect proportion by using the two like rates to form two ratios and then equating one of the ratios with the reciprocal of the other ratio.

22. To solve a proportion containing one unknown term, you can use the cross products to write a number sentence and then solve the number sentence.

23. A proportion can be used to solve a word problem if you are given a rate (ratio) and a fixed amount of one of the units of measure involved in the rate (ratio), and then asked to find a proportional amount of the other unit of measure in the rate (ratio).

24. A word problem that can be solved using a proportion is called a proportion problem.

25. To solve a proportion problem, you write either a direct or indirect proportion and then solve the proportion.

26. You can solve a proportion problem using a direct proportion when the two units of measure involved in the rate (ratio) act in the same way. That is, the number of one amount increases as the number of the other amount increases. Likewise, the number of one amount decreases as the number of the other amount decreases.

27. You can solve a proportion problem using an indirect proportion when the two units of measure involved in the rate act in opposite ways. That is, the number of one amount increases as the number of the other amount decreases. Likewise, the number of one amount decreases as the number of the other amount increases.

28. *Caution:* A direct proportion cannot be used to solve a proportion problem that requires an indirect proportion (and vice versa).

SOLVED PROBLEMS

PROBLEM 13-1   Find each ratio of each pair of like measures in lowest terms:

(a) 126 meters to 42 meters   **(b)** 32 wins to 48 wins   **(c)** $5\frac{1}{3}$ days to 24 days   **(d)** \$2.25 to \$15
(e) 225 people to 150 people

Solution: Recall that to find the ratio of two like measures in lowest terms, you use the first like measure as the numerator of the fraction and the second like measure as the denominator. Then you divide the like measures, and rename the fraction quotient in lowest terms [see Example 13-1]:

(a) 126 meters to 42 meters $= \dfrac{126 \text{ meters}}{42 \text{ meters}} = \dfrac{126}{42} = \dfrac{3}{1}$

(b) 32 wins to 48 wins $= \dfrac{32 \text{ wins}}{48 \text{ wins}} = \dfrac{32}{48} = \dfrac{2}{3}$

(c) $5\frac{1}{3}$ days to 24 days $= \dfrac{5\frac{1}{3}\text{ days}}{24\text{ days}} = \dfrac{5\frac{1}{3}}{24} = 5\frac{1}{3} \div 24 = \dfrac{16}{3} \div \dfrac{24}{1} = \dfrac{16}{3} \times \dfrac{1}{24} = \dfrac{2}{9}$

(d) \$2.25 to \$15 $= \dfrac{\$2.25}{\$15} = \dfrac{2.25}{15} = \dfrac{225}{1500} = \dfrac{3}{20}$

(e) 225 people to 150 people $= \dfrac{225\text{ people}}{150\text{ people}} = \dfrac{225}{150} = \dfrac{3}{2}$

PROBLEM 13-2   Rename each time or money measure as indicated:

(a) 3 decades = ? years   **(b)** 1 normal year = ? hours   **(c)** 1 day = ? seconds
(d) 9 quarters = ? cents   **(e)** 615 cents = ? nickels

Solution: Recall that to rename unlike time or money measures, you can use the unit conversions on page 263 [see Example 13-6]:

(a) 3 decades $= \dfrac{3\text{ decades}}{1} \times \dfrac{10\text{ years}}{1\text{ decade}} = (3 \times 10)\text{ years} = 30\text{ years}$

(b) 1 normal year $= \dfrac{1\text{ normal year}}{1} \times \dfrac{365\text{ days}}{1\text{ normal year}} \times \dfrac{24\text{ hours}}{1\text{ day}} = (365 \times 24)\text{ hours} = 8760\text{ hours}$

(c) 1 day $= \dfrac{1\text{ day}}{1} \times \dfrac{24\text{ hours}}{1\text{ day}} \times \dfrac{60\text{ minutes}}{1\text{ hour}} \times \dfrac{60\text{ seconds}}{1\text{ minute}} = (24 \times 60 \times 60)\text{ seconds} = 86{,}400$
seconds

(d) 9 quarters $= \dfrac{9\text{ quarters}}{1} \times \dfrac{25\text{ cents}}{1\text{ quarter}} = (9 \times 25)\text{ cents} = 225\text{ cents}$

(e) 615 cents $= \dfrac{615\text{ cents}}{1} \times \dfrac{1\text{ nickel}}{5\text{ cents}} = \dfrac{615}{5}\text{ nickels} = 123\text{ nickels}$

PROBLEM 13-3   Find each ratio of two unlike measures that belong to the same measurement family:

(a) 45 minutes to 6 hours   **(b)** 2 dimes to 3 quarters   **(c)** 42 nurses to 126 doctors
(d) 32 Mondays to 48 Fridays   **(e)** 48 feet to 12 yards

Solution: Recall that to find the ratio of two unlike measures that belong to the same measurement family, you first rename the unlike measures as like measures and then find the ratio of the like measures [see Example 13-7]:

(a) 6 hours $= \dfrac{6\text{ hours}}{1} \times \dfrac{60\text{ minutes}}{1\text{ hour}} = (6 \times 60)\text{ minutes} = 360\text{ minutes}$

45 minutes to 6 hours = 45 minutes to 360 minutes $= \dfrac{45\text{ minutes}}{360\text{ minutes}} = \dfrac{1}{8}$

(b) 2 dimes = 20 cents   3 quarters = 75 cents

2 dimes to 3 quarters = 20 cents to 75 cents $= \dfrac{20\text{ cents}}{75\text{ cents}} = \dfrac{4}{15}$

(c) 42 nurses = 42 people   126 doctors = 126 people

42 nurses to 126 doctors = 42 people to 126 people $= \dfrac{42\text{ people}}{126\text{ people}} = \dfrac{1}{3}$

(d) 32 Mondays = 32 days   48 Fridays = 48 days

32 Mondays to 48 Fridays = 32 days to 48 days $= \dfrac{32\text{ days}}{48\text{ days}} = \dfrac{2}{3}$

(e) 48 feet = 16 yards

$$48 \text{ feet to 12 yards} = 16 \text{ yards to 12 yards} = \frac{16 \text{ yards}}{12 \text{ yards}} = \frac{4}{3}$$

PROBLEM 13-4 Find each rate in fraction form in lowest terms:

(a) 500 miles to 30 gallons **(b)** 496 miles to 12 hours **(c)** $28 to 8 pounds
(d) 810 people to 60 rooms **(e)** 7.5 grams to 30¢

Solution: Recall that to find the rate in lowest terms of two unlike measures that cannot be renamed as like measures, you use the first given measure for a numerator of a fraction and the second given measure for the denominator. Then you divide the unlike measures, and rename the numerical part of the fractional quotient in lowest terms. Remember that the numerator is always plural and the denominator is always singular in a rate [see Example 13-9]:

(a) $500 \text{ miles to 30 gallons} = \dfrac{500 \text{ miles}}{30 \text{ gallons}} = \dfrac{500 \text{ miles}}{30 \text{ gallon}} = \dfrac{50 \text{ miles}}{3 \text{ gallon}}$

(b) $496 \text{ miles to 12 hours} = \dfrac{496 \text{ miles}}{12 \text{ hours}} = \dfrac{496 \text{ miles}}{12 \text{ hour}} = \dfrac{124 \text{ miles}}{3 \text{ hour}}$

(c) $28 \text{ dollars to 8 pounds} = \dfrac{28 \text{ dollars}}{8 \text{ pounds}} = \dfrac{28 \text{ dollars}}{8 \text{ pound}} = \dfrac{7 \text{ dollars}}{2 \text{ pound}}$

(d) $810 \text{ people to 60 rooms} = \dfrac{810 \text{ people}}{60 \text{ rooms}} = \dfrac{810 \text{ people}}{60 \text{ room}} = \dfrac{27 \text{ people}}{2 \text{ room}}$

(e) $7.5 \text{ grams to 30 cents} = \dfrac{7.5 \text{ grams}}{30 \text{ cents}} = \dfrac{7.5 \text{ grams}}{30 \text{ cent}} = \dfrac{1 \text{ gram}}{4 \text{ cent}}$

PROBLEM 13-5 Find the unit rate for each fractional rate found in Problem 13-14:

(a) $\dfrac{50 \text{ miles}}{3 \text{ gallon}}$ **(b)** $\dfrac{124 \text{ miles}}{3 \text{ hour}}$ **(c)** $\dfrac{7 \text{ dollars}}{2 \text{ pound}}$ **(d)** $\dfrac{27 \text{ people}}{2 \text{ room}}$ **(e)** $\dfrac{1 \text{ gram}}{4 \text{ cent}}$

Solution: Recall that every rate can be renamed as a unit rate by renaming the fractional amount as a whole number, mixed number, or decimal [see Example 13-13]:

(a) $\dfrac{50 \text{ miles}}{3 \text{ gallon}} = 16\frac{2}{3} \dfrac{\text{miles}}{\text{gallon}}$ or $16\frac{2}{3}$ miles/gallon or $16\frac{2}{3}$ miles per gallon or $16\frac{2}{3}$ mpg

(b) $\dfrac{124 \text{ miles}}{3 \text{ hour}} = 41\frac{1}{3} \dfrac{\text{miles}}{\text{hour}}$ or $41\frac{1}{3}$ miles/hour or $41\frac{1}{3}$ miles per hour on $41\frac{1}{3}$ mph

(c) $\dfrac{7 \text{ dollars}}{2 \text{ pound}} = 3\frac{1}{2} \dfrac{\text{dollars}}{\text{pound}}$ or 3.5 dollars/pound or $3.50 per pound or $3.50/pound

(d) $\dfrac{27 \text{ people}}{2 \text{ room}} = 13\frac{1}{2} \dfrac{\text{people}}{\text{room}}$ or 13.5 people/room or 13.5 people per room

(e) $\dfrac{1 \text{ gram}}{4 \text{ cent}} = 0.25 \dfrac{\text{gram}}{\text{cent}}$ or 0.25 gram/cent or 0.25 gram per cent or 0.25 gram/¢

PROBLEM 13-6 Write a proportion for each pair of ratios that are proportional:

(a) $\dfrac{10}{32}$ and $\dfrac{21}{63}$ **(b)** $\dfrac{150}{60}$ and $\dfrac{45}{18}$ **(c)** $\dfrac{12}{18}$ and $\dfrac{28}{42}$ **(d)** $\dfrac{11}{16}$ and $\dfrac{21}{32}$ **(e)** $\dfrac{2\frac{1}{2}}{3}$ and $\dfrac{\frac{5}{8}}{\frac{3}{4}}$

Solution: Recall that to determine if two given ratios are proportional, you can compare the two fractions using cross multiplication. If the cross products of the fractions are equal, then the ratios are proportional. If the cross products of the fractions are not equal, then the fractions are not proportional. Also recall that to write a proportion given two ratios that are proportional (equal), you just write an equality symbol (=) between them [see Examples 13-14 and 13-15]:

(a) $\dfrac{10}{32}$ and $\dfrac{21}{63}$ are not proportional because $63 \times 10 = 630$ and $32 \times 21 = 672$.

(b) $\dfrac{150}{60}$ and $\dfrac{45}{18}$ are proportional because $18 \times 150 = 60 \times 45 = 2700$: $\dfrac{150}{60} = \dfrac{45}{18}$.

(c) $\dfrac{12}{18}$ and $\dfrac{28}{42}$ are proportional because $42 \times 12 = 18 \times 28 = 504$: $\dfrac{12}{18} = \dfrac{28}{42}$.

(d) $\dfrac{11}{16}$ and $\dfrac{21}{32}$ are not proportional because $32 \times 11 = 352$ and $16 \times 21 = 336$.

(e) $\dfrac{2\frac{1}{2}}{3}$ and $\dfrac{\frac{5}{8}}{\frac{3}{4}}$ are proportional because $\dfrac{3}{4} \times 2\dfrac{1}{2} = 3 \times \dfrac{5}{8} = 1\dfrac{7}{8}$.

PROBLEM 13-7 Write a direct proportion for each pair of like rates that are directly proportional:

(a) $\dfrac{10}{6} \dfrac{\text{miles}}{\text{minutes}}$ and $\dfrac{5}{3} \dfrac{\text{miles}}{\text{minutes}}$ **(b)** $\dfrac{35}{45} \dfrac{\text{feet}}{\text{dollars}}$ and $\dfrac{4}{5} \dfrac{\text{feet}}{\text{dollars}}$ **(c)** $\dfrac{7}{8} \dfrac{\text{days}}{\text{pounds}}$ and $\dfrac{42}{48} \dfrac{\text{days}}{\text{pounds}}$

(d) $\dfrac{30}{20} \dfrac{\text{liters}}{\text{grams}}$ and $\dfrac{24}{15} \dfrac{\text{liters}}{\text{grams}}$ **(e)** $\dfrac{\frac{1}{2}}{\frac{2}{3}} \dfrac{\text{ounce}}{\text{cent}}$ and $\dfrac{12}{16} \dfrac{\text{ounces}}{\text{cents}}$

Solution: Recall that two like rates are directly proportional if the amount of one like rate is proportional (equal) to the amount of the other like rate. Also recall that when two like rates are directly proportional, you can always write a direct proportion by using the two like rates to form two ratios and then equating the ratios [see Examples 13-17 and 13-18]:

(a) $\dfrac{10}{6} \dfrac{\text{miles}}{\text{minutes}}$ and $\dfrac{5}{3} \dfrac{\text{miles}}{\text{minutes}}$ are directly proportional because $\dfrac{10}{6} = \dfrac{5}{3}$:

$$\begin{array}{l}\text{ratios} \qquad\qquad\qquad\qquad\qquad \text{direct proportion}\\[4pt] \dfrac{10\ \text{miles}}{5\ \text{miles}} \longleftarrow \dfrac{10\ \text{miles/6 minutes}}{5\ \text{miles/3 minutes}} \longrightarrow \dfrac{6\ \text{minutes}}{3\ \text{minutes}}\ \text{means}\ \dfrac{10}{5} = \dfrac{6}{3}\end{array}$$

(b) $\dfrac{35}{45} \dfrac{\text{feet}}{\text{dollars}}$ and $\dfrac{4}{5} \dfrac{\text{feet}}{\text{dollars}}$ are not directly proportional because $\dfrac{35}{45} \neq \dfrac{4}{5}$:

$$5 \times 35 = 175 \text{ and } 45 \times 4 = 180$$

(c) $\dfrac{7}{8} \dfrac{\text{days}}{\text{pounds}}$ and $\dfrac{42}{48} \dfrac{\text{days}}{\text{pounds}}$ are directly proportional because $\dfrac{7}{8} = \dfrac{42}{48}$:

$$\dfrac{7\ \text{days}}{42\ \text{days}} \longleftarrow \dfrac{7\ \text{days/8 pounds}}{42\ \text{days/48 pounds}} \longrightarrow \dfrac{8\ \text{pounds}}{48\ \text{pounds}}\ \text{means}\ \dfrac{7}{42} = \dfrac{8}{48}$$

(d) $\dfrac{30}{20} \dfrac{\text{liters}}{\text{grams}}$ and $\dfrac{24}{15} \dfrac{\text{liters}}{\text{grams}}$ are not directly proportional because $\dfrac{30}{20} \neq \dfrac{24}{15}$:

$$15 \times 30 = 450 \text{ and } 20 \times 24 = 480$$

(e) $\dfrac{\frac{1}{2}}{\frac{2}{3}} \dfrac{\text{ounce}}{\text{cent}}$ and $\dfrac{12}{16} \dfrac{\text{ounces}}{\text{cents}}$ are directly proportional because $\dfrac{\frac{1}{2}}{\frac{2}{3}} = \dfrac{12}{16}$:

$$\dfrac{\frac{1}{2}\ \text{ounce}}{12\ \text{ounces}} \longrightarrow \dfrac{\frac{1}{2}\ \text{ounce/}\frac{2}{3}\ \text{cent}}{12\ \text{ounces/16 cents}} \longrightarrow \dfrac{\frac{2}{3}\ \text{cent}}{16\ \text{cents}}\ \text{means}\ \dfrac{\frac{1}{2}}{12} = \dfrac{\frac{2}{3}}{16}$$

PROBLEM 13-8 Write an indirect proportion for each pair of like rates that are indirectly proportional: **(a)** $\dfrac{20}{15} \dfrac{\text{men}}{\text{days}}$ and $\dfrac{12}{25} \dfrac{\text{men}}{\text{days}}$ **(b)** $\dfrac{12}{210} \dfrac{\text{inches}}{\text{rpm}}$ and $\dfrac{14}{180} \dfrac{\text{inches}}{\text{rpm}}$ **(c)** $\dfrac{20}{480} \dfrac{\text{teeth}}{\text{rpm}}$ and $\dfrac{55}{180} \dfrac{\text{teeth}}{\text{rpm}}$

(d) $\dfrac{56}{33} \dfrac{\text{women}}{\text{hours}}$ and $\dfrac{82}{24} \dfrac{\text{women}}{\text{hours}}$ **(e)** $\dfrac{5}{\frac{1}{2}} \dfrac{\text{people}}{\text{hour}}$ and $\dfrac{3}{\frac{5}{6}} \dfrac{\text{people}}{\text{hour}}$

Solution: Recall that two like rates are indirectly proportional if the amount of one like rate is proportional (equal) to the amount of the other like rate after interchanging both numerators or both denominators. Also recall that when two like rates are indirectly proportional, you can always write an indirect proportion by using the two like rates to form two ratios and then equating one of the ratios with the reciprocal of the other ratio [see Examples 13-20 and 13-21]:

interchange numerators

(a) $\dfrac{20 \text{ men}}{15 \text{ days}}$ and $\dfrac{12 \text{ men}}{25 \text{ days}}$ are indirectly proportional because $\dfrac{12}{15} = \dfrac{20}{25}$:

ratio   indirect proportion

$\dfrac{20 \text{ men}}{12 \text{ men}} \longleftarrow \dfrac{20 \text{ men/15 days}}{12 \text{ men/25 days}} \longrightarrow \dfrac{15 \text{ days}}{25 \text{ days}}$ means $\dfrac{20}{12} = \dfrac{25}{15}$ $\longleftarrow$ one reciprocal ratio

(b) $\dfrac{12 \text{ inches}}{210 \text{ rpm}}$ and $\dfrac{14 \text{ inches}}{180 \text{ rpm}}$ are indirectly proportional because $\dfrac{14}{210} = \dfrac{12}{180}$:

$\dfrac{12 \text{ inches}}{14 \text{ inches}} \longleftarrow \dfrac{12 \text{ inches/210 rpm}}{14 \text{ inches/180 rpm}} \longrightarrow \dfrac{210 \text{ rpm}}{180 \text{ rpm}}$ means $\dfrac{12}{14} = \dfrac{180}{210}$

(c) $\dfrac{20 \text{ teeth}}{480 \text{ rpm}}$ and $\dfrac{55 \text{ teeth}}{180 \text{ rpm}}$ are not indirectly proportional because $\dfrac{55}{480} \neq \dfrac{20}{180}$:

$$180 \times 55 = 9900 \text{ and } 480 \times 20 = 9600$$

(d) $\dfrac{56 \text{ women}}{33 \text{ hours}}$ and $\dfrac{82 \text{ women}}{24 \text{ hours}}$ are not indirectly proportional because $\dfrac{82}{33} \neq \dfrac{56}{24}$:

$$24 \times 82 = 1968 \text{ and } 33 \times 56 = 1848$$

(e) $\dfrac{5 \text{ people}}{\frac{1}{2} \text{ hour}}$ and $\dfrac{3 \text{ people}}{\frac{5}{6} \text{ hour}}$ are indirectly proportional because $\dfrac{3}{\frac{1}{2}} = \dfrac{5}{\frac{5}{6}}$:

$\dfrac{5 \text{ people}}{3 \text{ people}} \longleftarrow \dfrac{5 \text{ people/}\frac{1}{2} \text{ hour}}{3 \text{ people/}\frac{5}{6} \text{ hour}} \longrightarrow \dfrac{\frac{1}{2} \text{ hour}}{\frac{5}{6} \text{ hour}}$ means $\dfrac{5}{3} = \dfrac{\frac{5}{6}}{\frac{1}{2}}$

PROBLEM 13-9 Solve each proportion containing one unknown term:

(a) $\dfrac{n}{51} = \dfrac{12}{18}$ (b) $\dfrac{3\frac{1}{2}}{a} = \dfrac{36}{24}$ (c) $\dfrac{5}{2.5} = \dfrac{7.2}{b}$ (d) $\dfrac{2}{7} = \dfrac{c}{35}$ (e) $\dfrac{y}{9} = \dfrac{100}{16\frac{2}{3}}$

Solution: Recall that to solve a proportion containing one unknown term, you can use the cross products of the fractions to write a number sentence and then solve the number sentence [see Examples 13-25 and 13-26]:

(a) $\dfrac{n}{51} = \dfrac{12}{18}$ means $18 \times n = 51 \times 12$ (b) $\dfrac{3\frac{1}{2}}{a} = \dfrac{36}{24}$ means $24 \times 3\frac{1}{2} = a \times 36$

$$18 \times n = 612$$
$$n = 612 \div 18$$
$$n = 34$$

Check: $18 \times 34 = 51 \times 12 = 612$

$$84 = a \times 36$$
$$a = 84 \div 36$$
$$a = 2\frac{1}{3}$$

Check: $24 \times 3\frac{1}{2} = 2\frac{1}{3} \times 36 = 84$

(c) $\dfrac{5}{2.5} = \dfrac{7.2}{b}$ means $b \times 5 = 2.5 \times 7.2$

$$b \times 5 = 18$$

$$b = 18 \div 5$$

$$b = 3\frac{3}{5} \text{ or } 3.6$$

Check: $3.6 \times 5 = 2.5 \times 7.2 = 18$

(d) $\dfrac{2}{7} = \dfrac{c}{35}$ means $35 \times 2 = 7 \times c$

$$70 = 7 \times c$$

$$c = 70 \div 7$$

$$c = 10$$

Check: $35 \times 2 = 7 \times 10 = 70$

(e) $\dfrac{y}{9} = \dfrac{100}{16\frac{2}{3}}$ means $16\frac{2}{3} \times y = 9 \times 100$

$$16\frac{2}{3} \times y = 900$$

$$y = 900 \div 16\frac{2}{3}$$

$$y = 54$$

Check: $16\frac{2}{3} \times 54 = 9 \times 100 = 900$

PROBLEM 13-10   Solve each proportion problem using a direct proportion:

(a) A photo is 8 inches long and 5 inches wide. If a reduced copy is made so that the width is 3 inches, then what will be the length of the copy?

(b) A women earns \$560 in 15 days. At that rate how much will she earn in 7 days?

(c) An 8-ounce jar of instant coffee contains 12 servings. At that rate, how many servings will a 22-ounce jar of the same coffee contain?

(d) At the rate given in problem **(c)**, how much instant coffee is needed for 22 servings?

(e) Concrete is made of cement, sand, and rock. The ratio of sand to rock is 2:4 in concrete. How much sand is needed for 125 pounds of rocks?

(f) At the rate given in problem **(e)**, how much rock is needed for 125 tons of sand?

Solution: Recall that to solve a proportion problem you write either a direct or an indirect proportion and then solve the proportion. Also recall that you solve a proportion problem using a direct proportion when the two units of measure in the rate (ratio) act in the same way. That is, the number of one amount increases as the number of the other amount increases or, the number of one amount decreases as the number of the other amount decreases [see Example 13-27]:

(a) *Understand:*   The problem asks you to compare two ratios. The two units of measure involved in any ratio are always the same (inches and inches).

Decide: To solve a proportion problem when the two units of measure are the same, you write and solve a direct proportion.

Form ratios:

$$\underbrace{\quad}_{\text{length}} \qquad \underbrace{\quad}_{\text{width}}$$

$$\frac{8 \text{ inches}}{n \text{ inches}} \longleftarrow \text{8 inches/5 inches} \longrightarrow \frac{5 \text{ inches}}{3 \text{ inches}}$$

Write the proportion: length ratio $\longrightarrow \dfrac{8}{n} = \dfrac{5}{3} \longleftarrow$ width ratio

Solve as before:

$3 \times 8 = n \times 5$	Equate the cross products.
$24 = n \times 5$	Simplify.
$n = 24 \div 5$	Divide the product by the known factor.
$n = 4\dfrac{4}{5}$	Check as before.

Interpret: $n = 4\frac{4}{5}$ means a 5 inch by 8 inch photo will reduce to a 3 inch by $4\frac{4}{5}$ inch photo.

(b) *Understand:*   In this problem, you compare two rates. The two units of measure involved in the given rate (dollars and days) act in the same way. That is, the number of dollars increases as the number of days increases. Or, the number of dollars decreases as the number of days decreases.

Decide: To solve a proportion when the two units of measure involved in the given rate act in the same way, you write and solve a direct proportion.

Form ratios: $\dfrac{560 \text{ dollars}}{n \text{ dollars}} \longleftarrow \begin{array}{c} \$560 \text{ in 15 days} \\ \$n \quad \text{in} \ 7 \text{ days} \end{array} \longrightarrow \dfrac{15 \text{ days}}{7 \text{ days}}$

Write the proportion: dollars ratio $\longrightarrow \dfrac{560}{n} = \dfrac{15}{7} \longleftarrow$ days ratio

Solve as before:

$7 \times 560 = n \times 15$	Equate cross products.
$3920 = n \times 15$	
$n = 3920 \div 15$	Divide the product by the known factor.
$n = 261.33\overline{3}$	Divide to the thousandths place.
$n \approx 261.33$	Round to the nearest hundredth.

Interpret: $n \approx 261.33$ means that, at the given rate, the woman will earn \$266.33 (to the nearest cent) in 7 days.

(c) The two units of measure is this rate (ounces and servings) act in the same way: as ounces increase, so do servings. The direct proportion is set up as follows:

$$\frac{8 \text{ ounces}}{22 \text{ ounces}} \longleftarrow \begin{array}{c} \text{8 ounces/12 servings} \\ \text{22 ounces/}n \text{ servings} \end{array} \longrightarrow \frac{12 \text{ servings}}{n \text{ servings}}$$

$$\text{ounces ratio} \longrightarrow \frac{8}{22} = \frac{12}{n} \longleftarrow \text{serving ratio}$$

$$n \times 8 = 22 \times 12$$

$$n \times 8 = 264$$

$$n = 264 \div 8$$

$$n = 33 \longleftarrow \text{solution}$$

This means that a 22-ounce jar of the same instant coffee contains 33 servings.

(d) Now the unknown term is ounces instead of servings.

$$\frac{8 \text{ ounces}}{n \text{ ounces}} \longleftarrow \begin{array}{c} \text{8 ounces/12 servings} \\ n \text{ ounces/22 servings} \end{array} \longrightarrow \frac{12 \text{ servings}}{22 \text{ servings}}$$

$$\text{ounces ratio} \longrightarrow \frac{8}{n} = \frac{12}{22} \longleftarrow \text{serving ratio}$$

$$22 \times 8 = n \times 12$$

$$176 = n \times 12$$

$$n = 176 \div 12$$

$$n = 14\frac{2}{3} \longleftarrow \text{solution}$$

This means that $14\frac{2}{3}$ ounces is needed to make 22 servings of instant coffee.

(e) The two units of measure (pounds of sand and pounds of rock) act in the same way, so you would set up the direct proportion as follows:

$$\frac{2 \text{ pounds sand}}{n \text{ pounds sand}} \begin{array}{c} \longleftarrow \; 2: \; 4 \; \longrightarrow \\ \longleftarrow \; n:125 \; \longrightarrow \end{array} \frac{4 \text{ pounds rock}}{125 \text{ pounds rock}}$$

$$\text{sand ratio} \longrightarrow \frac{2}{n} = \frac{4}{125} \longleftarrow \text{rock ratio}$$

$$125 \times 2 = n \times 4$$

$$250 = n \times 4$$

$$n = 250 \div 4$$

$$n = 62\frac{1}{2} \longleftarrow \text{solution}$$

This means that $62\frac{1}{2}$ pounds of sand is needed for 125 pounds of rocks.

(f) Now the unknown term is rocks, not sand:

$$\frac{2 \text{ tons sand}}{125 \text{ tons sand}} \begin{array}{c} \longleftarrow \quad 2:4 \; \longrightarrow \\ \longleftarrow \; 125:n \; \longrightarrow \end{array} \frac{4 \text{ tons rocks}}{n \text{ tons rocks}}$$

$$\text{sand ratio} \longrightarrow \frac{2}{125} = \frac{4}{n} \longleftarrow \text{rocks ratio}$$

$$n \times 2 = 125 \times 4$$

$$n \times 2 = 500$$

$$n = 500 \div 2$$

$$n = 250 \longleftarrow \text{solution}$$

This means that 250 tons of rock is needed for 125 tons of sand.

PROBLEM 13-11   Solve each proportion problem using an indirect proportion:

(a) It takes 6 men 14 days to do a job. At that rate, how long will it take 8 men to do the same job?

(b) At the rate given in problem **(a)**, what is the fewest number of men needed to do the same job in 9 days?

(c) A driven pulley with a diameter of 10 inches turns at 500 rpm. How many revolutions per minute does the driving pulley turn at if its diameter is 15 inches?

(d) At the rate given in problem **(c)**, what is the diameter of the driving pulley if it turns at 200 rpm?

Solution: Recall that you solve a proportion problem using an indirect proportion when the two units of measure involved in the rate act in opposite ways. That is, the number of one amount increases as the number of the other decreases. Or, the number of one amount decreases as the number of the other amount increases [see Example 13-29]:

(a) *Understand:*   The two units of measure in the rate (men and days) act in opposite ways. That is, the number of men increases as the number of days decreases. Or, the number of men decreases as the number of days increases.

Decide: To solve a proportion problem when the two units of measure act in opposite ways, you write and solve an indirect proportion.

Form ratios: $\dfrac{6 \text{ \sout{men}}}{8 \text{ \sout{men}}}$ ⟵— 6 men/14 days —⟶ $\dfrac{14 \text{ \sout{days}}}{n \text{ \sout{days}}}$

 8 men/n days

Write the proportion: men ratio —⟶ $\dfrac{6}{8} = \dfrac{n}{14}$ ⟵— reciprocal of days ratio

Solve as before: $14 \times 6 = 8 \times n$ Equate the cross products.

 $84 = 8 \times n$ Simplify.

 $n = 84 \div 8$ Divide the product by the known factor.

 $n = 10\dfrac{1}{2}$ ⟵— solution

Interpret: $n = 10\frac{1}{2}$ means that it will take 8 men $10\frac{1}{2}$ days to do the same job.

(b) Now the unknown term is *men*, not *days*. So you would set up the indirect proportion as follows:

$\dfrac{6 \text{ \sout{men}}}{n \text{ \sout{men}}}$ ⟵— 6 men/14 days —⟶ $\dfrac{14 \text{ \sout{days}}}{9 \text{ \sout{days}}}$

 n men/ 9 days

men ratio —⟶ $\dfrac{6}{n} = \dfrac{9}{14}$ ⟵— reciprocal of days ratio

$14 \times 6 = n \times 9$

$84 = n \times 9$

$n = 84 \div 9$

$n = 9\dfrac{1}{3}$ ⟵— solution

Interpret: $n = 9\frac{1}{3}$ means that it would take more than 9 men, or a minimum of 10 men, to do the same job in 9 days.

(c) The two units of measure in this rate (inches and rpm's) act in opposite ways: as the number of inches increases, the revolutions per minute decrease:

$\dfrac{10 \text{ \sout{inches}}}{15 \text{ \sout{inches}}}$ ⟵— 10 inches/500 rpm —⟶ $\dfrac{500 \text{ \sout{rpm}}}{n \text{ \sout{rpm}}}$

 15 inches/n rpm

inches ratio —⟶ $\dfrac{10}{15} = \dfrac{n}{500}$ ⟵— reciprocal of rpm ratio

$500 \times 10 = 15 \times n$

$5000 = 15 \times n$

$n = 5000 \div 15$

$n = 333\dfrac{1}{3}$ ⟵— solution

This means that a pulley with a diameter of 15 inches turns at $333\frac{1}{3}$ rpm.

(d) Now you must find out the diameter of the driving pulley if it turns at only 200 rpm:

$\dfrac{10 \text{ \sout{inches}}}{n \text{ \sout{inches}}}$ ⟵— 10 inches/500 rpm —⟶ $\dfrac{500 \text{ \sout{rpm}}}{200 \text{ \sout{rpm}}}$

 n inches/200 rpm

$$\text{inches ratio} \longrightarrow \frac{10}{n} = \frac{200}{500} \longleftarrow \text{reciprocal of rpm ratio}$$

$$500 \times 10 = n \times 200$$

$$5000 = n \times 200$$

$$n = 5000 \div 200$$

$$n = 25 \longleftarrow \text{solution}$$

This means that the pulley must be 25 inches in diameter if it turns at 200 rpm.

Supplementary Exercises

PROBLEM 13-12   Find each ratio: **(a)** 30 miles to 50 miles   **(b)** 14 hours to 16 hours
(c) 36 feet to $4\frac{1}{2}$ feet   **(d)** 48 meters to 0.6 meters   **(e)** \$25 to \$1.25   **(f)** 1 year to 1 month
(g) 6 days to 2 weeks   **(h)** 2 half-dollars to 3 nickels   **(i)** 2 quarters to 3 dimes
(j) 50 women to 30 men   **(k)** 14 wins to 30 games   **(l)** 3000 students to 120 instructors
(m) 0.5 grams to 500 kilograms   **(n)** 2 feet to 5 yards   **(o)** $1\frac{1}{2}$ pounds to 12 ounces

PROBLEM 13-13   Find each rate: **(a)** 55 miles to 60 minutes   **(b)** \$28 to 8 pounds
(c) 210 miles to 4 hours   **(d)** 280 miles to 16 gallons   **(e)** \$1740 to 4 weeks
(f) 12 gallons to $4\frac{1}{2}$ hours   **(g)** 2000 calories to 24 hours   **(h)** $31\frac{1}{4}$ hours to 5 people
(i) 5000 pages to 48 days   **(j)** 10,000 bricks to 12 walls   **(k)** 260 pounds to 8 cases
(l) \$19,011 to 12 months   **(m)** 730.8 miles to 14.5 hours   **(n)** 335 feet to 40 seconds
(o) \$102 to 8 people   **(p)** 425 houses to 200 TV sets   **(q)** 160 planes to 3 hours
(r) $6\frac{1}{4}$ pounds to \$2.50   **(s)** 1020 tons to 50 trucks

PROBLEM 13-14   Solve each proportion: **(a)** $\dfrac{5\frac{1}{2}}{1} = \dfrac{p}{24}$ **(b)** $\dfrac{50}{1} = \dfrac{m}{3}$ **(c)** $\dfrac{400}{20} = \dfrac{250}{n}$

(d) $\dfrac{3}{4} = \dfrac{b}{16}$ **(e)** $\dfrac{8}{c} = \dfrac{6}{30}$ **(f)** $\dfrac{9}{12} = \dfrac{18}{d}$ **(g)** $\dfrac{5}{f} = \dfrac{35}{14}$ **(h)** $\dfrac{g}{9} = \dfrac{4}{12}$ **(i)** $\dfrac{h}{36} = \dfrac{1}{3}$

(j) $\dfrac{10}{k} = \dfrac{15}{12}$ **(k)** $\dfrac{p}{3.5} = \dfrac{6}{15}$ **(l)** $\dfrac{4.5}{5} = \dfrac{27}{r}$ **(m)** $\dfrac{6}{9} = \dfrac{3\frac{1}{3}}{z}$ **(n)** $\dfrac{a}{15} = \dfrac{4}{5}$ **(o)** $\dfrac{900}{54} = \dfrac{b}{3}$

(p) $\dfrac{16}{c} = \dfrac{2}{7}$ **(q)** $\dfrac{50}{8} = \dfrac{150}{d}$ **(r)** $\dfrac{f}{32} = \dfrac{3}{4}$ **(s)** $\dfrac{12}{210} = \dfrac{g}{140}$ **(t)** $\dfrac{24}{h} = \dfrac{108}{18}$ **(u)** $\dfrac{6}{7} = \dfrac{54}{k}$

(v) $\dfrac{1.5}{0.3} = \dfrac{m}{8}$ **(w)** $\dfrac{0.04}{0.12} = \dfrac{0.6}{n}$ **(x)** $\dfrac{\frac{1}{2}}{10} = \dfrac{2\frac{1}{2}}{v}$ **(y)** $\dfrac{t}{3\frac{1}{2}} = \dfrac{8}{7}$ **(z)** $\dfrac{\frac{1}{2}}{\frac{1}{4}} = \dfrac{\frac{1}{3}}{u}$

PROBLEM 13-15   Solve each proportion problem:

(a) A map has a scale of $1\frac{1}{4}$ inch to 10 miles. If the map distance between two cities is $3\frac{1}{2}$ inches, then what is the actual distance between the two cities?

(b) Using the scale given in problem **(a)**, if the actual distance between the two cities is 500 miles, what is the map distance between the two cities?

(c) The average 150-pound person should eat 1800 calories each day to maintain that weight. How many calories should be eaten by a 110-pound person to maintain weight?

(d) Using the rate given in problem (c), how much should you weigh to eat 2000 calories a day to maintain weight?

(e) Three men take 5 hours to unload a train car. At that rate, how long will it take 10 men to unload the same train car?

(f) At the rate given in problem (e), how many men will be needed to unload the same train car in a maximum of 2 hours?

(g) Apples are on sale for $1.50 for 3 pounds. At that rate, how much will 5 pounds of apples cost?

(h) At the rate given in problem (g), how many pounds of apples can be purchased for $5?

(i) To feed 12 people, it takes 5 pounds of canned ham. At that rate, how many people can be given a full serving from 3 pounds of canned ham?

(j) At the rate given in problem (i), how many pounds of canned ham will it take to feed 20 people?

(k) A certain car travels 500 miles in 8 hours. At that rate, how far can the car travel in 12 hours?

(l) At the rate given in problem (i), how many hours will it take the car to travel 300 miles?

(m) A driving gear with 36 teeth turns at 90 rpm. How many teeth are in the driven gear if it turns at 60 rpm?

(n) Using the same driving gear in problem (m), how fast is the driven gear turning if it has 20 teeth?

(o) Two windows are to be constructed so that they are proportional. The height and width of one window is 5 feet by 3 feet, respectively. What should the be height of the other window if its width is 5 feet?

(p) Given the dimensions of the windows in problem (k), what should the width be of the other window if its height is 3 feet?

(q) For every $5 invested in a certain savings account, the bank returns a total of $5.30 at the end of one year. How much should be invested to earn $100 in one year?

(r) Using the same bank rate in problem (q) how much would be earned if $100 were invested?

(s) A driving pulley has a diameter of 12 inches and turns at 400 rpm. How fast does the driven pulley turn if it has a 20-inch diameter?

(t) Using the same driving pulley in problem (s), what is the diameter of the driven pulley if it turns at 600 rpm?

(u) Jose and Rosa are sitting on opposite ends of a seesaw. Jose weighs 160 pounds and is sitting 5 feet from the pivot (balance point). If Rosa weighs 120 pounds, how far must she sit from the pivot to balance the seesaw? (*Hint:* To balance a seesaw, the number of pounds must increase as the distance from the pivot decreases, or the number of pounds must decrease as the distance from the pivot increases.)

(v) Using the information given for Jose in problem (u), how much does Rosa weigh if she balances the seesaw at 6 feet from the pivot?

(w) A crowbar is built so that it provides its own pivot one inch from the end. If the handle is 30 inches long, how much force can a 120-pound woman exert with the crowbar if she applies all her body weight? (*Hint:* The same principle used for the seesaw in problem (u) can be used here.)

(x) A man wants to pry up a 750-pound weight. If he places the pivot one foot from the weight, how long must the total length of the lever be so that his weight of 150 pounds will just balance the weight? (*Hint:* The same principle used for the seesaw in problem (u) can be used here too. Also, the word *total* is important.)

Answers to Supplementary Exercises

(13-12)   **(a)** $\frac{3}{5}$ **(b)** $\frac{7}{8}$ **(c)** $\frac{8}{1}$ **(d)** $\frac{80}{1}$ **(e)** $\frac{20}{1}$ **(f)** $\frac{12}{1}$ **(g)** $\frac{3}{7}$ **(h)** $\frac{20}{3}$ **(i)** $\frac{5}{3}$

(j) $\frac{5}{3}$ **(k)** $\frac{7}{15}$ **(l)** $\frac{25}{1}$ **(m)** $\frac{1}{1,000,000}$ **(n)** $\frac{2}{15}$ **(o)** $\frac{2}{1}$

(13-13)   (Any one of the three different forms for the rate is acceptable.) **(a)** $\frac{11}{12}\frac{\text{miles}}{\text{minute}}$
(b) \$3.50/pound   **(c)** $52\frac{1}{2}$ mph   **(d)** $17\frac{1}{2}$ mpg   **(e)** \$435/week   **(f)** $2\frac{2}{3}$ gallons per hour
(g) $83\frac{1}{3}$ calories/hour   **(h)** $6\frac{1}{4}\frac{\text{hours}}{\text{person}}$ **(i)** $104\frac{1}{6}$ pages per day   **(j)** $833\frac{1}{3}$ bricks per wall
(k) $32\frac{1}{2}$ pounds/case   **(l)** \$1584.25 per month   **(m)** 50.4 mph   **(n)** $8\frac{3}{8}$ fps
(o) \$12.75/person   **(p)** $2\frac{1}{8}$ houses per TV set   **(q)** $53\frac{1}{3}\frac{\text{planes}}{\text{hour}}$ **(r)** $2\frac{1}{2}$ pounds/dollar
(s) $20\frac{2}{3}$ tons per truck

(13-14)   **(a)** 132   **(b)** 150   **(c)** $12\frac{1}{2}$ or 12.5   **(d)** 12   **(e)** 40   **(f)** 24   **(g)** 2   **(h)** 3
(i) 12   **(j)** 8   **(k)** 1.4 or $1\frac{2}{5}$ **(l)** 30   **(m)** 5   **(n)** 12   **(o)** 50   **(p)** 56   **(q)** 24
(r) 24   **(s)** 8   **(t)** 4   **(u)** 63   **(v)** 40   **(w)** 1.8   **(x)** 50   **(y)** 4   **(z)** $\frac{1}{6}$

(13-15)   **(a)** 28 miles   **(b)** $62\frac{1}{2}$ inches   **(c)** 1320 calories   **(d)** $166\frac{2}{3}$ pounds
(e) $1\frac{1}{2}$ hours   **(f)** 8 men   **(g)** \$2.50   **(h)** 10 pounds   **(i)** 7 people   **(j)** $8\frac{1}{3}$ pounds
(k) 750 miles   **(l)** $4\frac{4}{5}$ hours   **(m)** 54 teeth   **(n)** 162 rpm   **(o)** $8\frac{1}{3}$ feet   **(p)** $1\frac{4}{5}$ feet
(q) \$1666.67   **(r)** \$6   **(s)** 240 rpm   **(t)** 8 inches   **(u)** $6\frac{2}{3}$ feet   **(v)** $133\frac{1}{3}$ pounds
(w) 3600 pounds   **(x)** 6 feet

14 POWERS AND SQUARE ROOTS

THIS CHAPTER IS ABOUT

- ☑ **Writing Power Notation**
- ☑ **Evaluating Power Notation**
- ☑ **Finding Squares and Square Roots Using Paper and Pencil**
- ☑ **Finding Squares and Square Roots Using a Table**
- ☑ **Finding Squares and Square Roots Using a Calculator**

14-1. Writing Power Notation

Recall: Multiplication is a short way to write a sum of repeated addends.

EXAMPLE 14-1: Write the following as a multiplication problem: $2 + 2 + 2 + 2$

$$\text{repeated addend}$$
$$\text{number of repeated addends}$$

Solution: $2 + 2 + 2 + 2 = 2 \times 4 \longleftarrow$ multiplication problem (4 repeated addends of 2)

Recall: In 2×4, 2 and 4 are called factors.

Power notation or **exponential notation** is a short way to write a product of **repeated factors.**

EXAMPLE 14-2: Write the following as a power notation: $2 \times 2 \times 2 \times 2$

$$\text{repeated factor}$$
$$\text{number of repeated factors}$$

Solution: $2 \times 2 \times 2 \times 2 = 2^4 \longleftarrow$ power notation (4 repeated factors of 2)

Note: In 2^4, 2 is called the **base** and 4 is called the **power** or **exponent.** The power notation 2^4 is read as "two to the **fourth power**" or "two **raised** to the fourth power."

Power Notation Rule

To write a product of repeated factors as power notation, you use the repeated factor as the base and the number of repeated factors as the power (exponent):

$$\text{repeated factors} \longrightarrow \underbrace{a \times a \times a \times \cdots \times a}_{n \text{ repeated factors of } a} = a^n \longleftarrow \text{power notation}$$

EXAMPLE 14-3: Write the following repeated factors using power notation: **(a)** 10×10
(b) $1.2 \times 1.2 \times 1.2$

Solution
(a) $10 \times 10 = 10^2$ — Use the repeated factor 10 for the base.
Use the number of repeated factors (2) for the power (exponent).

(b) $1.2 \times 1.2 \times 1.2 = 1.2^3$ Use the repeated factor 1.2 for the base.
Use the number of repeated factors (3) for the power (exponent).

Note: The power notation 10^2 is read as "ten to the **second power**" or "ten **squared.**" The power notation 1.2^3 is real as "one and two tenths to the **third power**" or "one and two tenths **cubed.**"

To **write power notation** when the repeated factor is a fraction, you must write parentheses around the fraction before writing the power (exponent).

EXAMPLE 14-4: Write the following repeated factors using power notation:
(a) $\dfrac{3}{4} \times \dfrac{3}{4}$ **(b)** $\dfrac{3 \times 3}{4}$

Solution

(a) $\dfrac{3}{4} \times \dfrac{3}{4} = \left(\dfrac{3}{4}\right)^2$ Use parentheses to show that the whole fraction is being raised to the second power.

(b) $\dfrac{3 \times 3}{4} = \dfrac{3^2}{4}$ Do not use parentheses to show that only the numerator is being raised to the second power.

EXAMPLE 14-5: Show that $\dfrac{2^3}{5}$ does not mean $\left(\dfrac{2}{5}\right)^3$.

Solution: $\dfrac{2^3}{5} = \dfrac{2 \times 2 \times 2}{5}$ $\Bigg|$ $\left(\dfrac{2}{5}\right)^3 = \dfrac{2}{5} \times \dfrac{2}{5} \times \dfrac{2}{5}$

$= \dfrac{8}{5}$ $\Bigg|$ $= \dfrac{8}{125}$

14-2. Evaluating Power Notation

To **evaluate power notation,** you first write the power notation as a product of repeated factors and then multiply to get a whole number, or fraction, or decimal product.

EXAMPLE 14-6: Evaluating the following power notation: **(a)** 3^2 **(b)** $\left(\frac{2}{5}\right)^3$ **(c)** 1^5 **(d)** 0^6
(e) 0.1^4 **(f)** 10^7

Solution

(a) In 3^2, the base 3 has a power (exponent) of 2.
A power (exponent) of 2 means the base 3 is used as a repeated factor 2 times.

$3^2 = 3 \times 3$ Write 2 repeated factors of 3.

$= 9$ Multiply.

(b) In $\left(\dfrac{2}{5}\right)^3$, the base $\dfrac{2}{5}$ has a power (exponent) of 3.

A power (exponent) of 3 means the base $\dfrac{2}{5}$ is used as a repeated factor 3 times.

$\left(\dfrac{2}{5}\right)^3 = \dfrac{2}{5} \times \dfrac{2}{5} \times \dfrac{2}{5}$ Write 3 repeated factors of $\dfrac{2}{5}$.

$= \dfrac{8}{125}$ Multiply.

(c) In 1^5, the base 1 has a power (exponent) of 5.
A power (exponent) of 5 means the base 1 is used as a repeated factor 5 times.

$1^5 = 1 \times 1 \times 1 \times 1 \times 1$ Write 5 repeated factors of 1.

$= 1$ Multiply [see the following *Note 1*].

(d) In 0^6, the base 0 has a power (exponent) of 6.
A power (exponent) of 6 means the base 0 is used as a repeated factor 6 times.

$0^6 = 0 \times 0 \times 0 \times 0 \times 0 \times 0$ Write 6 repeated factors of 0.

$= 0$ Multiply [see the following *Note 2*].

(e) In 0.1^4, the base 0.1 has a power (exponent) of 4.
A power (exponent) of 4 means the base 0.1 is used as a repeated factor 4 times.

$0.1^4 = 0.1 \times 0.1 \times 0.1 \times 0.1$ Write 4 repeated factors of 0.1.

$= \underbrace{0.0001}_{4 \text{ zeros}}$ Multiply [see the following *Note 3*].

(f) In 10^7, the base 10 has a power (exponent) of 7.
A power (exponent) of 7 means the base 10 is used as a repeated factor 7 times.

$10^7 = 10 \times 10 \times 10 \times 10 \times 10 \times 10 \times 10$ Write 7 repeated factors of 10.

$= \underbrace{10,000,000}_{7 \text{ zeros}}$ Multiply [see the following *Note 4*].

Note 1: If n is any power (exponent), then $1^n = 1$.

Note 2: If n is any nonzero power (exponent), then $0^n = 0$ (0^0 is not defined).

Note 3: If n is any whole number power (exponent), then $0.1^n = \underbrace{0.00 \cdots 01}_{n \text{ zeros}}$.

Note 4: If n is any whole number power (exponent), then $10^n = \underbrace{1000 \cdots 0}_{n \text{ zeros}}$.

If you use 2 as a base with whole number exponents, an interesting pattern occurs. As each exponent decreases by 1, the amount is halved. This is illustrated in the following example.

The following base 2 example show a very important pattern.

EXAMPLE 14-7: Evaluate the following powers of 2: **(a)** 2^5 **(b)** 2^4 **(c)** 2^3 **(d)** 2^2
(e) 2^1 **(f)** 2^0

Solution: **(a)** $2^5 = 2 \times 2 \times 2 \times 2 \times 2 = 32$
            (b) $2^4 = 2 \times 2 \times 2 \times 2 = 16$ ⟵— half of 32
            (c) $2^3 = 2 \times 2 \times 2 = 8$ ⟵——— half of 16
            (d) $2^2 = 2 \times 2 = 4$ ⟵———— half of 8
            (e) $2^1 = 2$ ⟵————————— half of 4
            (f) $2^0 = 1$ ⟵————————— half of 2

Note 1: The power notation 2^1 is just another way of writing the base 2. The general rule for this power notation is

$$\text{If } a \text{ is any base, then } a^1 = a.$$

Note 2: The power notation 2^0 is just another way of writing the whole number 1. The general rule for this power notation is

$$\text{If } a \text{ is any nonzero base, then } a^0 = 1. \ (0^0 \text{ is not defined.})$$

EXAMPLE 14-8: Evaluate the following power notation: **(a)** 5^1 **(b)** $\left(\dfrac{3}{4}\right)^1$ **(c)** 2.5^1 **(d)** 1^1

(e) 0^1 **(f)** 10^0 **(g)** $\left(\dfrac{2}{3}\right)^0$ **(h)** 0.75^0 **(i)** 1^0 **(j)** 0^0

Solution

(a) $5^1 = 5$ **(b)** $\left(\dfrac{3}{4}\right)^1 = \dfrac{3}{4}$ **(c)** $2.5^1 = 2.5$ **(d)** $1^1 = 1$ **(e)** $0^1 = 0$

(f) $10^0 = 1$ **(g)** $\left(\dfrac{2}{3}\right)^0 = 1$ **(h)** $0.75^0 = 1$ **(i)** $1^0 = 1$ **(j)** 0^0 is not defined.

14-3. Finding Squares and Square Roots Using Paper and Pencil

If a and b are numbers and $a = b^2$, then a is called the **square** of b.

EXAMPLE 14-9: Find the square of the following numbers: **(a)** 3   **(b)** $\dfrac{3}{4}$ **(c)** 0.1

Solution
(a) The square of 3 is 9 because $3^2 = 3 \times 3 = 9$.

(b) The square of $\dfrac{3}{4}$ is $\dfrac{9}{16}$ because $\left(\dfrac{3}{4}\right)^2 = \dfrac{3}{4} \times \dfrac{3}{4} = \dfrac{9}{16}$.

(c) The square of 0.1 is 0.01 because $0.1^2 = \underset{\text{2 zeros}}{\underline{0.01}}$

Note: To find the square of a given number, you multiply the given number times itself.

You should know all the whole number squares up to 100 from memory. These are listed in the following example.

EXAMPLE 14-10: Name all the whole number squares up to 100.

Solution: 1 is a whole number square because $1^2 =$ 1.
   4 is a whole number square because $2^2 =$ 4.
   9 is a whole number square because $3^2 =$ 9.
   16 is a whole number square because $4^2 =$ 16.
   25 is a whole number square because $5^2 =$ 25.
   36 is a whole number square because $6^2 =$ 36.
   49 is a whole number square because $7^2 =$ 49.
   64 is a whole number square because $8^2 =$ 64.
   81 is a whole number square because $9^2 =$ 81.
   100 is a whole number square because $10^2 =$ 100.

If a and b are numbers and $a = b^2$, then b is called the **square root** of a.

EXAMPLE 14-11: Find the square root of the following numbers: **(a)** 9   **(b)** $\dfrac{9}{16}$ **(c)** 0.01

Solution
(a) The square root of 9 is 3 because $9 = 3 \times 3 = 3^2$ [see Example 14-9(a)].

(b) The square root of $\dfrac{9}{16}$ is $\dfrac{3}{4}$ because $\dfrac{9}{16} = \dfrac{3 \times 3}{4 \times 4} = \left(\dfrac{3}{4}\right)^2$ [see Example 14-9(b)].

(c) The square root of 0.01 is 0.1 because $0.01 = 0.1 \times 0.1 = 0.1^2$ [see Example 14-9(c)].

Note: To find the square root of a given number, you find the number that, when squared, will equal the given number.

"The square root of 9" can be written as $\sqrt{9}$. In $\sqrt{9}$, $\sqrt{}$ is called the **square root symbol** or **radical sign,** 9 is called the square or **radicand,** and $\sqrt{9}$ is called a **radical.**

To find the square root of a given whole number square, you can divide the given whole number by various whole number divisors until the whole number divisor and the quotient are the same.

EXAMPLE 14-12: Find $\sqrt{196}$ using division.

Solution: $196 \div 2 = 98$ $\qquad$ *Continue:* Since the divisor 2 and the quotient 98 are not the same, 2 is not the square root of 196.

$$196 \div 3 = 65\frac{1}{3}$$

$$196 \div 4 = 49$$

$$196 \div 5 = 39\frac{1}{5}$$

$$196 \div 6 = 32\frac{2}{3}$$

$$196 \div 7 = 28$$

$$196 \div 8 = 24\frac{1}{2}$$

$$196 \div 9 = 21\frac{7}{9}$$

$$196 \div 10 = 19\frac{3}{5}$$

$$196 \div 11 = 17\frac{9}{11}$$

$$196 \div 12 = 16\frac{1}{3}$$

$$196 \div 13 = 15\frac{1}{13}$$

$196 \div 14 = 14$ $\qquad$ *Stop!* Since the divisor 14 and the quotient 14 are the same, 14 is the square root of 196.

$\sqrt{196} = 14$ $\qquad$ Write square root notation.

Hint: The multiplication fact $9 \times 9 = 81$ or $10 \times 10 = 100$ tells you that the square root of 196 is larger than 9 or 10. This means you should start the division process in Example 14-12 with a whole number at least as large as 11.

To find the square root of a fraction, you can write the square root of the numerator over the square root of the denominator.

EXAMPLE 14-13: Find $\sqrt{\dfrac{25}{64}}$ using division.

same

Solution: $25 \div 5 = 5$ means $\sqrt{25} = 5$.

same

$64 \div 8 = 8$ means $\sqrt{64} = 8$.

$$\sqrt{\frac{25}{64}} = \frac{5}{8}$$ $\begin{matrix}\longleftarrow \text{ square root of the numerator} \\ \longleftarrow \text{ square root of the denominator}\end{matrix}$

14-4. Finding Squares and Square Roots Using a Table

To find the square of any whole number from 0 to 100, you can use Appendix Table 2 found in the back of this book. Part of Appendix Table 2 is displayed in the following example.

EXAMPLE 14-14: Find 47^2 using Appendix Table 2.

Solution: from Appendix Table 2

Number N	Square N^2	Square Root $\sqrt{N}$
35	1225	5.916
36	1296	6
37	1369	6.083
38	1444	6.164
39	1521	6.245
40	1600	6.325
41	1681	6.403
42	1764	6.481
43	1849	6.557
44	1936	6.633
45	2025	6.708
46	2116	6.782
47	2209	6.856
48	2304	6.928
49	2401	7

Place your left finger on the base 47 in the "Number N" column.
Place your right finger at the top of the "Square N^2" column.
Move your left finger straight across and your right finger straight down until they meet at 2209. Therefore $47^2 = 2209$.

Note: $47^2 = 47 \times 47 = 2209$

You can also use Appendix Table 2 to find the square root of any whole number from 0 to 100. Part of Appendix Table 2 is displayed in Example 14-15.

EXAMPLE 14-15: Find $\sqrt{47}$ using Appendix Table 2.

Solution: From Appendix Table 2

Number N	Square N^2	Square Root $\sqrt{N}$
35	1225	5.916
36	1296	6
37	1369	6.083
38	1444	6.164
39	1521	6.245
40	1600	6.325
41	1681	6.403
42	1764	6.481
43	1849	6.557
44	1936	6.633
45	2025	6.708
46	2116	6.782
47	2209	6.856
48	2304	6.928
49	2401	7

Place your left finger on the radicand 47 in the "Number N" column.
Place your right finger at the top of the "Square Root $\sqrt{N}$" column.
Move your left finger straight across and your right finger straight down until they meet at 6.856. Therefore $\sqrt{47} \approx 6.856$.

Note: $\sqrt{47} \approx 6.856$ because $6.856 \times 6.856 = 47.004736 \approx 47$.

14-5. Finding Squares and Square Roots Using a Calculator

To find the square of a whole number or decimal with four digits or less, you can use a standard 8-digit display **calculator**. To find the square of a whole number or decimal with 5 digits or less, you can use a standard 10-digit display calculator. Both versions are simulated in the following example.

EXAMPLE 14-16: Find the following squares using a calculator: **(a)** 276^2 **(b)** 915.8^2

Solution:

	Press	Display	Interpret
(a)	$\boxed{2}\,\boxed{7}\,\boxed{6}$	276	276
	$\boxed{\times}$	276	$276 \times$
	$\boxed{2}\,\boxed{7}\,\boxed{6}$	276	276×276
	$\boxed{=}$	76176	$276 \times 276 = 76{,}176$
(b)	$\boxed{9}\,\boxed{1}\,\boxed{5}\,\boxed{.}\,\boxed{8}$	915.8	915.8
	$\boxed{\times}$	915.8	$915.8 \times$
	$\boxed{9}\,\boxed{1}\,\boxed{5}\,\boxed{.}\,\boxed{8}$	915.8	915.8×915.8
	$\boxed{=}$	838689.64	$915.8 \times 915.8 = 838{,}689.64$

You can also find (or approximate) the square root of a whole number or decimal using a display calculator. To find or approximate the square root of a 8-digit whole number or decimal with 8 digits or less, you can use a standard 8-digit calculator. To find or approximate the square root of a whole number or decimal with 10 digits or less, you can use a standard 10-digit calculator. Both versions are simulated in the following example.

EXAMPLE 14-17: Approximate the following square roots using a calculator: **(a)** $\sqrt{785{,}273}$
(b) $\sqrt{14{,}906.385}$

Solution:

	Press	Display	Interpret
(a)	$\boxed{7}\,\boxed{8}\,\boxed{5}\,\boxed{2}\,\boxed{7}\,\boxed{3}$	785273	785,273
	$\boxed{\sqrt{\ }}$	886.15631	$\sqrt{785{,}273} \approx 886.15631$
(b)	$\boxed{1}\,\boxed{4}\,\boxed{9}\,\boxed{0}\,\boxed{6}\,\boxed{.}\,\boxed{3}\,\boxed{8}\,\boxed{5}$	14906.385	14,906.385
	$\boxed{\sqrt{\ }}$	122.09171	$\sqrt{14{,}906.385} \approx 122.09171$

To find or approximate the square root of a fraction using a calculator, you first divide the numerator by the denominator and then press the square root key.

EXAMPLE 14-18: Approximate $\sqrt{\dfrac{17}{32}}$ using a calculator.

Solution:

	Press	Display	Interpret
(a)	$\boxed{1}\,\boxed{7}$	17	17
	$\boxed{\div}$	17	$17 \div$
	$\boxed{3}\,\boxed{2}$	32	$17 \div 32$
	$\boxed{=}$	0.53125	$17 \div 32 = 0.53125$
	$\boxed{\sqrt{\ }}$	0.7288689	$\sqrt{0.53125} \approx 0.7288689$

Note: $\sqrt{\dfrac{17}{32}} \approx 0.7$ (nearest tenth), or 0.73 (nearest hundredth), or 0.729 (nearest thousandth), and so on, depending on what decimal place you round to.

RAISE YOUR GRADES

Can you . . . ?

☑ identify the base and power (exponent) given power notation (exponential notation)
☑ write a product of repeated factors as power notation
☑ write power notation when the repeated factor is a fraction
☑ evaluate power notation
☑ find the square of a number
☑ find the square root of a number
☑ identify the radicand, radical, and radical sign for a given square root
☑ find squares manually, using a table, and using a calculator
☑ find square roots manually, using a table, and using a calculator

SUMMARY

1. A short way to write a product of repeated factors is power notation, or exponential notation.
2. In a^n, a is called the base and n is called the power or exponent.
3. The power notation a^n is read as "a to the nth power" or "a raised to the nth power."
4. To write a product of repeated factors as power notation, you use the repeated factor as the base and the number of repeated factors as the power (exponent):

$$\text{repeated factors} \qquad \underbrace{a \times a \times a \times \cdots \times a}_{n \text{ repeated factors of } a} = a^n \longleftarrow \text{power notation}$$

5. To write power notation when the repeated factor is a fraction, you must write parentheses around the fraction before writing the power (exponent).
6. To evaluate power notation, you first write power notation as repeated factors and then multiply to get a whole number, or fraction, or decimal product.
7. *Caution:* $\dfrac{a^n}{b}$ does not mean $\left(\dfrac{a}{b}\right)^n$.
8. If n is any power (exponent), then $1^n = 1$.
9. If n is any nonzero power (exponent), then $0^n = 0$ (0^0 is not defined).
10. If n is any whole number power (exponent), then $0.1^n = \underbrace{0.00 \cdots 01}_{n \text{ zeros}}$
11. If n is any whole number power (exponent), then $10^n = \underbrace{1000 \cdots 0}_{n \text{ zeros}}$
12. If a is any base, then: $a^1 = a$.
13. If a is any nonzero base, then $a^0 = 1$ (0^0 is not defined).
14. If a and b are numbers and $a = b^2$, then a is called the square of b and b is called the square root of a.
15. To find the square of a given number, you multiply the given number times itself.
16. The whole number squares up to 100 are 1, 4, 9, 16, 25, 36, 49, 64, 81, and 100.
17. To find the square root of a given number, you find the number that, when squared, will equal the given number.
18. In $\sqrt{a}$, $\sqrt{}$ is called the square root symbol or radical sign, a is called the square or radicand, and $\sqrt{a}$ is called a radical.
19. To find the square root of a given whole number square, you can manually divide the given whole number by various whole number divisors until the whole number divisor and the quotient are the same.
20. To find the square root of a fraction, you write the square root of the numerator over the square root of the denominator.

21. To find the square or square root of a whole number from 1 to 100, you can use Appendix Table 2 as shown in Examples 14-14 and 14-15 (or in the back of this book).
22. To find the square of a whole number or decimal with 4 digits or less, you can use a standard 8-digit calculator as shown in Example 14-16(a). To find the square of a whole number or decimal with 5 digits or less, you can use a standard 10-digit calculator as shown in Example 14-16(b).
23. To find the square root of a whole number or decimal with 8 digits or less, you can use a standard 8-digit calculator as shown in Example 14-17(a). To find the square root of a whole number or decimal with 10 digits or less, you can use a standard 10-digit calculator as shown in Example 14-17(b).
24. To find the square root of a fraction using a calculator, you first divide the numerator by the denominator and then press the square root key.

SOLVED PROBLEMS

PROBLEM 14-1   Write each product of repeated factors as power notation:

(a) 5×5 (b) 0.1×0.1 (c) $\dfrac{4}{5} \times \dfrac{4}{5}$ (d) $2 \times 2 \times 2$ (e) $2.5 \times 2.5 \times 2.5$

(f) $\dfrac{1}{2} \times \dfrac{1}{2} \times \dfrac{1}{2}$ (g) $6 \times 6 \times 6 \times 6$ (h) $1.5 \times 1.5 \times 1.5 \times 1.5$ (i) $\dfrac{2}{3} \times \dfrac{2}{3} \times \dfrac{2}{3} \times \dfrac{2}{3}$

(j) $\dfrac{8 \times 8}{3}$ (k) $\dfrac{7}{9 \times 9 \times 9}$ (l) $\dfrac{2 \times 2}{3 \times 3 \times 3}$ (m) $10 \times 10 \times 10 \times 10 \times 10 \times 10$

Solution: Recall that to write a product of repeated factors as power notation, you use the repeated factor as the base and the number of repeated factors as the power (exponent) [see Examples 14-2, 14-3, and 14-4]:

repeated factor
number of repeated factors   use parentheses

(a) $5 \times 5 = 5^2$ (b) $0.1 \times 0.1 = 0.1^2$ (c) $\dfrac{4}{5} \times \dfrac{4}{5} = \left(\dfrac{4}{5}\right)^2$ (d) $2 \times 2 \times 2 = 2^3$

(e) $2.5 \times 2.5 \times 2.5 = 2.5^3$ (f) $\dfrac{1}{2} \times \dfrac{1}{2} \times \dfrac{1}{2} = \left(\dfrac{1}{2}\right)^3$ (g) $6 \times 6 \times 6 \times 6 = 6^4$

(h) $1.5 \times 1.5 \times 1.5 \times 1.5 = 1.5^4$ (i) $\dfrac{2}{3} \times \dfrac{2}{3} \times \dfrac{2}{3} \times \dfrac{2}{3} = \left(\dfrac{2}{3}\right)^4$ (j) $\dfrac{8 \times 8}{3} = \dfrac{8^2}{3}$

(k) $\dfrac{7}{9 \times 9 \times 9} = \dfrac{7}{9^3}$ (l) $\dfrac{2 \times 2}{3 \times 3 \times 3} = \dfrac{2^2}{3^3}$ (m) $10 \times 10 \times 10 \times 10 \times 10 \times 10 = 10^6$

PROBLEM 14-2   Evaluate each power notation: (a) 4^2 (b) 1.5^2 (c) $\left(\dfrac{1}{3}\right)^2$ (d) 2^3

(e) 0.1^3 (f) $\left(\dfrac{3}{4}\right)^3$ (g) 10^2 (h) 10^3 (i) 10^5 (j) 10^{12} (k) 0^3 (l) 1^{10} (m) $\dfrac{5^2}{4}$

(n) $\left(\dfrac{5}{4}\right)^2$ (o) 3^1 (p) $\left(\dfrac{7}{10}\right)^1$ (q) 0^1 (r) 8.25^1 (s) 1^0 (t) $\left(\dfrac{9}{10}\right)^0$ (u) 0^0

(v) 198.756^0 (w) $\dfrac{2}{3^2}$ (x) $\dfrac{5^2}{2^3}$ (y) $\dfrac{8^1}{6^0}$ (z) $\dfrac{0^5}{5^0}$

Solution: Recall that to evaluate power notation, you first write the power notation as a product of repeated factors and then multiply to get a whole number, or fraction, or decimal. Also, $a^1 = a\,(a \neq 0)$

and $a^0 = 1$ $(a \neq 0)$ [see Examples 14-5, 14-6, 14-7, and 14-8]:

repeated factor
number of repeated factors

(a) $4^2 = 4 \times 4 = 16$ (b) $1.5^2 = 1.5 \times 1.5 = 2.25$ (c) $\left(\dfrac{1}{3}\right)^2 = \dfrac{1}{3} \times \dfrac{1}{3} = \dfrac{1}{9}$

(d) $2^3 = 2 \times 2 \times 2 = 8$ (e) $0.1^3 = 0.1 \times 0.1 \times 0.1 = \overset{\text{3 zeros}}{0.001}$ (f) $\left(\dfrac{3}{4}\right)^3 = \dfrac{3}{4} \times \dfrac{3}{4} \times \dfrac{3}{4} = \dfrac{27}{64}$

(g) $10^2 = \overset{\text{2 zeros}}{100}$ (h) $10^3 = \overset{\text{3 zeros}}{1000}$ (i) $10^5 = \overset{\text{5 zeros}}{100,000}$ (j) $10^{12} = \overset{\text{12 zeros}}{1,000,000,000,000}$

(k) $0^3 = 0$ $(0^n = 0$ if $n \neq 0)$ (l) $1^{10} = 1$ $(1^n = 1)$ (m) $\dfrac{5^2}{4} = \dfrac{5 \times 5}{4} = \dfrac{25}{4}$ or $6\dfrac{1}{4}$

(n) $\left(\dfrac{5}{4}\right)^2 = \dfrac{5}{4} \times \dfrac{5}{4} = \dfrac{25}{16}$ or $1\dfrac{9}{16}$ (o) $3^1 = 3$ $(a^1 = a)$ (p) $\left(\dfrac{7}{10}\right)^1 \equiv \dfrac{7}{10}$ (q) $0^1 \equiv 0$

(r) $8.25^1 = 8.25$ (s) $1^0 = 1$ $(a^0 = 1$ if $a \neq 0)$ (t) $\left(\dfrac{9}{10}\right)^0 = 1$ (u) 0^0 is not defined

(v) $198.756^0 = 1$ (w) $\dfrac{2}{3^2} = \dfrac{2}{3 \times 3} = \dfrac{2}{9}$ (x) $\dfrac{5^2}{2^3} = \dfrac{5 \times 5}{2 \times 2 \times 2} = \dfrac{25}{8}$ or $3\dfrac{1}{8}$ (y) $\dfrac{8^1}{6^0} = \dfrac{8}{1} = 8$

(z) $\dfrac{0^5}{5^0} = \dfrac{0}{1} = 0$

PROBLEM 14-3 Find each square using paper and pencil: (a) 1^2 (b) 2^2 (c) 3^2 (d) 4^2

(e) 5^2 (f) 6^2 (g) 7^2 (h) 8^2 (i) 9^2 (j) 10^2 (k) $\left(\dfrac{1}{2}\right)^2$ (l) $\left(\dfrac{2}{3}\right)^2$ (m) $\left(\dfrac{3}{4}\right)^2$

(n) 0.2^2 (o) 0.03^2 (p) 0.004^2 (q) 11^2 (r) 12^2 (s) 20^2 (t) 300^2 (u) 1000^2
(v) $10,000^2$

Solution: Recall that to find the square of a given number, you multiply the given number times itself [see Examples 14-9 and 14-10]:

one times itself

(a) $1^2 = \overset{}{1 \times 1} = 1$ (b) $2^2 = 4$ (c) $3^2 = 9$ (d) $4^2 = 16$ (e) $5^2 = 25$ (f) $6^2 = 36$

(g) $7^2 = 49$ (h) $8^2 = 64$ (i) $9^2 = 81$ (j) $10^2 = 100$ (k) $\left(\dfrac{1}{2}\right)^2 = \dfrac{1}{4}$ (l) $\left(\dfrac{2}{3}\right)^2 = \dfrac{4}{9}$

(m) $\left(\dfrac{3}{4}\right)^2 = \dfrac{9}{16}$ (n) $0.2^2 = 0.04$ (o) $0.03^2 = 0.0009$ (p) $0.004^2 = 0.000016$

(q) $11^2 = 121$ (r) $12^2 = 144$ (s) $20^2 = 400$ (t) $300^2 = 90,000$ (u) $1000^2 = 1,000,000$

(v) $10,000^2 = 100,000,000$

PROBLEM 14-4 Find each square root using paper and pencil: (a) $\sqrt{1}$ (b) $\sqrt{100}$ (c) $\sqrt{49}$

(d) $\sqrt{4}$ (e) $\sqrt{16}$ (f) $\sqrt{81}$ (g) $\sqrt{25}$ (h) $\sqrt{9}$ (i) $\sqrt{36}$ (j) $\sqrt{64}$ (k) $\sqrt{\dfrac{4}{9}}$

(l) $\sqrt{\dfrac{25}{49}}$ (m) $\sqrt{\dfrac{1}{100}}$ (n) $\sqrt{\dfrac{1}{16}}$ (o) $\sqrt{121}$ (p) $\sqrt{225}$ (q) $\sqrt{400}$ (r) $\sqrt{144}$

Solution: Recall that to find the square root of a given whole number square, you can divide the given whole number by various whole number divisors until the whole number divisor and the quotient are the same [see Examples 14-12 and 14-13]:

(a) $1 \div 1 = 1$ means $\sqrt{1} = 1$ **(b)** $100 \div 10 = 10$ means $\sqrt{100} = 10$

(c) $49 \div 7 = 7$ means $\sqrt{49} = 7$ **(d)** $4 \div 2 = 2$ means $\sqrt{4} = 2$ **(e)** $16 \div 4 = 4$ means $\sqrt{16} = 4$

(f) $81 \div 9 = 9$ means $\sqrt{81} = 9$ **(g)** $25 \div 5 = 5$ means $\sqrt{25} = 5$ **(h)** $9 \div 3 = 3$ means $\sqrt{9} = 3$

(i) $36 \div 6 = 6$ means $\sqrt{36} = 6$ **(j)** $64 \div 8 = 8$ means $\sqrt{64} = 8$ **(k)** $\sqrt{\dfrac{4}{9}} = \dfrac{2}{3}$ **(l)** $\sqrt{\dfrac{25}{49}} = \dfrac{5}{7}$

(m) $\sqrt{\dfrac{1}{100}} = \dfrac{1}{10}$ **(n)** $\sqrt{\dfrac{1}{16}} = \dfrac{1}{4}$ **(o)** $121 \div 11 = 11$ means $\sqrt{121} = 11$

(p) $225 \div 15 = 15$ means $\sqrt{225} = 15$ **(q)** $400 \div 20 = 20$ means $\sqrt{400} = 20$

(r) $144 \div 12 = 12$ means $\sqrt{144} = 12$

PROBLEM 14-5   Find each square using Appendix Table 2: **(a)** 17^2 **(b)** 23^2 **(c)** 39^2
(d) 41^2 **(e)** 58^2 **(f)** 60^2 **(g)** 72^2 **(h)** 84^2 **(i)** 96^2

Solution: Recall that to find the square of any whole number from 0 to 100, you can use Appendix Table 2 by placing your left finger on the given base number in the "Number N" column and your right finger at the top of the "Number N^2" column and then moving your left finger straight across and your right finger straight down until they meet at the square of the given base number [see Example 14-14]:

(a) $17^2 = 289$ **(b)** $23^2 = 529$ **(c)** $39^2 = 1521$ **(d)** $41^2 = 1681$ **(e)** $58^2 = 3364$
(f) $60^2 = 3600$ **(g)** $72^2 = 5184$ **(h)** $84^2 = 7056$ **(i)** $96^2 = 9216$

PROBLEM 14-6   Find each square root using Appendix Table 2: **(a)** $\sqrt{2}$ **(b)** $\sqrt{3}$ **(c)** $\sqrt{8}$
(d) $\sqrt{26}$ **(e)** $\sqrt{30}$ **(f)** $\sqrt{59}$ **(g)** $\sqrt{72}$ **(h)** $\sqrt{87}$ **(i)** $\sqrt{99}$

Solution: Recall that to find the square root of any whole number from 0 to 100, you can use Appendix Table 2 by placing your left finger on the given radicand in the "Number N" column and your right finger at the top of the "Square Root $\sqrt{N}$" column and then moving your left finger straight across and your right finger straight down until they meet at the approximated square root of the given radicand [see Example 14-15]: **(a)** $\sqrt{2} \approx 1.414$ **(b)** $\sqrt{3} \approx 1.732$ **(c)** $\sqrt{8} \approx 2.828$
(d) $\sqrt{26} \approx 5.099$ **(e)** $\sqrt{30} \approx 5.477$ **(f)** $\sqrt{59} \approx 7.681$ **(g)** $\sqrt{72} \approx 8.485$ **(h)** $\sqrt{87} \approx 9.327$
(i) $\sqrt{99} \approx 9.950$

PROBLEM 14-7   Find each square using a calculator: **(a)** 101^2 **(b)** 300^2 **(c)** 4.56^2
(d) 82.1^2 **(e)** 1000^2 **(f)** 9508^2 **(g)** 27.82^2 **(h)** 502.7^2 **(i)** 0.095^2

Solution: Recall that to find the square of any whole number or decimal with 4 digits or less, you can use a standard 8-digit display calculator. To find the square of any whole number or decimal with 5 digits or less, you can use a standard 10-digit display calculator [see Example 14-16]:
(a) $101^2 = 10,201$ **(b)** $300^2 = 90,000$ **(c)** $4.56^2 = 20.7936$ **(d)** $82.1^2 = 6740.41$
(e) $1000^2 = 1,000,000$ **(f)** $9508^2 = 90,402,064$ **(g)** $27.82^2 = 773.9524$
(h) $502.7^2 = 252,707.29$ **(i)** $0.095^2 = 0.009025$

PROBLEM 14-8   Find or approximate each square root using a calculator: **(a)** $\sqrt{101}$
(b) $\sqrt{1000}$ **(c)** $\sqrt{72,361}$ **(d)** $\sqrt{265,271}$ **(e)** $\sqrt{9,205,000}$ **(f)** $\sqrt{38,521.619}$ **(g)** $\sqrt{3.5}$
(h) $\sqrt{31.36}$ **(i)** $\sqrt{0.002589}$

Solution: Recall that to find or approximate the square root of any 8-digit whole number or decimal, you can use a standard 8-digit calculator. To find or approximate the square root of any 10-digit number or decimal, you can use a standard 10-digit display calculator [see Example 14-17]:

(a) $\sqrt{101} \approx 10.049876$ **(b)** $\sqrt{1000} \approx 31.622777$ **(c)** $\sqrt{72,361} = 269$
(d) $\sqrt{265,271} \approx 515.04466$ **(e)** $\sqrt{9,205,000} \approx 3033.9743$ **(f)** $\sqrt{38,521.619} \approx 196.26925$
(g) $\sqrt{3.5} \approx 1.8708287$ **(h)** $\sqrt{31.36} = 5.6$ **(i)** $\sqrt{0.002589} \approx 0.0508822$

PROBLEM 14-9 Find or approximate each square root of a fraction using a calculator: **(a)** $\sqrt{\dfrac{1}{2}}$

(b) $\sqrt{\dfrac{1}{3}}$ **(c)** $\sqrt{\dfrac{1}{4}}$ **(d)** $\sqrt{\dfrac{2}{3}}$ **(e)** $\sqrt{\dfrac{3}{4}}$ **(f)** $\sqrt{\dfrac{7}{10}}$ **(g)** $\sqrt{\dfrac{9}{16}}$ **(h)** $\sqrt{\dfrac{5}{32}}$ **(i)** $\sqrt{\dfrac{1}{100}}$

Solution: Recall that to find or approximate the square root of a fraction using a calculator, you first divide the numerator by the denominator and then press the square root key [see Example 14-18]:

(a) $\sqrt{\dfrac{1}{2}} = \sqrt{0.5} \approx 0.7071067$ **(b)** $\sqrt{\dfrac{1}{3}} \approx \sqrt{0.3333333} \approx 0.5773502$ **(c)** $\sqrt{\dfrac{1}{4}} = \sqrt{0.25} = 0.5$

(d) $\sqrt{\dfrac{2}{3}} \approx \sqrt{0.66666666} \approx 0.8164965$ **(e)** $\sqrt{\dfrac{3}{4}} = \sqrt{0.75} \approx 0.8660254$

(f) $\sqrt{\dfrac{7}{10}} = \sqrt{0.7} \approx 0.8366600$ **(g)** $\sqrt{\dfrac{9}{16}} = \sqrt{0.5625} = 0.75$

(h) $\sqrt{\dfrac{5}{32}} = \sqrt{0.15625} \approx 0.3952847$ **(i)** $\sqrt{\dfrac{1}{100}} = \sqrt{0.01} = 0.01$

Supplementary Exercises

PROBLEM 14-10 Write each as power notations: **(a)** 8×8 **(b)** 2.3×2.3 **(c)** $\dfrac{9}{10} \times \dfrac{9}{10}$

(d) $7 \times 7 \times 7$ **(e)** $0.1 \times 0.1 \times 0.1$ **(f)** $\dfrac{3}{4} \times \dfrac{3}{4} \times \dfrac{3}{4}$ **(g)** $10 \times 10 \times 10 \times 10$

(h) $24.75 \times 24.75 \times 24.75 \times 24.75$ **(i)** $\dfrac{2}{3} \times \dfrac{2}{3} \times \dfrac{2}{3} \times \dfrac{2}{3}$ **(j)** $\dfrac{6 \times 6}{5}$ **(k)** $\dfrac{2}{5 \times 5 \times 5}$

(l) $\dfrac{9 \times 9 \times 9}{3 \times 3}$ **(m)** 0 **(n)** 1 **(o)** 2 **(p)** 10 **(q)** 12.5 **(r)** $\dfrac{3}{4}$ **(s)** $5 \times 5 \times 4$

(t) $3 \times 6 \times 6 \times 6 \times 6$ **(u)** 0×0 **(v)** $1 \times 1 \times 1$ **(w)** 2×2 **(x)** $3 \times 3 \times 3$
(y) $2 \times 3 \times 2$ **(z)** $5 \times 4 \times 4 \times 5 \times 4 \times 5 \times 5$

PROBLEM 14-11 Evaluate each power notation: **(a)** 9^2 **(b)** 0.2^2 **(c)** $\left(\dfrac{3}{4}\right)^2$ **(d)** 3^3

(e) 0.2^3 **(f)** $\left(\dfrac{1}{2}\right)^3$ **(g)** 10^1 **(h)** 10^4 **(i)** 10^6 **(j)** 10^{11} **(k)** 0^5 **(l)** 1^8 **(m)** $\dfrac{3^2}{2}$

(n) $\left(\dfrac{3}{2}\right)^2$ **(o)** 5^1 **(p)** $\left(\dfrac{2}{3}\right)^1$ **(q)** 0^1 **(r)** 25.3^1 **(s)** 10^0 **(t)** $\left(\dfrac{4}{5}\right)^0$ **(u)** 0^0

(v) 0.00856^0 **(w)** $\dfrac{18}{2^3}$ **(x)** $\left(\dfrac{3}{4}\right)^4$ **(y)** $\dfrac{9^2}{10}$ **(z)** $\dfrac{2^8}{2}$

PROBLEM 14-12 Find each square using paper and pencil and Appendix Table 2, or a calculator:
(a) 0^2 **(b)** 1^2 **(c)** 2^2 **(d)** 3^2 **(e)** 4^2 **(f)** 5^2 **(g)** 6^2 **(h)** 7^2 **(i)** 8^2
(j) 9^2 **(k)** 10^2 **(l)** 11^2 **(m)** 12^2 **(n)** 13^2 **(o)** 14^2 **(p)** 15^2 **(q)** 225^2

(r) $\left(\dfrac{1}{3}\right)^2$ **(s)** $\left(\dfrac{1}{4}\right)^2$ **(t)** $\left(\dfrac{2}{5}\right)^2$ **(u)** $\left(\dfrac{9}{10}\right)^2$ **(v)** $\left(\dfrac{9}{100}\right)^2$ **(w)** 0.1^2 **(x)** 2.5^2

(y) 0.25^2 **(z)** 75.85^2

PROBLEM 14-13 Find or approximate each square root using paper and pencil and Appendix Table 2, or a calculator. Round to the nearest thousandth when necessary: **(a)** $\sqrt{0}$ **(b)** $\sqrt{1}$

(c) $\sqrt{2}$ **(d)** $\sqrt{3}$ **(e)** $\sqrt{4}$ **(f)** $\sqrt{5}$ **(g)** $\sqrt{6}$ **(h)** $\sqrt{7}$ **(i)** $\sqrt{8}$ **(j)** $\sqrt{9}$ **(k)** $\sqrt{10}$

(l) $\sqrt{11}$ **(m)** $\sqrt{12}$ **(n)** $\sqrt{13}$ **(o)** $\sqrt{14}$ **(p)** $\sqrt{15}$ **(q)** $\sqrt{225}$ **(r)** $\sqrt{\dfrac{1}{3}}$ **(s)** $\sqrt{\dfrac{1}{4}}$

(t) $\sqrt{\dfrac{2}{5}}$ **(u)** $\sqrt{\dfrac{9}{10}}$ **(v)** $\sqrt{\dfrac{9}{100}}$ **(w)** $\sqrt{0.1}$ **(x)** $\sqrt{2.5}$ **(y)** $\sqrt{0.25}$ **(z)** $\sqrt{75.85}$

Answers to Supplementary Exercises

(14-10) **(a)** 8^2 **(b)** 2.3^2 **(c)** $\left(\dfrac{9}{10}\right)^2$ **(d)** 7^3 **(e)** 0.1^3 **(f)** $\left(\dfrac{3}{4}\right)^3$ **(g)** 10^4

(h) 24.75^4 **(i)** $\left(\dfrac{2}{3}\right)^4$ **(j)** $\dfrac{6^2}{5}$ **(k)** $\dfrac{2}{5^3}$ **(l)** $\dfrac{9^3}{3^2}$ **(m)** 0^1 **(n)** 1^1 or n^0 where $n \neq 0$

(o) 2^1 **(p)** 10^1 **(q)** 12.5^1 **(r)** $\left(\dfrac{3}{4}\right)^1$ **(s)** $5^2 \times 4$ **(t)** 3×6^4

(u) 0^2 or 0^n where $n \neq 0$ **(v)** 1^3 or 1^n **(w)** $(2^3)^2$ **(x)** $(3^2)^3$ **(y)** $2^2 \times 3$ or 3×2^2

(z) $5^4 \times 4^3$ or $4^3 \times 5^4$

(14-11) **(a)** 81   **(b)** 0.04   **(c)** $\dfrac{9}{16}$ **(d)** 27   **(e)** 0.008   **(f)** $\dfrac{1}{8}$ **(g)** 10   **(h)** 10,000

(i) 1,000,000   **(j)** 100,000,000,000   **(k)** 0   **(l)** 1   **(m)** $\dfrac{9}{2}$ **(n)** $\dfrac{9}{4}$ **(o)** 5   **(p)** $\dfrac{2}{3}$

(q) 0   **(r)** 25.3   **(s)** 1   **(t)** 1   **(u)** not defined   **(v)** 1   **(w)** $\dfrac{18}{8}$ or $\dfrac{9}{4}$ **(x)** $\dfrac{81}{256}$

(y) $\dfrac{81}{10}$ **(z)** $\dfrac{256}{2} = 128$

(14-12) **(a)** 0   **(b)** 1   **(c)** 4   **(d)** 9   **(e)** 16   **(f)** 25   **(g)** 36   **(h)** 49   **(i)** 64

(j) 81   **(k)** 100   **(l)** 121   **(m)** 144   **(n)** 169   **(o)** 196   **(p)** 225   **(q)** 50,625   **(r)** $\dfrac{1}{9}$

(s) $\dfrac{1}{16}$ **(t)** $\dfrac{4}{25}$ **(u)** $\dfrac{81}{100}$ **(v)** $\dfrac{81}{10,000}$ **(w)** 0.01   **(x)** 6.25   **(y)** 0.0625   **(z)** 5753.2225

(14-13) **(a)** 0   **(b)** 1   **(c)** 1.414   **(d)** 1.732   **(e)** 2   **(f)** 2.236   **(g)** 2.449   **(h)** 2.646
(i) 2.828   **(j)** 3   **(k)** 3.162   **(l)** 3.317   **(m)** 3.464   **(n)** 3.606   **(o)** 3.742   **(p)** 3.873
(q) 15   **(r)** 0.577   **(s)** 0.5   **(t)** 0.632   **(u)** 0.949   **(v)** 0.3   **(w)** 0.316   **(x)** 1.581
(y) 0.5   **(z)** 8.709

15 GEOMETRY

THIS CHAPTER IS ABOUT

- ☑ **Identifying Geometric Figures**
- ☑ **Finding the Perimeter**
- ☑ **Finding the Circumference**
- ☑ **Finding the Area**
- ☑ **Finding the Volume**

15-1. Identifying Geometric Figures

A. Identify points, lines, and line segments.

A **point** is an exact location in space. A **line** is an **infinite** (uncountable) collection of points that has no beginning and no ending. A **line segment** is a complete piece of a line that includes two **end points**.

EXAMPLE 15-1: Draw a representation of the following: **(a)** point   **(b)** line   **(c)** line segment

Solution

(a)   · ⟵   point (A simple dot can be used to represent a point.)

(b) ⟷ ⟵   line (The arrowheads show that the line continues on forever in both directions.)

(c) —— ⟵   line segment (No arrowheads show that the line segment
has a definite beginning and ending point.)

B. Identify two-dimensional figures.

The shape of a **2-dimensional figure** is formed by lines that are **straight, curved,** and/or **broken.** A two-dimensional figure in which you cannot get from the **inside** to the **outside** without crossing over a point **on** the figure itself is called a **closed figure.** A two-dimensional figure that is not a closed figure is called an **open figure.** An open figure does not have an inside or an outside.

EXAMPLE 15-2: Identify each of the following as an open or closed figure:

(a)   **(b)**

Solution

(a)   This is an open figure because it has no inside and no outside.

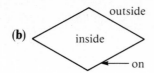

(b)   This is a closed figure because you cannot get from the inside to the outside without crossing over a point on the figure.

A closed figure with **sides** that are all straight line segments is called a **polygon.**

EXAMPLE 15-3: Draw a polygon.

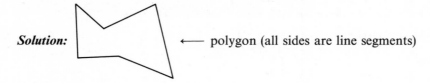

Solution: ⟵ polygon (all sides are line segments)

A polygon with three sides is called a **triangle.**

EXAMPLE 15-4: Draw a triangle.

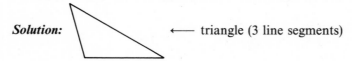

Solution: ⟵ triangle (3 line segments)

A triangle with a right angle (square corner) is called a **right triangle.**

EXAMPLE 15-5: Draw a right triangle.

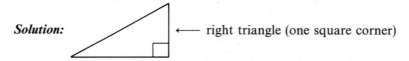

Solution: ⟵ right triangle (one square corner)

A triangle with three equal sides is called an **equilateral triangle.**

EXAMPLE 15-6: Draw an equilateral triangle.

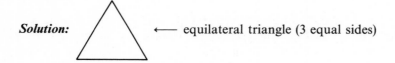

Solution: ⟵ equilateral triangle (3 equal sides)

A triangle with only two equal sides is called an **isosceles triangle.**

EXAMPLE 15-7: Draw an isosceles triangle.

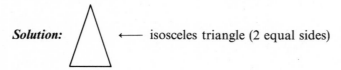

Solution: ⟵ isosceles triangle (2 equal sides)

A polygon with four sides is called a **quadrilateral.**

EXAMPLE 15-8: Draw a quadrilateral.

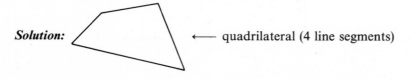

Solution: ⟵ quadrilateral (4 line segments)

A quadrilateral with four right angles is called a **rectangle.**

EXAMPLE 15-9: Draw a rectangle.

Solution: ⟵ rectangle (4 right angles)

A rectangle with four equal sides is called a **square.**

EXAMPLE 15-10: Draw a square.

Solution: ⟵ square (4 right angles and 4 equal sides)

A quadrilateral with two pairs of **parallel sides** (always the same distance apart, like railroad tracks) is called a **parallelogram.**

EXAMPLE 15-11: Draw a parallelogram.

Solution: ⟵ parallelogram (2 pairs of parallel sides)

A quadrilateral with one pair of parallel sides is called a **trapezoid.**

EXAMPLE 15-12: Draw a trapezoid.

Solution: ⟵ trapezoid (1 pair of parallel sides)

A polygon with five equal sides and five equal angles is called a **regular pentagon.**

EXAMPLE 15-13: Draw a regular pentagon.

Solution: ⟵ regular pentagon (5 equal sides and 5 equal angles)

Note: **"Regular"** means each side and each angle of the polygon is equal.

A polygon with six equal sides and six equal angles is called a **regular hexagon.**

EXAMPLE 15-14: Draw a regular hexagon.

Solution: ⟵ regular hexagon (6 equal sides and 6 equal angles)

A polygon with eight equal sides and eight equal angles is called a **regular octagon.**

EXAMPLE 15-15: Draw a regular octagon.

Solution: ⟵ regular octagon (8 equal sides and 8 equal angles)

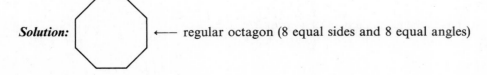

A **circle** is a closed figure in which every point is equally distanced from a fixed point inside the figure. The fixed point is called the **center** of the circle.

EXAMPLE 15-16: Draw a circle.

Solution: 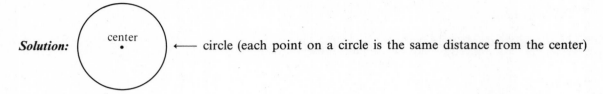 ⟵ circle (each point on a circle is the same distance from the center)

C. Identify three-dimensional figures.

A **three-dimensional figure** has length, width, and height. A three-dimensional figure shaped like a standard cardboard box is called a **rectangular prism.**

EXAMPLE 15-17: Draw a rectangular prism.

Solution: 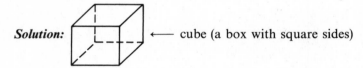 ⟵ rectangular prism (looks like a standard box)

A rectangular prism with all square sides is called a **cube.**

EXAMPLE 15-18: Draw a cube.

Solution: 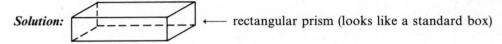 ⟵ cube (a box with square sides)

A three-dimensional figure shaped like a standard soup can is called a **cylinder.**

EXAMPLE 15-19: Draw a cylinder.

Solution: 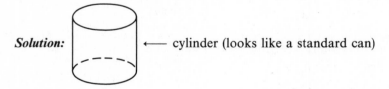 ⟵ cylinder (looks like a standard can)

A three-dimensional figure shaped like a round ball is called a **sphere.**

EXAMPLE 15-20: Draw a sphere.

Solution: 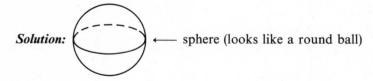 ⟵ sphere (looks like a round ball)

One-half of a sphere is called a **hemisphere.**

EXAMPLE 15-21: Draw a hemisphere.

Solution: 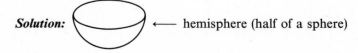 ⟵ hemisphere (half of a sphere)

15-2. Finding the Perimeter

A. Find the perimeter using addition.

The total distance around a polygon is called the **perimeter** *P*. To find the perimeter of any polygon, you add the lengths of all the sides together.

EXAMPLE 15-22: Find the perimeter of Figure 15-1.

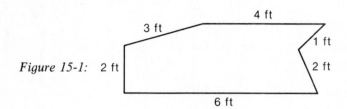

Figure 15-1:   2 ft

Solution: The figure in Figure 15-1 is a polygon with sides of 2 ft, 3 ft, 4 ft, 1 ft, 2 ft, and 6 ft. Add the lengths of all the sides together:

$P = 2\,\text{ft} + 3\,\text{ft} + 4\,\text{ft} + 1\,\text{ft} + 2\,\text{ft} + 6\,\text{ft}$

$\quad = 18\,\text{ft} \longleftarrow$ total distance around Figure 15-1

Note: The perimeter of Figure 15-1 is 18 feet because the total distance around Figure 15-1 is 18 feet.

B. Find the perimeter using a formula.

You can also find the perimeter of a polygon using a **perimeter formula.** Most of the more common polygons have specific perimeter formulas.

EXAMPLE 15-23: Write the common perimeter formulas.

Solution:

Common Perimeter Formulas

no sign means multiplication

Equilateral Triangle: $P = 3 \times \overbrace{\text{side}}^{s}$, or $P = 3 \times s$ or $\boldsymbol{P = 3s}$

Rectangle: $P = 2 \times \overbrace{\text{length}}^{l} + 2 \times \overbrace{\text{width}}^{w}$, or $\boldsymbol{P = 2l + 2w}$ or $\boldsymbol{P = 2(l + w)}$

Square: $P = 4 \times \text{side}$, or $\boldsymbol{P = 4s}$

Parallelogram: $P = 2 \times \overbrace{\text{one side}}^{a} + 2 \times \overbrace{\text{adjoining side}}^{b}$, or $\boldsymbol{P = 2a + 2b}$ or $\boldsymbol{P = 2(a + b)}$

Regular Pentagon: $P = 5 \times \text{side}$, or $\boldsymbol{P = 5s}$

Regular Hexagon: $P = 6 \times \text{side}$, or $\boldsymbol{P = 6s}$

Regular Octagon: $P = 8 \times \text{side}$, or $\boldsymbol{P = 8s}$

To find the perimeter of a given figure using its perimeter formula, you write the correct perimeter formula, then substitute the known measures for the corresponding letters, and compute the measures.

EXAMPLE 15-24: Find the perimeter of Figure 15-2 using the appropriate perimeter formula:

Figure 15-2:

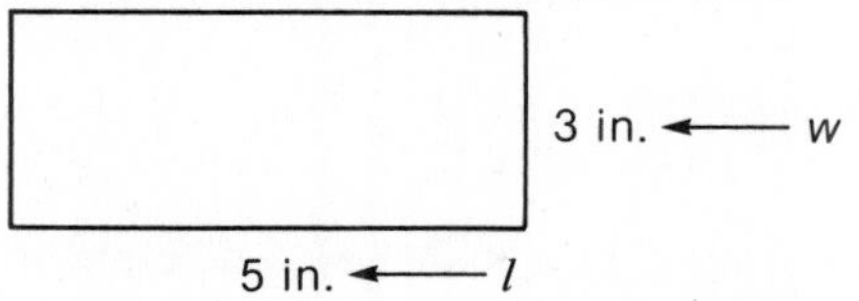

3 in. $\longleftarrow$ *w*

5 in. $\longleftarrow$ *l*

Solution: The figure in Figure 15-2 is a rectangle with a length *l* of 5 inches and a width *w* of 3 inches.

$P = 2(l + w)$ ⟵ perimeter formula for a rectangle

$\quad = 2(5 \text{ in.} + 3 \text{ in.})$ Substitute 5 in. for *l* (length) and 3 in. for *w* (width).

$\quad = 2(8 \text{ in.})$ Add inside the parentheses first.

$\quad = 16 \text{ in.}$ Then multiply.

Note 1: Whenever parentheses are present in a problem, you always compute inside the parentheses first.

Note 2: If you used addition to find the perimeter of Figure 15-2, you would see that the total distance around Figure 15-2 is 16 inches (5 in. + 3 in. + 5 in. + 3 in. = 16 in.).

EXAMPLE 15-25: Show that the perimeter of Figure 15-2 is 16 inches using the optional perimeter formula for a rectangle ($P = 2l + 2w$):

Solution: $P = 2l + 2w$ ⟵ optional perimeter formula for a rectangle

$\quad\quad\quad = 2 \times 5 \text{ in.} + 2 \times 3 \text{ in.}$ Substitute 5 in. for *l* (length) and 3 in. for *w* (width).

$\quad\quad\quad = 10 \text{ in.} + 6 \text{ in.}$ Multiply first.

$\quad\quad\quad = 16 \text{ in.}$ Then add.

15-3. Finding the Circumference

A. Rename the radius or diameter.

Any straight line segment that connects a point on a circle with its center is called a **radius *r***. More than one radius are called **radii.**

EXAMPLE 15-26: Draw a circle and one of its radii.

Solution:
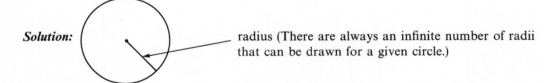
radius (There are always an infinite number of radii that can be drawn for a given circle.)

Any straight line segment that connects two points on a circle and also passes through its center is called a **diameter *d.***

EXAMPLE 15-27: Draw a circle and one of its diameters.

Solution:
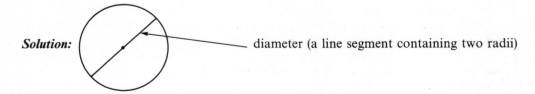
diameter (a line segment containing two radii)

The length of a diameter for a given circle is always twice the length of a radius.

EXAMPLE 15-28: Write a formula relating the length of a diameter for a circle given the length of a radius.

Solution: diameter = 2 × radius, or $d = 2r$ ⟵ **diameter/radius formula**

To find the length of a diameter of a circle given the length of a radius of the same circle, you can use the diameter/radius formula $d = 2r$.

EXAMPLE 15-29: Find the length of a diameter for the circle shown in Figure 15-3.

Figure 15-3: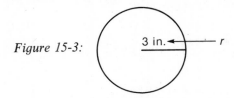

Solution: $d = 2r$ Write the diameter/radius formula.

       $= 2 \times 3$ in.   Substitute 3 in. for r (radius).

       $= 6$ in.   Multiply a measure by a number [see Example 12-38].

Note: The length of any diameter for Figure 15-3 is 6 inches.

The length of a radius for a given circle is always one-half the length of a diameter.

EXAMPLE 15-30: Write a formula relating the length of a radius for a cricle given the length of a diameter.

Solution: radius = one-half diameter or $r = \frac{1}{2}d$, or $r = d \div 2$ **radius/diameter formula**

To find the length of a radius of a circle given the length of a diameter for the same circle, you can compute the radius/diameter formula $r = d \div 2$.

EXAMPLE 15-31: Find the length of a radius for the circle and diameter shown in Figure 15-4.

Figure 15-4: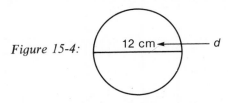

Solution: $r = d \div 2$ Write the radius/diameter formula.

       $= 12$ cm $\div 2$ Substitute 12 cm for d (diameter).

       $= 6$ cm   Divide a measure by a number [see Example 12-43.]

Note: The length of any radius for Figure 15-4 is 6 cm.

B. Find the circumference of a circle.

The distance around a circle is called its **circumference** C. The ratio of circumference to diameter for any given circle is always **pi** (π). The number π is a nonrepeating and nonterminating decimal that is approximately equal to $3\frac{1}{7}$, $\frac{22}{7}$, or 3.14. (One or the other of these forms of π is sometimes preferred when using certain measures.)

EXAMPLE 15-32: Write the approximate value of π that is usually used for
(a) U.S. Customary measures   **(b)** metric measures.

Solution
(a) When U.S. Customary measures are given, $\pi \approx 3\frac{1}{7}$ or $\frac{22}{7}$ is usually used.
(b) When metric measures are given, $\pi \approx 3.14$ is usually used.

To find the circumference C of a circle given the length of a radius or a diameter, you can compute the correct circumference formula.

EXAMPLE 15-33: Write the common circumference formulas.

Solution: Circle with a given radius length: Circumference = 2 × π × radius, or **C = 2πr**
Circle with a given diameter length: Circumference = π × diameter,   or **C = πd**

Note: If you memorize only $C = \pi d$, you can always derive $C = 2\pi r$ by using the diameter/radius formula $d = 2r$.

EXAMPLE 15-34: Find $C = 2\pi r$ using $C = \pi d$ and $d = 2r$.

Solution: $C = \pi d$ ◄——— memorized formula

$C = \pi(2r)$ Substitute $2r$ for d because $d = 2r$.

$C = 2\pi r$ ◄——— derived formula

To find the **approximate circumference** of a circle with a given length of a diameter, you first write the circumference formula $C = \pi d$, then substitute the given diameter length for d and either $3\frac{1}{7}$ or 3.14 for π, and then multiply.

EXAMPLE 15-35: Find the approximate circumference of the circle in Figure 15-5.

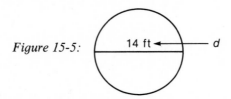

Figure 15-5:

Solution: The circle in Figure 15-5 has a diameter d of 14 feet.

$C = \pi d$ ◄——— circumference formula for a circle with a given diameter length

$\approx 3\frac{1}{7} \times 14$ ft   Substitute 14 ft for d and $3\frac{1}{7}$ for π because 14 feet is a U.S. Customary measure.

$= \dfrac{22}{\cancel{7}} \times \dfrac{2 \times \cancel{7}}{1}$ ft   Multiply.

$= 44$ ft ◄——— approximate circumference of Figure 15-5

Note: The circumference of the circle in 15-5 is only approximately equal to 44 feet ($C \approx 44$ ft) because π is only approximately equal to $3\frac{1}{7}$ ($\pi \approx 3\frac{1}{7}$).

To find the approximate circumference of a circle given the length of a radius, you first write the circumference formula $C = 2\pi r$, then substitute the given radius length for r and either $3\frac{1}{7}$ or 3.14 for π, and then multiply.

EXAMPLE 15-36: Find the approximate circumference of the circle in Figure 15-6.

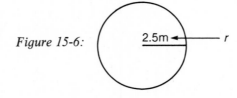

Figure 15-6:

Solution: The circle in Figure 15-6 has a radius r of 2.5 meters.

$C = 2\pi r$ ◄——— circumference formula for a circle with a given radius length

$\approx 2 \times 3.14 \times 2.5$ m   Substitute 2.5 m for r and 3.14 for π because 2.5 m is a metric measure.

$= 6.28 \times 2.5$ m   Multiply.

$= 15.7$ m ◄——— approximate circumference of Figure 15-6

15-4. Finding the Area

A. Identify common area units.

When the surface of a closed figure is measured by finding how many squares of a given size are needed to completely cover the surface, you are finding the **area** A of that surface in **square units of measure.**

Recall: The common square units of measure are
(**a**) square inches (in.^{2})   (**b**) square feet (ft^2)   (**c**) square yards (yd^2)
(**d**) square centimeters (cm^2)   (**e**) square meters (m^2).

B. Find the area of common polygons.

The **area formula for a rectangle** is Area = length × width, or $A = lw$.

EXAMPLE 15-37: Find the area of the rectangle in Figure 15-7.

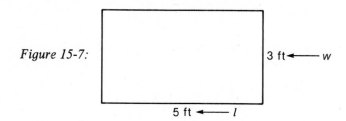

Figure 15-7:

3 ft ◄——— w

5 ft ◄——— l

Solution: The rectangle in Figure 15-7 has a length l of 5 feet and a width w of 3 feet.

$A = lw$ ◄——— area formula for a rectangle

   $= 5\,\text{ft} \times 3\,\text{ft}$ Substitute 5 ft for l (length) and 3 ft for w (width).

   $= (5 \times 3)\,\text{ft}^2$ Multiply two like length measures [see Example 12-40].

   $= 15\,\text{ft}^2$ ◄——— area of Figure 15-7

Note: It would take 15 squares each measuring 1 foot on a side to completely cover the surface of the rectangle in Figure 15-7.

EXAMPLE 15-38: Show by counting that it takes 15 squares each measuring 1 foot on a side to completely cover the surface of the rectangle in Figure 15-7.

Solution:

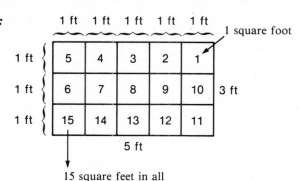

Subdivide Figure 15-7 into squares of 1 square foot each and then counting the number of square feet.

Note: The total number of square units it takes to completely cover a surface is the area of that surface.

The **area formula for a square** is Area = side × side or $A = s \times s$, or $A = s^2$.

EXAMPLE 15-39: Find the area of the square in Figure 15-8.

Figure 15-8:

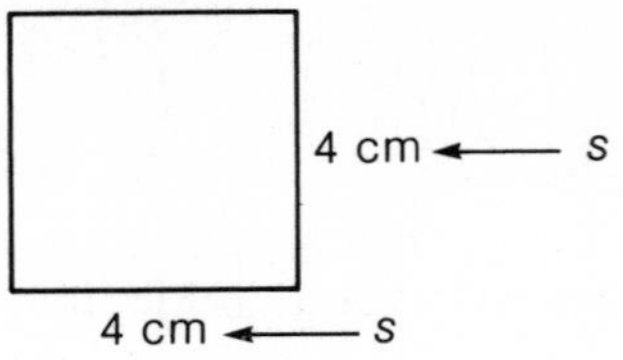

```
          ┌─────────┐
          │         │  4 cm ◄─── s
          │         │
          └─────────┘
           4 cm ◄─── s
```

Solution: The square in Figure 15-8 has sides s with a length of 4 centimeters.

$A = s^2$ ◄── area formula for a square

$\quad = (4 \text{ cm})^2$ $\qquad$ Substitute 4 cm for s (side).

$\quad = 4 \text{ cm} \times 4 \text{ cm}$ $\qquad$ Rename as a product of repeated factors.

$\quad = (4 \times 4) \text{ cm}^2$ $\qquad$ Multiply two like length measures.

$\quad = 16 \text{ cm}^2$ ◄── area of Figure 15-8

Note: It takes 16 squares each measuring 1 centimeter on a side to completely cover the surface of the square in Figure 15-8.

The **area formula for a parallelogram** is Area = base × height, or $A = bh$.

EXAMPLE 15-40: Find the area of the parallelogram in Figure 15-9.

Figure 15-9:

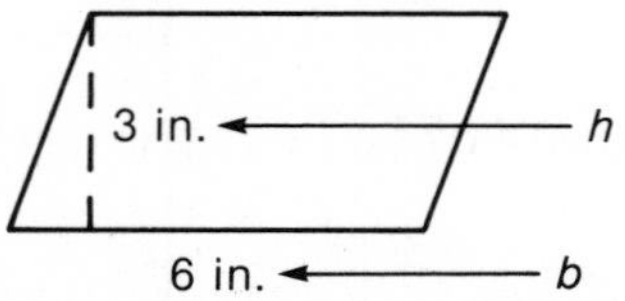

```
        ╱──────────╱
       ╱ | 3 in. ◄──── h
      ╱  |        ╱
     ╱___|_____╱
        6 in. ◄──── b
```

Solution: The parallelogram in Figure 15-9 has a base b of 6 inches and a height h of 3 inches.

$A = bh$ ◄── area formula for a parallelogram

$\quad = 6 \text{ in.} \times 3 \text{ in.}$ $\qquad$ Substitute 6 in. for b (base) and 3 in. for h (height).

$\quad = (6 \times 3) \text{ in.}^2$ $\qquad$ Multiply two like length measures.

$\quad = 18 \text{ in.}^2$ ◄── area of Figure 15-9

Note: It takes the equivalent of 18 squares each measuring 1 inch on a side to completely cover the surface of the parallelogram in Figure 15-9.

The **area formula for a triangle** is Area $= \frac{1}{2} \times$ base × height, or $A = \frac{1}{2} bh$.

EXAMPLE 15-41: Find the area of the triangle in Figure 15-10 (page 311).

Solution: The triangle in Figure 15-10 has a base b of 7 meters and a height h of 4 meters.

$A = \frac{1}{2} bh$ ◄── area formula for a triangle

$\quad = \frac{1}{2} \times 7 \text{ m} \times 4 \text{ m}$ $\qquad$ Substitute 7 m for b (base) and 4 m for h (height).

$\quad = \frac{1}{2} \times (7 \times 4) \text{ m}^2$ $\qquad$ Multiply two like length measures.

$\quad = \frac{1}{2} \times 28 \text{ m}^2$ $\qquad$ Multiply by $\frac{1}{2}$.

$\quad = 14 \text{ m}^2$ ◄── area of Figure 15-10

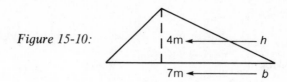

Figure 15-10:

Note: It takes the equivalent of 14 squares each measuring 1 meter on a side to completely cover the surface of the triangle in Figure 15-10.

The **area formula for a trapezoid** is $A = \frac{1}{2} \times$ (one base + other base) $\times$ height, or $A = \frac{1}{2}(b_1 + b_2)h.$

EXAMPLE 15-42: Find the area of the trapezoid

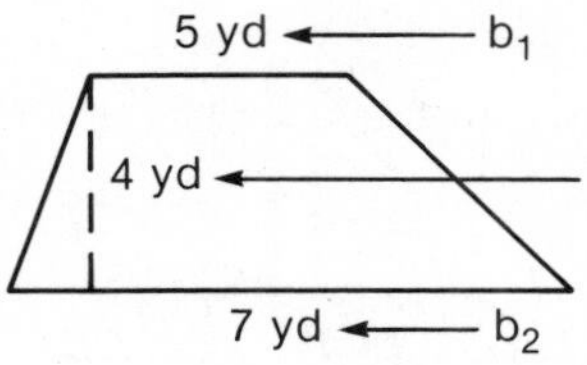

Figure 15-11:

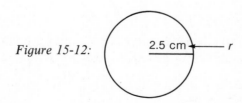

Solution: The trapezoid in Figure 15-11 has bases b_1 and b_2 of 5 yards and 7 yards and height h of 4 yards.

$A = \frac{1}{2}(b_1 + b_2)h \longleftarrow$ area formula for a trapezoid

$= \frac{1}{2}(5 \text{ yd} + 7 \text{ yd})\, 4 \text{ yd}$ Substitute 5 yd for b_1 (one base), 7 yd for b_2 (other base), and 4 yd for h (height).

$= \frac{1}{2} \times 12 \text{ yd} \times 4 \text{ yd}$ Add in parentheses first.

$= \frac{1}{2} \times (12 \times 4) \text{ yd}^2$ Then multiply two like length measures.

$= \frac{1}{2} \times 48 \text{ yd}^2$ Multiply by $\frac{1}{2}$.

$= 24 \text{ yd}^2 \longleftarrow$ area of Figure 15-11

Note: It takes the equivalent of 24 squares each measuring 1 yard on a side to completely cover the surface of the trapezoid in Figure 15-11.

C. Find the area of a circle.

The **area formula for a circle** is Area $= \pi \times$ radius $\times$ radius, or $A = \pi r r$ or $A = \pi r^2.$

EXAMPLE 15-43: Find the area of the circle in Figure 15-12.

Figure 15-12:

Solution: The circle in Figure 15-12 has a radius r of 2.5 centimeters.

$A = \pi r^2$ ⟵ area formula for a circle

$\approx 3.14\,(2.5\text{ cm})^2$ Substitute 2.5 cm for r (radius) and 3.14 for π because 2.5 cm is a metric measure.

$= 3.14 \times 2.5\text{ cm} \times 2.5\text{ cm}$ Rename as a product of repeated factors.

$= 3.14 \times (2.5 \times 2.5)\text{ cm}^2$ Multiply two like length measures.

$= 3.14 \times 6.25\text{ cm}^2$ Multiply by 3.14.

$= 19.625\text{ cm}^2$ ⟵ approximate area of Figure 15-12

Note: It takes the equivalent of about 19.625 ($19\frac{5}{8}$) squares each measuring 1 meter on a side to completely cover the surface of the circle in Figure 15-12.

To find the area of a circle with a given diameter, you first use $r = d \div 2$ to find the length of a radius and then compute $A = \pi r^2$.

EXAMPLE 15-44: Find the area of the circle in Figure 15-13.

Figure 15-13:

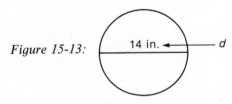

Solution: The circle in Figure 15-13 is a circle with diameter d 14 inches.

$r = d \div 2$ ⟵ radius/diameter formula

$= 14\text{ in.} \div 2$ Substitute 14 in. for d (diameter).

$= (14 \div 2)\text{ in.}$ Divide a measure by a number.

$= 7\text{ in.}$ ⟵ length of a radius r for the circle in Figure 15-13

$A = \pi r^2$ ⟵ area formula for a circle

$\approx 3\frac{1}{7}(7\text{ in.})^2$ Substitute 7 in. for r (radius) and $3\frac{1}{7}$ for π because 7 in. is a U.S. Customary measure.

$= 3\frac{1}{7} \times 7\text{ in.} \times 7\text{ in.}$ Rename as a product of repeated factors.

$= \dfrac{22}{\cancel{7}} \times \dfrac{\cancel{7}\text{ in.}}{1} \times \dfrac{7\text{ in.}}{1}$ Multiply.

$= 22 \times 7\text{ in.}^2$

$= 154\text{ in.}^2$ ⟵ approximate area of Figure 15-13

Note: It takes the equivalent of about 154 squares each measuring 1 inch on a side to completely cover the surface of the circle in Figure 15-13.

15-5. Finding the Volume

A. Identify common volume units.

When the space occupied by a three-dimensional object is measured by finding how many cubes of a given size are needed to completely fill that space, you are finding the **volume** V of that object in **cubic units of measure.**

Recall: The common cubic units of measure are
(a) cubic inches (in.^{3})   **(b)** cubic feet (ft^3)   **(c)** cubic yards (yd^3)   **(d)** cubic centimeters (cm^3)
(e) cubic meters (m^3).

B. Find the volume of 3-dimensional figures with flat surfaces.

The **volume formula for a rectangular prism** is Volume = length × width × height, or $V = lwh.$

EXAMPLE 15-45: Find the volume of the rectangular prism in Figure 15-14.

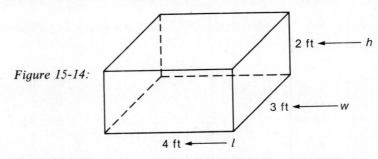

Figure 15-14:

Solution: The rectangular prism in Figure 15-14 has a length l of 4 feet, a width w of 3 feet, and a height h of 2 feet.

$V = lwh$ ⟵ volume formula for a rectangular prism

   $= 4\text{ ft} \times 3\text{ ft} \times 2\text{ ft}$ Substitute 4 ft for l (length), 3 ft for w (width), and 2 ft for h (height).

   $= (4 \times 3)\text{ ft}^2 \times 2\text{ ft}$ Multiply three like length measures [see Example 12-41].

   $= 12\text{ ft}^2 \times 2\text{ ft}$

   $= (12 \times 2)\text{ ft}^3$

   $= 24\text{ ft}^3$ ⟵ area of Figure 15-14

Note: It takes 24 cubes each measuring 1 foot on an edge to completely fill the space occupied by the rectangular prism in Figure 15-14.

EXAMPLE 15-46: Show by counting that it takes 24 cubes each measuring 1 foot on an edge to completely fill the space occupied by the rectangular prism in Figure 15-14.

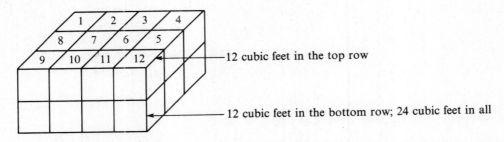

Note: The total number of cubic units it takes to completely fill the space occupied by an object is the volume of that object.

The **volume formula for a cube** is Volume = edge × edge × edge, or $V = (e)(e)(e)$, or $V = e^3.$

EXAMPLE 15-47: Find the volume of the cube in Figure 15-15.

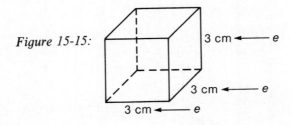

Figure 15-15:

Solution: The cube in Figure 15-15 has an edge (*e*) of 3 centimeters.

$V = e^3$ ◀— volume formula for a cube

$= (3 \text{ cm})^3$ Substitute 3 cm for *e* (edge).

$= 3 \text{ cm} \times 3 \text{ cm} \times 3 \text{ cm}$ Rename as a product of repeated factors.

$= (3 \times 3) \text{ cm}^2 \times 3 \text{ cm}$ Multiply three like length measures.

$= 9 \text{ cm}^2 \times 3 \text{ cm}$

$= (9 \times 3) \text{ cm}^3$

$= 27 \text{ cm}^3$ ◀— volume of Figure 15-15

Note: It takes 27 cubes each measuring 1 centimeter on an edge to completely fill the space occupied by the cube in Figure 15-15.

C. Find the volume of 3-dimensional figures with curved surfaces.

The **volume formula for a cylinder** is Volume = $\pi \times$ radius $\times$ radius $\times$ height, or $V = \pi rrh$ or $V = \pi r^2 h$.

EXAMPLE 15-48: Find the volume of the cylinder in Figure 15-16.

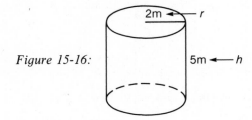

Figure 15-16:

Solution: The cylinder in Figure 15-16 has a radius *r* of 2 meters and a height *h* of 5 meters.

$V = \pi r^2 h$ ◀— volume formula for a cylinder

$\approx 3.14 (2 \text{ m})^2 \, 5 \text{ m}$ Substitute 2 m for *r* (radius), 5 m for *h* (height), and 3.14 for π since meters is a metric measure.

$= 3.14 \times 2 \text{ m} \times 2 \text{ m} \times 5 \text{ m}$ Rename as a product of repeated factors.

$= 3.14 \times (2 \times 2 \times 5) \text{m}^3$ Multiply three like length measures.

$= 3.14 \times 20 \text{ m}^3$ Multiply by 3.14.

$= 62.8 \text{ m}^3$ ◀— approximate volume of Figure 15-16

Note: It takes the equivalent of about 62.8 cubes each measuring 1 meter on an edge to completely fill the space occupied by the cylinder in Figure 15-16.

The **volume formula for a sphere** is Volume $= \dfrac{4}{3} \times \pi \times$ radius $\times$ radius $\times$ radius, or $V = \dfrac{4}{3} \pi rrr$ or $V = \dfrac{4}{3} \pi r^3$.

EXAMPLE 15-49: Find the volume of the sphere in Figure 15-17.

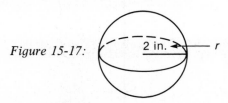

Figure 15-17:

Solution: The sphere in Figure 15-17 has a radius *r* of 2 inches.

$$V = \frac{4}{3}\pi r^3 \longleftarrow \text{volume formula for a sphere}$$

$$\approx \frac{4}{3} \times 3\frac{1}{7}(2 \text{ in.})^3 \qquad \text{Substitute 2 in. for } r \text{ (radius) and } 3\frac{1}{7} \text{ for } \pi \text{ because inches is a U.S. Customary measure.}$$

$$= \frac{4}{3} \times 3\frac{1}{7} \times 2 \text{ in.} \times 2 \text{ in.} \times 2 \text{ in.} \qquad \text{Rename as a product of repeated factors.}$$

$$= \frac{4}{3} \times \frac{22}{7} \times (2 \times 2 \times 2) \text{ in.}^3 \qquad \text{Multiply three like length measures.}$$

$$= \frac{88}{21} \times 8 \text{ in.}^3 \qquad \text{Multiply } \frac{4}{3} \text{ and } \frac{22}{7}.$$

$$= \frac{704}{21} \text{ in.}^3 \qquad \text{Multiply by } \frac{88}{21}.$$

$$= 33\frac{11}{21} \text{ in.}^3 \longleftarrow \text{approximate volume of Figure 15-17}$$

Note: It takes the equivalent of about $33\frac{11}{21}$ cubes each measuring 1 inch on an edge to completely fill the space occupied by the sphere in Figure 15-17.

RAISE YOUR GRADES

Can you . . . ?

- ☑ identify the representation of a point, line, and line segment
- ☑ identify open figures, closed figures, and polygons
- ☑ identify a triangle, right triangle, equilateral triangle, and an isosceles triangle
- ☑ identify a quadrilateral rectangle, square, parallelogram, and trapezoid
- ☑ identify a regular pentagon, regular hexagon, and regular octagon
- ☑ identify a circle
- ☑ identify a rectangular prism, cube, cylinder, sphere, and hemisphere
- ☑ find the perimeter of a polygon using addition
- ☑ find the perimeter of an equilateral triangle, rectangle, square, parallelogram, regular pentagon, regular hexagon, or regular octagon using a formula
- ☑ identify the radius and the diameter of a given circle
- ☑ find the length of a diameter for a circle given the length of the radius
- ☑ find the length of a radius for a circle given the length of the diameter
- ☑ find the circumference of a circle given a radius length or a diameter length
- ☑ identify common area units
- ☑ find the area of a rectangle, square, parallelogram, triangle, trapezoid, or circle using a formula
- ☑ identify common volume units
- ☑ find the volume of a rectangular prism, cube, cylinder, or sphere using a formula

SUMMARY

1. A point is an exact location in space.
2. A line is an infinite (uncountable) collection of points that has no beginning and no ending.
3. A line segment is a complete piece of a line that includes two end points.

4. The shape of a two-dimensional figure is formed by lines that are straight, curved, and/or broken.
5. A two-dimensional figure in which you cannot get from the inside to the outside without crossing over a point on the figure itself is called a closed figure.
6. A two-dimensional figure that is not a closed figure (with no inside nor outside) is called an open figure.
7. A closed figure with sides that are all straight line segments is called a polygon.
8. A polygon with three sides is called a triangle.
9. A triangle with a right angle (square corner) is called a right triangle.
10. A triangle with three equal sides is called an equilateral triangle.
11. A triangle with two equal sides is called an isosceles triangle.
12. A polygon with four sides is called a quadrilateral.
13. A quadrilateral with four right angles is called a rectangle.
14. A rectangle with four equal sides is called a square.
15. A quadrilateral with two pairs of parallel sides is called a parallelogram.
16. A quadrilateral with one pair of parallel sides is called a trapezoid.
17. A polygon with five equal sides and five equal angles is called a regular pentagon.
18. A polygon with six equal sides and six equal angles is called a regular hexagon.
19. A polygon with eight equal sides and eight equal angles is called a regular octagon.
20. A circle is a closed figure where each point on it is an equal distance from a point inside called its center.
21. A three-dimensional figure has length, width, and height.
22. A three-dimensional figure shaped like a standard cardboard box is called a rectangular prism.
23. A rectangular prism with all square sides is called a cube.
24. A three-dimensional figure shaped like a standard soup can is called a cylinder.
25. A three-dimensional figure shaped like a round ball is called a sphere.
26. One half of a sphere is called a hemisphere.
27. The total distance around a polygon is called the perimeter P.
28. To find the perimeter of any polygon, you can add the lengths of all the sides together.
29. You can also find the perimeter of a polygon using a perimeter formula. Most of the common polygons have a specific perimeter formula:
    (a) Equilateral Triangle: $P = 3s$ ⟵ s stands for side
    (b) Rectangle: $P = 2(l + w)$ ⟵ l stands for length and w stands for width
    (c) Square: $P = 4s$
    (d) Parallelogram: $P = 2(a + b)$ ⟵ a and b stand for adjoining sides
    (e) Regular Pentagon: $P = 5s$
    (f) Regular Hexagon: $P = 6s$
    (g) Regular Octagon: $P = 8s$
30. Any straight line segment that connects a point on a circle with its center is called a radius r. More than one radius are called radii.
31. Any straight line segment that connects two points on a circle and also passes through its center is called a diameter d.
32. The length of a diameter for a given circle is always twice the length of a radius: $d = 2r$.
33. The length of a radius for a given circle is always one-half the length of a diameter: $r = \frac{1}{2} d$ or $r = d \div 2$.
34. The distance around a circle is called the circumference C.
35. The ratio of circumference to diameter for any given circle is always pi (π).
36. The number π is a nonrepeating and nonterminating decimal that is approximately equal to $3\frac{1}{7}, \frac{22}{7}$, or 3.14.
37. When U.S. Customary measures are given, $\pi \approx 3\frac{1}{7}$ is usually used.
38. When metric measures are given, $\pi \approx 3.14$ is usually used.
39. To find the circumference C of a circle given the length of a radius r or a diameter d, you can compute the circumference formula $C = 2\pi r$, or $C = \pi d$.

40. When the surface of a closed figure is measured by finding how many squares of a given size are needed to completely cover the surface, you are finding the area A of that surface in square units of measure.

41. The common square units of measure are (**a**) in.^{2} (**b**) ft^2 (**c**) yd^2 (**d**) cm^2 (**e**) m^2

42. The area formula for a rectangle is $A = lw$ ⟵——————— Area = length × width

43. The area formula for a square is $A = s^2$ ⟵——————— Area = side × side

44. The area formula for a parallelogram is $A = bh$ ⟵——————— Area = base × height

45. The area formula for a triangle is $A = \dfrac{1}{2}bh$ ⟵——————— Area = $\dfrac{1}{2}$ × base × height

46. The area formula for a trapezoid is

$$A = \frac{1}{2}(b_1 + b_2)h \;\longleftarrow\; \text{Area} = \frac{1}{2}(\text{one base} + \text{other base}) \times \text{height}$$

47. The area formula for a circle is $A = \pi r^2$ ⟵——————— Area = π × radius × radius

48. When the space occupied by a three-dimensional object is measured by finding how many cubes of a given size are needed to completely fill that space, you are finding the volume (V) of that object in cubic units of measure.

49. The common cubic units of measure are (**a**) in.^{3} (**b**) ft^3 (**c**) yd^3 (**d**) cm^3 (**e**) m^3.

50. The volume formula for a rectangular prism is $V = lwh$ ⟵— Volume = length × width × height.

51. The volume formula for a cube is $V = e^3$ ⟵——————— Volume = edge × edge × edge.

52. The volume formula for a cylinder is $V = \pi r^2 h$ ⟵— Volume = π × radius × radius × height.

53. The volume formula for a sphere is $V = \dfrac{4}{3}\pi r^3$ ⟵— Volume = $\dfrac{4}{3}$ × π × radius × radius × radius.

SOLVED PROBLEMS

PROBLEM 15-1   Identify each geometric figure:

(a)

(b)

(c)

(d)

(e)

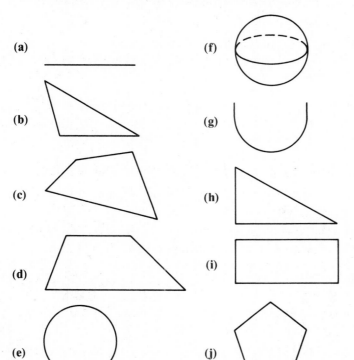

(f)

(g)

(h)

(i)

(j)

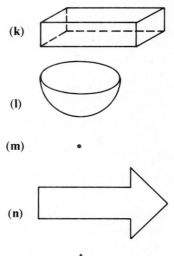

(k)

(l)

(m)

(n)

(o)

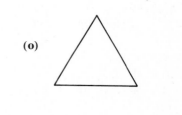

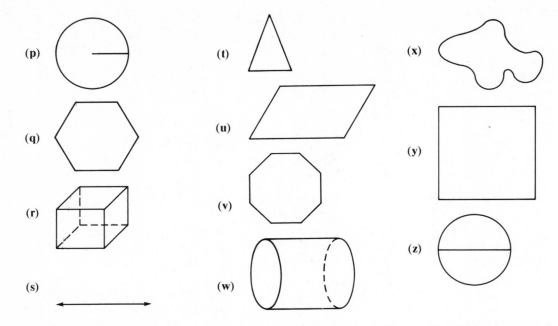

(p)   **(t)**   **(x)**

(q)   **(u)**   **(y)**

(r)   **(v)**

(s)   **(w)**   **(z)**

Solution: **(a)** line segment   **(b)** triangle   **(c)** quadrilateral   **(d)** trapezoid   **(e)** circle
(f) sphere   **(g)** open figure   **(h)** right triangle   **(i)** rectangle   **(j)** regular pentagon
(k) rectangular prism   **(l)** hemisphere   **(m)** point   **(n)** polygon   **(o)** equilateral triangle
(p) circle with a radius   **(q)** regular hexagon   **(r)** cube   **(s)** line   **(t)** isosceles triangle
(u) parallelogram   **(v)** octagon   **(w)** cylinder   **(x)** closed figure   **(y)** square
(z) circle with a diameter   [see Examples 15-1 through 15-21].

PROBLEM 15-2   Find the perimeter for **(a)** an equilateral triangle with side 5 in.
(b) a square with side 3 cm   **(c)** a regular pentagon with side 8 ft
(d) a regular hexagon with side 2.5 m   **(e)** a regular octagon with side $2\frac{1}{2}$ yd
(f) a rectangle with length 8 in. and width 5 in.   **(g)** a parallelogram with sides 7 cm and 6 cm
(h) a polygon with sides 2 ft, 3 ft, 8 ft, $6\frac{1}{2}$ ft, 10 ft, and $12\frac{3}{4}$ ft.

Solution: Recall that to find the perimeter of a polygon, you can either add the lengths of all the
sides together or use the appropriate perimeter formula [see Examples 15-22 to 15-25]:
(a) $P = 3s = 3 \times 5$ in. $= 15$ in.   **(b)** $P = 4s = 4 \times 3$ cm $= 12$ cm
(c) $P = 5s = 5 \times 8$ ft $= 40$ ft   **(d)** $P = 6s = 6 \times 2.5$ m $= 15$ m
(e) $P = 8s = 8 \times 2\frac{1}{2}$ yd $= 20$ yd   **(f)** $P = 2(l + w) = 2(8$ in. $+ 5$ in.$) = 2 \times 13$ in. $= 26$ in.
(g) $P = 2(a + b) = 2(7$ cm $+ 6$ cm$) = 2 \times 13$ cm $= 26$ cm
(h) $P = 2$ ft $+ 3$ ft $+ 8$ ft $+ 6\frac{1}{2}$ ft $+ 10$ ft $+ 12\frac{3}{4}$ ft $= 42\frac{1}{4}$ ft

PROBLEM 15-3   Find the length of a diameter of a circle given the length of a radius for the same
circle:

(a) $r = 5$ in.   **(b)** $r = 7$ cm   **(c)** $r = 2\frac{1}{4}$ ft   **(d)** $r = 1.3$ m

Solution: Recall that to find the length of a diameter of a circle given the length of a radius for the
same circle, you can use the diameter/radius formula $d = 2r$ [see Examples 15-28 and 15-29]:
(a) $d = 2r = 2 \times 5$ in. $= 10$ in.   **(b)** $d = 2r = 2 \times 7$ cm $= 14$ cm
(c) $d = 2r = 2 \times 2\frac{1}{4}$ ft $= 4\frac{1}{2}$ ft   **(d)** $d = 2r = 2 \times 1.3$ m $= 2.6$ m

PROBLEM 15-4   Find the length of a radius of a circle given the length of a diameter for the same
circle: **(a)** $d = 5$ in.   **(b)** $d = 7$ cm   **(c)** $d = 2\frac{1}{4}$ ft   **(d)** $d = 1.3$ m

Solution: Recall that to find the length of a radius of a circle given the length of a diameter for the
same circle, you can use the radius/diameter formula $r = d \div 2$ [see Examples 15-30 and 15-31]:
(a) $r = d \div 2 = 5$ in. $\div 2 = 2\frac{1}{2}$ in.   **(b)** $r = d \div 2 = 7$ cm $\div 2 = 3.5$ cm
(c) $r = \frac{1}{2}d = \frac{1}{2} \times 2\frac{1}{4}$ ft $= 1\frac{1}{8}$ ft   **(d)** $r = d \div 2 = 1.3$ m $\div 2 = 0.65$ m

PROBLEM 15-5   Find the approximate circumference for a circle given:
(a) $r = 3\frac{1}{2}$ ft   **(b)** $r = 2m$ **(c)** $d = 7$ yd   **(d)** $d = 1.5$ cm

Solution: Recall that to find the circumference of a circle given the length of a radius or a diameter, you use the circumference formula $C = 2\pi r$ or $C = \pi d$ using $\pi \approx 3\frac{1}{7}$ for U.S. customary measure and $\pi \approx 3.14$ for metric measures [see Examples 15-36 and 15-37]:

(a) $C = 2\pi r \approx 2 \times 3\frac{1}{7} \times 3\frac{1}{2}$ ft $= \frac{2}{1} \times \frac{22}{7} \times \frac{7}{2}$ ft $= 22$ ft

(b) $C = 2\pi r \approx 2 \times 3.14 \times 2$ m $= 6.28 \times 2$ m $= 12.56$ m

(c) $C = \pi d \approx 3\frac{1}{7} \times 7$ yd $= \frac{22}{7} \times \frac{7}{1}$ yd $= 22$ yd

(d) $C = \pi d \approx 3.14 \times 1.5$ cm $= 4.71$ cm

PROBLEM 15-6   Find the area of a: **(a)** rectangle with length 9 m and width 5 m
(b) square with side 3 in.   **(c)** parallelogram with base 6 ft. and height 4 ft
(d) triangle with base 2 cm and height 8 cm   **(e)** trapezoid with bases 2 in. and 3 in. and height 4 in.
(f) circle with radius $2\frac{1}{3}$ yd   **(g)** circle with diameter 6 m.

Solution: Recall that to find the area of a square, rectangle, parallelogram, triangle, or circle with a given radius or diameter, you can use the appropriate area formula [see Examples 15-37, 15-39, 15-40, 15-41, 15-42, 15-43, and 15-44, respectively]:

(a) $A = lw = 9$ m $\times 5$ m $= (9 \times 5)$ m$^2 = 45$ m^2
(b) $A = s^2 = (3$ in.$)^2 = 3$ in. $\times 3$ in. $= (3 \times 3)$ in.$^2 = 9$ in.^{2}
(c) $A = bh = 6$ ft $\times 4$ ft $= (6 \times 4)$ ft$^2 = 24$ ft^2

(d) $A = \frac{1}{2} bh = \frac{1}{2} \times 2$ cm $\times 8$ cm $= \left(\frac{1}{2} \times 2 \times 8\right)$ cm$^2 = 8$ cm^2

(e) $A = \frac{1}{2}(b_1 + b_2)h = \frac{1}{2}(2$ in. $+ 3$ in.$) 4$ in. $= \frac{1}{2} \times 5$ in. $\times 4$ in. $= \left(\frac{1}{2} \times 5 \times 4\right)$ in.$^2 = 10$ in.^{2}

(f) $A = \pi r^2 \approx 3\frac{1}{7}(2\frac{1}{3}$ yd$)^2 = 3\frac{1}{7} \times 2\frac{1}{3}$ yd $\times 2\frac{1}{3}$ yd $= \frac{22}{7} \times \frac{7}{3}$ yd $\times \frac{7}{3}$ yd $= (\frac{22}{7} \times \frac{7}{3} \times \frac{7}{3})$ yd$^2 = 17\frac{1}{9}$ yd^2
(g) $r = d \div 2 = 6$ m $\div 2 = 3$ m and
$A = \pi r^2 \approx 3.14 (3$ m$)^2 = 3.14 \times 3$ m $\times 3$ m $= (3.14 \times 3 \times 3)$ m$^2 = 28.26$ m^2

PROBLEM 15-7   Find the volume of a **(a)** rectangular prism with length 3 in., width 5 in., and height 2 in.   **(b)** cube with edge 4 cm   **(c)** cylinder with radius $1\frac{3}{4}$ ft and height 6 ft
(d) sphere with radius 3 m

Solution: Recall that to find the volume of a rectangular prism, cube, cylinder, or sphere, you can use the appropriate volume formula [see Examples 15-45, 15-47, 15-48, and 15-49, respectively]:

(a) $V = lwh = 3$ in. $\times 5$ in. $\times 2$ in. $= (3 \times 5 \times 2)$ in.$^3 = 30$ in.^{3}
(b) $V = e^3 = (4$ cm$)^3 = 4$ cm $\times 4$ cm $\times 4$ cm $= (4 \times 4 \times 4)$ cm$^3 = 64$ cm^3
(c) $V = \pi r^2 h \approx 3\frac{1}{7}(1\frac{3}{4}$ ft$)^2 6$ ft $= \frac{22}{7} \times \frac{7}{4}$ ft $\times \frac{7}{4}$ ft $\times 6$ ft $= \frac{22}{7} \times (\frac{7}{4} \times \frac{7}{4} \times 6)$ ft$^3 = 57\frac{3}{4}$ ft^3
(d) $V = \frac{4}{3}\pi r^3 \approx \frac{4}{3} \times 3.14 (3$ m$)^3$
$= \frac{4}{3} \times 3.14 \times 3$ m $\times 3$ m $\times 3$ m
$= \frac{4}{3} \times 3.14 \times (3 \times 3 \times 3)$ m^3
$= 113.04$ m^3

Supplementary Exercises

PROBLEM 15-8   Find the approximate perimeter for **(a)** an equilateral triangle with side 8 cm
(b) a square with side 6 in.   **(c)** a regular pentagon with side 3 ft
(d) a regular hexagon with side 5.2 m   **(e)** a regular octagon with side $1\frac{1}{2}$ yd
(f) a rectangle with length 5 cm and width 6 cm   **(g)** a parallelogram with sides 2.5 cm and 3.5 cm
(h) a polygon with sides 2.3 m, 1.8 m, 6.4 m. 0.25 m, 3.18 m, and 0.08 m.

PROBLEM 15-9   Find the circumference for a circle given **(a)** $r = 3\frac{1}{2}$ in.   **(b)** $r = 4$ cm
(c) $r = 14$ yd   **(d)** $r = 5$ mm   **(e)** $r = 7$ in.   **(f)** $r = 1.5$ m   **(g)** $r = 8$ cm
(h) $r = 1\frac{2}{5}$ ft   **(i)** $r = 1.4$ m   **(j)** $r = 2\frac{4}{5}$ in.   **(k)** $r = 2.8$ mm   **(l)** $r = 1\frac{1}{2}$ ft
(m) $r = 3.25$ cm   **(n)** $d = 2.8$ m   **(o)** $d = 7$ ft   **(p)** $d = 21$ yd   **(q)** $d = 3$ cm
(r) $d = 28$ in.   **(s)** $d = 1.3$ cm   **(t)** $d = 4\frac{2}{3}$ ft   **(u)** $d = 4.7$ m   **(v)** $d = 10\frac{1}{2}$ in.
(w) $d = 10.5$ cm   **(x)** $d = 2\frac{2}{3}$ yd   **(y)** $d = 5.625$ mm   **(z)** $d = 35$ ft.

PROBLEM 15-10   Find the area of **(a)** a rectangle with length 4 ft and width 2 ft
(b) a square with side 5 cm   **(c)** a parallelogram with base 3.4 mm and height 5.6 mm
(d) a triangle with base 1 ft 6 in. and height 6 in.   **(e)** a trapezoid with bases 8 m and 1.5 m and
height 3.2 m   **(f)** a circle with radius $3\frac{1}{2}$ yd   **(g)** a circle with diameter 2.5 m.

PROBLEM 15-11   Find the volume of: **(a)** a rectangular prism with length 6 in., width 8 in., and
height 4 in.   **(b)** a cube with edge 1.5 mm   **(c)** a cylinder with radius 7 cm and height 10 cm
(d) a cylinder with diameter 2 ft and height 7 ft   **(e)** a sphere with radius 1.5 m   **(f)** a sphere
with diameter 7 yd.

Answers to Supplementary Exercises

(15-8) **(a)** 24 cm   **(b)** 24 in.   **(c)** 15 ft   **(d)** 31.2 m   **(e)** 12 yd   **(f)** 22 cm   **(g)** 12 cm
(h) 14.01 m

(15-9) **(a)** 22 in.   **(b)** 25.12 cm   **(c)** 88 yd   **(d)** 31.4 mm   **(e)** 44 in.   **(f)** 9.42 m
(g) 50.24 cm   **(h)** $8\frac{4}{5}$ ft   **(i)** 8.792 m   **(j)** $17\frac{3}{5}$ in.   **(k)** 17.584 mm   **(l)** $9\frac{3}{7}$ ft
(m) 20.41 cm   **(n)** 8.792 m   **(o)** 22 ft   **(p)** 66 yd   **(q)** 9.42 cm   **(r)** 88 in.
(s) 4.082 cm   **(t)** $14\frac{2}{3}$ ft   **(u)** 14.758 m   **(v)** 33 in.   **(w)** 32.97 cm   **(x)** $8\frac{8}{21}$ in.
(y) 17.6625 mm   **(z)** 110 ft

(15-10) **(a)** 8 ft^2 **(b)** 25 cm^2 **(c)** 19.04 mm^2 **(d)** 54 in.^{2} or $4\frac{1}{2}$ ft^2 **(e)** 15.2 m^2
(f) $38\frac{1}{2}$ yd^2 **(g)** 4.90625 m^2

(15-11) **(a)** 192 in.^{3} **(b)** 3.375 mm^3 **(c)** 1538.6 cm^3 **(d)** 22 ft^3 **(e)** 14.13 m^3
(f) $179\frac{2}{3}$ yd^3

16 REAL NUMBERS

16-1. Identifying Real Numbers

A. Identify positive and negative numbers.

Recall: The whole numbers are listed as 0, 1, 2, 3, $\cdots$

To graph whole numbers, you can use a **number line.**

EXAMPLE 16-1: Graph whole numbers on a number line.

Solution:

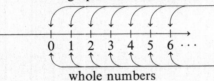

On a number line **(a)** the point that represents zero is called the **origin.**
(b) all points to the right of the origin represent **positive numbers.**
(c) all points to the left of the origin represent **negative numbers.**

Note: To write a negative number, you use a **negative sign** $(-)$.

EXAMPLE 16-2: Write some negative numbers.

Solution: $-1, -\frac{3}{4}, -1.5, -\sqrt{6}, -\pi$ $\longleftarrow$ negative numbers

Note: To write a positive number, you do *not* have to write a **positive sign** $(+)$, although you may include one.

EXAMPLE 16-3: Write some positive numbers.

Solution: $+1$ or 1, $+\frac{3}{4}$ or $\frac{3}{4}$, $+1.5$ or 1.5, $+\sqrt{6}$ or $\sqrt{6}$, $+\pi$ or π $\longleftarrow$ positive numbers

To graph positive and negative numbers, you can use a number line.

EXAMPLE 16-4: Graph some positive and negative numbers.

Solution

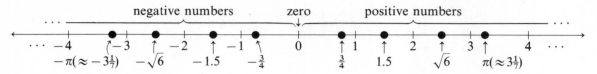

B. Identify integers.

Note: On a number line, the points representing
(a) $-\frac{3}{4}$ and $\frac{3}{4}$ are the same distance from the origin.
(b) -1 and 1 are the same distance from the origin.
(c) -1.5 and 1.5 are the same distance from the origin.
(d) $-\sqrt{6}$ or $\sqrt{6}$ are the same distance from the origin.
(e) $-\pi$ and π are the same distance from the origin.

Two numbers with **unlike signs** $(+, -)$ that are the same distance from the origin (0) on a number line are called **opposites.** To write **the opposite of a number,** you just change the number's sign to the **opposite sign.**

EXAMPLE 16-5: Write the opposite of each of the following numbers:
(a) $+4$ (b) $\frac{1}{2}$ (c) -2.3 (d) 0

Solution
(a) The opposite of $+4$ is -4 because $+4$ and -4 have opposite signs and they are both the same distance from 0 on a number line.
(b) The opposite of $\frac{1}{2}$ is $-\frac{1}{2}$ because $\frac{1}{2}$ and $-\frac{1}{2}$ have opposite signs $(\frac{1}{2} = +\frac{1}{2})$ and they are both the same distance from 0 on a number line.
(c) The opposite of -2.3 is $+2.3$ (or just 2.3) because -2.3 and $+2.3$ have opposite signs and they are both the same distance from 0 on a number line.
(d) The opposite of 0 is **0** because 0 can be written as two numbers with opposite signs $(0 = +0 = -0)$ and 0 is at the origin.

Note: The sum of two opposites is always zero: $+4 + (-4) = 0$

The whole numbers $(0, 1, 2, 3, \cdots)$ and their opposites $(\cdots, -3, -2, -1, 0)$ are called **integers.**

EXAMPLE 16-6: List the set of integers.

Solution: $\cdots, -3, -2, -1, 0, 1, 2, 3, \cdots$ ◀— integers

Note 1: In Example 16-6, the number collection
(a) $1, 2, 3, \cdots$ is called the **positive integers.**
(b) $\cdots, -3, -2, -1$ is called the **negative integers.**

Note 2: The ellipsis notation used in Example 16-6 indicates that the positive integers continue on forever in a positive counting pattern, and that the negative integers continue on forever in a negative counting pattern.

EXAMPLE 16-7: List the (a) first ten positive integers   (b) first ten negative integers.

Solution: $1, \quad 2, \quad 3, \quad 4, \quad 5, \quad 6, \quad 7, \quad 8, \quad 9, \quad 10$ ◀— first ten positive integers
$-10, -9, -8, -7, -6, -5, -4, -3, -2, -1$ ◀— first ten negative integers

Note: The negative numbers $-\frac{3}{4}$, -1.5, $-\sqrt{6}$, and $-\pi$ are not integers because $\frac{3}{4}$, 1.5, $\sqrt{6}$, and π are not whole numbers. *Only* the whole numbers and their opposites are integers.

C. Identify rational numbers.

When you draw a number line, you will usually need to graph several integers as reference points.

EXAMPLE 16-8: Draw a number line graphing several integers as reference points.

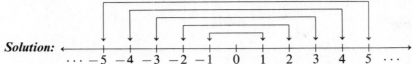

same distance from the origin (0)

Solution:

$$\cdots \; -5 \;\; -4 \;\; -3 \;\; -2 \;\; -1 \;\;\; 0 \;\;\; 1 \;\;\; 2 \;\;\; 3 \;\;\; 4 \;\;\; 5 \;\; \cdots$$

If a and b are integers ($b \neq 0$), then the collection of all numbers that can be written in the fractional form $\dfrac{a}{b}$ are called **rational numbers.**

Note: Integers and whole numbers are rational numbers.

EXAMPLE 16-9: Identify which of the following numbers are rational numbers: **(a)** $\dfrac{1}{2}$ **(b)** $\dfrac{-3}{4}$

(c) $\dfrac{2}{-5}$ **(d)** $1\dfrac{1}{2}$ **(e)** 5   **(f)** -2 **(g)** 0.1   **(h)** -1.3 **(i)** $0.\overline{3}$ **(j)** $\sqrt{2}$ **(k)** π

Solution

(a) $\dfrac{1}{2}$ is a rational number because 1 and 2 are integers.

(b) $\dfrac{-3}{4}$ is a rational number because -3 and 4 are integers.

(c) $\dfrac{2}{-5}$ is a rational number because 2 and -5 are integers.

(d) $1\dfrac{1}{2}$ is a rational number because $1\dfrac{1}{2}$ can be written as $\dfrac{3}{2}$, and 3 and 2 are integers.

(e) 5 is a rational number because 5 can be written as $\dfrac{5}{1}$, and 5 and 1 are integers.

(f) -2 is a rational number because -2 can be written as $\dfrac{-2}{1}$, and -2 and 1 are integers.

(g) 0.1 is a rational number because 0.1 can be written as $\dfrac{1}{10}$, and 1 and 10 are integers.

(h) -1.3 is a rational number because -1.3 can be written as $\dfrac{-13}{10}$, and -13 and 10 are integers.

(i) $0.\overline{3}$ is a rational number because $0.\overline{3}$ can be written as $\dfrac{1}{3}$, and 1 and 3 are integers.

(j) $\sqrt{2}$ is not a rational number because $\sqrt{2}$ cannot be written in the form $\dfrac{a}{b}$ where a and b are integers: $\sqrt{2} = 1.41421356237\cdots$ and is not a terminating or repeating decimal.

(k) π is not a rational number because π cannot be written in the form $\dfrac{a}{b}$ where a and b are integers: $\pi = 3.1415926536\cdots$ and is not a terminating or repeating decimal.

Note: If the decimal form of a number does *not* terminate or repeat, then that number is *not* a rational number. If the decimal form of a number does terminate or repeat, then that number is a rational number.

D. Identify irrational numbers.

Numbers whose decimal form does not terminate and does not repeat are called **irrational numbers.**

EXAMPLE 16-10: List some irrational numbers.

Solution: $\sqrt{2}, -\sqrt{3}, \sqrt{5}, -\sqrt{6}, \pi, -\pi$ ⟵ irrational numbers

Note: The positive or negative square root of any whole number that is not a square is an irrational number ($\cdots, -\sqrt{5}, -\sqrt{3}, -\sqrt{2}, \sqrt{2}, \sqrt{3}, \sqrt{5}, \cdots$).

E. Identify real numbers.

The collection of all rational and irrational numbers is called the set of **real numbers.**

Note 1: The real numbers include whole numbers, integers, fractions, and decimals.

Note 2: The only numbers that will be used in this text are real numbers.

EXAMPLE 16-11: Draw a diagram that shows the relationships between real numbers, irrational numbers, rational numbers, integers, and whole numbers.

Solution:

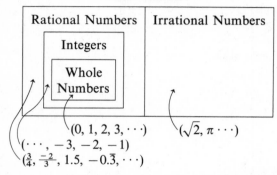

Note: Because every real number can be represented by a point on a number line, a number line is sometimes called a **real number line.** Every point on a number line represents a real number; that is, a number line is a pictorial representation of the collection of all real numbers.

16-2. Comparing Real Numbers

To compare two real numbers a and b using a number line, you write
(a) $a < b$ if a is to the **left** of b;
(b) $a > b$ if a is to the **right** of b;
(c) $a = b$ if a and b are represented by the **same** point.

EXAMPLE 16-12: Compare each of the following pairs of real numbers using either $<$, $>$, or $=$:
(a) $\frac{3}{4}$ and 0.7 **(b)** -5 and 2 **(c)** -3 and -4 **(d)** $-\frac{1}{3}$ and $-0.\overline{3}$

Solution

(a) $\frac{3}{4} > 0.7$ because $0.7 = \frac{7}{10}$, and $\frac{3}{4} > \frac{7}{10}$ (overbrace cross products: $10 \times 3 = 30 > 28 = 4 \times 7$).
(b) $-5 < 2$ because every negative number is to the left of every positive number on a number line.
(c) $-3 > -4$ because -3 is to the right of -4 on a number line.
(d) $-\frac{1}{3} = -0.\overline{3}$ because $0.\overline{3} = \frac{1}{3}$.

16-3. Finding Absolute Values

The distance between a given real number and its origin (0) on a number line is called its **absolute value.** The **absolute value symbol** is $|\ \ |$. Read $|-5|$ as "the absolute value of negative five."

EXAMPLE 16-13: Find each absolute value: **(a)** $|+5|$ **(b)** $|-5|$ **(c)** $|\frac{3}{4}|$ **(d)** $|0|$

Solution
(a) $|+5| = 5$ because $+5$ (or just 5) is 5 units from 0 on a number line.
(b) $|-5| = 5$ because -5 is 5 units from 0 on a number line.
(c) $|\frac{3}{4}| = \frac{3}{4}$ because $\frac{3}{4}$ is $\frac{3}{4}$ of a unit from 0 on a number line.
(d) $|0| = 0$ because 0 is 0 units from 0 on a number line.

Note: The absolute value of any real number is *never* negative.

16-4. Computing with Integers

A. Add integers.
To add two or more integers with like signs, you use the following Addition Rules for Two or More Real Numbers with Like Signs:

Addition Rules for Two or More Real Numbers with Like Signs
1. Find the sum of the absolute values of the addends.
2. Write the same like sign on the sum.

Note: You can use this rule for all integers because all integers are real numbers.

EXAMPLE 16-14: Add the following integers with like signs: **(a)** $+5 + (+4)$
(b) $-4 + (-3)$ **(c)** $8 + 5 + 9 + 7$

Solution:

                                  absolute values

(a) $+5 + (+4) = ?9$ *Think:* $5 + 4 = 9 \longleftarrow$ sum of absolute values

          $= +9 \text{ or } 9$ *Think:* Because both addends are positive, the sum is positive.

                                  absolute values

(b) $-4 + (-3) = ?7$ *Think:* $4 + 3 = 7 \longleftarrow$ sum of absolute values

          $= -7$ *Think:* Because both addends are negative, the sum is negative.

(c) $8 + 5 + 9 + 7 = ?29$ *Think:* $8 + 5 + 9 + 7 = 29$

          $= +29 \text{ or } 29$ *Think:* No sign on a number means it is positive.

To add two integers with unlike signs, you can use the following Addition Rules for Two Real Numbers with Unlike Signs:

Addition Rules for Two Real Numbers with Unlike Signs
1. Find the difference between the absolute values of the addends.
2. Write the sign of the number with the larger absolute value on the sum.

Note: You can use this rule for all integers because all integers are real numbers.

EXAMPLE 16-15: Add the following integers with unlike signs: **(a)** $-3 + 7$ **(b)** $2 + (-8)$
(c) $+6 + (-6)$

Solution

                              larger absolute value

(a) $-3 + 7 = ?4$ *Think:* $7 - 3 = 4 \longleftarrow$ difference between absolute values

          $= +4 \text{ or } 4$ *Think:* Because 7 has the larger absolute value, the sum is positive.

larger absolute value
↓

(b) $2 + (-8) = ?\ 6$ *Think:* $8 - 2 = 6$ ⟵ difference between absolute values

         $= -6$ *Think:* Because -8 has the larger absolute value, the sum is negative.

(c) $+6 + (-6) = 0$ *Think:* The sum of two numbers that are opposites is always 0.

Note: The sum of any given real number and zero is always the given real number.

EXAMPLE 16-16: Add the following numbers:
(a) $0 + 7$ **(b)** $0 + (-8)$ **(c)** $5 + 0$ **(d)** $-2 + 0$

Solution: **(a)** $0 + 7 = 7$ **(b)** $0 + (-8) = -8$ **(c)** $5 + 0 = 5$ **(d)** $-2 + 0 = -2$

B. Subtract integers.

To subtract two integers, you can use the following Subtraction Rules for Two Real Numbers:

Subtraction Rules for Two Real Numbers
1. Change subtraction to addition and write the opposite of the subtrahend.
2. Follow the rules for adding real numbers.
3. Check by adding the proposed difference to the original subtrahend to see if you get the original minuend.

Note: You can use these rules for all integers because all integers are real numbers.

EXAMPLE 16-17: Subtract the following integers:
(a) $5 - 8$ **(b)** $-2 - 7$ **(c)** $6 - (-4)$ **(d)** $-3 - (-9)$

Solution

change to +

(a) $5 - 8 = 5 + (-8)$ Add the opposite of the subtrahend.

opposite of 8

       $= -3$ Add real numbers with unlike signs [see Example 16-15].

   Check: $-3 + 8 = +5$ ⟵ -3 checks

change to +

(b) $-2 - 7 = -2 + (-7)$ Add the opposite of the subtrahend.

opposite of 7

       $= -9$ Add real numbers with like signs [see Example 16-14].

   Check: $-9 + 7 = -2$ ⟵ -9 checks

change to +

(c) $6 - (-4) = 6 + (+4)$ Add the opposite of the subtrahend.

opposite of -4

       $= +10$ or 10 Add real numbers with like signs.

   Check: $10 + (-4) = 6$ ⟵ 10 checks

change to +

(d) $-3 - (-9) = -3 + 9$ Add the opposite of the subtrahend.

opposite of -9

$= 6$ Add real numbers with unlike signs.

Check: $6 + (-9) = -3$ ←— 6 checks

C. Multiply integers.

There are several different ways to write the multiplication problem "3 times 4."

EXAMPLE 16-18: Write "3 times 4" in five different ways.

Solution: 3 times 4 $= 3 \times 4$

$= 3 \cdot 4$

$= 3(4)$ } all equivalent ways of writing "3 times 4"

$= (3)4$

$= (3)(4)$

To multiply two integers, you can use the following Multiplication Rules for Two Real Numbers:

Multiplication Rules for Two Real Numbers
1. Find the product of the absolute values.
2. Make the product
    (a) positive if the two factors have like signs;
    (b) negative if the two factors have unlike signs;
    (c) zero if either one of the factors is zero.

EXAMPLE 16-19: Multiply the following pairs of integers: **(a)** $9 \times (-7)$ **(b)** $-5 \cdot 2$
(c) $(9)(3)$ **(d)** $(-2)0$ **(e)** $-4(-6)$

Solution

absolute values

(a) $9 \times (-7) = ?\ 63$ *Think:* $9 \times 7 = 63$ ←— product of the absolute values

$= -63$ The factors 9 and -7 have unlike signs so the product is negative.

(b) $-5 \cdot 2 = ?\ 10$ *Think:* $5 \times 2 = 10$

$= -10$ The factors -5 and 2 have unlike signs so the product is negative.

(c) $(9)(3) = ?\ 27$ *Think:* $9 \times 3 = 27$

$= +27$ or 27   The factors 9 and 3 have like signs so the product is positive.

(d) $(-2)0 = 0$ *Think:* When one of the factors is zero the product is always zero.

(e) $-4(-6) = ?\ 24$ *Think:* $4 \times 6 = 24$

$= +24$ or 24   *Think:* The factors -4 and -6 have like signs so the product
is positive.

Note: The product of two negative numbers is always positive!

To multiply more than two integers you can use the following Multiplication Rules for Two or More
Real Numbers:

Multiplication Rules for Two or More Real Numbers
1. Find the product of all the absolute values.
2. Make the product
  (a) positive if there are an **even number** (0, 2, 4, · · ·) of negative factors;
  (b) negative if there are an **odd number** (1, 3, 5, · · ·) of negative factors;
  (c) zero if one or more factors are zero.

Note: You can use these rules for all integers because all integers are real numbers.

EXAMPLE 16-20: Multiply the following groups of integers: **(a)** $5 \cdot 3 \cdot 2$ **(b)** $4(-1)3$
(c) $-5(-2)2$ **(d)** $-1(-6)(-3)$ **(e)** $2(-3)(-1)(0)(-3)(-2)$

Solution
(a) $5 \cdot 3 \cdot 2 = ? \, 30$ *Think:* $5 \times 3 \times 2 = 30$ ⟵ product of the absolute values

   $= +30$ or 30 There is an even number (0) of negative signs
                so the product is positive.

(b) $4(-1)3 = ? \, 12$ *Think:* $4 \times 1 \times 3 = 12$

   $= -12$ There is an odd number (1) of negative
                signs so the product is negative.

(c) $-5(-2)2 = ? \, 20$ *Think:* $5 \times 2 \times 2 = 20$

   $= +20$ or 20 There is an even number (2) of negative
                signs so the product is positive.

(d) $-1(-6)(-3) = ? \, 18$ *Think:* $1 \times 6 \times 3 = 18$

   $= -18$ There is an odd number (3) of negative
                signs so the product is negative.

(e) $2(-3)(-1)(0)(-3)(-2) = 0$ *Think:* When one or more factors are zero, the
                product is always zero.

D. Divide integers.
There are several different ways to write the division problem "8 divided by 3."

EXAMPLE 16-21: Write "8 divided by 3" in six different ways.

Solution: 8 divided by 3 $= 8 \div 3$

   $= 8/3$

   $= \frac{8}{3}$

   $= 8 \cdot \frac{1}{3}$ } all equivalent ways of writing "8 divided by 3"

   $= \frac{1}{3} \cdot 8$

   $= 3\overline{)8}$

To divide two integers, you can use the following Division Rules for Two Real Numbers:

Division Rules for Two Real Numbers
1. Find the quotient of the absolute values.
2. Use the sign rules for multiplication to write the correct sign on the quotient.
3. Check by multiplying the proposed quotient by the original divisor to see if you get the original dividend.

Note: You can use these rules for all integers because all integers are real numbers.

Recall: Zero divided by any nonzero number is always zero.
    Division by zero is not defined.

EXAMPLE 16-22: Divide the following pairs of integers: (a) $15 \div 3$ (b) Divide -10 by 2.
(c) $\frac{-8}{-2}$ (b) $-3\overline{)-18}$ (e) $\frac{0}{3}$ (f) $\frac{0}{3}$ (g) $\frac{0}{0}$

Solution

(a) $15 \div 3 = +5$ or 5   Use the sign rules for multiplication.

    Check: $5 \times 3 = 15$ ⟵ 5 checks

(b) Divide -10 by $2 = -10 \div 2$ Find the quotient of the absolute values.

               $= -5$ Use the sign rules for multiplication.

    Check: $-5 \times 2 = -10$ ⟵ -5 checks

(c) $\frac{-8}{-2} = 8 \div (-2)$ Find the quotient of the absolute values.

      $= -4$ Use the sign rules for multiplication.

    Checks: $-4(-2) = +8$ or 8 ⟵ -4 checks

(d) $-3\overline{)-18} = -18 \div (-3)$ Find the quotient of the absolute values.

           $= +6$ or 6   Use the sign rules for multiplication.

    Check: $6(-3) = -18$ ⟵ 6 checks

(e) $\frac{0}{3} = 0$ Zero divided by any nonzero real number
                       is always zero.

    Check: $0 \times 3 = 0$ ⟵ 0 checks

(f) $\frac{3}{0}$ is not defined.

(g) $\frac{0}{0}$ is not defined.

16-5. Computing with Rational Numbers

A. Rename rational numbers.

Recall: If a and b are integers ($b \neq 0$), then any number that can be written in the form $\frac{a}{b}$ is called a rational number.

Every rational number in fraction form has three signs: the sign of the fraction, the sign of the numerator, and the sign of the denominator.

EXAMPLE 16-23: Identify the three signs for each of the following rational numbers:

(a) $\frac{3}{4}$ (b) $-\frac{-3}{4}$

Solution

(a) $\dfrac{3}{4} = +\dfrac{+3}{+4}$

    sign of the fraction → (leading +)
    sign of the numerator → ($+3$)
    sign of the denominator → ($+4$)

(b) $-\dfrac{-3}{4} = -\dfrac{-3}{+4}$

    sign of the fraction → (leading $-$)
    sign of the numerator → (-3)
    sign of the denominator → ($+4$)

If exactly two of a fraction's three signs are changed, you will always get an equal fraction.

The following are all equivalent ways to write a fraction with form $\frac{a}{b}$.

$$\frac{a}{b} = -\frac{-a}{b} = -\frac{a}{-b} = \frac{-a}{-b}$$

EXAMPLE 16-24: Rename $\dfrac{3}{4}$ in three equivalent ways using sign changes.

Solution: $\dfrac{3}{4} = -\dfrac{-3}{4}$ *Think:* $+\dfrac{+3}{+4} = -\dfrac{-3}{+4}$ ⟵ two sign changes

$\dfrac{3}{4} = -\dfrac{3}{-4}$ *Think:* $+\dfrac{+3}{+4} = -\dfrac{+3}{-4}$ ⟵ two sign changes

$\dfrac{3}{4} = \dfrac{-3}{-4}$ *Think:* $+\dfrac{+3}{+4} = +\dfrac{-3}{-4}$ ⟵ two sign changes

Note: The fractions $\dfrac{3}{4}$, $-\dfrac{-3}{4}$, $-\dfrac{3}{-4}$, and $\dfrac{-3}{-4}$ are equal fractions.

Caution: To get an equal fraction using sign changes, you must change exactly two of a fraction's three signs.

The following are all equivalent ways to write a fraction with form $-\dfrac{a}{b}$:

$$-\frac{a}{b} = \frac{-a}{b} = \frac{a}{-b} = -\frac{-a}{-b}$$

EXAMPLE 16-25: Rename $-\dfrac{3}{4}$ in three equivalent ways using sign changes.

Solution: $-\dfrac{3}{4} = \dfrac{-3}{4}$ *Think:* $-\dfrac{+3}{+4} = +\dfrac{-3}{+4}$ ⟵ two sign changes

$-\dfrac{3}{4} = \dfrac{3}{-4}$ *Think:* $-\dfrac{+3}{+4} = +\dfrac{+3}{-4}$ ⟵ two sign changes

$-\dfrac{3}{4} = -\dfrac{-3}{-4}$ *Think:* $-\dfrac{+3}{+4} = -\dfrac{-3}{-4}$ ⟵ two sign changes

Note: The fractions $-\dfrac{3}{4}$, $\dfrac{-3}{4}$, $\dfrac{3}{-4}$, and $-\dfrac{-3}{-4}$ are equal fractions.

B. Simplify rational numbers.

A rational number in fraction form for which the numerator and denominator do not share a common integer factor other than 1 or -1 is called a **rational number in lowest terms**.

EXAMPLE 16-26: Which of the following rational numbers are in lowest terms?

(a) $\dfrac{3}{4}$ (b) $\dfrac{-5}{2}$ (c) $\dfrac{-6}{-10}$

Solution

(a) $\dfrac{3}{4}$ is a rational number in lowest terms because the only common integer factors of 3 and 4 are 1

or -1: $\dfrac{3}{4} = \dfrac{(3)}{1(4)}$ and $\dfrac{3}{4} = \dfrac{-3}{-4} = \dfrac{-1(3)}{-1(4)}$.

(b) $\dfrac{-5}{2}$ is a rational number in lowest terms because the only common integer factors of -5 and 2

are 1 and -1: $\dfrac{-5}{2} = \dfrac{1(-5)}{1(2)}$ and $\dfrac{-5}{2} = \dfrac{5}{-2} = \dfrac{-1(-5)}{-1(2)}$.

(c) $\dfrac{-6}{-10}$ is not a rational number in lowest terms because 6 and 10 share a common integer factor

of 2 or -2: $\dfrac{-6}{-10} = \dfrac{2(-3)}{2(-5)}$ or $\dfrac{-6}{-10} = \dfrac{-2(3)}{-2(5)}$.

To **simplify a rational number in fraction form,** you write the rational number using as few negative signs as possible and then **reduce the rational number in fraction form** to lowest terms using the following rule:

Fundamental Rule for Rational Numbers

If a, b, and c are integers ($b \neq 0$ and $c \neq 0$), then $\dfrac{a \cdot c}{b \cdot c} = \dfrac{a}{b}$.

Note: By the Fundamental Rule for Rational Numbers, the value of a rational number in fraction form will not change when you divide both the numerator and denominator by the same nonzero

number: $\dfrac{a \cdot \cancel{c}}{b \cdot \cancel{c}} = \dfrac{a}{b}$

EXAMPLE 16-27: Simplify the following fractions: **(a)** $\dfrac{6}{10}$ **(b)** $-\dfrac{6}{3}$ **(c)** $\dfrac{-12}{8}$ **(d)** $-\dfrac{-10}{-25}$

Solution:

(a) $\dfrac{6}{10} = \dfrac{2 \cdot 3}{2 \cdot 5}$ Factor both the numerator and denominator.

$= \dfrac{\cancel{2} \cdot 3}{\cancel{2} \cdot 5}$ Use the Fundamental Rule for Rational Numbers.

$= \dfrac{3}{5}$ ⟵ simplest form

(b) $-\dfrac{6}{3} = -\dfrac{2 \cdot \cancel{3}}{1 \cdot \cancel{3}}$ Use the Fundamental Rule for Rational Numbers.

$= -\dfrac{2}{1}$

$= -2$ ⟵ simplest form

(c) $\dfrac{-12}{8} = -\dfrac{12}{8}$

$= -\dfrac{\cancel{2} \cdot \cancel{2} \cdot 3}{\cancel{2} \cdot \cancel{2} \cdot 2}$ or $-\dfrac{\cancel{4} \cdot 3}{\cancel{4} \cdot 2}$ Use the Fundamental Rule for Rational Numbers.

$= -\dfrac{3}{2}$ ⟩ simplest form

or $-1\dfrac{1}{2}$

(d) $-\dfrac{-10}{-25} = -\dfrac{10}{25}$ Write as few negative signs as possible: $-\dfrac{-10}{-25} = -\dfrac{+10}{+25}$

$= -\dfrac{2 \cdot \cancel{5}}{5 \cdot \cancel{5}}$ Use the Fundamental Rule for Rational Numbers.

$= -\dfrac{2}{5}$ ⟵ simplest form

C. Add rational numbers.

To **add rational numbers,** you use the Addition Rules for Real Numbers with Like or Unlike Signs given in Section 16-4, part A.

EXAMPLE 16-28: Add the following rational numbers: **(a)** $\frac{1}{4} + 0.5$ **(b)** $-2.5 + 1.75$
(c) $\frac{3}{4} + \left(-\frac{1}{2}\right)$ **(d)** $-3.2 + \left(-1\frac{1}{3}\right)$

Solution

(a) Rename to get fractions   or   Rename to get decimals

$$\frac{1}{4} + 0.5 = \frac{1}{4} + \frac{5}{10}$$

$$= \frac{1}{4} + \frac{1}{2}$$

$$= \frac{1}{4} + \frac{2}{4}$$

$$= \frac{3}{4}$$

$$\frac{1}{4} + 0.5 = 0.25 + 0.5$$

$$= 0.75$$

(b) $-2.5 + 1.75 = \ ? \ 0.75$

$$= -0.75$$

larger absolute value
$\downarrow$
Think: $2.5 - 1.75 = 0.75 \longleftarrow$ difference between absolute values

Think: Because -2.5 has the larger absolute value, the sum is negative.

(c) $\frac{3}{4} + \left(-\frac{1}{2}\right) = \ ? \ \frac{1}{4}$

$$= +\frac{1}{4} \text{ or } \frac{1}{4}$$

larger absolute value
$\downarrow$
Think: $\frac{3}{4} - \frac{1}{2} = \frac{1}{4} \longleftarrow$ difference between the absolute values

Think: Because $\frac{3}{4}$ has the larger absolute value, the sum is positive.

(d) $-3.2 + \left(-1\frac{1}{3}\right) = -3\frac{2}{10} + \left(-1\frac{1}{3}\right)$ Rename 3.2 as a mixed number $\left(3\frac{2}{10}\right)$.

$$= -3\frac{1}{5} + \left(-1\frac{1}{3}\right)$$ Simplify $3\frac{2}{10}$ to $3\frac{1}{5}$.

$$= -\frac{16}{5} + \left(-\frac{4}{3}\right)$$ Rename as fractions.

$$= -\frac{48}{15} + \left(-\frac{20}{15}\right)$$ Use the LCD 15 to get like fractions.

$$= \ ? \ \frac{68}{15}$$ *Think:* $\frac{48}{15} + \frac{20}{15} = \frac{68}{15}$

$$= -\frac{68}{15}$$ *Think:* Because both addends are negative, the sum is negative.

D. Subtract rational numbers.

To subtract rational numbers, you use the Subtraction Rules for Real Numbers given in Section 16-4, part B.

EXAMPLE 16-29: Subtract the following rational numbers: $-4.5 - 2\frac{3}{4}$

Solution: You can subtract by first renaming as decimals, or first renaming as fractions.

Rename to get decimals

$$-4.5 - 2\frac{3}{4} = -4.5 - 2.75 \qquad \text{Rename } 2\frac{3}{4} \text{ as } 2.75.$$

$$= -4.5 + (-2.75) \qquad \text{Add the opposite of the subtrahend.}$$

$$= -7.25$$

Rename to get fractions

$$-4.5 - 2\frac{3}{4} = -4\frac{5}{10} - 2\frac{3}{4} \qquad \text{Rename } -4.5 \text{ as } -4\frac{5}{10}.$$

$$= -4\frac{5}{10} + \left(-2\frac{3}{4}\right) \qquad \text{Add the opposite of the subtrahend.}$$

$$= -4\frac{1}{2} + \left(-2\frac{3}{4}\right) \qquad \text{Rename } 4\frac{5}{10} \text{ as } 4\frac{1}{2}.$$

$$= -4\frac{2}{4} + \left(-2\frac{3}{4}\right) \qquad \text{Use the LCD 4 to get like fractions.}$$

$$= -6\frac{5}{4} \qquad \text{Add rational numbers.}$$

$$= -7\frac{1}{4} \qquad \text{Rename } \frac{5}{4} \text{ as } 1\frac{1}{4}.$$

E. Multiply rational numbers.

To **multiply rational numbers,** you use the Multiplication Rules for Real Numbers given in Section 16-4, part C.

EXAMPLE 16-30: Multiply the following rational numbers: **(a)** $\frac{1}{2}(0.25)$ **(b)** $-\frac{1}{4} \cdot \frac{2}{3}$

(c) $1.5(-2.3)$ **(d)** $-0.5\left(-\frac{2}{3}\right)$

Solution

(a) Rename to get fractions or Rename to get decimals

$$\frac{1}{2}(0.25) = \frac{1}{2} \cdot \frac{1}{4} \qquad\qquad \frac{1}{2}(0.25) = 0.5(0.25)$$

$$= \frac{1}{8} \qquad\qquad\qquad\qquad = 0.125$$

(b) $-\frac{1}{4} \cdot \frac{2}{3} = -\frac{1}{2 \cdot \cancel{2}} \cdot \frac{\cancel{2}}{3}$ Eliminate common factors.

$$= ?\frac{1}{6} \qquad \textit{Think: } \frac{1}{2} \cdot \frac{1}{3} = \frac{1}{6} \longleftarrow \text{product of the absolute values}$$

$$= -\frac{1}{6} \qquad \textit{Think: } \text{Because the factors have unlike signs, the product is negative.}$$

(c) $1.5(-2.3) = ?\ 3.45$ *Think:* $1.5(2.3) = 3.45$ ⟵ product of the absolute values

$= -3.45$ *Think:* Because the fractions have unlike signs, the product is negative.

(d) $-0.5\left(-\dfrac{2}{3}\right) = -\dfrac{1}{2}\left(-\dfrac{2}{3}\right)$ Rename 0.5 as $\dfrac{1}{2}$.

$= -\dfrac{1}{2}\left(-\dfrac{2}{3}\right)$ Multiply fractions.

$= ?\ \dfrac{1}{3}$ *Think:* $\dfrac{1}{\cancel{2}} \cdot \dfrac{\cancel{2}}{3} = \dfrac{1}{3}$ ⟵ product of the absolute values

$= +\dfrac{1}{3} \text{ or } \dfrac{1}{3}$ *Think:* Because both factors are negative, the product is positive.

F. Divide rational numbers.

To divide rational numbers, you use the Division Rules for Real Numbers in Section 16-4, part D.

EXAMPLE 16-31: Divide the following rational numbers: $\dfrac{3}{4} \div (-0.25)$

Solution: You can divide by first renaming as decimals, or first renaming as fractions.

Rename to get decimals

$\dfrac{3}{4} \div (-0.25) = 0.75 \div (-0.25)$ Rename $\dfrac{3}{4}$ as 0.75.

$= ?\ 3$ Divide decimals.

$= -3$ *Think:* Because the decimals have unlike signs, the quotient is negative.

Rename to get fractions

$\dfrac{3}{4} \div (-0.25) = \dfrac{3}{4} \div \left(-\dfrac{1}{4}\right)$ Rename (-0.25) as $\left(-\dfrac{1}{4}\right)$.

$= \dfrac{3}{4} \cdot \left(-\dfrac{4}{1}\right)$ Multiply by the reciprocal of the divisor.

$= -\dfrac{3}{1}$ Multiply fractions and simplify.

$= -3$

Check: $-3(-0.25) = +0.75 \text{ or } 0.75$ ⟵ -3 checks

16-6. Using the Order of Operations

To avoid getting a wrong answer when computing with real numbers, you must use the following rules:

Order of Operations
· First, **clear grouping symbols** such as ( ), [ ], { }, —, and $\sqrt{}$ by performing the operations inside of them.
· Then, evaluate each power notation and radical.
· Next, multiply or divide in order from left to right.
· Last, add or subtract in order from left to right.

Note: The saying *Please Excuse My Dear Aunt Sally* can help you remember the Order of Operations: **P**arentheses (grouping symbols), **E**xponents (and radicals), **M**ultiply and **D**ivide, **A**dd and **S**ubtract.

EXAMPLE 16-32: Evaluate the following expressions: **(a)** $3 + 5 \cdot 2$ **(b)** $(1 - 5)^2 - 2\sqrt{9} + \dfrac{12}{4}$

(c) $\dfrac{5 - \sqrt{5^2 - 4(6)(-4)}}{2(6)}$

Solution

(a) $3 + 5 \cdot 2 = 3 + 10$ *Think:* Multiply first.

            $= 13$ Then add.

(b) $(1 - 5)^2 - 2\sqrt{9} + \frac{12}{4} = (-4)^2 - 2\sqrt{9} + 3$ *Think:* Clear grouping symbols first.

                    $= 16 - 2 \cdot 3 + 3$ Evaluate each power notation and radical.

                    $= 16 - 6 + 3$ Multiply and divide in order from left to right.

                    $= 10 + 3$ Add and subtract in order from left to right.

                    $= 13$

(c) $\dfrac{5 - \sqrt{5^2 - 4(6)(-4)}}{2(6)} = \dfrac{5 - \sqrt{25 - 4(6)(-4)}}{2(6)}$ *Think:* $5^2 = 25$

             $= \dfrac{5 - \sqrt{25 - 24(-4)}}{2(6)}$ $4(6) = 24$

             $= \dfrac{5 - \sqrt{25 - (-96)}}{2(6)}$ $24(-4) = -96$

             $= \dfrac{5 - \sqrt{25 + 96}}{2(6)}$ $25 - (-96) = 25 + (+96)$

             $= \dfrac{5 - \sqrt{121}}{2(6)}$ $25 + 96 = 121$

             $= \dfrac{5 - 11}{2(6)}$ $\sqrt{121} = 11$ because $11 \cdot 11 = 121$

             $= \dfrac{-6}{2(6)}$ $5 - 11 = -6$

             $= \dfrac{-1(\cancel{6})}{2(\cancel{6})}$ $-6 = -1(6)$

             $= \dfrac{-1}{2}$ or $-\dfrac{1}{2}$ $\dfrac{a \cdot c}{b \cdot c} = \dfrac{a}{b}$

Caution: To avoid getting a wrong answer when there are no grouping symbols present, always multiply and divide *before* adding and subtracting.

EXAMPLE 16-33: Evaluate the following expressions: **(a)** $8 + 6 \div 2$ **(b)** $18 - 2 \cdot 3$

Solution

Correct Method	Wrong Method
(a) $8 + \overbrace{6 \div 2} = 8 + 3$ Divide first.	$\overbrace{8 + 6} \div 2 = 14 \div 2$ No! Divide first.
$= 11 \longleftarrow$ correct answer	$= 7 \longleftarrow$ wrong answer
(b) $18 - \overbrace{2 \cdot 3} = 18 - 6$ Multiply first.	$\overbrace{18 - 2} \cdot 3 = 16 \cdot 3$ No! Multiply first.
$= 12 \longleftarrow$ correct answer	$= 48 \longleftarrow$ wrong answer

Caution: Be sure to work from left to right when adding, subtracting, multiplying, or dividing.

EXAMPLE 16-34: Evaluate the following expressions: (a) $3 - 5 + 7$ (b) $12 \div 2 \times 3$

Solution

Correct Method	Wrong Method
(a) $\overbrace{3 - 5}^{} + 7 = -2 + 7$ Work in order from left to right.	$3 - \overbrace{5 + 7}^{} = 3 - 12$ No! Never add and subtract from right to left.
$= 5 \longleftarrow$ correct answer	$= -9 \longleftarrow$ wrong answer
(b) $\overbrace{12 \div 2}^{} \times 3 = 6 \times 3$ Work in order from left to right.	$12 \div \overbrace{2 \times 3}^{} = 12 \div 6$ No! Never multiply and divide from right to left.
$= 18 \longleftarrow$ correct answer	$= 2 \longleftarrow$ wrong answer

RAISE YOUR GRADES

Can you . . . ?

☑ identify positive and negative numbers
☑ graph positive and negative numbers
☑ write the opposite of any positive or negative number
☑ list the positive integers, negative integers, and integers
☑ identify rational numbers, irrational numbers, and real numbers
☑ draw a diagram showing the relationships between real numbers, rational numbers, irrational numbers, integers, and whole numbers
☑ compare any two real numbers using either $<$, $>$, or $=$
☑ find the absolute value of any given real number
☑ add, subtract, multiply, and divide integers
☑ identify when division is not defined
☑ identify the three signs of a rational number in fraction form
☑ write rational numbers in the form $\frac{a}{b}$ in three equivalent ways using sign changes
☑ simplify rational numbers in fraction form
☑ add, subtract, multiply, and divide rational numbers
☑ evaluate an expression using the Order of Operations

SUMMARY

1. On a number line
   (a) the point that represents zero is called the origin.
   (b) all points to the right of the origin represent positive numbers.
   (c) all points to the left of the origin represent negative numbers.
2. To write a negative number, you use a negative sign $(-)$.
3. To write a positive number, you do not need to write a positive sign $(+)$.
4. Two numbers with unlike signs $(+, -)$ that are the same distance from the origin (0) on a number line are called opposites.
5. To find the opposite of a real number, you just change the number's sign to the opposite sign.
6. The sum of two opposites is always zero.
7. The whole numbers $(0, 1, 2, 3, \cdots)$ and their opposites $(\cdots, -3, -2, -1, 0)$ are called integers $(\cdots, -3, -2, -1, 0, 1, 2, 3, \cdots)$.
8. The positive integers are $1, 2, 3, \cdots$

9. The negative integers are $\cdots, 3, 2, 1$.

10. If a and b are integers ($b \neq 0$), then the collection of all numbers that can be written in fractional form $\frac{a}{b}$ are called rational numbers.

11. Numbers whose decimal form does not terminate and does not repeat are called irrational numbers.

12. The collection of all rational and irrational numbers is called the set of real numbers.

13. A diagram showing the relationships between real numbers, rational numbers, irrational numbers, integers, and whole numbers can be drawn as follows:

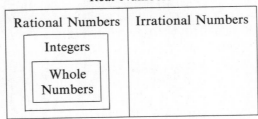

14. Every real number is represented by a point on a number line and every point on a number line represents a real number. That is, a real number line is a pictorial representation of the collection of all real numbers.

15. To compare two real numbers a and b on a number line, you write
    (a) $a < b$ if a is to the left of b.
    (b) $a > b$ if a is to the right of b.
    (c) $a = b$ if a and b are represented by the same point.

16. The distance between a given real number and its origin (0) on a number line is its absolute value.

17. The absolute value of any real number is never negative.

18. **Addition Rules for Two or More Real Numbers with Like Signs**
    (a) Find the sum of the absolute values.
    (b) Write the same like sign on the sum.

19. **Addition Rules for Two Real Numbers with Unlike Signs**
    (a) Find the difference between the absolute values.
    (b) Write the sign of the number with the larger absolute value on the sum.

20. **Subtraction Rules for Two Real Numbers**
    (a) Change subtraction to addition by writing the opposite of the subtrahend.
    (b) Follow the rules for adding real numbers.
    (c) Check by adding the proposed difference to the original subtrahend to see if you get the original minuend.

21. A multiplication problem like "3 times 4" can be written as 3×4, $3 \cdot 4$, $3(4)$, $(3)4$, or $(3)(4)$.

22. **Multiplication Rules for Two or More Real Numbers**
    (a) Find the product of the absolute values.
    (b) Make the product
        (a) positive if there are an even number $(0, 2, 4, \cdots)$ of negative factors.
        (b) negative if there are an odd number $(1, 3, 5, \cdots)$ of negative factors.
        (c) zero if one or more factors are zero.

23. A division problem like "8 divided by 3" can be written as $8 \div 3$, $8/3$, $\frac{8}{3}$, $8 \cdot \frac{1}{3}$, $\frac{1}{3} \cdot 8$, or $3\overline{)8}$.

24. **Division Rules for Two Real Numbers**
    (a) Find the quotient of the absolute values.
    (b) Use the sign rules for multiplication to write the correct sign on the quotient.
    (c) Check by multiplying the proposed quotient by the original divisor to see if you get the original dividend.

25. Every rational number in fraction form has three signs: the sign of the fraction, the sign of the numerator, and the sign of the denominator.

26. The following are all equivalent ways to write a fraction with form $\frac{a}{b}$:

$$\frac{a}{b} = -\frac{-a}{b} = -\frac{a}{-b} = \frac{-a}{-b}$$

27. *Caution:* To get an equal fraction using changes, you must change exactly two of a fraction's three signs.

28. The following are all equivalent ways to write a fraction with form $-\dfrac{a}{b}$:

$$-\frac{a}{b} = \frac{-a}{b} = \frac{a}{-b} = -\frac{-a}{-b}$$

29. A rational number in fraction form for which the numerator and denominator do not share a common integer factor other than 1 or -1 is called a rational number in lowest terms.

30. To simplify a rational number in fraction form, you write the rational number using as few negative signs as possible and then reduce the rational number in fraction form to lowest terms using the Fundamental Rule for Rational Numbers.

31. **Fundamental Rule for Rational Numbers**

    If a, b, and c are integers ($b \neq 0$ and $c \neq 0$), then $\dfrac{a \cdot c}{b \cdot c} = \dfrac{a}{b}$.

32. To avoid getting a wrong answer when computing with real numbers, you must use the Order of Operations rules in the order given below.

33. **Order of Operations**
    (a) Perform operations inside grouping symbols such as ( ), [ ], { }, ─, and $\sqrt{}$.
    (b) Evaluate power notation and radicals.
    (c) Multiply and divide in order from left to right.
    (d) Add and subtract in order from left to right.

34. The saying "Please Excuse My Dear Aunt Sally" can help you remember the Order of Operations rules: **P**arentheses (enclosure symbols), **E**xponents (and radicals), **M**ultiply and **D**ivide, **A**dd and **S**ubtract.

35. *Caution:* To avoid getting a wrong answer when there are no grouping symbols present, always multiply and divide *before* adding and subtracting.

36. *Caution:* Be sure to work from left to right when adding, subtracting, multiplying, or dividing.

SOLVED PROBLEMS

PROBLEM 16-1   Write the opposite of each real number: (a) $+8$ (b) $-\frac{2}{3}$ (c) 0.6 (d) 0

Solution: Recall that to find the opposite of a real number, you just change the number's sign to the opposite sign [see Example 16-5]:
(a) The opposite of $+8$ is -8.   (b) The opposite of $-\frac{2}{3}$ is $\frac{2}{3}$.   (c) The opposite of 0.6 is -0.6.
(d) The opposite of 0 is 0.

PROBLEM 16-2:   Identify which of the following numbers are rational numbers and which are irrational numbers: (a) 0 (b) 2 (c) -3 (d) $\frac{3}{4}$ (e) $-\frac{5}{3}$ (f) 2.3 (g) -0.25
(h) $0.\overline{6}$ (i) $\sqrt{9}$ (j) $-\sqrt{10}$ (k) π

Solution: Recall that the rational numbers include all the whole numbers, integers, fractions, terminating decimals, and repeating decimals. The irrational numbers include all numbers whose decimal form does not terminate and does not repeat [see Examples 16-9 and 16-10]:
(a) $-$(i) The numbers $0, 2, -3, \frac{3}{4}, \frac{-5}{3}, 2.3, -0.25, 0.\overline{6}$, and $\sqrt{9}$ because $\sqrt{9} = 3$) are all rational numbers.
(j) $-\sqrt{10}$ is an irrational number because $-\sqrt{10} = -3.1622766016\cdots$, and does not terminate or repeat.   (k) π is an irrational number because $\pi = 3.1415926536\cdots$, and does not terminate or repeat.

PROBLEM 16-3   Compare each pair of real numbers using $<$, $>$, or $=$:   (a) $2, 5$ (b) $2, -5$
(c) $-2, 5$ (d) $-2, -5$ (e) $\frac{8}{12}, \frac{12}{18}$ (f) $0, \frac{1}{2}$ (g) $0, -\frac{3}{4}$ (h) $-\frac{2}{3}, -0.\overline{6}$

Solution: Recall that to compare two real numbers a and b using a number line, you write $a < b$ if a is to the left of b, $a > b$ if a is to the right of b, and $a = b$ if a and b represent the same point [see Example 16-12]:

(a) $2 < 5$ because 2 is to the left of 5 on a number line.

(b) $2 > -5$ because every positive number is to the right of every negative number.

(c) $-2 < 5$ because every negative number is to the left of every positive number.

(d) $-2 > -5$ because -2 is to the right of -5 on a number line.

(e) $\dfrac{8}{12} = \dfrac{12}{18}$ because $\dfrac{8}{12} = \dfrac{2 \cdot \cancel{4}}{3 \cdot \cancel{4}} = \dfrac{2}{3}$ and $\dfrac{12}{18} = \dfrac{2 \cdot \cancel{6}}{3 \cdot \cancel{6}} = \dfrac{2}{3}$.

(f) $0 < \frac{1}{2}$ because zero is to the left of every positive number.

(g) $0 > -\frac{3}{4}$ because zero is to the right of every negative number.

(h) $-\frac{2}{3} = -0.\overline{6}$ because $2 \div 3 = 0.666 \cdots = 0.\overline{6}$.

PROBLEM 16-4   Find each absolute value: (a) $|+2|$ (b) $\left|-\frac{2}{3}\right|$ (c) $|2.75|$ (d) $|0|$

Solution: Recall that the distance that a given real number is from the origin (0) on a number line is the given real number's absolute value [see Example 16-13]:

(a) $|+2|$ $= 2$ because 2 is 2 units from 0 on a number line.

(b) $\left|-\frac{2}{3}\right|$ $= \frac{2}{3}$ because $-\frac{2}{3}$ is $\frac{2}{3}$ of a unit from 0 on a number line.

(c) $|2.75|$ $= 2.75$ because 2.75 is 2.75 units from 0 on a number line.

(d) $|0|$ $\;\; = 0$ because 0 is 0 units from 0 on a number line.

PROBLEM 16-5   Add integers with like signs: (a) $+5 + (+8)$ (b) $-6 + (-3)$ (c) $4 + 7$
(d) $+8 + (+7) + (+2)$ (e) $-5 + (-4) + (-7) + (-8)$

Solution: Recall that to add two or more real numbers with like signs, you first find the sum of the absolute values and then write the same like sign on the sum [see Example 16-14]:

same like sign

sum of the absolute values

(a) $+5 + (+8) = +(5 + 8) = +13$ or 13   (b) $-6 + (-3) = -(6 + 3) = -9$
(c) $4 + 7 = 11$ or $+11$ (d) $+8 + (+7) + (+2) = +(8 + 7 + 2) = +17$ or 17
(e) $-5 + (-4) + (-7) + (-8) = -(5 + 4 + 7 + 8) = -24$

PROBLEM 16-6   Add integers with unlike signs: (a) $+8 + (-6)$ (b) $2 + (-9)$
(c) $-7 + (+8)$ (d) $-8 + 1$ (e) $0 + 6$ (f) $0 + (-7)$ (g) $+9 + 0$ (h) $-4 + 0$

Solution: Recall that to add two integers with unlike signs, you first find the difference between the absolute values and then write the sign of the number with the larger absolute value on the sum [see Examples 16-15 and 16-16]:

sign of the number with the larger absolute value

difference of the absolute values

(a) $+8 + (-6) = +(8 - 6) = +2$ or 2   (b) $2 + (-9) = -(9 - 2) = -7$
(c) $-7 + (+8) = +(8 - 7) = +1$ or 1   (d) $-8 + 1 = -(8 - 1) = -7$
(e) $0 + 6 = 6$ (f) $0 + (-7) = -7$ (g) $+9 + 0 = +9$ or 9   (h) $-4 + 0 = -4$

PROBLEM 16-7   Subtract integers: (a) $+5 - (+2)$ (b) $6 - 8$ (c) $-6 - (+4)$
(d) $-2 - 9$ (e) $+8 - (-3)$ (f) $5 - (-8)$ (g) $-5 - (-2)$ (h) $-3 - (-7)$

Solution: Recall that to subtract two integers, you first change subtraction to addition while writing the opposite of the subtrahend and then follow the rules for adding real numbers [see Example 16-17]:

change to addition

write opposite of subtrahend

(a) $+5 - (+2) = +5 + (-2) = +3$ or 3   (b) $6 - 8 = 6 + (-8) = -2$
(c) $-6 - (+4) = -6 + (-4) = -10$ (d) $-2 - 9 = -2 + (-9) = -11$

(e) $+8 - (-3) = +8 + (+3) = +11$ or 11 **(f)** $5 - (-8) = 5 + (+8) = +13$ or 13
(g) $-5 - (-2) = -5 + (+2) = -3$ **(h)** $-3 - (-7) = -3 + (+7) = +4$ or 4

PROBLEM 16-8   Multiply integers: **(a)** $5 \cdot 2$ **(b)** $+4(+3)$ **(c)** $(-2)3$ **(d)** $5(-3)$
(e) $+7(-4)$ **(f)** $(-5)(+6)$ **(g)** $-2(-8)$ **(h)** $(-3)(-2)(-5)$ **(i)** $5(-1)2(-1)(-2)(-1)$

Solution: Recall that to multiply two or more real numbers, you first find the product of all the absolute values and then make the product positive if there are an even number of negative factors; negative if there are an odd number of negative factors; zero if one or more factors are zero [see Examples 16-19 and 16-20]:

even number (0) of negative factors
positive product

odd number (1) of negative factors
negative product

(a) $\overbrace{5 \cdot 2} = 10$ **(b)** $+4(+3) = +12$ or 12 **(c)** $\overbrace{(-2)3} = -6$ **(d)** $5(-3) = -15$
(e) $+7(-4) = -28$ **(f)** $(-5)(+6) = -30$ **(g)** $-2(-8) = +16$ or 16
(h) $(-3)(-2)(-5) = -30$ **(i)** $5(-1)2(-1)(-2)(-1) = +20$ or 20

PROBLEM 16-9   Divide integers: **(a)** $12 \div 3$ **(b)** $-18 \div 2$ **(c)** $20 \div (-5)$
(d) $-10 \div (-2)$ **(e)** Divide $+8$ by $+2$ **(f)** $\frac{+15}{-3}$ **(g)** $-6\overline{)-12}$ **(h)** $0 \div (-4)$
(i) $-4 \div 0$ **(j)** $0 \div 0$

Solution: Recall that to divide two integers, you first find the quotient of the absolute values and then use the sign rules for multiplication to write the correct sign on the product [see Example 16-22]:

two positive numbers
positive quotient

one negative number
negative quotient

(a) $\overbrace{12 \div 3} = 4$ **(b)** $\overbrace{-18 \div 2} = -9$

two negative numbers
positive quotient

(c) $20 \div (-5) = -4$ **(d)** $\overbrace{-10 \div (-2)} = +5$ or 5
(e) Divide $+8$ by $+2 = +8 \div (+2) = +4$ or 4 **(f)** $\frac{+15}{-3} = +15 \div (-3) = -5$
(g) $-6\overline{)-12} = -12 \div (-6) = +2$ or 2 **(h)** $0 \div (-4) = 0$
(i) $-4 \div 0$ is not defined.   **(j)** $0 \div 0$ is not defined.

PROBLEM 16-10   Rename each fraction in three equivalent ways using sign changes:

(a) $\frac{1}{2}$ **(b)** $-\frac{2}{3}$ **(c)** $-\frac{5}{-8}$ **(d)** $\frac{-1}{-4}$ **(e)** $-\frac{3}{5}$ **(f)** $\frac{-5}{6}$ **(g)** $\frac{2}{-9}$ **(h)** $-\frac{-1}{-3}$

Solution: Recall that if exactly two of a fraction's three signs are changed, you will always get an equal fraction [see Examples 16-24 and 16-25]:

(a) $\frac{1}{2} = -\frac{-1}{2} = -\frac{1}{-2} = \frac{-1}{-2}$ **(b)** $-\frac{-2}{3} = -\frac{2}{-3} = \frac{-2}{-3} = \frac{2}{3}$

(c) $-\frac{5}{-8} = -\frac{-5}{8} = \frac{-5}{-8} = \frac{5}{8}$ **(d)** $\frac{-1}{-4} = \frac{1}{4} = -\frac{-1}{4} = -\frac{1}{-4}$

(e) $-\frac{3}{5} = \frac{-3}{5} = \frac{3}{-5} = -\frac{-3}{-5}$ **(f)** $\frac{-5}{6} = -\frac{5}{6} = \frac{5}{-6} = -\frac{-5}{-6}$

(g) $\frac{2}{-9} = -\frac{2}{9} = \frac{-2}{9} = -\frac{-2}{-9}$ **(h)** $-\frac{-1}{-3} = -\frac{1}{3} = \frac{-1}{3} = \frac{1}{-3}$

PROBLEM 16-11   Simplify each rational number: **(a)** $\frac{12}{18}$ **(b)** $\frac{-10}{12}$ **(c)** $-\frac{42}{-14}$

(d) $\frac{-22}{-8}$ **(e)** $-\frac{12}{3}$ **(f)** $\frac{-15}{10}$ **(g)** $\frac{24}{-36}$ **(h)** $-\frac{-12}{-30}$

Solution: Recall the Fundamental Rule for Rational Numbers: If a, b, and c are integers ($b \neq 0$ and $c \neq 0$), then $\dfrac{a \cdot c}{b \cdot c} = \dfrac{a}{b}$ [see Example 16-27]:

(a) $\dfrac{12}{18} = \dfrac{2 \cdot 6}{3 \cdot 6} = \dfrac{2}{3}$ **(b)** $-\dfrac{-10}{12} = \dfrac{10}{12} = \dfrac{\not{2} \cdot 5}{\not{2} \cdot 6} = \dfrac{5}{6}$ **(c)** $-\dfrac{42}{-14} = \dfrac{42}{14} = \dfrac{3 \cdot \not{14}}{1 \cdot \not{14}} = \dfrac{3}{1} = 3$

(d) $\dfrac{-22}{-8} = \dfrac{22}{8} = \dfrac{\not{2} \cdot 11}{\not{2} \cdot 4} = 2\dfrac{3}{4}$ **(e)** $-\dfrac{12}{3} = -\dfrac{\not{3} \cdot 4}{1 \cdot \not{3}} = -\dfrac{4}{1} = -4$

(f) $\dfrac{-15}{10} = -\dfrac{15}{10} = -\dfrac{3 \cdot \not{5}}{2 \cdot \not{5}} = -\dfrac{3}{2}$ or $-1\dfrac{1}{2}$ **(g)** $\dfrac{24}{-36} = -\dfrac{24}{36} = -\dfrac{2 \cdot \not{12}}{3 \cdot \not{12}} = -\dfrac{2}{3}$

(h) $-\dfrac{-12}{-30} = -\dfrac{12}{30} = -\dfrac{2 \cdot \not{6}}{5 \cdot \not{6}} = -\dfrac{2}{5}$

PROBLEM 16-12   Add rational numbers: **(a)** $3 + \frac{3}{8}$ **(b)** $-0.5 + 0.2$ **(c)** $0.75 + \left(-\frac{3}{8}\right)$
(d) $-1\frac{1}{2} + (-1.25)$ **(e)** $6 + \frac{1}{2} + 0.2 + \left(-3\frac{1}{4}\right) + (-2.75) + (-1)$

Solution: Recall that to add rational numbers, you use the Addition Rules for Real Numbers with Like (or Unlike) Signs given in Section 16-4, part A [see Example 16-28]:

(a) $3 + \frac{3}{8} = 3\frac{3}{8}$ or 3.375 **(b)** $-0.5 + 0.2 = -0.3$ or $-\frac{3}{10}$
(c) $0.75 + \left(-\frac{3}{8}\right) = 0.75 + (-0.375) = 0.375$ or $\frac{3}{8}$
(d) $-1\frac{1}{2} + (-1.25) = -1\frac{1}{2} + \left(-1\frac{1}{4}\right) = -2\frac{3}{4}$ or -2.75
(e) $6 + \frac{1}{2} + 0.2 + \left(-3\frac{1}{4}\right) + (-2.75) + (-1) = 6 + 0.5 + 0.2 + (-3.25) + (-2.75) + (-1)$
$$= 6.7 + (-7)$$
$$= -0.3 \text{ or } -\tfrac{3}{10}$$

PROBLEM 16-13   Subtract rational numbers: **(a)** $5 - \frac{3}{4}$ **(b)** $1.3 - 2.9$ **(c)** $0.25 - \left(-\frac{4}{5}\right)$
(d) $-2\frac{1}{2} - (-3.75)$

Solution: Recall that to subtract rational numbers, you use the Subtraction Rules for Real Numbers given in Section 16-4, part B [see Example 16-29]:

(a) $5 - \frac{3}{4} = 4\frac{4}{4} - \frac{3}{4} = 4\frac{1}{4}$ or 4.25 **(b)** $1.3 - 2.9 = -1.6$ or $-1\frac{3}{5}$
(c) $0.25 - \left(-\frac{4}{5}\right) = 0.25 + \left(+\frac{4}{5}\right) = 0.25 + 0.8 = 1.05$ or $1\frac{1}{20}$
(d) $-2\frac{1}{2} - (-3.75) = -2\frac{1}{2} + (+3.75) = -2.5 + 3.75 = 1.25$ or $1\frac{1}{4}$

PROBLEM 16-14   Multiply rational numbers: **(a)** $\frac{1}{2}\left(-\frac{3}{4}\right)$ **(b)** $-0.5(2.6)$ **(c)** $\frac{1}{4}(0.25)$
(d) $-1.25\left(-2\frac{3}{4}\right)$ **(e)** $-3(0.5)\left(-\frac{1}{2}\right)(-2.5)(-2)(0)(-8)(-1.75)$

Solution: Recall that to multiply rational numbers, you use the Multiplication Rules for Real Numbers given in Section 16-4, part C [see Example 16-30]:

(a) $\frac{1}{2}\left(-\frac{3}{4}\right) = -\frac{3}{8}$ or -0.375 **(b)** $-0.5(2.6) = -1.3$ or $-1\frac{3}{10}$
(c) $\frac{1}{4}(0.25) = \frac{1}{4} \cdot \frac{1}{4} = \frac{1}{16}$ or 0.0625 **(d)** $-1.25\left(-2\frac{3}{4}\right) = -1.25(-2.75) = 3.4375$ or $3\frac{7}{16}$
(e) $-3(0.5)\left(-\frac{1}{2}\right)(-2.5)(-2)(0)(-8)(-1.75) = 0$

PROBLEM 16-15   Divide rational numbers: **(a)** $-\frac{2}{3} \div \frac{3}{4}$ **(b)** $1.2 \div (-0.5)$ **(c)** $\frac{3}{4} \div 0.25$
(d) $-2.5 \div \left(-1\frac{1}{2}\right)$ **(e)** $0 \div 0.5$ **(f)** $-3.6 \div 0$

Solution: Recall that to divide rational numbers, you use the Division Rules for Real Numbers given in Section 16-4, part D [see Example 16-31]:

(a) $-\frac{2}{3} \div \frac{3}{4} = -\frac{2}{3} \cdot \frac{4}{3} = -\frac{8}{9}$ or $-0.\overline{8}$ **(b)** $1.2 \div (-0.5) = -2.4$ or $-2\frac{2}{5}$
(c) $\frac{3}{4} \div 0.25 = \frac{3}{4} \div \frac{1}{4} = \frac{3}{\not{4}} \cdot \frac{\not{4}}{1} = \frac{3}{1} = 3$ **(d)** $-2.5 \div \left(-1\frac{1}{2}\right) = -2.5 \div (-1.5) = 1.\overline{6}$ or $1\frac{2}{3}$
(e) $0 \div 0.5 = 0$ **(f)** $-3.6 \div 0$ is not defined

PROBLEM 16-16 Evaluate using the Order of Operations: **(a)** $8 - 6 + 2$ **(b)** $8 \div 2 \cdot 4$

(c) $9 - 3 \cdot 5$ **(d)** $9 + 3 \div 4$ **(e)** $4 + 6(-2)$ **(f)** $\dfrac{4 + 6}{2}$ **(g)** $10 \div (2 + 3)$

(h) $5 - \dfrac{8}{1 + 3} \cdot 3 + 6$ **(i)** $2 + [6 - (-4)(-2 + 8)] \div 6$

(j) $-3 + \{-2[16 + (-4 - (-3))] \div (-5)\}(-2)$ **(k)** $3 \cdot 2^3$ **(l)** $(-5 \cdot 2)^2$
(m) $-2 \cdot 3^2 + 10$ **(n)** $3^2 \cdot 2 \div 3\sqrt{16}$ **(o)** $-\sqrt{5^2 - 4^2}$ **(p)** $3^2(1 - 2) + 5\sqrt{4}$

(q) $2^3(3 - 7) \div 16\sqrt{9}$ **(r)** $\dfrac{11 + \sqrt{(-11)^2 - 4(6)(-10)}}{2(6)}$

Solution: Recall that to evaluate using the Order of Operations, you:
1. Perform operations inside grouping symbols such as ( ), [ ], { }, —, and $\sqrt{}$
2. Evaluate each power notation and radical.
3. Multiply or divide in order from left to right.
4. Add or subtract in order from left to right.
   [See Examples 16-32, 16-33, and 16-34]:

(a) $8 - 6 + 2 = 2 + 2 = 4$ **(b)** $8 \div 2 \cdot 4 = 4 \cdot 4 = 16$ **(c)** $9 - 3 \cdot 5 = 9 - 15 = -6$

(d) $9 + 3 \div 4 = 9 + \dfrac{3}{4} = 9\dfrac{3}{4}$ **(e)** $4 + 6(-2) = 4 + (-12) = -8$ **(f)** $\dfrac{4 + 6}{2} = \dfrac{10}{2} = 5$

(g) $10 \div (2 + 3) = 10 \div 5 = 2$

(h) $5 - \dfrac{8}{1 + 3} \cdot 3 + 6 = 5 - \dfrac{8}{4} \cdot 3 + 6 = 5 - 2 \cdot 3 + 6 = 5 - 6 + 6 = -1 + 6 = 5$

(i) $2 + [6 - (-4)(-2 + 8)] \div 6 = 2 + [6 - (-4)(6)] \div 6$

$\qquad\qquad\qquad\qquad\qquad\quad = 2 + [6 - (-24)] \div 6$

$\qquad\qquad\qquad\qquad\qquad\quad = 2 + 30 \div 6$

$\qquad\qquad\qquad\qquad\qquad\quad = 2 + 5$

$\qquad\qquad\qquad\qquad\qquad\quad = 7$

(j) $-3 + \{-2[16 + (-4 - (-3))] \div (-5)\}(-2) = -3 + \{-2[16 + (-1)] \div (-5)\}(-2)$

$\qquad\qquad\qquad\qquad\qquad\qquad\qquad\qquad\quad = -3 + \{-2[15] \div (-5)\}(-2)$

$\qquad\qquad\qquad\qquad\qquad\qquad\qquad\qquad\quad = -3 + \{-30 \div (-5)\}(-2)$

$\qquad\qquad\qquad\qquad\qquad\qquad\qquad\qquad\quad = -3 + \{6\}(-2)$

$\qquad\qquad\qquad\qquad\qquad\qquad\qquad\qquad\quad = -3 + (-12)$

$\qquad\qquad\qquad\qquad\qquad\qquad\qquad\qquad\quad = -15$

(k) $3 \cdot 2^3 = 3 \cdot 8 = 24$ **(l)** $(-5 \cdot 2)^2 = (-10)^2 = (-10)(-10) = 100$
(m) $-2 \cdot 3^2 + 10 = -2 \cdot 9 + 10 = -18 + 10 = -8$
(n) $3^2 \cdot 2 \div 3\sqrt{16} = 9 \cdot 2 \div 3(4) = 18 \div 3(4) = 6(4) = 24$
(o) $-\sqrt{5^2 - 4^2} = -\sqrt{25 - 16} = -\sqrt{9} = -(3) = -3$
(p) $3^2(1 - 2) + 5\sqrt{4} = 3^2(-1) + 5\sqrt{4} = 9(-1) + 5(2) = -9 + 5(2) = -9 + 10 = 1$
(q) $2^3(3 - 7) \div 16\sqrt{9} = 2^3(-4) \div 16\sqrt{9} = 8(-4) \div 16(3) = -32 \div 16(3) = -2(3) = -6$

(r)
$$\frac{11 + \sqrt{(-11)^2 - 4(6)(-10)}}{2(6)} = \frac{11 + \sqrt{121 - 4(6)(-10)}}{2(6)}$$

$$= \frac{11 + \sqrt{121 - 24(-10)}}{2(6)}$$

$$= \frac{11 + \sqrt{121 + 240}}{2(6)}$$

$$= \frac{11 + \sqrt{361}}{2(6)}$$

$$= \frac{11 + 19}{2(6)}$$

$$= \frac{30}{2(6)}$$

$$= \frac{5(\cancel{6})}{2(\cancel{6})}$$

$$= \frac{5}{2} \text{ or } 2\frac{1}{2} \text{ or } 2.5$$

Supplementary Exercises

PROBLEM 16-17   Write the opposite of each real number: **(a)** 6   **(b)** -8 **(c)** $+9$
(d) 0   **(e)** 1   **(f)** -1 **(g)** $\frac{3}{4}$ **(h)** $-\frac{2}{3}$ **(i)** $+\frac{1}{8}$ **(j)** 1.25   **(k)** -2.5 **(l)** $+0.75$
(m) $0.\overline{6}$ **(n)** $-0.\overline{3}$ **(o)** $\sqrt{2}$ **(p)** $-\sqrt{5}$ **(q)** $-\pi$ **(r)** π **(s)** $2\frac{1}{2}$ **(t)** $-3\frac{3}{4}$
(u) $0.1\overline{6}$ **(v)** $-0.8\overline{3}$ **(w)** 2^3 **(x)** -3^2 **(y)** $(-4)^2$ **(z)** $-(-5)^3$

PROBLEM 16-18   Compare each pair of real numbers using $<$, $>$, or $=$: **(a)** 5, 8
(b) $-3, 4$ **(c)** $5, -9$ **(d)** $-6, -8$ **(e)** 0, 7   **(f)** $0, -5$ **(g)** $\frac{1}{2}, \frac{1}{3}$ **(h)** $-\frac{3}{4}, \frac{1}{2}$
(i) $-\frac{15}{20}, -\frac{2}{3}$ **(j)** $\frac{0}{-5}, 0$ **(k)** 1.5, 1.6   **(l)** $0.3, -\frac{1}{3}$ **(m)** $-0.75, -\frac{3}{4}$ **(n)** $\frac{15}{16}, \frac{31}{32}$
(o) 0.023, 0.032   **(p)** $-0.12, -0.21$ **(q)** $-1.0101, 1.0101$ **(r)** $\frac{7}{16}, 0.4375$ **(s)** $1\frac{1}{2}, \frac{3}{2}$
(t) $2.5, -2\frac{1}{2}$ **(u)** $-1.75, -1\frac{1}{4}$ **(v)** $\frac{300}{7}, 42\frac{6}{7}$ **(w)** $-\frac{18}{40}, -\frac{12}{32}$ **(x)** $-0.0001, -0.000101$
(y) $2.125, \frac{17}{8}$ **(z)** $\pi, 3.14$

PROBLEM 16-19   Find each absolute value: **(a)** $|2|$ **(b)** $|-9|$ **(c)** $|+7|$ **(d)** $|0|$ **(e)** $|1|$
(f) $|-1|$ **(g)** $|\frac{1}{2}|$ **(h)** $|-\frac{2}{3}|$ **(i)** $|+\frac{1}{8}|$ **(j)** $|1.25|$ **(k)** $|2.5|$ **(l)** $|0.75|$ **(m)** $|0.\overline{6}|$
(n) $|0.\overline{3}|$ **(o)** $|\sqrt{2}|$ **(p)** $|-\sqrt{5}|$ **(q)** $|-\pi|$ **(r)** $|\pi|$ **(s)** $|2\frac{1}{2}|$ **(t)** $|-3\frac{3}{4}|$ **(u)** $|0.1\overline{6}|$
(v) $|0.8\overline{3}|$ **(w)** $|2^3|$ **(x)** $|-3^2|$ **(y)** $|(-4)^2|$ **(z)** $|-(-5)^3|$

PROBLEM 16-20   Compute with integers: **(a)** $+4 + (+2)$ **(b)** $7 + (-7)$ **(c)** $4 - (-7)$
(d) $-6 + 5$ **(e)** $-8 + (-9)$ **(f)** $-8 - 6$ **(g)** $-7 - (-8)$ **(h)** $2 - 9$
(i) $2 + 3 + 5$ **(j)** $-3 + 5 - 4$ **(k)** $4 - (-3) + 5$ **(l)** $-1 - 3 + (-4)$
(m) $2 - 3 + 1 + (-2)$ **(n)** $-2 - 3 + (-4) - (-3)$ **(o)** 4×8 **(p)** $36 \div 6$ **(q)** $-6(7)$
(r) $-21 \div 3$ **(s)** $8(-3)$ **(t)** $\frac{18}{6}$ **(u)** $-7(-7)$ **(v)** $\frac{-35}{-7}$ **(w)** $5 \cdot 3 \cdot 2$
(x) $50 \div 5 \div (-2)$ **(y)** $0 \times 8 \div (-6)$ **(z)** $3(-2)(-6) \div 4 \div 0$

PROBLEM 16-21 Simplify each rational number: (a) $\frac{24}{40}$ (b) $\frac{-20}{50}$ (c) $\frac{16}{-80}$

(d) $\frac{-75}{-90}$ (e) $-\frac{-20}{120}$ (f) $-\frac{84}{-120}$ (g) $-\frac{8}{10}$ (h) $-\frac{-3}{-9}$ (i) $\frac{3}{3}$ (j) $\frac{-12}{12}$

(k) $\frac{9}{-3}$ (l) $\frac{-18}{-2}$ (m) $-\frac{10}{-6}$ (n) $-\frac{-14}{4}$ (o) $-\frac{48}{24}$ (p) $-\frac{-56}{-32}$ (q) $\frac{256}{32}$

(r) $\frac{-125}{40}$ (s) $\frac{15}{-25}$ (t) $\frac{-20}{-45}$ (u) $-\frac{12}{18}$ (v) $-\frac{-9}{27}$ (w) $\frac{20}{-1}$ (x) $-\frac{-90}{1}$

(y) $-\frac{0}{-8}$ (z) $-\frac{-6}{0}$

PROBLEM 16-22 Compute with rational numbers: (a) $0.5 + 0.9$ (b) $1\frac{1}{5} + (-0.9)$
(c) $-0.8 + 2\frac{3}{5}$ (d) $-3\frac{1}{2} + (-2.1)$ (e) $\frac{1}{2} + \frac{7}{16}$ (f) $\frac{-3}{4} + 0.75$ (g) $-3\frac{1}{2} + 2.5$
(h) $-5\frac{1}{8} + (-7.25)$ (i) $0.2 - 0.6$ (j) $-0.8 - 1\frac{1}{2}$ (k) $-\frac{3}{8} - (-0.25)$ (l) $4\frac{3}{16} - (2.125)$
(m) $0.5(0.2)$ (n) $\frac{1}{2} \cdot \frac{3}{4}$ (o) $-4.8(-\frac{3}{10})$ (p) $-\frac{1}{4}(-0.\overline{3})$ (q) $0.8(-0.7)$ (r) $-\frac{3}{8} \cdot \frac{2}{3}$
(s) $0.7 \div 2\frac{2}{5}$ (t) $\frac{3}{8} \div 0.5$ (u) $-1.44 \div (-4.8)$ (v) $-\frac{1}{12} \div (-0.25)$ (w) $-4.5 \div 1.8$
(x) $\frac{1}{4} \div 1.8$ (y) $-6.125 \div (-1\frac{5}{16})$ (z) $-8\frac{3}{4} \div 2.5$

PROBLEM 16-23 Evaluate using the Order of Operations: (a) $4 \cdot 6(-3)$ (b) $\frac{-4}{-6} + 3$
(c) $4 - 6 \div (-3)$ (d) $-4(-6) - 3$ (e) $[4 + (-6)](-3)$ (f) $(-4 - 6) \div 3$
(g) $-5[-7 + 2(-3)]$ (h) $-20 \div [10 - (-3)(-2)]$ (i) $7[-4 - (-10) \div 2] + 2$
(j) $5 - [12 + (-2)(-4)] \div (-5)$ (k) $-6 + \{-4[2 - (\frac{-20}{10} - 2)] - (-15)\} \div 9$
(l) $4 - \{3 + [-2 + (-8 + \frac{-24}{3}) \div (-4)](-2)\}2$ (m) $4 \cdot 2^3$ (n) $3(-5^2)$ (o) $5(1 - 2)^2$

(p) $\frac{(4 - 1)^2}{6}$ (q) $4(5)^2 + 8(5) - 6$ (r) $(5 - 1)^2 + \frac{8}{2} - (3 + 5)^2$

(s) $2 - [5(6 - 3^2)^2 + 3] \div 8$ (t) $\sqrt{\frac{2(-4)}{-5(40)}}$ (u) $\frac{-13 + \sqrt{13^2 - 4(6)(6)}}{2(6)}$

(v) $\sqrt{8^2 - 15}$ (w) $-\sqrt{49} + 1^8$ (x) $\sqrt{3^2 + 4^2}$ (y) $\sqrt{5^2 - 3^2}$ (z) $-2(3^2 - \sqrt{9})^2$

Answers to Supplementary Exercises

(16-17) (a) -6 (b) 8 (c) -9 (d) 0 (e) -1 (f) 1 (g) $-\frac{3}{4}$ (h) $\frac{2}{3}$ (i) $-\frac{1}{8}$
(j) -1.25 (k) 2.5 (l) -0.75 (m) $-0.\overline{6}$ (n) $0.\overline{3}$ (o) $-\sqrt{2}$ (p) $\sqrt{5}$ (q) π
(r) $-\pi$ (s) $-2\frac{1}{2}$ (t) $3\frac{3}{4}$ (u) $-0.1\overline{6}$ (v) $0.8\overline{3}$ (w) -2^3 (x) 3^2 (y) $-(-4)^2$
(z) $(-5)^3$

(16-18) (a) $5 < 8$ (b) $-3 < 4$ (c) $5 > -9$ (d) $-6 > -8$ (e) $0 < 7$ (f) $0 > -5$
(g) $\frac{1}{2} > \frac{1}{3}$ (h) $-\frac{3}{4} < \frac{1}{2}$ (i) $-\frac{15}{20} < -\frac{2}{3}$ (j) $\frac{-0}{-5} = 0$ (k) $1.5 < 1.6$ (l) $0.\overline{3} > -\frac{1}{3}$
(m) $-0.75 = -\frac{3}{4}$ (n) $\frac{15}{16} < \frac{31}{32}$ (o) $0.023 < 0.032$ (p) $-0.12 > -0.21$
(q) $-1.0101 < 1.0101$ (r) $\frac{7}{16} = 0.4375$ (s) $1\frac{1}{2} = \frac{3}{2}$ (t) $2.5 > -2\frac{1}{2}$ (u) $-1.75 < -1\frac{1}{4}$
(v) $\frac{300}{7} = 42\frac{6}{7}$ (w) $-\frac{18}{40} < -\frac{12}{32}$ (x) $-0.0001 > -0.000101$ (y) $2.125 = \frac{17}{8}$
(z) $\pi > 3.14$

(16-19) (a) 2 (b) 9 (c) 7 (d) 0 (e) 1 (f) 1 (g) $\frac{1}{2}$ (h) $\frac{2}{3}$ (i) $\frac{1}{8}$ (j) 1.25
(k) 2.5 (l) 0.75 (m) $0.\overline{6}$ (n) $0.\overline{3}$ (o) $\sqrt{2}$ (p) $\sqrt{5}$ (q) π (r) π (s) $2\frac{1}{2}$
(t) $3\frac{3}{4}$ (u) $0.1\overline{6}$ (v) $0.8\overline{3}$ (w) 2^3 (x) 3^2 (y) 4^2 (z) 5^3

(16-20)   **(a)** 6   **(b)** 0   **(c)** 11   **(d)** -1 **(e)** -17 **(f)** -14 **(g)** 1   **(h)** -7 **(i)** 10   **(j)** -2 **(k)** 12   **(l)** -8 **(m)** -2 **(n)** -6 **(o)** 32   **(p)** 6   **(q)** -42 **(r)** -7 **(s)** -24 **(t)** 3   **(u)** 49   **(v)** 5   **(w)** 30   **(x)** -5 **(y)** 0   **(z)** not defined

(16-21)   **(a)** $\frac{3}{5}$ **(b)** $-\frac{2}{5}$ **(c)** $-\frac{1}{5}$ **(d)** $\frac{5}{6}$ **(e)** $\frac{1}{6}$ **(f)** $\frac{7}{10}$ **(g)** $-\frac{4}{5}$ **(h)** $-\frac{1}{3}$ **(i)** 1   **(j)** -1 **(k)** -3 **(l)** 9   **(m)** $\frac{5}{3}$ or $1\frac{2}{3}$ **(n)** $\frac{7}{2}$ or $3\frac{1}{2}$ **(o)** -2 **(p)** $-\frac{7}{4}$ **(q)** 8   **(r)** $-\frac{25}{8}$ or $-3\frac{1}{8}$ **(s)** $-\frac{3}{5}$ **(t)** $\frac{4}{9}$ **(u)** $-\frac{2}{3}$ **(v)** $\frac{1}{3}$ **(w)** -20 **(x)** 90   **(y)** 0   **(z)** not defined

(16-22)   **(a)** 1.4 or $1\frac{2}{5}$ **(b)** 0.3 or $\frac{3}{10}$ **(c)** 1.8 or $1\frac{4}{5}$ **(d)** -5.6 or $-5\frac{3}{5}$ **(e)** $\frac{15}{16}$ or 0.9375   **(f)** 0   **(g)** -1 **(h)** -12.375 or $-12\frac{3}{8}$ **(i)** -0.4 or $-\frac{2}{5}$ **(j)** -2.3 or $-2\frac{3}{10}$ **(k)** -0.125 or $-\frac{1}{8}$ **(l)** 2.0625 or $2\frac{1}{16}$ **(m)** 0.1 or $\frac{1}{10}$ **(n)** $\frac{3}{8}$ or 0.375   **(o)** 1.44 or $1\frac{11}{25}$ **(p)** $\frac{1}{12}$ or $0.08\overline{3}$ **(q)** -0.56 or $-\frac{14}{25}$ **(r)** $-\frac{1}{4}$ or -0.25 **(s)** 0.2916 or $\frac{7}{24}$ **(t)** $\frac{3}{4}$ or 0.75   **(u)** 0.3 or $\frac{3}{10}$ **(v)** $\frac{1}{3}$ or $0.\overline{3}$ **(w)** -2.5 or $-2\frac{1}{2}$ **(x)** $\frac{5}{36}$ or $0.13\overline{8}$ **(y)** $4\frac{2}{3}$ or $4.\overline{6}$ **(z)** $-3\frac{1}{2}$ or -3.5

(16-23)   **(a)** -72 **(b)** $\frac{11}{3}$ or $3\frac{2}{3}$ **(c)** 6   **(d)** 21   **(e)** 6   **(f)** $-\frac{10}{3}$ or $-3\frac{1}{3}$ **(g)** 65   **(h)** -5 **(i)** 9   **(j)** 9   **(k)** -7 **(l)** -10 **(m)** 32   **(n)** -75 **(o)** 5   **(p)** $\frac{3}{2}$ or $1\frac{1}{2}$ **(q)** 134   **(r)** -44 **(s)** -4 **(t)** $\frac{1}{5}$ **(u)** $-\frac{2}{3}$ **(v)** 7   **(w)** -6 **(x)** 5   **(y)** 4   **(z)** -72

17 PROPERTIES AND EXPRESSIONS

THIS CHAPTER IS ABOUT

☑ **Identifying Properties**
☑ **Using the Distributive Properties**
☑ **Clearing Parentheses**
☑ **Combining Like Terms**
☑ **Evaluating Expressions**
☑ **Evaluating Formulas**

17-1. Identifying Properties

A. Write indicated products and quotients that involve variables.

It is often necessary in algebra to write **indicated products** of real numbers and/or **variables** (letters).

Recall: There are five correct ways to write the indicated product of two real numbers. For example, the real number expression "2 times 5" can be written as

$$2 \times 5 \qquad 2 \cdot 5 \qquad 2(5) \qquad (2)5 \qquad (2)(5)$$

There are also five correct ways to write the indicated product of variables and real numbers, or the indicated product of variables and variables.

EXAMPLE 17-1: Write five correct ways to indicate the products of the following expressions:
(a) 2 times x **(b)** -3 times y **(c)** m times n

Solution
(a) 2 times x $\quad = 2 \cdot x \quad = 2(x) \quad = (2)x \quad = (2)(x) \quad = 2x$
(b) -3 times $y = -3 \cdot y = -3(y) = (-3)y = (-3)(y) = -3y \longrightarrow$ preferred form
(c) m times $n \quad = m \cdot n \quad = m(n) \quad = (m)n \quad = (m)(n) \quad = mn$

Note: When a variable is used as a factor, the traditional "times" symbol $\times$ is never used because it looks too much like the letter x.

It is often necessary in algebra to write **indicated quotients** of real numbers and/or variables.

Recall: There are six correct ways to write the indicated quotient of two real numbers. For example, the real number expression "8 divided by -3" can be written as

$$8 \div (-3) \qquad 8/(-3) \qquad 8 \cdot \frac{1}{-3} \qquad \frac{1}{-3} \cdot 8 \qquad \frac{8}{-3}$$

There are five correct ways to write the indicated quotient of variables and real numbers, or the indicated quotient of variables and variables.

EXAMPLE 17-2: Write five correct ways to indicate the quotients of the following expressions:
(a) x divided by -3 **(b)** 8 divided by y **(c)** h divided by k

Solution

(a) x divided by $-3 = x \div (-3) = x/(-3) = x \cdot \dfrac{1}{-3} = \dfrac{1}{-3} \cdot x = \dfrac{x}{-3}$

(b) 8 divided by y $= 8 \div y$ $= 8/y$ $= 8 \cdot \dfrac{1}{y}$ $= \dfrac{1}{y} \cdot 8$ $= \dfrac{8}{y}$ ⟵ preferred form

(c) h divided by k $= h \div k$ $= h/k$ $= h \cdot \dfrac{1}{k}$ $= \dfrac{1}{k} \cdot h$ $= \dfrac{h}{k}$

Note: When variables are used in algebra, the traditional division symbol $\div$ and the slash division symbol $/$, although acceptable, are seldom used to write an indicated quotient. The preferred form of division notation is $\dfrac{a}{b}$.

EXAMPLE 17-3: Write the following expressions using the preferred form of division notation:
(a) x divided by $(x + 3)$ **(b)** $(y - 2)$ divided by 6 **(c)** $(m + n)$ divided by $(m - n)$

Solution

(a) x divided by $(x + 3)$ $= \dfrac{x}{x + 3}$

(b) $(y - 2)$ divided by 6 $= \dfrac{y - 2}{6}$ ⟵ preferred division form

(c) $(m + n)$ divided by $(m - n) = \dfrac{m + n}{m - n}$

Note: If you choose to use a form other than the preferred form of division notation, you must use parentheses around the divisor to avoid any errors in computation, as in the expressions $x \div (x + 3)$ or $x/(x + 3)$.

Caution: Never write an expression like x divided by $(x + 3)$ as $x \div x + 3$ or $x/x + 3$. Omitting the parentheses will cause you to get a wrong answer when you solve the expression.

B. Identify the commutative properties.

The set of real numbers follows certain rules called **properties.**

When you change the **order** of real number addends, the sum of the addends does not change.

EXAMPLE 17-4: Show that the sum does not change when the order of the addends is changed in $5 + 2$.

Solution: different order
$5 + 2 = 7$ same
$2 + 5 = 7$ sum

Example 17-4 illustrates the following property:

Commutative Property of Addition
If a and b are real numbers, then $a + b = b + a$.

When you change the order of real number factors, the product of the factors does not change.

EXAMPLE 17-5: Show that the product does not change when the order of the factors is changed in $2 \cdot 3$.

Solution: different order
$2 \cdot 3 = 6$ same
$3 \cdot 2 = 6$ product

Example 17-5 illustrates the following property:

Commutative Property of Multiplication
If *a* and *b* are real numbers, then $ab = ba$.

The Commutative Property of Addition and the Commutative Property of Multiplication are jointly called the **commutative properties.**

Caution: Subtraction of real numbers is not commutative.

EXAMPLE 17-6: Show that the difference changes when the order of the minuend and subtrahend are changed in $5 - 2$.

Solution: different order
$$5 - 2 = 3 \qquad \text{different}$$
$$2 - 5 = -3 \qquad \text{differences}$$

Caution: Division of real numbers is not commutative.

EXAMPLE 17-7: Show that the quotient changes when the order of the dividend and divisor are changed in $2 \div 1$.

Solution: different order
$$2 \div 1 = 2 \qquad \text{different}$$
$$1 \div 2 = \tfrac{1}{2} \qquad \text{quotients}$$

C. Identify the associative properties.

When you change the **grouping** of real number addends, the sum of the addends does not change.

EXAMPLE 17-8: Show that the sum does not change when the grouping of the addends is changed in $(2 + 3) + 4$.

Solution: different groupings
$$(2 + 3) + 4 = 5 + 4 = 9 \qquad \text{same}$$
$$2 + (3 + 4) = 2 + 7 = 9 \qquad \text{sum}$$

Note: Use the Order of Operations first to clear grouping symbols.

Example 17-8 illustrates the following property:

Associative Property of Addition
If *a*, *b*, and *c* are real numbers, then $(a + b) + c = a + (b + c)$.

When you change the grouping of real number factors, the product of the factors does not change.

EXAMPLE 17-9: Show that the product does not change when the grouping of the factors is changed in $(2 \cdot 3)4$.

Solution: different groupings
$$(2 \cdot 3)4 = 6 \cdot 4 = 24 \qquad \text{same}$$
$$2(3 \cdot 4) = 2 \cdot 12 = 24 \qquad \text{product}$$

Note: Use the Order of Operations first to clear grouping symbols.

Example 17-9 illustrates the following property:

Associative Property of Multiplication
If *a*, *b*, and *c* are real numbers, then $(ab)c = a(bc)$.

The Associative Property of Addition and the Associative Property of Multiplication are jointly called the **associative properties.**

Caution: Subtraction of real numbers is not associative.

EXAMPLE 17-10: Show that the difference changes when the grouping is changed in $(9 - 5) - 3$.

Solution: different $\longrightarrow (9 - 5) - 3 = 4 - 3 = 1 \longleftarrow$ different

groupings $\longrightarrow 9 - (5 - 3) = 9 - 2 = 7 \longleftarrow$ answers

Note: Use the Order of Operations first to clear grouping symbols.

Caution: Division of real numbers is not associative.

EXAMPLE 17-11: Show that the quotient changes when the grouping is changed in $(24 \div 6) \div 2$.

Solution: different $\longrightarrow (24 \div 6) \div 2 = 4 \div 2 = 2 \longleftarrow$ different

groupings $\longrightarrow 24 \div (6 \div 2) = 24 \div 3 = 8 \longleftarrow$ answers

D. Identify the distributive properties.

To multiply a sum by a real number, you can first add and then multiply, or you can first multiply and then add.

EXAMPLE 17-12: Compute $2(4 + 3)$ by (**a**) first adding and then multiplying, (**b**) first multiplying and then adding.

Solution

(**a**) First add, then multiply: $2(4 + 3) = 2 \cdot 7 = 14 \longleftarrow$ ┐ same

(**b**) First multiply, then add: $2(4 + 3) = 2 \cdot 4 + 2 \cdot 3 = 8 + 6 = 14 \longleftarrow$ ┘ answer

Example 17-12, part (**b**) illustrates the following property:

Distributive Property of Multiplication Over Addition

If a, b, and c are real numbers, then $a(b + c) = ab + ac$

and $(b + c)a = ba + ca$.

To multiply a difference by a real number, you can first subtract and then multiply, or you can first multiply and then subtract.

EXAMPLE 17-13: Compute $4(5 - 2)$ by (**a**) first subtracting and then multiplying, (**b**) first multiplying and then subtracting.

Solution: (**a**) First subtract, then multiply: $4(5 - 2) = 4 \cdot 3 = 12 \longleftarrow$ ┐ same

(**b**) First multiply, then subtract: $4(5 - 2) = 4 \cdot 5 - 4 \cdot 2 = 20 - 8 = 12 \longleftarrow$ ┘ answer

Example 17-13, part **b** illustrates the following property:

Distributive Property of Multiplication Over Subtraction

If a, b, and c are real numbers, then $a(b - c) = ab - ac$

and $(b - c)a = ba - ca$.

Note: The Distributive Property of Multiplication Over Addition and the Distributive Property of Multiplication Over Subtraction are jointly called the **distributive properties.**

E. Identify properties involving zero or one.

There are several important properties of real numbers that involve zero.

EXAMPLE 17-14: List the important properties of real numbers that involve zero.

Solution: Properties Involving Zero

If a and b are real numbers, then:

$a + 0 = 0 + a = a$ **Identity Property for Addition**

Because $a + 0 = a$ for every real number a, 0 is called the *identity element for addition.*

$$a + (-a) = -a + a = 0$$

Additive Inverse Property

Because $a + (-a) = 0$ for every real number a, the symbol $-a$ is called the *opposite* of a or the *additive inverse* of a.

$$a - 0 = a + 0 = a$$
$$0 - a = 0 + (-a) = -a$$
$$a - b = 0 \text{ means } a = b$$

$$a \cdot 0 = 0 \cdot a = 0$$

Zero-Factor Property

The Zero-Factor Property states that a zero factor always gives a zero product.

$$ab = 0 \text{ means } a = 0 \text{ or } b = 0$$

Zero-Product Property

The Zero-Product Property states that the only way for the product of two real numbers to be zero is for one or both of the real numbers to be zero.

$$0 \div a = \frac{0}{a} = 0 \ (a \neq 0)$$

Zero-Dividend Property

The Zero-Dividend Property states that zero divided by any nonzero real number equals zero.

$$a \div 0 \text{ and } \frac{a}{0} \text{ are not defined.}$$

Zero-Divisor Property

The Zero-Divisor Property states that dividing by zero is not defined.

There are several important properties of real numbers that involve the number one.

EXAMPLE 17-15: List the important properties of real numbers that involve one.

Solution:
If a and b are real numbers, then:

Properties Involving One

$$a \cdot 1 = 1 \cdot a = a$$

Identity Property for Multiplication

Because $a \cdot 1 = a$ for every real number a, one is called the *identity element for multiplication.*

$$a \cdot \frac{1}{a} = \frac{1}{a} \cdot a = 1 \ (a \neq 0)$$

Multiplicative Inverse Property

Because $a \cdot \dfrac{1}{a} = 1$ for every nonzero real number a, $\dfrac{1}{a} (a \neq 0)$ is called the reciprocal of a or the *multiplicative inverse* of a.

$$a \div 1 = \frac{a}{1} = a$$

$$a \div b = \frac{a}{b} = 1 \text{ means } a = b$$
$$(a \neq 0 \text{ and } b \neq 0)$$

Unit-Fraction Property

The Unit-Fraction Property states that the only way for the quotient of two real numbers to equal one is for the two real numbers to be the same.

F. Identify properties involving negative signs.

There are several important properties of real numbers that involve negative signs.

EXAMPLE 17-16: List the important properties of real numbers that involve negative signs.

Solution: **Properties Involving Negative Signs**
If a and b are real numbers, then:

$$-a = -1 \cdot a = a(-1)$$

Read $-a$ as "the opposite of a."

$$-(-a) = a$$

Read $-(-a)$ as "the opposite of the opposite of a."

$$ab = -a(-b)$$

$$-(ab) = -a(b) = a(-b) = -(-a)(-b)$$

G. Identify the Subtraction and Division Properties.

There are two important properties of real numbers that involve subtraction and division. You have already been introduced to each of these properties in previous sections of this book.

The subtraction property of real numbers states that to subtract two real numbers, you change subtraction to addition while writing the opposite of the subtrahend.

EXAMPLE 17-17: State the subtraction property algebraically.

Solution: **Subtraction Property**
If a and b are real numbers, then $a - b = a + (-b)$.

The **division property** of real numbers states that to divide fractions, you change division to multiplication while writing the reciprocal of the divisor.

EXAMPLE 17-18: State the division property algebraically.

Solution: **Division Property**
If a and b are real numbers, then $\dfrac{a}{b} = a \cdot \dfrac{1}{b}$ or $\dfrac{1}{b} \cdot a$.

17-2. Using the Distributive Properties

A. Compute using the distributive properties.

Caution: To compute with real numbers using the distributive properties, you must multiply each number inside the parentheses separately by the number outside the parentheses.

EXAMPLE 17-19: Evaluate the following expressions using the distributive properties: **(a)** $(3 + 1)5$ **(b)** $3(7 - 2)$

Solution

(a) Correct Method

$$(3 + 1)5 = 3 \cdot 5 + 1 \cdot 5$$
$$= 15 + 5$$
$$= 20 \longleftarrow \text{correct answer}$$

Wrong Method

$$(3 + 1)5 = 3 + 1 \cdot 5 \quad \text{No! (Multiply both 3 and 1 by 5.)}$$
$$= 3 + 5$$
$$= 8 \longleftarrow \text{wrong answer}$$

(b) Correct Method

$$3(7 - 2) = 3 \cdot 7 - 3 \cdot 2$$
$$= 21 - 6$$
$$= 15 \longleftarrow \text{correct answer}$$

Wrong Method

$$3(7 - 2) = 3 \cdot 7 - 2 \quad \text{No! (Multiply both 7 and 2 by 3.)}$$
$$= 21 - 2$$
$$= 19 \longleftarrow \text{wrong answer}$$

B. Compute without using the distributive properties.

Sometimes it is easier to compute with real numbers without using the distributive properties. To compute without using the distributive properties, you use the Order of Operations.

EXAMPLE 17-20: Evaluate the following without using the distributive properties: **(a)** $(3 + 1)5$ **(b)** $3(7 - 2)$

Solution

(a) $(3 + 1)5 = 4 \cdot 5$ Use the Order of Operations by first adding inside the parentheses.

$\qquad\qquad = 20 \longleftarrow$ same answer as found in Example 17-19
$\qquad\qquad\qquad\quad$ using a distributive property

(b) $3(7 - 2) = 3 \cdot 5$ Use the Order of Operations by first subtracting inside the parentheses.

$= 15 \longleftarrow$ same answer as found in Example 17-19
using a distributive property

17-3. Clearing Parentheses

A. Identify algebraic expressions.

A number, variable, or the sum, difference, product, quotient, or square root of numbers and variables is called an **algebraic expression.**

EXAMPLE 17-21: List several different algebraic expressions.

Solution: The following are all algebraic expressions:

(a) 3   **(b)** x **(c)** $2y$ **(d)** $\dfrac{abc}{5}$ **(e)** $\dfrac{u}{v}$ **(f)** $2w + 3$ **(g)** $6z - \sqrt{z}$

(h) $m^2 - 2mn + n^2$

B. Identify terms.

In an algebraic expression that does not contain grouping symbols, the addition symbols separate the **terms.**

EXAMPLE 17-22: Identify the terms of each algebraic expression listed in Example 17-21.

Solution
(a) The number 3 is called a **constant term.**
(b) The variable x is called a **letter term.**
(c) Terms with only products and/or quotients of numbers and variables like $2y$, $\dfrac{abc}{5}$, and $\dfrac{u}{v}$ are called **general terms.**
(d) $2w + 3$ has two terms ($2w$ and 3) which are separated by an addition sign
(e) $6z - \sqrt{z}$ has two terms ($6z$ and $-\sqrt{z}$) because $6z - \sqrt{z} = 6z + (-\sqrt{z})$.
(f) $m^2 - 2mn + n^2$ has three terms—m^2, $-2mn$, and n^2.

Caution: To identify the terms of algebraic expressions like $-2(x + 3)$ or $\dfrac{w - 2}{5}$, you must first **clear grouping symbols.** Often this can be done by using the distributive properties.

C. Clear parentheses using the distributive properties.

Recall that to **clear parentheses** by evaluating an indicated product like $(3 + 1)5$ or $3(7 - 2)$, it is almost always easier to compute without using the distributive properties [see Example 17-20].

However, to clear parentheses in an algebraic expression like $-2(x + 3)$ or $(m - n)3$, you must use the distributive properties because it is otherwise impossible to add different terms like x and 3 or to subtract different terms like m and n.

EXAMPLE 17-23: Clear parentheses and simplify the following expressions: **(a)** $-2(x + 3)$
(b) $(m - n)3$ **(c)** $5(2 + y)$ **(d)** $(-3r - 2s)(-4)$

Solution
(a) $-2(x + 3) = -2 \cdot x + (-2)3$ **(b)** $(m - n)3 = m \cdot 3 - n \cdot 3$

$= -2x + (-6)$ $= 3 \cdot m - 3 \cdot n$

$= -2x - 6$ $= 3m - 3n$

(c) $5(2 + y) = 5 \cdot 2 + 5 \cdot y$

$\qquad\qquad = 10 + 5y$

$\qquad\qquad$ or $5y + 10$

(d) $(-3r - 2s)(-4) = -3r(-4) - 2s(-4)$

$\qquad\qquad\qquad\quad = -3(-4)r - 2(-4)s$

$\qquad\qquad\qquad\quad = +12r - (-8s)$

$\qquad\qquad\qquad\quad = 12r + (+8s)$

$\qquad\qquad\qquad\quad = 12r + 8s$

Note: The terms of **(a)** $-2(x + 3)$ are $-2x$ and -6 because $-2(x + 3) = -2x - 6$.

$\qquad\qquad\quad$ **(b)** $(m - n)3$ are $3m$ and $-3n$ because $(m - n)3 = 3m - 3n$.

$\qquad\qquad\quad$ **(c)** $5(2 + y)$ are $5y$ and 10 because $5(2 + y) = 5y + 10$.

$\qquad\qquad\quad$ **(d)** $(-3r - 2s)(-4)$ are $12r$ and $8s$ because $(-3r - 2s)(-4) = 12r + 8s$.

When parentheses have a positive sign (or no sign) in front of them, you can clear parentheses by just writing the same algebraic expression that is inside of the parentheses.

EXAMPLE 17-24: Clear parentheses in $+(3a + 2)$.

same

Solution: $+(3a + 2) = 3a + 2$ because $+(3a + 2) = +1(3a + 2)$

$\qquad\qquad\qquad\qquad\qquad\qquad\qquad\qquad\quad = (+1)3a + (+1)2$

$\qquad\qquad\qquad\qquad\qquad\qquad\qquad\qquad\quad = 3a + 2$

Note: The terms of $+(3a + 2)$ are $3a$ and 2 because $+(3a + 2) = 3a + 2$.

When parentheses have a negative sign in front of them, you can clear parentheses by just writing the opposite of each term inside the parentheses.

EXAMPLE 17-25: Clear parentheses in $-(7b - 3)$.

opposite

Solution: $-(7b - 3) = -7b + 3$ because $-(7b - 3) = -1(7b - 3)$

opposite

$\qquad\qquad\qquad\qquad\qquad\qquad\qquad\qquad\quad = (-1)7b - (-1)3$

$\qquad\qquad\qquad\qquad\qquad\qquad\qquad\qquad\quad = -7b - (-3)$

$\qquad\qquad\qquad\qquad\qquad\qquad\qquad\qquad\quad = -7b + (+3)$

$\qquad\qquad\qquad\qquad\qquad\qquad\qquad\qquad\quad = -7b + 3$

Note: The terms of $-(7b - 3)$ are $-7b$ and 3 because $-(7b - 3) = -7b + 3$.

17-4. Combining Like Terms

A. Identify the numerical coefficient and the literal part of a term.

In a term, the number that multiplies the variable(s) is called the **numerical coefficient**.

EXAMPLE 17-26: Identify the numerical coefficient in each of the following: **(a)** $2y$ **(b)** x **(c)** $-w^2$ **(d)** $\dfrac{abc}{5}$

Solution

(a) In $2y$, the numerical coefficient is 2.

(b) In x, the numerical coefficient is 1 because $x = 1x$.

(c) In $-w^2$, the numerical coefficient is -1 because $-w^2 = -1w^2$.

(d) In $\dfrac{abc}{5}$, the numerical coefficient is $\dfrac{1}{5}$ because $\dfrac{abc}{5} = \dfrac{1}{5}abc$.

The part of a term that is not the numerical coefficient is called the **literal part** of the term.

EXAMPLE 17-27: Identify the literal part in each of the following:

(a) $2y$ **(b)** x **(c)** $-w^2$ **(d)** $\dfrac{abc}{5}$

Solution

(a) In $2y$, the literal part is the variable y.
(b) In x, the literal part is the variable x.
(c) In $-w^2$, the literal part is the variable w^2.
(d) In $\dfrac{abc}{5}$, the literal part is the product of variables abc.

B. Identify like terms.

Terms that have the same literal part are called **like terms.** Terms that have different literal parts are called **unlike terms.**

EXAMPLE 17-28: Identify the like terms in each of the following: **(a)** $2x + 5 - x$

(b) $y - 5y + 3x$ **(c)** $\dfrac{mn}{2} - 3 + 5mn$ **(d)** $5w + \dfrac{3}{w}$ **(e)** $-5y + 3x$ **(f)** $mn + 2m$

(g) $2x^2 + 3x$

Solution

(a) In $2x + 5 - x$, the like terms are $2x$ and $-x$ because they have the same literal part x.
(b) In $y - 5y + 3x$, the like terms are y and $-5y$ because they have the same literal part y.
(c) In $\dfrac{mn}{2} - 3 + 5mn$, the like terms are $\dfrac{mn}{2}$ and $5mn$ because they have the same literal part mn.
(d) In $5w + \dfrac{3}{w}$, there are no like terms because the literal parts of $5w$ and $\dfrac{3}{w}$ are different (w and $\dfrac{1}{w}$, respectively).
(e) In $-5y + 3x$, there are no like terms because the literal parts of $-5y$ and $3x$ are different (y and x, respectively).
(f) In $mn + 2m$, there are no like terms because the literal parts of mn and $2m$ are different (mn and m, respectively).
(g) In $2x^2 + 3x$, there are no like terms because the literal parts of $2x^2$ and $3x$ are different (x^2 and x, respectively).

C. Combine like terms.

To **combine like terms** in an algebraic expression, you use the distributive properties to add or subtract the numerical coefficients of like terms and then write the same like literal part on the sum or difference.

EXAMPLE 17-29: Combine like terms in the following expressions: **(a)** $2w + 3w$ **(b)** $8ab - 5ab$

(c) $2x + 5 - x$ **(d)** $y - 5y + 3x$ **(e)** $\dfrac{mn}{2} - 3 + 5mn$ **(f)** $8u - 5v - 6u + v$

Solution

(a) $2w + 3w = (2 + 3)w$ Combine like terms using the distributive property of addition:
$= 5w$ $ac + bc = (a + b)c$

(b) $8ab - 5ab = (8 - 5)ab$ Combine the like terms using the distributive property of subtraction:
$= 3ab$ $ac - bc = (a - b)c$

like terms

(c) $2x + 5 - x = \overbrace{2x - x} + 5$ Group like terms.

$ = 2x - 1x + 5$ *Think:* $-x = -1x$

$ = (2 - 1)x + 5$ Combine like terms.

$ = 1x + 5$

$ = x + 5$ Simplify.

(d) $y - 5y + 3x = 1y - 5y + 3x$ *Think:* $y = 1y$

$ = (1 - 5)y + 3x$ Combine like terms.

$ = -4y + 3x$

$ = 3x - 4y$ Write terms in alphabetical order.

like terms

(e) $\dfrac{mn}{2} - 3 + 5mn = \overbrace{\dfrac{mn}{2} + 5mn} - 3$ Group like terms.

$\phantom{\dfrac{mn}{2} - 3 + 5mn} = \dfrac{1}{2}mn + 5mn - 3$ *Think:* $\dfrac{mn}{2} = \dfrac{1}{2}mn$

$\phantom{\dfrac{mn}{2} - 3 + 5mn} = \left(\dfrac{1}{2} + 5\right)mn - 3$ Combine like terms.

$\phantom{\dfrac{mn}{2} - 3 + 5mn} = 5\dfrac{1}{2}mn - 3 \text{ or } \dfrac{11}{2}mn - 3 \text{ or } \dfrac{11mn}{2} - 3$

(f) $8u - 5v - 6u + v = (8u - 6u) + (-5v + 1v)$ Group like terms together.

$ = (8 - 6)u + (-5 + 1)v$ Combine like terms.

$ = 2u + (-4)v$

$ = 2u - 4v$ Simplify.

Note 1: In $3x - 4y$, $3x$ and $-4y$ cannot be combined because $3x$ and $-4y$ are not like terms.

Note 2: In $2u - 4v$, $2u$ and $-4v$ cannot be combined because $2u$ and $-4v$ are not like terms.

Caution: Only like terms can be combined.

D. Clear parentheses and combine like terms.

To clear parentheses and combine like terms, you first combine any like terms inside the parentheses, then clear the parentheses, and then combine any like terms that remain.

EXAMPLE 17-30: Clear parentheses and combine like terms in the following expressions:
(a) $2(x + 5) - 5x$ **(b)** $-(2y - 3) + (3y - 5)$ **(c)** $3(2a - b + 5a) - 8b$
(d) $6m - 2(m + n)$

Solution
(a) $2(x + 5) - 5x = \mathbf{2} \cdot x + \mathbf{2} \cdot 5 - 5x$ Clear parentheses.

$ = 2x + 10 - 5x$ Simplify.

$ = 2x - 5x + 10$ Group like terms.

$ = -3x + 10$ Combine like terms.

$ \text{or } 10 - 3x$

(b) $-(2y - 3) + (3y - 5) = -2y + 3 + 3y - 5$ Clear parentheses.

$\qquad\qquad\qquad\qquad\quad = -2y + 3y + 3 - 5$ Group like terms.

$\qquad\qquad\qquad\qquad\quad = 1y + (-2)$ Combine like terms.

$\qquad\qquad\qquad\qquad\quad = y - 2$ Simplify.

(c) $3(2a - b + 5a) - 8b = 3(2a + 5a - b) - 8b$ Combine like terms inside the parentheses.

$\qquad\qquad\qquad\qquad\quad = 3(7a - b) - 8b$

$\qquad\qquad\qquad\qquad\qquad = 3 \cdot 7a - 3 \cdot b - 8b$ Clear parentheses.

$\qquad\qquad\qquad\qquad\qquad = 21a - 3b - 8b$ Simplify.

$\qquad\qquad\qquad\qquad\qquad = 21a - 11b$ Combine like terms.

(d) $6m - 2(m + n) = 6m + (-2)(m + n)$ Rename terms.

$\qquad\qquad\qquad\quad = 6m + (-2)m + (-2)n$ Clear parentheses.

$\qquad\qquad\qquad\quad = 6m - 2m - 2n$ Simplify.

$\qquad\qquad\qquad\quad = 4m - 2n$ Combine like terms.

Caution: To avoid making an error in Example 17-30, part **(d)**, you first rename $6m - 2(m + n)$ as $6m + (-2)(m + n)$.

EXAMPLE 17-31: Show how it is possible to make an error when $6m - 2(m + n)$ is not first renamed as $6m + (-2)(m + n)$.

Solution

$\qquad$ <u>Wrong Method</u>

$6m - 2(m + n) = 6m - 2 \cdot m + 2 \cdot n$ No! Multiply each term inside the parentheses by -2, not 2.

$\qquad\qquad\qquad = 6m - 2m + 2n$ Simplify.

$\qquad\qquad\qquad = (6 - 2)m + 2n$ Combine like terms.

$\qquad\qquad\qquad = 4m + 2n \longleftarrow$ wrong answer (The correct answer is $4m - 2n$.)

17-5. Evaluating Expressions

A. Evaluate algebraic expressions in one variable.

To **evaluate an algebraic expression in one variable** given a numerical value for that variable, you first substitute the given value for the variable in each term that the variable appears in, and then evaluate using the Order of Operations.

EXAMPLE 17-32: Evaluate $2x^2 - 5x - 6$ for **(a)** $x = 3$ **(b)** $x = -2$.

Solution

(a) $2x^2 - 5x - 6 = 2(3)^2 - 5(3) - 6$ Substitute 3 for x.

$\qquad\qquad\qquad = 2 \cdot 9 - 5(3) - 6$ Evaluate using the Order of Operations

$\qquad\qquad\qquad = 18 - 5(3) - 6$ [see Example 16-32].

$\qquad\qquad\qquad = 18 - 15 - 6$

$\qquad\qquad\qquad = 3 - 6$

$\qquad\qquad\qquad = -3$

(b) $2x^2 - 5x - 6 = 2(-2)^2 - 5(-2) - 6$ Substitute -2 for x.

$$= 2 \cdot 4 - 5(-2) - 6$$ Evaluate using the Order of Operations.

$$= 8 - 5(-2) - 6$$

$$= 8 + 10 - 6$$

$$= 18 - 6$$

$$= 12$$

Note 1: For $x = 3$: $2x^2 - 5x - 6 = -3$

Note 2: For $x = -2$: $2x^2 - 5x - 6 = 12$

Caution: To avoid making an error, always use parentheses when substituting a numerical value for a variable, as in Example 17-32.

B. Evaluate algebraic expressions in two or more variables.

To **evaluate an algebraic expression in two or more variables** given a numerical value for each different variable, you first substitute each given value for the associated variable and then evaluate using the Order of Operations.

EXAMPLE 17-33: Evaluate $\dfrac{-b + \sqrt{b^2 - 4ac}}{2a}$ for $a = 2$, $b = -1$, and $c = -3$.

Solution: $\dfrac{-b + \sqrt{b^2 - 4ac}}{2a} = \dfrac{-(-1) + \sqrt{(-1)^2 - 4(2)(-3)}}{2(2)}$ Substitute 2 for a, -1 for b, and -3 for c.

$$= \frac{1 + \sqrt{1 - 4(2)(-3)}}{2(2)}$$ Evaluate using the Order of Operations.

$$= \frac{1 + \sqrt{1 + 24}}{2(2)}$$

$$= \frac{1 + \sqrt{25}}{2(2)}$$

$$= \frac{1 + 5}{2(2)}$$

$$= \frac{6}{2(2)}$$

$$= \frac{2(3)}{2(2)}$$

$$= \frac{3}{2} \text{ or } 1\frac{1}{2} \text{ or } 1.5$$

Note: For $a = 2$, $b = -1$, and $c = -3$, $\dfrac{-b + \sqrt{b^2 - 4ac}}{2a} = \dfrac{3}{2}$ or $1\frac{1}{2}$ or 1.5.

17-6. Evaluating Formulas

A. Identify the parts of an equation.

An **equation** is a **mathematical sentence** that contains an **equality symbol.** Every equation has three parts: a **left member,** an equality symbol, and a **right member.**

EXAMPLE 17-34: Identify the three different parts of the following equations: **(a)** $m + 3 = 2$
(b) $2n = 6$　　**(c)** $4w - 3 = 1$　　**(d)** $2x + 3y = -5$　　**(e)** $C = \frac{5}{9}(F - 32)$

Solution

equality symbol

(a)　　　　　　　　　　　$m + 3 = 2$

(b)　　　　　　　　　　$2n = 6$

(c) left member　　　　$4w - 3 = 1$　　right member

(d)　　　　　　　　$2x + 3y = -5$

(e)　　　　　　　　　$C = \frac{5}{9}(F - 32)$

B. Identify the parts of a formula

A **formula** is an equation that contains two or more variables and represents a known phenomenon. A **formula is solved for a given variable** when that variable is isolated in one member of the equation. When a formula is solved for a given variable, the algebraic expression in the other member of the equation is called the **solution.**

EXAMPLE 17-35: Identify the variable that each formula is solved for and the solution in each of the following: **(a) Temperature Formula:** $C = \frac{5}{9}(F - 32)$　　**(b) Distance Formula:** $d = rt$
(c) Volume Formula: $V = lwh$

Solution
(a) The formula $C = \frac{5}{9}(F - 32)$ is solved for C and the solution is $\frac{5}{9}(F - 32)$.
(b) The formula $d = rt$ is solved for d and the solution is rt.
(c) The formula $V = lwh$ is solved for V and the solution is lwh.

C. Evaluate formulas to solve problems.

In Chapter 15, you evaluated geometry formulas like $P = 4s$, $C = \pi d$, $A = \frac{1}{2}bh$, and $V = lwh$. To **evaluate a formula that is solved for a specific variable** given a measurement value for each of the other variables, you substitute each given value for the associated variable and then evaluate using the Order of Operations.

EXAMPLE 17-36: Water boils at 212 degrees Fahrenheit F. Find the temperature in degrees Celsius C at which water boils by evaluating the temperature formula $C = \frac{5}{9}(F - 32)$.

Solution: The question asks you to evaluate $C = \frac{5}{9}(F - 32)$ for $F = 212$.

$C = \frac{5}{9}(F - 32)$　　　　Write the given formula.

　$= \frac{5}{9}(\mathbf{212} - 32)$　　　　Substitute 212 for F.

　$= \frac{5}{9} \cdot 180$　　　　Evaluate using the Order of Operations.

　$= 100$　　　　*Think:* $\frac{5}{9} \cdot 180 = \frac{5}{9} \cdot \frac{180}{1} = \frac{5}{9} \cdot \frac{9 \cdot 20}{1} = 100$

Note: $C = 100$ means the Celsius temperature at which water boils is **100°C.**

D. Evaluate a formula when one of the variables represents a rate.

To **evaluate a formula when one of the variables represents a rate,** you may first need to rename to get a common unit of measure that can be eliminated.

EXAMPLE 17-37: What distance d can a car travel in a time t of 20 minutes at a constant rate r of 45 mph (miles per hour) using the distance formula $d = rt$?

different time units

Solution: The question asks you to evaluate $d = rt$ for $r = 45$ miles/hour and $t = 20$ minutes.

Caution: Before evaluating $d = rt$, you must have a common time unit.

20 minutes $= \frac{20}{60}$ hour $= \frac{1}{3}$ hour	Rename to get a common time unit [see Example 11-15].
$d = rt$	Write the given formula.
$= 45 \dfrac{\text{miles}}{\text{hour}} \cdot \dfrac{1}{3} \text{ hour}$	Substitute 45 mph for r and $\frac{1}{3}$ hour for t.
$= 45 \dfrac{\text{miles}}{\cancel{\text{hour}}} \cdot \dfrac{1}{3} \dfrac{\cancel{\text{hour}}}{1}$	Eliminate the common time unit.
$= (45 \cdot \frac{1}{3})$ miles	Multiply a measure by a number [see Example 12-38].
$= 15$ miles	Simplify.

Note: In 20 minutes, at a constant rate of 45 mph, a car can travel 15 miles.

RAISE YOUR GRADES

Can you . . . ?

☑ write indicated products of real numbers and/or variables
☑ write indicated quotients of real numbers and/or variables
☑ identify the commutative properties
☑ identify the associative properties
☑ identify the distributive properties
☑ identify properties involving zero or one
☑ identify properties involving negative signs
☑ identify the subtraction and division properties
☑ compute using the distributive properties
☑ compute using the Order of Operations
☑ identify algebraic expressions and terms
☑ clear parentheses using the distributive properties
☑ identify numerical coefficients and literal parts in terms
☑ identify like terms and unlike terms
☑ combine like terms
☑ clear parentheses and combine like terms
☑ evaluate algebraic expressions in one or more variables
☑ identify the parts of an equation
☑ identify the parts of a formula
☑ evaluate formulas to solve problems
☑ evaluate formulas when one of the variables represents a rate

SUMMARY

1. The following are five correct ways to write the indicated product of the real number expression
   (a) 3 times w **(b)** -8 times z **(c)** r times s:

   (a) 3 times w $= 3 \cdot w$ $= 3(w)$ $= (3)w$ $= (3)(w)$ $= 3w$ ↖
   (b) -8 times $z = -8 \cdot z = -8(z) = (-8)z = (-8)(z) = -8z$ ← preferred form
   (c) r times s $= r \cdot s$ $= r(s)$ $= (r)s$ $= (r)(s)$ $= rs$ ↙

2. The following are five correct ways to write the indicated quotient of the following variables and real numbers: **(a)** y divided by 2   **(b)** -5 divided by w **(c)** r divided by s
   (d) 3 divided by $(x + 3)$

(a) y divided by 2 $\quad = y \div 2 \quad = y/2 \quad = y \cdot \dfrac{1}{2} \quad = \dfrac{1}{2} \cdot y \quad = \dfrac{y}{2}$ ← preferred form

(b) -5 divided by $w = -5 \div w = -5/w = -5 \cdot \dfrac{1}{w} = \dfrac{1}{w}(-5) = \dfrac{-5}{w}$

(c) r divided by $s \quad = r \div s \quad = r/s \quad = r \cdot \dfrac{1}{s} \quad = \dfrac{1}{s} \cdot r \quad = \dfrac{r}{s}$

(d) 3 divided by $(x + 3) = 3 \div (x + 3) = 3/(x + 3) = 3 \cdot \dfrac{1}{x + 3} = \dfrac{1}{x + 3} \cdot 3 = \dfrac{3}{x + 3}$

3. Commutative Property of Addition
If a and b are real numbers, then $a + b = b + a$.

4. Commutative Property of Multiplication
If a and b are real numbers, then $ab = ba$.

5. Real numbers are not commutative with respect to subtraction or division.

6. Associative Property of Addition
If a, b, and c are real numbers, then $(a + b) + c = a + (b + c)$.

7. Associative Property of Multiplication
If a, b, and c are real numbers, then $(ab)c = a(bc)$.

8. Real numbers are not associative with respect to subtraction or division.

9. Distributive Property of Multiplication Over Addition
If a, b, and c are real numbers, then $a(b + c) = ab + ac$
and $(b + c)a = ba + ca$.

10. Distributive Property of Multiplication Over Subtraction
If a, b, and c are real numbers, then $a(b - c) = ab - ac$
and $(b - c)a = ba - ca$.

11. The common properties involving zero or one are

Identity Property for Addition: $\qquad a + 0 = 0 + a = a$

Identity Property for Multiplication: $a \cdot 1 = 1 \cdot a = a$

Additive Inverse Property: $\qquad a + (-a) = -a + a = 0$

Multiplicative Inverse Property: $\qquad a \cdot \dfrac{1}{a} = \dfrac{1}{a} \cdot a = 1$

Zero-Factor Property: $\qquad a \cdot 0 = 0 \cdot a = 0$

Zero-Product Property: $\qquad ab = 0$ means $a = 0$ or $b = 0$

Unit-Fraction Property: $\qquad \dfrac{a}{b} = 1$ means $a = b$

Zero-Dividend Property: $\qquad 0 \div a = \dfrac{0}{a} = 0 \, (a \neq 0)$

Zero-Divisor Property: $\qquad a \div 0$ and $\dfrac{a}{0}$ are not defined.

12. Properties Involving Negative Signs
If a and b are real numbers, then
(a) $-a = 1 \cdot a = a(-1)$
(b) $-(-a) = a$
(c) $ab = -a(-b)$
(d) $-(ab) = -a \cdot b = a(-b) = -(-a)(-b)$

13. Subtraction Property
If a and b are real numbers, then $a - b = a + (-b)$.

14. Division Property
If a and b are real numbers, then $\dfrac{a}{b} = a \cdot \dfrac{1}{b}$ or $\dfrac{1}{b} \cdot a$.

15. To compute using the distributive properties, you must multiply each term inside the parentheses separately by the term outside the parentheses.

16. To compute without using the distributive properties, you use the Order of Operations.
17. A number, variable, or the sum, difference, product, quotient, or square root of numbers and variables is called an algebraic expression.
18. In an algebraic expression that does not contain grouping symbols, the addition symbols separate the terms.
19. To clear parentheses by evaluating a product like $(3 + 1)5$ or $3(7 - 2)$, it is almost always easier to compute without using the distributive properties. However, to clear parentheses in an algebraic expression like $-2(x + 3)$ or $(m - n)3$, you must use the distributive properties because it is otherwise impossible to add different terms like x and 3 or to subtract different terms like m and n.
20. When parentheses have a positive sign (or no sign) in front of them, you can clear parentheses by just writing the same algebraic expression that is inside the parentheses.
21. When parentheses have a negative sign in front of them, you can clear parentheses by writing the opposite of each term inside the parentheses.
22. In a term, the number that multiplies the variable(s) is called the numerical coefficient.
23. The part of a term that is not the numerical coefficient is called the literal part of the term.
24. Terms that have the same literal part are called like terms.
25. Terms that have different literal parts are called unlike terms.
26. To combine like terms in an algebraic expression, you use the distributive properties to add or subtract the numerical coefficients of like terms and then write the same like literal part on the sum or difference.
27. Only like terms can be combined.
28. To clear parentheses and combine like terms, you first combine any like terms inside the parentheses, then clear the parentheses, and then combine like terms.
29. To evaluate an algebraic expression in one variable given a numerical value for that variable, you first substitute the given value for the variable in each term that the variable appears in, and then evaluate using the Order of Operations.
30. To avoid making an error, always use parentheses when substituting a numerical value for a variable.
31. To evaluate an algebraic expression in two or more variables given a numerical value for each different variable, you first substitute each given value for the associated variable and then evaluate using the Order of Operations.
32. An equation is a mathematical sentence that contains an equality symbol.
33. Every equation has three parts; a left member, an equality symbol, and a right member.
34. A formula is an equation that contains two or more variables and represents a known phenomenon.
35. A formula is solved for a given variable when that variable is isolated on one side of the equation.
36. When a formula is solved for a given variable, the algebraic expression on the other side of the equation is called the solution for the given variable.
37. To evaluate a formula that is solved for a specific variable, given a measurement value for each of the other variables, you substitute each given value for the associated variable and then evaluate using the Order of Operations.
38. To evaluate a formula when one of the variables represents a rate, you may first need to rename to get a common unit of measure that can be eliminated.

SOLVED PROBLEMS

PROBLEM 17-1 Write each indicated product in five ways for the following expressions:
(a) 4 times m **(b)** -5 times n **(c)** x times y

Solution: Recall that the traditional "times" symbol $\times$ is never used with variables because it looks too much like the letter x [see Example 17-1]:

(a) 4 times $m = 4 \cdot m = 4(m) = (4)m = (4)(m) = 4m$

(b) -5 times $n = -5 \cdot n = -5(n) = (-5)n = (-5)(n) = -5n$ ⟶ preferred form

(c) x times $y = x \cdot y = x(y) = (x)y = (x)(y) = xy$

PROBLEM 17-2   Write each indicated quotient in five ways for the following expressions:
(a) m divided by 8   **(b)** -3 divided by n **(c)** x divided by y

Solution: Recall that the traditional division symbol $\div$ and slash division symbol /, although acceptable, are seldom used to write an indicated quotient. The preferred form of division notation is $\dfrac{a}{b}$ [see Example 17-2]:

(a) m divided by 8 $\quad = \quad m \div 8 = \quad m/8 = \quad m \cdot \dfrac{1}{8} = \quad \dfrac{1}{8} \cdot m = \dfrac{m}{8}$

(b) -3 divided by $n = -3 \div n = -3/n = -3 \cdot \dfrac{1}{n} = \dfrac{1}{n}(-3) = \dfrac{-3}{n}$ preferred form

(c) x divided by $y \quad = \quad x \div y = \quad x/y =. \quad x \cdot \dfrac{1}{y} = \quad \dfrac{1}{y} \cdot x = \dfrac{x}{y}$

PROBLEM 17-3   Identify each missing number and identify the property by name when possible:
(a) $3 + ? = 2 + 3$ **(b)** $?(-5) = -5 \cdot 9$ **(c)** $(? + 3) + 6 = 8 + (3 + 6)$
(d) $(4 \cdot 6)? = 4(6 \cdot 9)$ **(e)** $?(5 + 3) = 2 \cdot 5 + 2 \cdot 3$ **(f)** $(5 - 2)3 = 5 \cdot 3 - 2 \cdot ?$
(g) $? + 0 = 4$ **(h)** $5 + ? = 0$ **(i)** $6 - ? = 6$ **(j)** $0 - ? = -5$ **(k)** $5 - ? = 0$
(l) $7 \cdot 0 = ?$ **(m)** $5 \cdot ? = 0$ **(n)** $0 \div 6 = ?$ **(o)** $2 \div 0 = ?$ **(p)** $8 \cdot ? = 1$ **(q)** $3 \cdot ? = 3$
(r) $7 \div ? = 7$ **(s)** $? \div 5 = 1$ **(t)** $-? = -1 \cdot 5$ **(u)** $-(-2) = ?$ **(v)** $3 \cdot 4 = -3 \cdot ?$
(w) $-(2 \cdot 5) = ?(-5)$ **(x)** $3 - 5 = 3 + ?$ **(y)** $\frac{3}{4} = 3 \cdot ?$

Solution
(a) $3 + 2 = 2 + 3$ by the Commutative Property of Addition [see Example 17-4].
(b) $9(-5) = -5 \cdot 9$ means by the Commutative Property of Multiplication [see Example 17-5].
(c) $(8 + 3) + 6 = 8 + (3 + 6)$ by the Associative Property of Addition [see Example 17-8].
(d) $(4 \cdot 6)9 = 4(6 \cdot 9)$ by the Associative Property of Multiplication [see Example 17-9].
(e) $2(5 + 3) = 2 \cdot 5 + 2 \cdot 3$ by the Distributive Property of Multiplication Over Addition [see Example 17-12].
(f) $(5 - 2)3 = 5 \cdot 3 - 2 \cdot 3$ by the Distributive Property of Multiplication over Subtraction [see Example 17-13].
(g) $4 + 0 = 4$ by the Identity Property for Addition [see Example 17-14].
(h) $5 + (-5) = 0$ by the Additive Inverse Property [see Example 17-14].
(i) $6 - 0 = 6$ [see Example 17-14].
(j) $0 - 5 = -5$ [see Example 17-14].
(k) $5 - 5 = 0$ [see Example 17-14].
(l) $7 \cdot 0 = 0$ by the Zero-Factor Property [see Example 17-14].
(m) $5 \cdot 0 = 0$ by the Zero-Product Property [see Example 17-14].
(n) $0 \div 6 = 0$ by the Zero-Dividend Property [see Example 17-14].
(o) $2 \div 0$ is not defined by the Zero-Divisor Property [see Example 17-14].
(p) $8 \cdot \frac{1}{8} = 1$ by the Multiplicative Inverse Property [see Example 17-15].
(q) $3 \cdot 1 = 3$ by the Identity Property for Multiplication [see Example 17-15].
(r) $7 \div 1 = 7$ [see Example 17-15].
(s) $5 \div 5 = 1$ by the Unit-Fraction Property [see Example 17-15].
(t) $-5 = -1 \cdot 5$ [see Example 17-16].
(u) $-(-2) = 2$ [see Example 17-16].
(v) $3 \cdot 4 = -3(-4)$ [see Example 17-16].
(w) $-(2 \cdot 5) = 2(-5)$ [see Example 17-16].
(x) $3 - 5 = 3 + (-5)$ by the Subtraction Property [see Example 17-17].
(y) $\frac{3}{4} = 3 \cdot \frac{1}{4}$ by the Division Property [see Example 17-18].

PROBLEM 17-4   Identify each term in **(a)** $x + 3$ **(b)** $y - 2$ **(c)** mn **(d)** $\dfrac{w}{2}$

(e) $2a - 3b + 2$ **(f)** $-4h + \dfrac{3}{h} - 2\sqrt{h}$ **(g)** $2(x + 5)$ **(h)** $3(y - 4)$ **(i)** $+(a^2 - a - 1)$

(j) $-(x - y)$ **(k)** 3 **(l)** w **(m)** $\dfrac{m + 1}{4}$ **(n)** $\dfrac{a - b}{c}$

Solution: Recall that in an algebraic expression that does not contain grouping symbols, the addition symbols separate the terms. Also, to identify terms when grouping symbols are present, you must first clear the grouping symbols [see Example 17-22]:

(a) In $x + 3$, the terms are x and 3.

(b) In $y - 2$, the terms are y and -2 because $y - 2 = y + (-2)$.

(c) The product of two or more numbers and variables like mn is a term.

(d) The quotient of numbers and variables like $\dfrac{w}{2}$ is a term.

(e) In $2a - 3b + 2$, the terms are $2a$, $-3b$, and 2.

(f) In $-4h + \dfrac{3}{h} - 2\sqrt{h}$, the terms are $-4h$, $\dfrac{3}{h}$, and $-2\sqrt{h}$.

(g) In $2(x + 5)$, the terms are $2x$ and 10 because $2(x + 5) = 2 \cdot x + 2 \cdot 5 = 2x + 10$.

(h) In $3(y - 4)$, the terms are $3y$ and -12 because $3(y - 4) = 3 \cdot y - 3 \cdot 4 = 3y - 12$.

(i) In $+(a^2 - a - 1)$, the terms are a^2, $-a$, and -1 because $+(a^2 - a - 1) = a^2 - a - 1$.

(j) In $-(x - y)$, the terms are $-x$ and y because $-(x - y) = -x + y$.

(k) A number like 3 is a constant term.

(l) A variable like w is a letter term.

(m) In $\dfrac{m + 1}{4}$, the terms are $\dfrac{m}{4}$ and $\dfrac{1}{4}$ because $\dfrac{m + 1}{4} = \dfrac{m}{4} + \dfrac{1}{4}$.

(n) In $\dfrac{a - b}{c}$, the terms are $\dfrac{a}{c}$ and $-\dfrac{b}{c}$ because $\dfrac{a - b}{c} = \dfrac{a}{c} - \dfrac{b}{c}$.

PROBLEM 17-5 Clear parentheses in **(a)** $5(u + 2)$ **(b)** $(v - 3)(-4)$ **(c)** $-2(x + y)$
(d) $(2m - 5n)3$ **(e)** $+(8h - 5k)$ **(f)** $-(y^2 - 2y + 5)$.

Solution: Recall that to clear parentheses in algebraic expressions like those in Problem 17-5, you must use the distributive properties because it is otherwise impossible to combine the unlike terms inside the parentheses [see Examples 17-23, 17-24, and 17-25]:

(a) $5(u + 2) = 5 \cdot u + 5 \cdot 2 = 5u + 10$ **(b)** $(v - 3)(-4) = v(-4) - 3(-4) = -4v + 12$

(c) $-2(x + y) = (-2)x + (-2)y = -2x - 2y$ **(d)** $(2m - 5n)3 = 2m \cdot 3 - 5n \cdot 3 = 6m - 15n$

(e) $+(8h - 5k) = 8h - 5k$ **(f)** $-(y^2 - 2y + 5) = -y^2 + 2y - 5$

PROBLEM 17-6 Combine like terms in **(a)** $2x + 5x$ **(b)** $3y - y$ **(c)** $w - 5w$

(d) $6ab - 5ab$ **(e)** $3u + 2 - 4u$ **(f)** $m - 5n + m$ **(g)** $\dfrac{c}{2} + c - \dfrac{1}{2}$

(h) $6s - 5r + 2s + 3r$.

Solution: Recall that to combine like terms in an algebraic expression, you use the distributive properties to add or subtract the numerical coefficients of like terms and then write the same literal part on the sum or difference [see Example 17-29]:

(a) $2x + 5x = (2 + 5)x = 7x$ **(b)** $3y - y = 3y - 1y = (3 - 1)y = 2y$

(c) $w - 5w = 1w - 5w = (1 - 5)w = -4w$ **(d)** $6ab - 5ab = (6 - 5)ab = 1ab = ab$

(e) $3u + 2 - 4u = 3u - 4u + 2 = (3 - 4)u + 2 = -1u + 2 = -u + 2$ or $2 - u$

(f) $m - 5n + m = m + m - 5n = 1m + 1m - 5n = (1 + 1)m - 5n = 2m - 5n$

(g) $\dfrac{c}{2} + c - \dfrac{1}{2} = \dfrac{1}{2}c + 1c - \dfrac{1}{2} = \left(\dfrac{1}{2} + 1\right)c - \dfrac{1}{2} = \dfrac{3}{2}c - \dfrac{1}{2}$ or $\dfrac{3c - 1}{2}$

(h) $6s - 5r + 2s + 3r = 6s + 2s + (-5r) + 3r = (6 + 2)s + (-5 + 3)r = 8s + (-2)r = 8s - 2r$

PROBLEM 17-7 Clear parentheses and combine like terms in **(a)** $3(x + 2) + 2x$
(b) $-2(3y - 4) - 5$ **(c)** $-(m - n) + (m + n)$ **(d)** $2(3a - b + 5a) - 8b$ **(e)** $2u - 3(u - v)$.

Solution: Recall that to clear parentheses and combine like terms, you first combine any like terms inside the parentheses, then clear the parentheses, and then combine like terms [see Examples 17-30 and 17-31]:

(a) $3(x + 2) + 2x = 3 \cdot x + 3 \cdot 2 + 2x = 3x + 6 + 2x = 3x + 2x + 6 = 5x + 6$
(b) $-2(3y - 4) - 5 = (-2)3y - (-2)4 - 5 = -6y + 8 - 5 = -6y + 3$
(c) $-(m - n) + (m + n) = -m + n + m + n = -m + m + n + n = 0 + 2n = 2n$
(d) $2(3a - b + 5a) - 8b = 2(8a - b) - 8b = 2 \cdot 8a - 2 \cdot b - 8b = 16a - 2b - 8b = 16a - 10b$
(e) $2u - 3(u - v) = 2u + (-3)(u - v) = 2u + (-3)u - (-3)v = -u + 3v$ or $3v - u$

PROBLEM 17-8 Evaluate the following algebraic expressions: (a) $5x$ for $x = 2$
(b) $-2y$ for $y = -4$ (c) $w + 3$ for $w = -6$ (d) $u - 2$ for $u = 1$ (e) $2v + 3$ for $v = 8$
(f) $3a - 5$ for $a = -2$ (g) $x^2 + 5x - 6$ for $x = 2$ (h) $x^2 + 5x - 6$ for $x = -2$

(i) $\dfrac{-b - \sqrt{b^2 - 4ac}}{2a}$ for $a = 6$, $b = 1$, and $c = -12$

Solution: Recall that to evaluate an algebraic expression given a numerical value for each different variable, you first substitute each given value for the associated variable and then evaluate using the Order of Operations [see Examples 17-32 and 17-33]:

(a) $5x = 5(2) = 10$ (b) $-2y = -2(-4) = 8$ (c) $w + 3 = (-6) + 3 = -3$
(d) $u - 2 = (1) - 2 = -1$ (e) $2v + 3 = 2(8) + 3 = 16 + 3 = 19$
(f) $3a - 5 = 3(-2) - 5 = -6 - 5 = -11$
(g) $x^2 + 5x - 6 = (2)^2 + 5(2) - 6 = 4 + 10 - 6 = 8$
(h) $x^2 + 5x - 6 = (-2)^2 + 5(-2) - 6 = 4 - 10 - 6 = -12$

(i) $\dfrac{-b - \sqrt{b^2 - 4ac}}{2a} = \dfrac{-(1) - \sqrt{(1)^2 - 4(6)(-12)}}{2(6)} = \dfrac{-1 - \sqrt{289}}{12} = \dfrac{-1 - 17}{12} = \dfrac{-18}{12} = -\dfrac{3}{2}$

PROBLEM 17-9 Solve each problem by evaluating the given formula:

(a) Water freezes at 0 degree Celsius C. Find the temperature in degrees Fahrenheit F at which water freezes by evaluating the **temperature formula** $F = \frac{9}{5}C + 32$.

(b) Find the ideal weight w for a man whose height h is 72 inches using the **male ideal weight formula** $w = 5\frac{1}{2}h - 231$.

(c) Find the ideal weight w for a woman whose height h is 60 inches using the **female ideal weight formula** $w = 5\frac{1}{4}h - 216$.

(d) What distance d can a car travel in a time t of 40 minutes at a constant rate r of 60 mph using the **distance formula** $d = rt$.

(e) What is the altitude a in kilometers when the average annual temperature t is 0 degrees Celsius using the **altitude/temperature formula** $a = 0.16(15 - t)$

(f) What is the amount A at the end of a 9 month time period t if \$500 principal P is invested at a simple interest rate r of 6% per year using the **amount formula** $A = P(1 + rt)$.

Solution: Recall that to evaluate a formula that is solved for a specific variable given a measurement value for each of the other variables, you substitute each given value for the associated variable and then evaluate using the Order of Operations. To evaluate a formula when one of the variables represents a rate, you may first need to rename to get a common unit of measure that can be eliminated [see Examples 17-36 and 17-37]:

(a) The question asks you to evaluate $F = \frac{9}{5}C + 32$ for $C = 0$:

$$F = \frac{9}{5}C + 32 = \frac{9}{5}(0) + 32 = 0 + 32 = 32 \text{ degrees Fahrenheit}$$

(b) The question asks you to evaluate $w = 5\frac{1}{2}h - 231$ for $h = 72$:

$$w = 5\frac{1}{2}h - 231 = 5\frac{1}{2}(72) - 231 = 396 - 231 = 165 \text{ pounds}$$

(c) The question asks you to evaluate $w = 5\frac{1}{4}h - 216$ for $h = 60$:

$$w = 5\frac{1}{4}h - 216 = 5\frac{1}{4}(60) - 216 = 315 - 216 = 99 \text{ pounds}$$

(d) The question asks you to evaluate $d = rt$ for $r = 60\dfrac{\text{miles}}{\text{hour}}$ and $t = 40$ minutes:

40 minutes $= \frac{40}{60}$ hour $= \frac{2}{3}$ hour

$$d = rt = 60\,\frac{\text{miles}}{\cancel{\text{hour}}} \cdot \frac{2\,\cancel{\text{hour}}}{3}\cdot\frac{1}{1} = \left(60\cdot\frac{2}{3}\right)\text{miles} = 40\text{ miles}$$

(e) The question asks you to evaluate $a = 0.16(15 - t)$ for $t = 0$:

$a = 0.16(15 - t) = 0.16(15 - 0) = 0.16(15) = 2.4$ kilometers

(f) The question asks you to evaluate $A = P(1 + rt)$ for $P = \$500$, $r = 6\%$, and $t = 9$ months:

9 months $= \dfrac{9}{12}$ year $= \dfrac{3}{4}$ year

$$A = P(1 + rt) = \$500\left(1 + \frac{6\%}{\text{year}}\cdot\frac{3\,\text{year}}{4}\cdot\frac{1}{1}\right) = \$500\left(1 + 0.06\cdot\frac{3}{4}\right) = \$500(1 + 0.045)$$
$$= \$500(1.045) = \$522.50$$

Supplementary Exercises

PROBLEM 17-10   Write five correct ways to indicate the following expressions: **(a)** 7 times u
(b) -3 times v **(c)** x times y **(d)** w divided by -2 **(e)** 9 divided by z **(f)** a divided by b

PROBLEM 17-11   Identify each missing number and identify the property by name when possible:

(a) $? + 3 = 3 + (-5)$ **(b)** $2 \cdot ? = -3 \cdot 2$ **(c)** $(4 + 5) + 2 = 4 + (? + 2)$
(d) $(2 \cdot 3)5 = 2(? \cdot 5)$ **(e)** $8 + ? = 8$ **(f)** $? + (-2) = 0$ **(g)** $? - 0 = 5$ **(h)** $? - 4 = -4$
(i) $? - 3 = 0$ **(j)** $? \cdot 5 = 0$ **(k)** $0 \cdot ? = 0$ **(l)** $? \div (-2) = 0$ **(m)** $2 \div ?$ is not defined
(n) $? \cdot 1 = 6$ **(o)** $? \cdot \frac{2}{3} = 1$ **(p)** $? \cdot 7 = -7$ **(q)** $? \div 1 = -6$ **(r)** $-2 \div ? = 1$
(s) $3(8 + 9) = ? \cdot 8 + ? \cdot 9$ **(t)** $(6 + 2)5 = 6 \cdot ? + 2 \cdot ?$ **(u)** $-(-5) = ?$ **(v)** $-2 \cdot 9 = 2 \cdot ?$
(w) $3 \cdot 4 = -3 \cdot ?$ **(x)** $2 - 8 = 2 + ?$ **(y)** $\frac{2}{3} = 2 \cdot ?$ **(z)** $?(-1) = -7$

PROBLEM 17-12   Identify each term in the following: **(a)** $u + 7$ **(b)** $v - 5$
(c) $3x^2 - 5x - 4$ **(d)** $2\sqrt{y} - y\sqrt{2} + y - 2$ **(e)** $-(m + n)$ **(f)** $(a - 6)$ **(g)** $-3(b - 5)$

(h) $2(9 - c)$ **(i)** 2 **(j)** p **(k)** $-3hk$ **(l)** $\dfrac{xyz}{5}$ **(m)** $\dfrac{x + 3}{x}$ **(n)** $\dfrac{m - n}{n}$

(o) $w\sqrt{5^2 - 4^2}$ **(p)** $\dfrac{a + b - c}{2}$ **(q)** $-5(2x^2 + 3x - 2)$

PROBLEM 17-13   Clear parentheses and/or combine like terms in the following: **(a)** $-3(x + 2)$
(b) $(2 - y)8$ **(c)** $5(m + n)$ **(d)** $(x - y)(-5)$ **(e)** $3h^2 - 2h + 5$ **(f)** $-(5 - 8k)$
(g) $6u + u$ **(h)** $3v - 7v$ **(i)** $9p - p$ **(j)** $3mn - 8mn$ **(k)** $2x - 5x + x$

(l) $5y - 8 + y - 5$ **(m)** $\dfrac{w}{4} - \dfrac{3w}{4}$ **(n)** $5b - 6a + b + 3a - 2b$ **(o)** $2(h - 2) + 2h$

(p) $m - 4(m - n) - n$ **(q)** $-(a - b + c) + 2a - 4c$ **(r)** $3 + 2(v - 5)$ **(s)** $u - (u + v)$
(t) $2x - 3(4x - 5) + 6$ **(u)** $a + (a - b)$ **(v)** $x^2 + 2x + 1 - (x^2 - 3x + 5)$
(w) $2y^2 - 6y + 5 + (3y^2 + 5y - 4)$

PROBLEM 17-14   Evaluate the following algebraic expressions: **(a)** $3v$ for $v = 8$
(b) $-5u$ for $u = 2$ **(c)** $x + 5$ for $x = 2$ **(d)** $y - 7$ for $y = -8$ **(e)** $5w + 6$ for $w = -1$

(f) $2z - 9$ for $z = 6$ **(g)** $a^2 + 3a - 9$ for $a = 3$ **(h)** $a^2 + 3a - 9$ for $a = -3$

(i) $\dfrac{-b + \sqrt{b^2 - 4ac}}{2a}$ for $a = 6, b = -13$ and $c = 6$

(j) $\dfrac{-b - \sqrt{b^2 - 4ac}}{2a}$ for $a = 6, b = -13$ and $c = 6$ **(k)** $2u - 8v$ for $u = \dfrac{1}{2}$ and $v = -\dfrac{1}{4}$

(l) $-6a - 9b$ for $a = -\dfrac{1}{3}$ and $b = \dfrac{2}{3}$

PROBLEM 17-15   Solve each problem by evaluating each given formula:

(a) Find the shoe-size number n for a man who has feet each measuring 12 inches in length l when standing using the **male shoe-size formula** $n = 3l - 25$.

(b) Find the shoe-size number n for a woman who has feet each measuring $9\frac{1}{3}$ inches in length l when standing using the **female shoe-size formula** $n = 3l - 22$.

(c) Find the average temperature t in degrees Celsius at an altitude a of 12 kilometers using the **temperature/altitude formula** $t = 15 - 6.25a$.

(d) Find the Fahrenheit F temperature for the answer to problem C using the **temperature formula** $F = \dfrac{9}{5}C + 32$.

(e) What distance d can a car travel in a time t of 15 minutes at a constant rate r of 15 mph using the **distance formula** $d = rt$?

(f) What is the amount A at the end of an 18-month time period t if \$800 principal P is invested at a simple interest rate r of 8% per year using the **amount formula** $A = P(1 + rt)$?

(g) How much time t will it take to travel a distance d of 300 miles at a constant rate r of 45 mph using the **time formula** $t = d \div r$?

(h) What is the constant rate r that is necessary to travel a distance d of 100 miles in a time t of $1\frac{1}{2}$ hours using the **rate formula** $r = d \div t$?

Answers to Supplementary Exercises

(17-10)   **(a)** 7 times $u = \quad 7 \cdot u = \quad 7(u) = \quad (7)u = \quad (7)(u) = 7u$

(b) -3 times $v = -3 \cdot v = -3(v) = (-3)v = (-3)(v) = -3v$

(c) x times $y = \quad x \cdot y = \quad x(y) = \quad (x)y = \quad (x)(y) = xy$

(d) w divided by $-2 = w \div (-2) = w/(-2) = w \cdot \dfrac{1}{-2} = \dfrac{1}{-2} \cdot w = \dfrac{w}{-2}$

(e) 9 divided by $z = \quad 9 \div z = \quad 9/z = \quad 9 \cdot \dfrac{1}{z} = \quad \dfrac{1}{z} \cdot 9 = \dfrac{9}{z}$

(f) a divided by $b = \quad a \div b = \quad a/b = \quad a \cdot \dfrac{1}{b} = \quad \dfrac{1}{b} \cdot a = \dfrac{a}{b}$

(17-11)   **(a)** -5; Commutative Property of Addition
(b) -3; Commutative Property of Multiplication   **(c)** 5; Associative Property of Addition
(d) 3; Associative Property of Multiplication   **(e)** 0; Identity Property for Addition
(f) 2; Additive Inverse Property   **(g)** 5   **(h)** 0   **(i)** 3   **(j)** 0; Zero-Factor Property
(k) any real number; Zero-Product Property   **(l)** 0; Zero-Dividend Property
(m) 0; Zero-Divisor Property   **(n)** 6; Identity Property for Multiplication
(o) $\frac{3}{2}$; Multiplicative Inverse Property   **(p)** -1 **(q)** -6 **(r)** -2; Unit-Fraction Property

(s) 3; Distributive Property of Multiplication Over Addition
(t) 5; Distributive Property of Multiplication Over Addition   **(u)** 5   **(v)** -9 **(w)** -4
(x) -8; Subtraction Property   **(y)** $\frac{1}{3}$; Division Property   **(z)** 7

(17-12) **(a)** $u, 7$ **(b)** $v, -5$ **(c)** $3x^2, -5x, -4$ **(d)** $2\sqrt{y}, -y\sqrt{2}, y, -2$ **(e)** $-m, -n$

(f) $a, -6$ **(g)** $-3b, 15$ **(h)** $18, -2c$ **(i)** 2   **(j)** p **(k)** $-3hk$ **(l)** $\frac{xyz}{5}$ **(m)** $1, \frac{3}{x}$

(n) $\frac{m}{n}, -1$ **(o)** $3w$ **(p)** $\frac{a}{2}, \frac{b}{2}, -\frac{c}{2}$ **(q)** $-10x^2, -15x, 10$

(17-13) **(a)** $-3x - 6$ **(b)** $16 - 8y$ **(c)** $5m + 5n$ **(d)** $5y - 5x$ **(e)** $3h^2 - 2h + 5$
(f) $8k - 5$ **(g)** $7u$ **(h)** $-4v$ **(i)** $8p$ **(j)** $-5mn$ **(k)** $-2x$ **(l)** $6y - 13$
(m) $-\frac{w}{2}$ or $-\frac{1}{2}w$ **(n)** $4b - 3a$ or $-3a + 4b$ **(o)** $4h - 4$ **(p)** $-3m + 3n$ or $3n - 3m$
(q) $a + b - 5c$ **(r)** $2v - 7$ **(s)** $-v$ **(t)** $-10x + 21$ or $21 - 10x$ **(u)** $2a - b$
(v) $5x - 4$ **(w)** $5y^2 - y + 1$

(17-14) **(a)** 24   **(b)** -10 **(c)** 7   **(d)** -15 **(e)** 1   **(f)** 3   **(g)** 9   **(h)** -9 **(i)** $\frac{3}{2}$

(j) $\frac{2}{3}$ **(k)** 3   **(l)** -4

(17-15) **(a)** size 11   **(b)** size 6   **(c)** $-60°$ **(d)** $-76°F$ **(e)** $3\frac{3}{4}$ miles   **(f)** \$896
(g) $6\frac{2}{3}$ hours   **(h)** $66\frac{2}{3}$ mph

18 LINEAR EQUATIONS IN ONE VARIABLE

THIS CHAPTER IS ABOUT

☑ **Identifying Linear Equations in One Variable**
☑ **Solving Linear Equations Using Rules**
☑ **Solving Linear Equations Containing Like Terms and/or Parentheses**
☑ **Solving Linear Equations Containing Fractions, Decimals, or Percents**
☑ **Solving Literal Equations and Formulas for a Given Variable**

18-1. Identifying Linear Equations in One Variable

A. Identify linear equations in one variable.

An equation in one variable that can be written in standard form as $Ax + B = C$, where A, B, and C are real numbers ($A \neq 0$) and x is any variable is called a **linear equation in one variable.**

EXAMPLE 18-1: Which of the following are linear equations in one variable? **(a)** $y + 5 = -2$

(b) $w - 4 = 5$ **(c)** $2z = 6$ **(d)** $\dfrac{u}{5} = -2$ **(e)** $-3v - 6 = -2$ **(f)** $x\sqrt{2} + 3 = 5$

(g) $\dfrac{a}{2} + 0.3 = 1.8$ **(h)** $\dfrac{b}{-3} - 4 = 5$ **(i)** $2m + 3n = 6$ **(j)** $2 + 3 = 5$

(k) $2x^2 + 4 = 3$ **(l)** $2\sqrt{y} - 6 = 4$ **(m)** $\dfrac{2}{w} + 5 = 7$

Solution
(a) $y + 5 = -2$ or $1y + 5 = -2$ is a linear equation in one variable.
(b) $w - 4 = 5$ or $1w + (-4) = 5$ is a linear equation in one variable.
(c) $2z = 6$ or $2z + 0 = 6$ is a linear equation in one variable.

(d) $\dfrac{u}{5} = -2$ or $\dfrac{1}{5}u + 0 = -2$ is a linear equation in one variable.

(e) $-3v - 6 = -2$ or $-3v + (-6) = -2$ is a linear equation in one variable.
(f) $x\sqrt{2} + 3 = 5$ or $(\sqrt{2})x + 3 = 5$ is a linear equation in one variable.

(g) $\dfrac{a}{2} + 0.3 = 1.8$ or $\dfrac{1}{2}a + 0.3 = 1.8$ is a linear equation in one variable.

(h) $\dfrac{b}{-3} - 4 = 5$ or $-\dfrac{1}{3}b + (-4) = 5$ is a linear equation in one variable.

(i) $2m + 3n = 6$ is not a linear equation in one variable because it has two variables.
(j) $2 + 3 = 5$ is not a linear equation in one variable because it has no variables.
(k) $2x^2 + 4 = 3$ is not a linear equation in one variable because the variable has an exponent greater than 1.

(l) $2\sqrt{y} - 6 = 4$ is not a linear equation in one variable because the variable is in the radicand.

(m) $\dfrac{2}{w} + 5 = 7$ is not a linear equation in one variable because the variable is in the denominator.

B. Identify the eight special types of linear equations in one variable.

Every linear equation in one variable that does not contain like terms or parentheses can be easily categorized as one of **eight special types of linear equations in one variable.**

EXAMPLE 18-2: List the eight special types of linear equations in one variable.

Solution: If A, B, and C are real numbers ($A \neq 0$) and x is any variable, then
1. $x + B = C$ is called an **addition equation** [see Example 18-5].
2. $x - B = C$ is called a **subtraction equation** [see Example 18-6].
3. $Ax = C$ is called a **multiplication equation** [see Example 18-7].

4. $\dfrac{x}{A} = C$ is called a **division equation** [see Example 18-8].

5. $Ax + B = C$ is called a **multiplication-addition equation** [see Example 18-9].
6. $Ax - B = C$ is called a **multiplication-subtraction equation** [see Example 18-10].

7. $\dfrac{x}{A} + B = C$ is called a **division-addition equation** [see Example 18-11].

8. $\dfrac{x}{A} - B = C$ is called a **division-subtraction equation** [see Example 18-12].

Given a linear equation in one variable that does not contain like terms or parentheses, you may need to first write the equation in standard form before it can be categorized as one of the eight special types of linear equations in one variable.

EXAMPLE 18-3: Which special type of linear equation in one variable given by is given by each of the following equations?

(a) $y + 5 = -2$ **(b)** $w - 4 = 5$ **(c)** $2z = 6$ **(d)** $\dfrac{u}{5} = -2$ **(e)** $2x + 3 = 9$

(f) $-3v - 6 = -2$ **(g)** $\dfrac{a}{2} + 3 = -1$ **(h)** $\dfrac{b}{-3} - 4 = 5$ **(i)** $-c + 2 = -3$

(j) $5 - 2r = 0$ **(k)** $2 = \dfrac{s}{3} + 5$ **(l)** $-1 = -2 - \dfrac{t}{8}$

Solution
(a) $y + 5 = -2$ is an addition equation.
(b) $w - 4 = 5$ is a subtraction equation.
(c) $2z = 6$ is a multiplication equation.

(d) $\dfrac{u}{5} = -2$ is a division equation.

(e) $2x + 3 = 9$ is a multiplication-addition equation.
(f) $-3v - 6 = -2$ is a multiplication-subtraction equation.

(g) $\dfrac{a}{2} + 3 = -1$ is a division-addition equation.

(h) $\dfrac{b}{-3} - 4 = 5$ is a division-subtraction equation.

(i) $-c + 2 = -3$ or $-1c + 2 = -3$ is a multiplication-addition equation.

(j) $-5 - 2r = 0$ or $-2r - 5 = 0$ is a multiplication-subtraction equation.

(k) $2 = \dfrac{s}{3} + 5$ or $\dfrac{s}{3} + 5 = 2$ is a division-addition equation.

(l) $-1 = -2 - \dfrac{t}{8}$ or $\dfrac{t}{-8} - 2 = -1$ is a division-subtraction equation.

18-2. Solving Linear Equations Using Rules

A. Check a proposed solution of a linear equation in one variable.

Every linear equation in one variable has exactly one solution. The **solution of a linear equation in one variable** is the real number that can replace the variable to make both members of the equation equal. To **check a proposed solution of a linear equation in one variable,** you first substitute the proposed solution for the variable and then compute using the Order of Operations to get a number sentence. If you get a **true number sentence,** then the proposed solution **checks** and is the one and only solution of the original linear equation in one variable. If you get a **false number sentence,** then the proposed solution does not check and is not the solution of the original linear equation in one variable.

EXAMPLE 18-4: Which one of the following is a solution of $3x + 2 = 5$: 0 or 1?

Solution: Check $x = 0$ in $3x + 2 = 5$. | Check $x = 1$ in $3x + 2 = 5$.

$3x + 2 = 5$		⟵ given equation	$3x + 2 = 5$		⟵ given equation
$3(0) + 2$	5	Substitute 0 for x.	$3(1) + 2$	5	Substitute 1 for x.
$0 + 2$	5	Compute.	$3 + 2$	5	Compute.
2	5	⟵ false $(2 \neq 5)$	5	5	⟵ true $(5 = 5)$

$x = 0$ is not a solution because $2 \neq 5$. | $x = 1$ is a solution because $5 = 5$.

Note: Because every linear equation in one variable has exactly one solution, the one and only solution of $3x + 2 = 5$ is $x = 1$.

B. Solve an addition equation ($x + B = C$).

To **solve an addition equation** in the form $x + B = C$, you subtract using the following Subtraction Rule for Equations.

Subtraction Rule for Equations
If a, b, and c are real numbers and $a = b$, then $a - c = b - c$.

Note: The Subtraction Rule for Equations states that if you subtract the same term from both members of an equation, the solution(s) will not change.

To solve an addition equation using the Subtraction Rule for Equations, you subtract the numerical addend from both members of the addition equation to isolate the variable in one member.

EXAMPLE 18-5: Solve $y + 5 = -2$.

Solution:

variable addend
numerical addend
sum

$y + 5 = -2$ *Think:* In $y + 5 = -2$, the numerical addend is 5.

$y + 5 \mathbf{- 5} = -2 \mathbf{- 5}$ Subtract the numerical addend 5 from both members to isolate the variable y in one member.

$y + 0 = -2 - 5$

y is isolated ⟶ $y = -2 - 5$

$y = -7$ ⟵ proposed solution

Check: $\underline{y + 5 = -2}$ ⟵ original equation

$\dfrac{-7 + 5\ \ |\ \ -2}{\qquad\qquad}$ Substitute the proposed solution -7 for y in the original equation to see if you get a true number sentence.

$-2\ \ |\ \ -2$ ⟵ $y = -7$ checks

Note: The one and only solution of $y + 5 = -2$ is $y = -7$.

C. Solve a subtraction equation ($x - B = C$).

To **solve a subtraction equation** in the form $x - B = C$, you add using the Addition Rule for Equations.

Addition Rule for Equations

If a, b, and c are real numbers and $a = b$, then $a + c = b + c$.

Note: The Addition Rule for Equations states that if you add the same term to both members of an equation, the solution(s) will not change.

To solve a subtraction equation using the Addition Rule for Equations, you add the subtrahend to both members of the subtraction equation to isolate the variable in one member.

EXAMPLE 18-6: Solve $w - 4 = 5$.

$$
\begin{array}{ccc}
\text{minuend} & & \\
\big| & \text{subtrahend} & \\
\big| & \big| & \text{difference} \\
\downarrow & \downarrow & \downarrow
\end{array}
$$

Solution: $w - 4 = 5$ *Think:* In $w - 4 = 5$, 4 is the subtrahend.

$w - 4 + 4 = 5 + 4$ Add the subtrahend 4 to both members
to isolate the variable w in one member.

$w + 0 = 5 + 4$

w is isolated ⟶ $w = 5 + 4$

$w = 9$ ⟵ proposed solution

Check: $w - 4 = 5$ ⟵ original equation

$\dfrac{9 - 4\ \ |\ \ 5}{\qquad}$ Substitute the proposed solution 9 for w.

$5\ \ |\ \ 5$ ⟵ $w = 9$ checks

Note: The one and only solution of $w - 4 = 5$ is $w = 9$.

D. Solve a multiplication equation ($Ax = C$).

To **solve a multiplication equation** in the form $Ax = C$, you divide using the Division Rule for Equations.

Division Rule for Equations

If a, b, and c are real numbers ($c \neq 0$) and $a = b$, then $\dfrac{a}{c} = \dfrac{b}{c}$.

Note: The Division Rule for Equations states that if you divide both members of an equation by the same nonzero term, the solution(s) will not change.

To solve a multiplication equation using the Division Rule for Equations, you divide both members of the multiplication equation by the numerical factor to isolate the variable in one member.

EXAMPLE 18-7: Solve $2z = 6$.

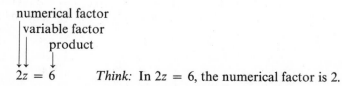

Solution: $2z = 6$ *Think:* In $2z = 6$, the numerical factor is 2.

$$\frac{2z}{2} = \frac{6}{2}$$ Divide both members by the numerical factor 2 to isolate the variable z in one member.

$$\frac{\cancel{2}z}{\cancel{2}} = \frac{6}{2}$$

z is isolated $\longrightarrow$ $z = \dfrac{6}{2}$

$z = 3$ $\longleftarrow$ proposed solution

Check: $2z = 6$ $\longleftarrow$ original equation

$\begin{array}{c|c} 2(3) & 6 \\ \hline 6 & 6 \end{array}$ Substitute the proposed solution 3 for z.

$\longleftarrow$ $z = 3$ checks

Note: The one and only solution of $2z = 6$ is $z = 3$.

E. Solve a division equation $\left(\dfrac{x}{A} = C\right)$.

To **solve a division equation** in the form $\dfrac{x}{A} = C$, you multiply using the Multiplication Rule for Equations.

Multiplication Rule for Equations
If a, b, and c are real numbers ($c \neq 0$) and $a = b$, then $ac = bc$.

Note: The Multiplication Rule for Equations states that if you multiply both members of an equation by the same nonzero term, the solution(s) will not change.

To solve a division equation using the Multiplication Rule for Equations, you multiply both members of the division equation by the numerical divisor to isolate the variable in one member.

EXAMPLE 18-8: Solve $\dfrac{u}{5} = -2$.

Solution

numerical divisor
  variable dividend
          quotient

$\dfrac{u}{5} = -2$ *Think:* In $\dfrac{u}{5} = -2$, the numerical divisor is 5.

$\mathbf{5} \cdot \dfrac{u}{5} = \mathbf{5}(-2)$ Multiply both members by the numerical divisor 5 to isolate the variable u in one member.

$\dfrac{\cancel{5}}{1} \cdot \dfrac{u}{\cancel{5}} = 5(-2)$

u is isolated $\longrightarrow$ $u = 5(-2)$

$u = -10$ $\longleftarrow$ proposed solution

Check: $\dfrac{u}{5} = -2$ $\longleftarrow$ original equation

$\begin{array}{c|c} \dfrac{-10}{5} & -2 \\ \hline -2 & -2 \end{array}$ Substitute the proposed solution -10 for u.

$\longleftarrow$ $u = -10$ checks

Note: The one and only solution of $\dfrac{u}{5} = -2$ is $u = -10$.

The Addition, Subtraction, Multiplication, and Division Rules for Equations are collectively called the **rules for linear equations.**

F. Solve a multiplication-addition equation ($Ax + B = C$).

To **solve a multiplication-addition equation** in the form $Ax + B = C$, you first use the Subtraction Rule for Equations to isolate the **variable term** in one member and then use the Division Rule for Equations to isolate the variable itself.

EXAMPLE 18-9: Solve $2x + 3 = 9$.

Solution

variable term

$$2x + 3 = 9 \qquad \text{Identify the variable term.}$$

$$2x + 3 - 3 = 9 - 3 \qquad \text{Use the Subtraction Rule for Equations to isolate the variable term in one member.}$$

$2x$ is isolated $\longrightarrow$ $2x = 6$

$$\frac{2x}{2} = \frac{6}{2} \qquad \text{Use the Division Rule for Equations to isolate the variable } x.$$

x is isolated $\longrightarrow$ $x = 3$ $\longleftarrow$ proposed solution

Check: $2x + 3 = 9$ $\longleftarrow$ original equation

$2(3) + 3$	9	Substitute the proposed solution 3 for x.
$6 + 3$	9	Compute.
9	9	$\longleftarrow$ $x = 3$ checks

Note: The one and only solution of $2x + 3 = 9$ is $x = 3$.

G. Solve a multiplication-subtraction equation ($Ax - B = C$).

To **solve a multiplication-subtraction equation** in the form $Ax - B = C$, you first use the Addition Rule for Equations to isolate the variable term in one member and then use the Division Rule for Equations to isolate the variable itself.

EXAMPLE 18-10: Solve $-3v - 6 = -2$.

Solution

variable term

$$-3v - 6 = -2 \qquad \text{Identify the variable term.}$$

$$-3v - 6 + 6 = -2 + 6 \qquad \text{Use the Addition Rule for Equations to isolate the variable term in one member.}$$

$-3v$ is isolated $\longrightarrow$ $-3v = 4$

$$\frac{-3v}{-3} = \frac{4}{-3} \qquad \text{Use the Division Rule for Equations to isolate the variable } v.$$

v is isolated $\longrightarrow$ $v = \dfrac{4}{-3}$ or $\dfrac{-4}{3}$ or $-\dfrac{4}{3}$ $\longleftarrow$ proposed solution

Check: $-3v - 6 = -2$ $\longleftarrow$ original equation

$-3\left(-\dfrac{4}{3}\right) - 6$	-2	Substitute the proposed solution $-\dfrac{4}{3}$ for v.
$4 - 6$	-2	Compute.
-2	-2	$\longleftarrow$ $v = -\dfrac{4}{3}$ checks

Note: The one and only solution of $-3v - 6 = -2$ is $v = -\dfrac{4}{3}$.

H. Solve a division-addition equation $\left(\dfrac{x}{A} + B = C\right)$.

To **solve a division-addition equation** in the form $\dfrac{x}{A} + B = C$, you first use the Subtraction Rule for Equations to isolate the variable term in one member and then use the Multiplication Rule for Equations to isolate the variable itself.

EXAMPLE 18-11: Solve $\dfrac{a}{2} + 3 = -1$.

Solution

$$\overset{\text{variable term}}{\underset{\downarrow}{}}$$

$$\dfrac{a}{2} + 3 = -1 \qquad \text{Identify the variable term.}$$

$$\dfrac{a}{2} + 3 - 3 = -1 - 3 \qquad \begin{array}{l}\text{Use the Subtraction Rule for Equations to isolate the variable} \\ \text{term in one member.}\end{array}$$

$\dfrac{a}{2}$ is isolated $\longrightarrow \dfrac{a}{2} = -4$

$$2 \cdot \dfrac{a}{2} = 2(-4) \qquad \begin{array}{l}\text{Use the Multiplication Rule for Equations to isolate the} \\ \text{variable } a.\end{array}$$

a is isolated $\longrightarrow a = -8 \longleftarrow$ proposed solution

Check: $\dfrac{a}{2} + 3 = -1 \longleftarrow$ original equation

$$\dfrac{-8}{2} + 3 \;\Big|\; -1 \qquad \text{Substitute the proposed solution } -8 \text{ for } a.$$

$$-4 + 3 \;\Big|\; -1 \qquad \text{Compute.}$$

$$-1 \;\Big|\; -1 \longleftarrow a = -8 \text{ checks}$$

Note: The one and only solution of $\dfrac{a}{2} + 3 = -1$ is $a = -8$.

I. Solve a division-subtraction equation $\left(\dfrac{x}{A} - B = C\right)$.

To **solve a division-subtraction equation** in the form $\dfrac{x}{A} - B = C$, you first use the Addition Rule for Equations to isolate the variable term in one member and then use the Multiplication Rule for Equations to isolate the variable itself.

EXAMPLE 18-12: Solve $\dfrac{b}{-3} - 4 = 5$.

Solution

$$\overset{\text{variable term}}{\underset{\downarrow}{}}$$

$$\dfrac{b}{-3} - 4 = 5 \qquad \text{Identify the variable term.}$$

$$\dfrac{b}{-3} - 4 + 4 = 5 + 4 \qquad \begin{array}{l}\text{Use the Addition Rule for Equations to} \\ \text{isolate the variable term in one member.}\end{array}$$

$\dfrac{b}{-3}$ is isolated $\longrightarrow \dfrac{b}{-3} = 9$

$$-3 \cdot \frac{b}{-3} = -3 \cdot 9$$

Use the Multiplication Rule for Equations to isolate the variable b.

b is isolated $\longrightarrow$ $b = -27$ $\longleftarrow$ proposed solution

Check: $\dfrac{b}{-3} - 4 = 5$ $\longleftarrow$ original equation

$$\frac{-27}{-3} - 4 \;\Big|\; 5$$

Substitute the proposed solution -27 for b.

$$9 - 4 \;\Big|\; 5$$

Compute.

$$5 \;\Big|\; 5 \longleftarrow b = -27 \text{ checks}$$

Note: The one and only solution of $\dfrac{b}{-3} - 4 = 5$ is $b = -27$.

J. Solve linear equations in one variable that do not contain like terms or parentheses.

Given a linear equation in one variable that does not contain like terms or parentheses and that is not in the form of one of the eight special types of linear equations, you can always rename the given linear equation as one of the special types and then solve using the rules for linear equations.

EXAMPLE 18-13: Solve (a) $-x + 2 = -3$ (b) $5 - 2y = 0$ (c) $2 = \dfrac{m}{3} + 5$

(d) $-1 = -3 + \dfrac{n}{-5}$

Solution

(a) $-x + 2 = -3$ Rename using $-x = -1x$.

$-1x + 2 = -3$ $\longleftarrow$ multiplication-addition equation

$-1x + 2 - 2 = -3 - 2$ Use the Subtraction Rule for Equations.

$-1x = -5$

$\dfrac{-1x}{-1} = \dfrac{-5}{-1}$ Use the Division Rule for Equations.

$x = 5$ $\longleftarrow$ solution

(b) $5 - 2y = 0$

$-1(5 - y) = -1(0)$ Multiply both members by -1.

$-5 + 2y = 0$

$2y - 5 = 0$ $\longleftarrow$ multiplication-subtraction equation

$2y - 5 + 5 = 0 + 5$ Use the Addition Rule for Equations.

$y = 5$

$\dfrac{2y}{2} = \dfrac{5}{2}$ Use the Division Rule for Equations.

$y = \dfrac{5}{2}$ or $2\dfrac{1}{2}$ or 2.5 $\longleftarrow$ solution

(c) $2 = \dfrac{m}{3} + 5$ Interchange the left and right members.

$\dfrac{m}{3} + 5 = 2$ $\longleftarrow$ division-addition equation

$\dfrac{m}{3} + 5 - 5 = 2 - 5$ Use the Subtraction Rule for Equations.

$$\frac{m}{3} = -3$$

$$3 \cdot \frac{m}{3} = 3(-3) \qquad \text{Use the Multiplication Rule for Equations.}$$

$$m = -9 \longleftarrow \text{solution}$$

(d)
$$-1 = -3 + \frac{n}{-5}$$

$$-3 + \frac{n}{-5} = -1 \qquad \text{Interchange the left and right members.}$$

$$\frac{n}{-5} + (-3) = -1 \qquad \text{Use the commutative property of addition.}$$

$$\frac{n}{-5} - 3 = -1 \longleftarrow \text{division-subtraction equation}$$

$$\frac{n}{-5} - 3 + 3 = -1 + 3 \qquad \text{Use the Addition Rule for Equations.}$$

$$\frac{n}{-5} = 2$$

$$-5 \cdot \frac{n}{-5} = -5(2) \qquad \text{Use the Multiplication Rule for Equations.}$$

$$n = -10 \longleftarrow \text{solution}$$

18-3. Solving Linear Equations Containing Like Terms and/or Parentheses

A. Solve linear equations in one variable containing like terms in only one member.

To **solve a linear equation in one variable that contains like terms in only one member** use the following steps:
1. Combine like terms.
2. Rename the equation from Step 1 when necessary to get one of the eight special types of linear equations.
3. Solve the equation using the rules for linear equations.

EXAMPLE 18-14: Solve $8 = 3x - 4 + x$.

Solution

$$\overset{\text{like terms}}{8 = 3x - 4 + x} \qquad \text{Identify like terms.}$$

$$8 = 4x - 4 \qquad \text{Combine like terms.}$$

$$4x - 4 = 8 \longleftarrow \text{multiplication-subtraction equation}$$

$$4x = 12 \qquad \text{Use the Addition Rule: } 4x - 4 + 4 = 8 + 4$$

$$x = 3 \qquad \text{Use the Division Rule: } \frac{4x}{4} = \frac{12}{4}$$

Check: $8 = 3x - 4 + x \longleftarrow$ original equation

8	$3(3) - 4 + (3)$	Substitute the proposed solution 3 for x.
8	$9 - 4 + 3$	Compute.
8	$5 + 3$	
8	$8 \longleftarrow x = 3$ checks	

Note: The one and only solution of $8 = 3x - 4 + x$ is $x = 3$.

B. Solve linear equations in one variable containing like terms in both members.

To **solve a linear equation in one variable that contains like terms in both members,** use the following steps:
1. Combine like terms in each member when possible.
2. Collect and combine like terms all in one member.
3. Rename the equation to get one of the eight special types of linear equations in one variable when necessary.
4. Solve the equation using the rules for linear equations.

EXAMPLE 18-15: Solve $-y + 8 - 2y = 3y - 6 + y$.

Solution

like terms

$-y + 8 - 2y = 3y - 6 + y$	Identify like terms
$-3y + 8 = 4y - 6$	Combine like terms in each member.
$-3y + 3y + 8 = 4y + 3y - 6$	Collect like terms all in one member.
$8 = 4y + 3y - 6$ ⟵ like terms are all in one member
$8 = 7y - 6$	Combine like terms.
$7y - 6 = 8$ ⟵ multiplication-subtraction equation
$7y = 14$	Use the Addition Rule: $7y - 6 + 6 = 8 + 6$
$y = 2$	Use the Division Rule: $\dfrac{7y}{7} = \dfrac{14}{7}$

Check: $-y + 8 - 2y = 3y - 6 + y$ ⟵ original equation

$-(2) + 8 - 2(2)$	$3(2) - 6 + (2)$	Substitue the proposed solution 2 for y.
$-2 + 8 - 4$	$6 - 6 + 2$	Compute.
$6 - 4$	$0 + 2$	
2	2 ⟵ $y = 2$ checks	

Note: The one and only solution of $-y + 8 - 2y = 3y - 6 + y$ is $y = 2$.

C. Solve linear equations in one variable containing parentheses.

To **solve a linear equation in one variable containing parentheses,** use the following steps:
1. Combine like terms inside the parentheses when possible.
2. Clear parentheses using the distributive properties.
3. Solve the equation using the rules for linear equations.

EXAMPLE 18-16: Solve $2(7 - 5u - 4) = u - 4(5u - 6 - 3u)$.

Solution

like terms   like terms

$2(7 - 5u - 4) = u - 4(5u - 6 - 3u)$	Identify like terms inside parentheses.
$2(3 - 5u) = u - 4(2u - 6)$	Combine like terms inside parentheses.
$2(3 - 5u) = u + (-4)(2u - 6)$	
$2(3) - 2(5u) = u + (-4)(2u) - (-4)(6)$	Use the distributive property to clear parentheses.
$6 - 10u = u - 8u + 24$ ⟵ parentheses cleared
$6 - 10u = -7u + 24$	Solve as before [see Example 18-15].
$6 - 10u + 7u = -7u + 7u + 24$	
$6 - 3u = 24$	
$-3u + 6 = 24$	
$-3u = 18$	
$u = -6$ ⟵ proposed solution

Check:

$$2(7 - 5u - 4) = u - 4(5u - 6 - 3u) \longleftarrow \text{ original equation}$$

$2(7 - 5(-6) - 4)$	$(-6) - 4(5(-6) - 6 - 3(-6))$	Substitute the proposed solution 6 for u.
$2(7 + 30 - 4)$	$-6 - 4(-30 - 6 + 18)$	Compute.
$2(37 - 4)$	$-6 - 4(-36 + 18)$	
$2(33)$	$-6 - 4(-18)$	
66	$-6 + 72$	
66	$66 \longleftarrow u = -6 \text{ checks}$	

Note: The one and only solution of $2(7 - 5u - 4) = u - 4(5u - 6 - 3u)$ is $u = -6$.

D. **Solve equations in one variable that simplify as true number sentences.**

If an equation in one variable simplifies as a true number sentence then every real number is a solution of that equation.

EXAMPLE 18-17: Solve $2 - 5w = 4w + 2 - 9w$.

Solution

$$\overbrace{2 - 5w = 4w + 2 - 9w}^{\text{like terms}} \qquad \text{Identify like terms.}$$

$$2 - 5w = -5w + 2 \qquad \text{Combine like terms.}$$

$$2 - 5w + 5w = -5w + 5w + 2 \qquad \text{Collect like terms.}$$

$$2 = 2 \qquad \textit{Stop!} \text{ The true number sentence } 2 = 2 \text{ means that}$$
every real number is a solution of
$2 - 5w = 4w + 2 - 9w$.

Note: The equation $2 - 5w = 4w + 2 - 9w$ is not a linear equation in one variable because it simplifies as $2 = 2$ or $0w + 2 = 2$ and therefore cannot be written in the form $Aw + B = C$ where $A \neq 0$.

E. **Solve equations in one variable that simplify as false number sentences.**

If an equation in one variable simplifies as a false number sentence then there are no solutions of that equation.

EXAMPLE 18-18: Solve $1 - z = 2 + 3z - 3 - 4z$.

Solution

$$1 - z = 2 + 3z - 3 - 4z \qquad \text{Identify like terms.}$$

$$1 - z = -1 - z \qquad \text{Combine like terms.}$$

$$1 - z + z = -1 - z + z \qquad \text{Collect like terms.}$$

$$1 = -1 \qquad \textit{Stop!} \text{ The false number sentence } 1 = -1 \text{ means}$$
that there is no solution of
$1 - z = 2 + 3z - 3 - 4z$.

Note: The equation $1 - z = 2 + 3z - 3 - 4z$ is not a linear equation in one variable because it simplifies as $1 = -1$ or $0z + 1 = -1$ and therefore cannot be written in the form $Az + B = C$ where $A \neq 0$.

18-4. Solving Linear Equations Containing Fractions, Decimals, or Percents

A. Solve linear equations in one variable containing fractions.

To **solve a linear equation in one variable containing fractions,** you first **clear the fractions** and then use the rules for linear equations.

To clear fractions, you
1. Find the least common denominator (LCD) of the fractions in the given equation.
2. Multiply both members of the given equation by the LCD.
3. Clear parentheses in the equation to clear all fractions.

EXAMPLE 18-19: Solve $\frac{2}{3}x - \frac{1}{2} = \frac{1}{4}$.

Solution: The LCD of $\frac{2}{3}$, $\frac{1}{2}$, and $\frac{1}{4}$ is 12 [see Section 6-2].

$$12(\tfrac{2}{3}x - \tfrac{1}{2}) = 12(\tfrac{1}{4}) \qquad \text{Multiply both members by the LCD 12.}$$

$$12(\tfrac{2}{3}x) - 12(\tfrac{1}{2}) = 12(\tfrac{1}{4}) \qquad \text{Clear parentheses to clear fractions.}$$

$$8x - 6 = 3 \longleftarrow \text{fractions cleared}$$

$$8x = 9 \qquad \text{Solve using the Addition Rule.}$$

$$x = \tfrac{9}{8} \text{ or } 1\tfrac{1}{8} \text{ or } 1.125 \longleftarrow \text{proposed solution}$$

Check: $\frac{2}{3}x - \frac{1}{2} = \frac{1}{4} \longleftarrow$ original equation

$\quad \frac{2}{3}(\frac{9}{8}) - \frac{1}{2} \ \Big|\ \frac{1}{4} \qquad$ Substitute the proposed solution $\frac{9}{8}$ for x.

$\qquad \frac{3}{4} - \frac{1}{2} \ \Big|\ \frac{1}{4} \qquad$ Compute.

$\qquad\quad \frac{1}{4} \ \Big|\ \frac{1}{4} \longleftarrow x = \frac{9}{8}$ checks

Note: The one and only solution of $\frac{2}{3}x - \frac{1}{2} = \frac{1}{4}$ is $x = \frac{9}{8}$ or $1\frac{1}{8}$ or 1.125.

B. Solve linear equations in one variable containing decimals.

To **solve a linear equation in one variable containing decimals,** you can use the rules for linear equations.

EXAMPLE 18-20: Solve $0.5y + 0.2 = 0.3$

Solution

$$\overset{\text{variable term}}{\overbrace{0.5y}} + 0.2 = 0.3$$

$$0.5y + 0.2 - \mathbf{0.2} = 0.3 - \mathbf{0.2} \qquad \text{Use the Subtraction Rule for Equations to isolate the variable term } 0.5y \text{ in one member.}$$

$$0.5y \text{ is isolated} \longrightarrow 0.5y = 0.1$$

$$\frac{0.5y}{\mathbf{0.5}} = \frac{0.1}{\mathbf{0.5}} \qquad \text{Use the Division Rule for Equations isolate the variable } y.$$

$$y \text{ is isolated} \longrightarrow y = 0.2 \text{ or } \frac{1}{5} \longleftarrow \text{proposed solution}$$

Check: $0.5y + 0.2 = 0.3 \longleftarrow$ original equation

$\quad 0.5(\mathbf{0.2}) + 0.2 \ \Big|\ 0.3 \qquad$ Substitute the proposed solution 0.2 for y.

$\qquad 0.1 + 0.2 \ \Big|\ 0.3 \qquad$ Compute.

$\qquad\qquad 0.3 \ \Big|\ 0.3 \longleftarrow y = 0.2 \text{ or } \frac{1}{5}$ checks

Note: The one and only solution of $0.5y + 0.2 = 0.3$ is $y = 0.2$ or $\frac{1}{5}$.

An equation containing decimals can also be solved by first **clearing decimals** and then using the rules for linear equations.

To clear decimals, you
1. Find the LCD of the fraction form of the decimals from the given equations.
2. Multiply both members of the given equation by the LCD.
3. Clear parentheses in the equation to clear all decimals.

EXAMPLE 18-21: Solve $0.5y + 0.2 = 0.3$ by first clearing decimals.

Solution: The LCD of $\frac{5}{10}$ (0.5), $\frac{2}{10}$ (0.2), and $\frac{3}{10}$ (0.3) is 10 [see Section 6-2].

$$10(0.5y + 0.2) = 10(0.3) \qquad \text{Multiply both members by the LCD 10.}$$

$$10(0.5y) + 10(0.2) = 10(0.3) \qquad \text{Clear parentheses to clear decimals.}$$

$$5y + 2 = 3 \longleftarrow \text{decimals cleared}$$

$$5y = 1 \qquad \text{Solve using the rules for linear equations.}$$

$$y = \frac{1}{5} \text{ or } 0.2 \longleftarrow \text{same solution as found in Example 18-20}$$

Note: To solve a linear equation in one variable containing decimals, you can either use the rules for linear equations directly without clearing decimals or you can first clear decimals and then use the rules for linear equations. In practice, the best method to use is the method that is easiest for you.

C. Solve linear equations containing percents.

To **solve a linear equation in one variable containing percents,** you
1. **Clear percents** by renaming each percent as either a fraction or a decimal.
2. Clear fractions or decimals in the equation.
3. Solve the equation using the rules for linear equations.

EXAMPLE 18-22: Solve $x + 25\%x = 20$.

Solution

	Renaming the percent as a decimal	Renaming the percent as a fraction
	$x + 25\%x = 20$	$x + 25\%x = 20$
Clear percents.	$x + 0.25x = 20$	$x + \frac{1}{4}x = 20$
Clear decimals or fractions.	$100(x + 0.25x) = 100(20)$	$4(x + \frac{1}{4}x) = 4(20)$
	$100(x) + 100(0.25x) = 100(20)$	$4(x) + 4(\frac{1}{4}x) = 4(20)$
	$100x + 25x = 2000$	$4x + 1x = 80$
Solve as before	$125x = 2000$	$5x = 80$ — Solve as before.
	$x = \frac{2000}{125}$	$x = \frac{80}{5}$
	$x = 16$	$x = 16$

Check:	$x + 25\%x = 20$		*Check:*	$x + 25\%x = 20$	
	$16 + 25\% (16)$	20		$16 + 25\% (16)$	20
	$16 + 0.25(16)$	20		$16 + \frac{1}{4}(16)$	20
	$16 + 4$	20		$16 + 4$	20
	20	20		20	20

Note: The one and only solution of $x + 25\%x = 20$ is $x = 16$.

18-5. Solving Literal Equations and Formulas for a Given Variable

A. Identify literal equations.

An equation containing more than one different variable is called a **literal equation.** Every formula is a literal equation because every formula contains two or more different variables.

EXAMPLE 18-23: Which of the following are literal equations? **(a)** $P = 2(l + w)$
(b) $2x + 3y = 5$ **(c)** $2x + 3x = 5$ **(d)** $2x + 3 = 5$ **(e)** $2 + 3 = 5$

Solution
(a) $P = 2(l + w)$ is a formula (perimeter of a rectangle) and therefore is a literal equation.
(b) $2x + 3y = 5$ is a literal equation because it contains more than one variable.
(c) $2x + 3x = 5$ is not a literal equation because it contains only one variable x.
(d) $2x + 3 = 5$ is not a literal equation because it contains only one variable x.
(e) $2 + 3 = 5$ is not a literal equation because it contains no variables.

B. Solve a literal equation for a given variable that is contained in only one term.

To **solve a literal equation for a given variable that is contained in only one term,** you first isolate the term containing the given variable in one member of the literal equation and then isolate the given variable itself. When a literal equation is solved for a given variable, the algebraic expression in the other member is called the **solution of the literal equation with respect to the given variable.**

EXAMPLE 18-24: Which variable is the literal equation $y = mx + b$ solved for and what is the solution with respect to that variable?

Solution: The literal equation $y = mx + b$ is solved for y and the solution is $mx + b$.

EXAMPLE 18-25: Solve $2x + 3y = 5$ for x.

Solution

term containing x

$2x + 3y = 5$ Identify the term containing the given variable x.

$2x + 3y - 3y = 5 - 3y$ Use the Subtraction Rule for Equations to isolate the term containing the given variable x.

$2x$ is isolated ⟶ $2x = 5 - 3y$

$$\frac{2x}{2} = \frac{5 - 3y}{2}$$ Use the Division Rule for Equations to isolate the given variable x.

x is isolated ⟶ $x = \dfrac{5 - 3y}{2}$ or $\dfrac{5}{2} - \dfrac{3y}{2}$ or $\dfrac{5}{2} - \dfrac{3}{2}y$ or $-\dfrac{3}{2}y + \dfrac{5}{2}$ ⟵ proposed solution

Check: $2x + 3y = 5$ ⟵ original equation

$2\left(\dfrac{5 - 3y}{2}\right) + 3y$ | 5   Substitute the proposed solution $\dfrac{5 - 3y}{2}$ for x.

$5 - 3y + 3y$ | 5   Simplify.

$5 + 0$ | 5

5 | 5 ⟵ $\dfrac{5 - 3y}{2}$ checks

C. Solve a literal equation for a given variable that is contained in two or more like terms.

To **solve a literal equation for a given variable that is contained in two or more like terms,** you
1. Collect and combine any like terms.
2. Isolate the term containing the given variable in one member of the literal equation.
3. Isolate the variable itself.

EXAMPLE 18-26: Solve $-3m + 7n = 2n + 10$ for n.

Solution

like terms containing n

$$-3m + 7n = 2n + 10 \qquad \text{Identify the like terms containing the given variable.}$$

$-3m + 7n - 2n = 2n - 2n + 10 \qquad$ Collect like terms.

$-3m + 5n = 10 \qquad$ Combine like terms.

$-3m + 3m + 5n = 10 + 3m \qquad$ Use the Subtraction Rule for Equations to isolate the term containing the given variable n.

$5n$ is isolated $\longrightarrow 5n = 10 + 3m$

$$\frac{5n}{5} = \frac{10 + 3m}{5} \qquad \text{Use the Division Rule for Equations to isolate the given variable } n.$$

n is isolated $\longrightarrow n = \dfrac{10 + 3m}{5}$ or $\dfrac{3m + 10}{5}$ or $\dfrac{3m}{5} + 2$ or $\dfrac{3}{5}m + 2 \longleftarrow$ solution (Check as before.)

D. Solve a literal equation for a given variable that is contained in parentheses.

To **solve a literal equation for a given variable that is contained in parentheses,** you:
1. Combine any like terms inside the parentheses.
2. Clear parentheses if the given variable is contained inside the parentheses.
3. Collect and combine any like terms.
4. Isolate the term containing the given variable in one member of the literal equation.
5. Isolate the given variable itself.

EXAMPLE 18-27: Solve $P = 2(l + w)$ for w.

Solution

term containing w

$$P = 2(l + w) \qquad \text{Identify the given variable } w \text{ as being contained inside the parentheses.}$$

$$P = 2(l) + 2(w) \qquad \text{Use a distributive property to clear parentheses.}$$

$$P = 2l + 2w \longleftarrow \text{parentheses cleared}$$

$2w$ is in the left member $\longrightarrow 2l + 2w = P \qquad$ Rename to get the term containing the given variable w in the left member.

$2l - 2l + 2w = P - 2l \qquad$ Use the Subtraction Rule for Equations to isolate the term containing the given variable w.

$2w$ is isolated $\longrightarrow 2w = P - 2l$

$$\frac{2w}{2} = \frac{P - 2l}{2} \qquad \text{Use the Division Rule for Equations to isolate the given variable } w.$$

w is isolated $\longrightarrow w = \dfrac{P - 2l}{2}$ or $\dfrac{P}{2} - l$ or $\dfrac{1}{2}P - l \longleftarrow$ solution (Check as before.)

Note: The formula $w = \dfrac{1}{2}P - l$ means that the width w of a rectangle always equals one-half the perimeter P minus the length l.

RAISE YOUR GRADES

Can you ...?

☑ identify linear equations in one variable
☑ identify the eight special types of linear equations in one variable
☑ check a proposed solution of a linear equation in one variable
☑ identify the rules for linear equations
☑ solve linear equations using the rules for linear equations
☑ solve linear equations in one variable containing like terms and/or parentheses
☑ solve an equation in one variable that simplifies as a true or false number sentence
☑ solve linear equations in one variable containing fractions, decimals, or percents
☑ solve literal equations for a given variable

SUMMARY

1. An equation in one variable that can be written in standard form as $Ax + B = C$, where A, B, and C are real numbers ($A \neq 0$) and x is any variable is called a linear equation in one variable.
2. If A, B, and C are real numbers ($A \neq 0$) and x is any variable, then the eight special types of linear equations in one variable are
   (a) $x + B \ = C$ (addition equation)
   (b) $x - B \ = C$ (subtraction equation)
   (c) $\qquad Ax = C$ (multiplication equation)

   (d) $\qquad \dfrac{x}{A} = C$ (division equation)

   (e) $Ax + B = C$ (multiplication-addition equation)
   (f) $Ax - B = C$ (multiplication-subtraction equation)

   (g) $\dfrac{x}{A} + B = C$ (division-addition equation)

   (h) $\dfrac{x}{A} - B = C$ (division-subtraction equation)

3. Every linear equation in one variable that does not contain like terms or parentheses can be categorized as one of the eight special types of linear equations in one variable.
4. Every linear equation in one variable has exactly one solution.
5. The solution of a linear equation in one variable is the real number that can replace the variable to make both members of the equation equal.
6. To check a proposed solution of a linear equation in one variable, you first substitute the proposed solution for the variable and then compute using the Order of Operations to get a number sentence. If you get a true number sentence then the proposed solution checks and is the one and only solution to the original linear equation in one variable. If you get a false number sentence, then the proposed solution does not check and is not a solution of the original linear equation in one variable.
7. To solve a linear equation in one variable that does not contain like terms or parentheses, you use the following rules for linear equations:
   (a) **Addition Rule for Equations**
       If a, b, and c are real numbers and $a = b$, then $a + c = b + c$.
   (b) **Subtraction Rule for Equations**
       If a, b, and c are real numbers and $a = b$, then $a - c = b - c$.
   (c) **Multiplication Rule for Equations**
       If a, b, and c are real numbers ($c \neq 0$) and $a = b$, then $ac = bc$.
   (d) **Division Rule for Equations**
       If a, b, and c are real numbers ($c \neq 0$) and $a = b$, then $\dfrac{a}{c} = \dfrac{b}{c}$.

8. To solve an addition equation in the form $x + B = C$, you use the Subtraction Rule for Equations to subtract the numerical addend from both members of the addition equation to isolate the variable in one member.

9. To solve a subtraction equation in the form $x - B = C$, you use the Addition Rule for Equations to add the subtrahend to both members of the subtraction equation to isolate the variable in one member.

10. To solve a multiplication equation in the form $Ax = C$, you use the Division Rule for Equations to divide both members of the multiplication equation by the numerical factor to isolate the variable in one member.

11. To solve a division equation in the form $\dfrac{x}{A} = C$, you use the Multiplication Rule for Equations to multiply both members of the division equation by the numerical divisor to isolate the variable in one member.

12. To solve a multiplication-addition equation in the form $Ax + B = C$, you first use the Subtraction Rule for Equations to isolate the variable term in one member and then use the Division Rule for Equations to isolate the variable itself.

13. To solve a multiplication-subtraction equation in the form $Ax - B = C$, you first use the Addition Rule for Equations to isolate the variable term in one member and then use the Division Rule for Equations to isolate the variable itself.

14. To solve a division-addition equation in the form $\dfrac{x}{A} + B = C$, you first use the Subtraction Rule for Equations to isolate the variable term in one member and then use the Multiplication Rule for Equations to isolate the variable itself.

15. To solve a division-subtraction equation in the form $\dfrac{x}{A} - B = C$, you first use the Addition Rule for Equations to isolate the variable term in one member and then use the Multiplication Rule for Equations to isolate the variable itself.

16. To solve a linear equation in one variable containing like terms and/or parentheses, you
  (a) Combine any like terms that are inside the parentheses.
  (b) Clear any parentheses using the distributive properties.
  (c) Combine any like terms in each member.
  (d) Collect all like terms in one member.
  (e) Combine any like terms from Step **(d)**.
  (f) Rename the equation (when necessary) to get one of the eight special types of linear equations.
  (g) Solve the equation using the rules for linear equations.

17. If an equation in one variable simplifies as a true number sentence then every real number is a solution of that equation.

18. If an equation in one variable simplifies as a false number sentence then there are no solutions of that equation.

19. A given equation that simplifies as a true or false number sentence is not a linear equation in one variable because it cannot be written as $Ax + B = C$ where $A \neq 0$.

20. To solve a linear equation in one variable containing fractions, you first clear the fractions and then use the rules for linear equations.

21. To clear fractions, you:
  (a) Find the LCD of the fractions in the given equation.
  (b) Multiply both members of the given equation by the LCD.
  (c) Clear parentheses in the equation to clear all fractions.

22. To solve a linear equation in one variable containing decimals, you can either use the rules for linear equations directly without clearing decimals or you can first clear decimals and then use the rules for linear equations. In practice, the best method to use is the method that is easiest for you.

23. To solve a linear equation in one variable containing percents, you
  (a) Clear percents by renaming each percent as either a fraction or a decimal.
  (b) Clear fractions or decimals in the equation.
  (c) Solve the equation using the rules for linear equations.

24. An equation containing more than one different variable is called a literal equation.

25. Every formula is a literal equation because every formula contains two or more different variables.

26. To solve a literal equation for a given variable that is contained in only one term, you first isolate the term containing the given variable in one member of the literal equation and then isolate the given variable itself.

27. When a literal equation is solved for a given variable, the algebraic expression in the other member is called the solution of the literal equation with respect to the given variable.

28. To solve a literal equation for a given variable, you:
  (a) Combine any like terms inside parentheses.
  (b) Clear parentheses if the given variable is contained inside the parentheses.
  (c) Collect and combine any like terms.
  (d) Isolate the term containing the given variable in one member of the literal equation.
  (e) Isolate the given variable itself.

SOLVED PROBLEMS

PROBLEM 18-1   Which of the following are linear equations in one variable? **(a)** $2x = 3$

(b) $y - 8 = 4$ **(c)** $\dfrac{z}{5} = -8$ **(d)** $\dfrac{w}{2} - 3 = 7$ **(e)** $m + 3 = 3 + m$ **(f)** $2 + n = -5$

(g) $3x + 2y = 6$ **(h)** $0.3x + 0.2 = 0.6$ **(i)** $3 + 2 = 5$ **(j)** $1 - 2r = 2 + 2r$

(k) $\dfrac{a}{2} + 5 = 4$ **(l)** $\dfrac{2}{a} + 5 = 4$ **(m)** $b\sqrt{5} - 2 = 3$ **(n)** $5\sqrt{b} + 2 = 3$ **(o)** $x^3 = 1$

(p) $2m + 3n + 6 = 8 + 3n$

Solution: Recall that a linear equation in one variable is any equation in one variable that can be written in standard form as $Ax + B = C$, where A, B, and C are real numbers ($A \neq 0$) and x is any variable [see Example 18-1]:

$$\text{standard form}$$

(a) $2x = 3$ or $\overbrace{2x + 0 = 3}$ is a linear equation in one variable.
(b) $y - 8 = 4$ or $1y + (-8) = 4$ is a linear equation in one variable.

(c) $\dfrac{z}{5} = -8$ or $\dfrac{1}{5}z + 0 = -8$ is a linear equation in one variable.

(d) $\dfrac{w}{2} - 3 = 7$ or $\dfrac{1}{2}w + (-3) = 7$ is a linear equation in one variable.

(e) $m + 3 = 3 + m$ or $3 = 3$ is not a linear equation in one variable because $3 = 3$ cannot be written as $Ax + B = C$ where $A \neq 0$.
(f) $2 + n = -5$ or $1n + 2 = -5$ is a linear equation in one variable.
(g) $3x + 2y = 6$ is not a linear equation in one variable because there are two different variables.
(h) $0.3x + 0.2 = 0.6$ is a linear equation in one variable in standard form.
(i) $3 + 2 = 5$ is not a linear equation in one variable because there are no variables.
(j) $1 - 2r = 2 + 2r$ or $-4r + 1 = 2$ is a linear equation in one variable.

(k) $\dfrac{a}{2} + 5 = 4$ or $\dfrac{1}{2}a + 5 = 4$ is a linear equation in one variable.

(l) $\dfrac{2}{a} + 5 = 4$ is not a linear equation because there is a variable in the denominator.

(m) $b\sqrt{5} - 2 = 3$ or $(\sqrt{5})b + (-2) = 3$ is a linear equation in one variable.
(n) $5\sqrt{b} + 2 = 3$ is not a linear equation in one variable because there is a variable under the radical symbol.

(o) $x^3 = 1$ is not a linear equation in one variable because the variable has an exponent greater than one.

(p) $2m + 3n + 6 = 8 + 3n$ or $2m + 6 = 8$ is a linear equation in one variable.

PROBLEM 18-2   List and name the eight special types of linear equations in one variable.

Solution: Recall that if A, B, and C are real numbers ($A \neq 0$) and x is any variable, then

(a) $x + B \quad = C$ is called an addition equation [see Example 18-5].
(b) $x - B \quad = C$ is called a subtraction equation [see Example 18-6].
(c) $\qquad Ax = C$ is called a multiplication equation [see Example 18-7].

(d) $\qquad \dfrac{x}{A} = C$ is called a division equation [see Example 18-8].

(e) $Ax + B = C$ is called a multiplication-addition equation [see Example 18-9].
(f) $Ax - B = C$ is called a multiplication-subtraction equation [see Example 18-10].

(g) $\dfrac{x}{A} + B = C$ is called a division-addition equation [see Example 18-11].

(h) $\dfrac{x}{A} - B = C$ is called a division-subtraction equation [see Example 18-12].

PROBLEM 18-3   Identify the special type of linear equation in one variable given by each of the following:

(a) $2 - x = 0$ **(b)** $-2 - x = 0$ **(c)** $5y = 1$ **(d)** $m + 2 = -5$ **(e)** $4 = n - 2$

(f) $-6 = \dfrac{u}{4}$ **(g)** $-3 + \dfrac{v}{5} = 6$ **(h)** $-15 = 2 + \dfrac{w}{3}$

Solution: Recall that a given linear equation in one variable that does not contain like terms or parentheses may need to be written in standard form before it can be categorized as one of the eight special types of linear equations in one variable [see Example 18-3]:

$$\overbrace{\qquad\qquad}^{\text{standard form}}$$

(a) $2 - x = 0$ or $-1x + 2 = 0$ is a multiplication-addition equation.
(b) $-2 - x = 0$ or $-1x - 2 = 0$ is a multiplication-subtraction equation.
(c) $5y = 1$ is a multiplication equation.
(d) $m + 2 = -5$ is an addition equation.
(e) $4 = n - 2$ or $n - 2 = 4$ is a subtraction equation.

(f) $-6 = \dfrac{u}{4}$ or $\dfrac{u}{4} = -6$ is a division equation.

(g) $-3 + \dfrac{v}{5} = 6$ or $\dfrac{v}{5} - 3 = 6$ is a division-subtraction equation.

(h) $-15 = 2 + \dfrac{w}{3}$ or $\dfrac{w}{3} + 2 = -15$ is a division-addition equation.

PROBLEM 18-4   Determine if the given real number is a solution of the given linear equation in one variable in each of the following: **(a)** $\dfrac{3}{2}$; $2x = 3$ **(b)** -3; $y + 5 = 2$ **(c)** 6; $\dfrac{w}{-2} = 3$

(d) 0; $z - 4 = -5$ **(e)** $-\dfrac{9}{2}$; $2m + 3 = -6$ **(f)** 3; $3n - 5 = 4$ **(g)** -4; $\dfrac{u}{4} + 3 = 2$

(h) -10; $\dfrac{v}{-2} - 5 = -1$ **(i)** 9; $2a + 5 = 3a - 4$ **(j)** 0; $2(b - 3) = -5 - (2 - b)$

(k) -2; $\dfrac{1}{2}x + \dfrac{3}{4} = -\dfrac{1}{4}$ **(l)** -2.4; $-0.5y - 0.2 = 1$ **(m)** 6; $w - 50\%w = 2.5$

Solution: Recall that to check a proposed solution of a given linear equation in one variable, you first substitute the proposed solution for the variable and then compute using the Order of Operations to get a number sentence. If you get a true number sentence, then the proposed solution checks and it is the one and only solution of the given equation. If you get a false number sentence, then the proposed solution does not check and it is not the solution of the given equation [see Example 18-4]:

(a)

$$2x = 3$$

$$2\left(\frac{3}{2}\right) \quad \Big| \quad 3$$

$$\rightarrow 3 \quad \Big| \quad 3$$

true $(3 = 3)$

$\frac{3}{2}$ is the solution of $2x = 3$.

(b)

$$y + 5 = 2$$

$$-3 + 5 \quad \Big| \quad 2$$

$$\rightarrow 2 \quad \Big| \quad 2$$

true $(2 = 2)$

-3 is the solution of $y + 5 = 2$.

(c)

$$\frac{w}{-2} = 3$$

$$\frac{6}{-2} \quad \Big| \quad 3$$

$$\rightarrow -3 \quad \Big| \quad 3$$

false $(-3 \neq 3)$

6 is not the solution of $\frac{w}{-2} = 3$.

(d)

$$z - 4 = -5$$

$$0 - 4 \quad \Big| \quad -5$$

$$\rightarrow -4 \quad \Big| \quad -5$$

false $(-4 \neq -5)$

0 is not the solution of $z - 4 = -5$.

(e)

$$2m + 3 = -6$$

$$2\left(-\frac{9}{2}\right) + 3 \quad \Big| \quad -6$$

$$-9 + 3 \quad \Big| \quad -6$$

$$\rightarrow -6 \quad \Big| \quad -6$$

true $(-6 = -6)$

$-\frac{9}{2}$ is the solution of $2m + 3 = -6$.

(f)

$$3n - 5 = 4$$

$$3(3) - 5 \quad \Big| \quad 4$$

$$9 - 5 \quad \Big| \quad 4$$

$$\rightarrow 4 \quad \Big| \quad 4$$

true $(4 = 4)$

3 is the solution of $3n - 5 = 4$.

(g)

$$\frac{u}{4} + 3 = 2$$

$$\frac{-4}{4} + 3 \quad \Big| \quad 2$$

$$-1 + 3 \quad \Big| \quad 2$$

$$\rightarrow 2 \quad \Big| \quad 2$$

true $(2 = 2)$

-4 is the solution of $\frac{u}{4} + 3 = 2$.

(h)

$$\frac{v}{-2} - 5 = -1$$

$$\frac{-10}{-2} - 5 \quad \Big| \quad -1$$

$$5 - 5 \quad \Big| \quad -1$$

$$\rightarrow 0 \quad \Big| \quad -1$$

false $(0 \neq -1)$

-10 is not the solution of $\frac{v}{-2} - 5 = -1$.

(i)

$$2a + 5 = 3a - 4$$

$$2(9) + 5 \quad \Big| \quad 3(9) - 4$$

$$18 + 5 \quad \Big| \quad 27 - 4$$

$$\rightarrow 23 \quad \Big| \quad 23$$

true $(23 = 23)$

9 is the solution of $2a + 5 = 3a - 4$.

(j)

$$2(b - 3) = -5 - (2 - b)$$

$$2(0 - 3) \quad \Big| \quad -5 - (2 - 0)$$

$$2(-3) \quad \Big| \quad -5 - 2$$

$$-6 \quad \Big| \quad -7 \quad \longleftarrow \quad \text{false } (-6 \neq -7)$$

0 is not the solution of $2(b - 3) = -5 - (2 - b)$.

(k)

$$\frac{1}{2}x + \frac{3}{4} = -\frac{1}{4}$$

$$\frac{1}{2}(-2) + \frac{3}{4} \quad \Big| \quad -\frac{1}{4}$$

$$-1 + \frac{3}{4} \quad \Big| \quad -\frac{1}{4}$$

$$-\frac{1}{4} \quad \Big| \quad -\frac{1}{4} \quad \longleftarrow \quad \text{true}\left(-\frac{1}{4} = -\frac{1}{4}\right)$$

-2 is the solution of $\frac{1}{2}x + \frac{3}{4} = -\frac{1}{4}$.

(l)

$$\frac{-0.5y - 0.2 = 1}{-0.5(-2.4) - 0.2 \quad | \quad 1}$$

$$1.2 - 0.2 \quad | \quad 1$$

$$1 \quad | \quad 1 \longleftarrow \text{ true } (1 = 1)$$

(m)

$$\frac{w - 50\%w = 2.5}{6 - 50\%(6) \quad | \quad 2.5}$$

$$6 - \frac{1}{2}(6) \quad | \quad 2.5$$

$$6 - 3 \quad | \quad 2.5$$

$$3 \quad | \quad 2.5 \longleftarrow \text{ false } (3 \neq 2.5)$$

-2.4 is the solution of $-0.5y - 0.2 = 1$.   6 is not the solution of $w - 50\%w = 2.5$.

PROBLEM 18-5   Solve the following equations using the rules for linear equations: **(a)** $x + 2 = 5$

(b) $y - 2 = -7$ **(c)** $3c = 12$ **(d)** $\dfrac{d}{2} = -4$ **(e)** $4r + 2 = 14$ **(f)** $3s - 1 = -10$

(g) $\dfrac{t}{2} + 3 = 5$ **(h)** $\dfrac{u}{-5} - 3 = 1$ **(i)** $-5 = 5 - z$ **(j)** $3 = 2 - 4g$ **(k)** $1 = \dfrac{m}{-3} + 1$

(l) $-3 = -5 + \dfrac{y}{-2}$

Solution: Recall that to solve a linear equation in one variable that does not contain like terms or parentheses, you
1. Rename when necessary as one of the eight special types of linear equation in one variable.
2. Use the rules for linear equations to isolate the variable in one member of the linear equation.

(a) $x + 2 = 5$

$$x + 2 - 2 = 5 - 2$$

$$x = 3 \text{ [See Example 18-5.]}$$

(b) $y - 2 = -7$

$$y - 2 + 2 = -7 + 2$$

$$y = -5 \text{ [See Example 18-6.]}$$

(c) $3c = 12$

$$\frac{3c}{3} = \frac{12}{3}$$

$$c = 4 \text{ [See Example 18-7.]}$$

(d) $\dfrac{d}{2} = -4$

$$2 \cdot \frac{d}{2} = 2(-4)$$

$$d = -8 \text{ [See Example 18-8.]}$$

(e) $4r + 2 = 14$

$$4r + 2 - 2 = 14 - 2$$

$$4r = 12$$

$$\frac{4r}{4} = \frac{12}{4}$$

$$r = 3 \text{ [See Example 18-9.]}$$

(f) $3s - 1 = -10$

$$3s - 1 + 1 = -10 + 1$$

$$3s = -9$$

$$\frac{3s}{3} = \frac{-9}{3}$$

$$s = -3 \text{ [See Example 18-10.]}$$

(g) $\dfrac{t}{2} + 3 = 5$

$$\frac{t}{2} + 3 - 3 = 5 - 3$$

$$\frac{t}{2} = 2$$

$$2 \cdot \frac{t}{2} = 2 \cdot 2$$

$$t = 4 \text{ [See Example 18-11.]}$$

(h) $\dfrac{u}{-5} - 3 = 1$

$$\frac{u}{-5} - 3 + 3 = 1 + 3$$

$$\frac{u}{-5} = 4$$

$$-5 \cdot \frac{u}{-5} = -5 \cdot 4$$

$$u = -20 \text{ [See Example 18-12.]}$$

(i)
$$-5 = 5 - z$$
$$5 - z = -5$$
$$-z + 5 = -5$$
$$-1z + 5 = -5$$
$$-1z + 5 - 5 = -5 - 5$$
$$-1z = -10$$
$$\frac{-1z}{-1} = \frac{-10}{-1}$$
$$z = 10 \text{ [See Example 18-13,}$$
part (**a**).]

(j)
$$3 = 2 - 4g$$
$$2 - 4g = 3$$
$$-4g + 2 = 3$$
$$-4g + 2 - 2 = 3 - 2$$
$$-4g = 1$$
$$\frac{-4g}{-4} = \frac{1}{-4}$$
$$g = -\frac{1}{4} \text{ [See Example 18-13,}$$
part (**b**).]

(k)
$$1 = \frac{m}{-3} + 1$$
$$\frac{m}{-3} + 1 = 1$$
$$\frac{m}{-3} + 1 - 1 = 1 - 1$$
$$\frac{m}{-3} = 0$$
$$-3 \cdot \frac{m}{-3} = -3 \cdot 0$$
$$m = 0 \text{ [See Example 18-13,}$$
part (**c**).]

(l)
$$-3 = -5 + \frac{y}{-2}$$
$$-5 + \frac{y}{-2} = -3$$
$$\frac{y}{-2} - 5 = -3$$
$$\frac{y}{-2} - 5 + 5 = -3 + 5$$
$$\frac{y}{-2} = 2$$
$$-2 \cdot \frac{y}{-2} = -2(2)$$
$$y = -4 \text{ [See Example 18-13,}$$
part (**d**).]

PROBLEM 18-6   Solve by first combining like terms: **(a)** $x - 5 - 6x = -10$
(b) $4y + 8 = 2y - 6$ **(c)** $-w + 8 - 2w = 3w - 6 + w$ **(d)** $2 - 5m = 4m + 2 - 9m$
(e) $1 - n = 2 + 3n - 3 - 4n$ **(f)** $5(3 - 2u) = -5$ **(g)** $3(v + 2) = 2(3 + v - 2v)$

Solution: Recall that to solve a linear equation in one variable that contains like terms and/or parentheses, you:
1. Combine any like terms inside the parentheses.
2. Clear any parentheses using the distributive properties.
3. Combine any like terms in each member.
4. Collect and combine the like terms in one member.
5. Rename the equation when necessary to get one of the eight special types of linear equations in one variable.
6. Solve the equation in one variable using the rules for linear equations.

(a)
$$x - 5 - 6x = -10$$
$$1x - 6x - 5 = -10$$
$$-5x - 5 = -10$$
$$-5x = -5$$
$$x = 1$$
[See Example 18-15.]

(b)
$$4y + 8 = 2y - 6$$
$$4y - 2y + 8 = 2y - 2y - 6$$
$$2y + 8 = -6$$
$$2y = -14$$
$$y = -7$$
[See Example 18-15.]

(c)
$$-w + 8 - 2w = 3w - 6 + w$$
$$-1w - 2w + 8 = 3w + 1w - 6$$
$$-3w + 8 = 4w - 6$$
$$-3w - 4w + 8 = 4w - 4w - 6$$
$$-7w = -14$$
$$w = 2$$
[See Example 18-15.]

(d)
$$2 - 5m = 4m + 2 - 9m$$
$$2 - 5m = 4m - 9m + 2$$
$$2 - 5m = -5m + 2$$
$$2 - 5m + 5m = -5m + 5m + 2$$
$$2 = 2 \longleftarrow \text{true number sentence}$$

Every real number is a solution of
$$2 - 5m = 4m + 2 - 9m$$
[See Example 18-17.]

(e)
$$1 - n = 2 + 3n - 3 - 4n$$
$$1 - n = 2 - 3 + 3n - 4n$$
$$1 - n = -1 - n$$
$$1 - n + n = -1 - n + n$$
$$1 = -1 \longleftarrow \text{false number sentence}$$

There are no solutions of
$$1 - n = 2 + 3n - 3 - 4n$$
[See Example 18-18.]

(f)
$$5(3 - 2u) = -5$$
$$5(3) - 5(2u) = -5$$
$$15 - 10u = -5$$
$$-10u + 15 = -5$$
$$-10u = -20$$
$$u = 2$$

(g)
$$3(v + 2) = 2(3 + v - 2v)$$
$$3(v + 2) = 2(3 - v)$$
$$3 \cdot v + 3 \cdot 2 = 2 \cdot 3 - 2 \cdot v$$
$$3v + 6 = 6 - 2v$$
$$3v + 2v + 6 = 6 - 2v + 2v$$
$$5v + 6 = 6$$
$$5v = 0$$
$$v = 0$$

[See Example 18-16.]

PROBLEM 18-7 Solve by first clearing fractions, decimals, or percents:

(a) $\frac{1}{2}x - \frac{1}{2} = \frac{3}{4}x + \frac{1}{2}$ **(b)** $\dfrac{y - 2}{6} = \dfrac{y + 3}{4} - 1$ **(c)** $\frac{1}{2}(2 - w) + 1 = \frac{2}{3}(3w - 1) + \frac{1}{6}$

(d) $0.03m - 1.2 = 0.06$ **(e)** $0.5(2 - n) = 0.3$ **(f)** $u - 25\%u = 1.5$

Solution: Recall that to solve linear equations containing fractions, decimals, or percents, you
1. Clear all fractions, decimals, and percents.
2. Combine like terms and/or clear parentheses.
3. Rename as one of the eight special types of linear equations in one variable.
4. Use the rules for linear equations. [See Examples 18-19, 18-21, and 18-22.]

(a) $\frac{1}{2}x - \frac{1}{2} = \frac{3}{4}x + \frac{1}{2}$

The LCD for $\frac{1}{2}$ and $\frac{3}{4}$ is 4.
$$4\left(\tfrac{1}{2}x - \tfrac{1}{2}\right) = 4\left(\tfrac{3}{4}x + \tfrac{1}{2}\right)$$
$$4\left(\tfrac{1}{2}x\right) - 4\left(\tfrac{1}{2}\right) = 4\left(\tfrac{3}{4}x\right) + 4\left(\tfrac{1}{2}\right)$$
$$2x - 2 = 3x + 2$$
$$2x - 3x - 2 = 3x - 3x + 2$$
$$-1x - 2 = 2$$
$$-1x = 4$$
$$x = -4$$

(b) $\dfrac{y - 2}{6} = \dfrac{y + 3}{4} - 1$

The LCD for $\dfrac{y - 2}{6}$ and $\dfrac{y + 3}{4}$ is 12.
$$12 \cdot \frac{y - 2}{6} = 12\left(\frac{y + 3}{4} - 1\right)$$
$$12 \cdot \frac{y - 2}{6} = 12 \cdot \frac{y + 3}{4} - 12(1)$$
$$\tfrac{12}{6}(y - 2) = \tfrac{12}{4}(y + 3) - 12$$
$$2(y - 2) = 3(y + 3) - 12$$
$$2y - 4 = 3y + 9 - 12$$
$$2y - 4 = 3y - 3$$
$$2y - 3y - 4 = 3y - 3y - 3$$
$$-1y - 4 = -3$$
$$-1y = 1$$
$$y = -1$$

(c) $\frac{1}{2}(2 - w) + 1 = \frac{2}{3}(3w - 1) + \frac{1}{6}$

The LCD for $\frac{1}{2}$, $\frac{2}{3}$, and $\frac{1}{6}$ is 6.

$6[\frac{1}{2}(2 - w) + 1] = 6[\frac{2}{3}(3w - 1) + \frac{1}{6}]$

$6 \cdot \frac{1}{2}(2 - w) + 6(1) = 6 \cdot \frac{2}{3}(3w - 1) + 6 \cdot \frac{1}{6}$

$3(2 - w) + 6 = 4(3w - 1) + 1$

$6 - 3w + 6 = 12w - 4 + 1$

$-3w + 12 = 12w - 3$

$-3w - 12w + 12 = 12w - 12w - 3$

$-15w + 12 = -3$

$-15w = -15$

$w = 1$

(d) $0.03m - 1.2 = 0.06$

The LCD for $\frac{3}{100}(0.03)$, $\frac{12}{10}(1.2)$, and $\frac{6}{100}(0.06)$ is 100.

$100(0.03 - 1.2) = 100(0.06)$

$100(0.03m) - 100(1.2) = 100(0.06)$

$3m - 120 = 6$

$3m = 126$

$m = 42$

(e) $0.5(2 - n) = 0.3$

The LCD for $\frac{5}{10}(0.5)$ and $\frac{3}{10}(0.3)$ is 10.

$10[0.5(2 - n)] = 10(0.3)$

$10(0.5)(2 - n) = 10(0.3)$

$5(2 - n) = 3$

$10 - 5n = 3$

$-5n + 10 = 3$

$-5n = -7$

$n = \frac{7}{5}$ or 1.4

(f) $u - 25\%u = 1.5$

$1u - 0.25u = 1.5$

$0.75u = 1.5$

$u = 2$

PROBLEM 18-8   Solve the following literal equations and formulas:
(a) Solve for c: $s = c + m$ [retail sales formula]
(b) Solve for x: $x - 4y = -2x + 5$ [literal equation]
(c) Solve for r: $d = rt$ [distance formula]   **(d)** Solve for l: $n = 3l - 25$ [shoe-size formula]
(e) Solve for a: $P = 2(a + b)$ [perimeter formula]   **(f)** Solve for b: $A = \frac{1}{2}bh$ [area formula]

Solution: Recall that to solve a literal equation for a given variable, you
1. Clear fractions, decimals, and percents.
2. Combine like terms and/or clear parentheses.
3. Isolate the term containing the given variable.
4. Isolate the given variable itself.

(a) $s = c + m$

$c + m = s$

$c + m - m = s - m$

$c = s - m$

[See Example 18-25.]

(b) $x - 4y = -2x + 5$

$x + 2x - 4y = -2x + 2x + 5$

$3x - 4y = 5$

$3x - 4y + 4y = 4y + 5$

$3x = 4y + 5$

$x = \dfrac{4y + 5}{3}$

or $x = \dfrac{4y}{3} + \dfrac{5}{3}$

or $x = \dfrac{4}{3}y + \dfrac{5}{3}$

[See Example 18-26.]

(c) $d = rt$

$rt = d$

$\dfrac{rt}{t} = \dfrac{d}{t}$

$r = \dfrac{d}{t}$

[See Example 18-25.]

(d)

$$n = 3l - 25$$
$$3l - 25 = n$$
$$3l - 25 + 25 = n + 25$$
$$3l = n + 25$$
$$l = \frac{n + 25}{3}$$

or

$$l = \frac{n}{3} + \frac{25}{3}$$

or

$$l = \frac{1}{3}n + \frac{25}{3}$$

[See Example 18-25.]

(e)

$$P = 2(a + b)$$
$$2(a + b) = P$$
$$2a + 2b = P$$
$$2a + 2b - 2b = P - 2b$$
$$2a = P - 2b$$
$$a = \frac{P - 2b}{2}$$

or

$$a = \frac{P}{2} - b$$

or

$$a = \frac{1}{2}P - b$$

[See Example 18-27.]

(f)

$$A = \frac{1}{2}bh$$
$$\frac{1}{2}bh = A$$
$$2\left(\frac{1}{2}bh\right) = 2(A)$$
$$2 \cdot \frac{1}{2}bh = 2A$$
$$bh = 2A$$
$$\frac{b\cancel{h}}{\cancel{h}} = \frac{2A}{h}$$
$$b = \frac{2A}{h}$$

[See Example 18-25.]

Supplementary Exercises

PROBLEM 18-9   Determine if the given real number is a solution of the given linear equation in one variable: **(a)** -8; $x + 6 = -2$ **(b)** 9; $y - 5 = 4$ **(c)** $-\frac{4}{3}$; $-3w = 4$

(d) -12; $\frac{z}{2} = -5$ **(e)** -1; $-3u + 2 = 5$ **(f)** $\frac{3}{2}$; $2v - 4 = -1$ **(g)** -2; $\frac{a}{-2} + 1 = 3$

(h) 15; $\frac{b}{5} - 2 = 1$ **(i)** $\frac{9}{4}$; $2 - 3m = m - 2 + 5m$ **(j)** 0; $-2(5 - n) + 5 = 3(n - 2)$

(k) $\frac{1}{3}$; $\frac{3}{4}h - \frac{1}{2} = \frac{1}{4}$ **(l)** 1; $0.2k + 0.3 = 0.5$ **(m)** -8; $r - 75\%r = -2$

PROBLEM 18-10   Solve the following linear equations: **(a)** $x + 5 = -2$ **(b)** $y - 4 = 2$

(c) $8 = w + 5$ **(d)** $7 = z - 2$ **(e)** $3m = 24$ **(f)** $\frac{n}{4} = 5$ **(g)** $-12 = 6a$

(h) $-9 = \frac{b}{-2}$ **(i)** $3n + 7 = -5$ **(j)** $-6 = 3v + 3$ **(k)** $2h - 10 = 6$ **(l)** $5 = 2k - 5$

(m) $\frac{r}{5} + 6 = 2$ **(n)** $3 = \frac{s}{2} + 5$ **(o)** $\frac{c}{-5} - 3 = -5$ **(p)** $-5 = \frac{d}{2} - 3$

(q) $5p - 12 - 3p = -7$ **(r)** $-q + 4q + 6 = -3q - 4 + q$ **(s)** $-2(3w - 5) = -8$

(t) $3(1 + x) + 2 = 5(x + 1) - 2x$ **(u)** $5y - 2 = 8y + 3(1 - y)$ **(v)** $\frac{1}{2}z + \frac{5}{6} = -\frac{2}{3}$

(w) $\frac{2m + 3}{2} + 1 = \frac{5 - m}{5}$ **(x)** $\frac{3}{4}(n - 4) = 2 - \frac{1}{2}(2n + 3)$ **(y)** $2.5a + 6 = 0.25$

(z) $b + 40\%b = 0.7$

PROBLEM 18-11   Solve the following literal equations and formulas:

(a) Solve for r: $s = r - d$ [retail sales formula]   **(b)** Solve for d: $s = r - d$

(c) Solve for x: $x - y = 5$ [literal equation]   **(d)** Solve for y: $x - y = 5$

(e) Solve for m: $4m - 5n = 2$ [literal equation]   **(f)** Solve for n: $4m - 5n = 2$

(g) Solve for x: $Ax + By = C$ [literal equation]   **(h)** Solve for y: $Ax + By = C$

(i) Solve for b: $y = mx + b$ [slope-intercept formula]   **(j)** Solve for m: $y = mx + b$

(k) Solve for P: $PB = A$ [percent formula]   **(l)** Solve for t: $d = rt$ [distance formula]

(m) Solve for l: $V = lwh$ [volume formula]   **(n)** Solve for r: $I = Prt$ [interest formula]

(o) Solve for h: $A = \frac{1}{2}bh$ [area formula]   **(p)** Solve for B: $V = \frac{1}{3}Bh$ [volume formula]

(q) Solve for P: $A = P(l + rt)$ [amount formula]   **(r)** Solve for r: $A = P(l + rt)$

(s) Solve for l: $n = 3l - 22$ [shoe-size formula]   **(t)** Solve for h: $w = 5\frac{1}{2}(h - 40)$ [weight formula]

(u) Solve for h: $A = \frac{1}{2}h(b_1 + b_2)$ [area formula]   **(v)** Solve for b_1: $A = \frac{1}{2}h(b_1 + b_2)$

(w) Solve for F: $C = \frac{5}{9}(F - 32)$ [temperature formula]

(x) Solve for t: $a = 0.16(15 - t)$ [altitude formula]

(y) Solve for v_0: $v = v_0 - 32t$ [velocity formula]   **(z)** Solve for t: $v = v_0 - 32t$

Answers to Supplementary Exercises

(18-9)   **(a)** yes   **(b)** yes   **(c)** yes   **(d)** no   **(e)** yes   **(f)** yes   **(g)** no   **(h)** yes
(i) no   **(j)** no   **(k)** no   **(l)** yes   **(m)** yes

(18-10)   **(a)** -7 **(b)** 6   **(c)** 3   **(d)** 9   **(e)** 8   **(f)** 20   **(g)** -2 **(h)** 18   **(i)** -4
(j) -3 **(k)** 8   **(l)** 5   **(m)** -20 **(n)** -4 **(o)** 10   **(p)** -4 **(q)** $\frac{5}{2}$ or $2\frac{1}{2}$ or 2.5
(r) -2 **(s)** 3   **(t)** every real number   **(u)** no solutions   **(v)** -3 **(w)** $-\frac{5}{4}$ or $-1\frac{1}{4}$ or
-1.25 **(x)** 2   **(y)** $-\frac{23}{10}$ or $-2\frac{3}{10}$ or -2.3 **(z)** $\frac{1}{2}$ or 0.5

(18-11)   **(a)** $r = s + d$ **(b)** $d = r - s$ **(c)** $x = y + 5$ **(d)** $y = x - 5$

(e) $m = \dfrac{5n + 2}{4}$ or $\dfrac{5n}{4} + \dfrac{1}{2}$ or $\dfrac{5}{4}n + \dfrac{1}{2}$ **(f)** $n = \dfrac{2 - 4m}{-5}$ or $-\dfrac{2}{5} + \dfrac{4m}{5}$ or $\dfrac{4}{5}m - \dfrac{2}{5}$

(g) $x = \dfrac{C - By}{A}$ or $\dfrac{C}{A} - \dfrac{By}{A}$ or $\dfrac{C}{A} - \dfrac{B}{A}y$ or $-\dfrac{B}{A}y + \dfrac{C}{A}$

(h) $y = \dfrac{C - Ax}{B}$ or $\dfrac{C}{B} - \dfrac{Ax}{B}$ or $\dfrac{C}{B} - \dfrac{A}{B}x$ or $-\dfrac{A}{B}x + \dfrac{C}{B}$ **(i)** $b = y - mx$

(j) $m = \dfrac{y - b}{x}$ or $\dfrac{y}{x} - \dfrac{b}{x}$ **(k)** $P = \dfrac{A}{B}$ **(l)** $t = \dfrac{d}{r}$ **(m)** $l = \dfrac{V}{wh}$ **(n)** $r = \dfrac{I}{Pt}$ **(o)** $h = \dfrac{2A}{b}$

(p) $B = \dfrac{3V}{h}$ **(q)** $P = \dfrac{A}{l + rt}$ **(r)** $r = \dfrac{A - Pl}{Pt}$ or $\dfrac{A}{Pt} = \dfrac{l}{t}$

(s) $1 = \dfrac{n + 22}{3}$ or $\dfrac{n}{3} + \dfrac{22}{3}$ or $\dfrac{1}{3}n + \dfrac{22}{3}$ **(t)** $h = \dfrac{w + 220}{5\frac{1}{2}}$ or $\dfrac{w}{5\frac{1}{2}} + 40$ or $\dfrac{2}{11}w + 40$

(u) $h = \dfrac{2A}{b_1 + b_2}$ **(v)** $b_1 = \dfrac{2A - hb_2}{h}$ or $\dfrac{2A}{h} - b_2$ **(w)** $F = \dfrac{9C + 160}{5}$ or $\dfrac{9C}{5} + 32$ or $\dfrac{9}{5}C + 32$

(x) $t = \dfrac{2.4 - a}{0.16}$ or $15 - \dfrac{a}{0.16}$ or $15 - 6.25a$ **(y)** $v_0 = v + 32t$

(z) $t = \dfrac{v - v_0}{-32}$ or $\dfrac{v}{-32} + \dfrac{v_0}{32}$ or $-\dfrac{1}{32}v + \dfrac{1}{32}v_0$ or $\dfrac{1}{32}v_0 - \dfrac{1}{32}v$ or $\dfrac{v_0 - v}{32}$

FINAL EXAMINATION

Chapters 10-18

Part 1: Skills and Concepts (80 questions)

1. 12% renamed as an equal decimal is
    (a) 12.0   **(b)** 1.2   **(c)** 0.12   **(d)** 120   **(e)** none of these

2. $66\frac{2}{3}\%$ renamed as an equal fraction is
    (a) $\frac{2}{3}$ **(b)** $\frac{2}{300}$ **(c)** $\frac{1}{3}$ **(d)** $\frac{200}{3}$ **(e)** none of these

3. 0.08 renamed as an equal percent is
    (a) 80%   **(b)** 8%   **(c)** 800%   **(d)** 0.8%   **(e)** none of these

4. $\frac{1}{3}$ renamed as an equal percent is
    (a) $\frac{1}{3}\%$ **(b)** $\frac{100}{3}$ **(c)** $\frac{3}{100}\%$ **(d)** 33%   **(e)** none of these

5. 25% of 40 is
    (a) 0.625   **(b)** 100   **(c)** 1.6   **(d)** 10   **(e)** none of these

6. 12 is 75% of
    (a) 16   **(b)** 9   **(c)** 0.0625   **(d)** 0.16   **(e)** none of these

7. The percent that 1 is of 4 is
    (a) 400%   **(b)** 250%   **(c)** 4%   **(d)** 25%   **(e)** none of these

8. The percent increase from 3 to 4 is
    (a) $133\frac{1}{3}\%$ **(b)** $33\frac{1}{3}\%$ **(c)** 25%   **(d)** 75%   **(e)** none of these

9. The percent decrease from 4 to 3 is
    (a) $133\frac{1}{3}\%$ **(b)** $33\frac{1}{3}\%$ **(c)** 25%   **(d)** 75%   **(e)** none of these

10. The height of a tall milk glass might be
    (a) 6 in.   **(b)** 6 ft   **(c)** 6 yd   **(d)** 6 mi   **(e)** none of these

11. The amount of milk that a tall milk glass might contain is
    (a) 1 c   **(b)** 1 pt   **(c)** 1 qt   **(d)** 1 gal   **(e)** none of these

12. The weight of a tall milk glass full of milk might be
    (a) 1 oz   **(b)** 1 lb   **(c)** 1 T   **(d)** all of these   **(e)** none of these

13. 54 inches renamed as an equal amount of yards is
    (a) $4\frac{1}{2}$ yd   **(b)** 18 yd   **(c)** 6 yd   **(d)** $1\frac{1}{2}$ yd   **(e)** none of these

14. The height of a tall milk glass might be
    (a) 15 mm   **(b)** 15 cm   **(c)** 15 m   **(d)** 15 km   **(e)** none of these

15. The amount of milk that a tall milk glass might contain is
    (a) 470 mL   **(b)** 470 L   **(c)** 470 kL   **(d)** all of these   **(e)** none of these

16. The mass of a tall milk glass full of milk might be
    (a) 470 mg   **(b)** 470 g   **(c)** 470 kg   **(d)** 470 t   **(e)** none of these

17. 1.5 km renamed as an equal amount of centimeters is
  (a) 150 cm   **(b)** 1500 cm   **(c)** 15,000 cm   **(d)** 150,000 cm   **(e)** none of these

18. The difference of 2 yd 1 ft − 16 in. is
  (a) 20 in.   **(b)** 5 ft 6 in.   **(c)** 5 ft 8 in.   **(d)** 65 in.   **(e)** none of these

19. The quotient of 2 yd 1 ft ÷ 16 in. is
  (a) $5\frac{1}{4}$ **(b)** $2\frac{1}{4}$ **(c)** $\frac{3}{16}$ **(d)** 3   **(e)** none of these

20. The ratio in lowest terms of 30 seconds to 2 minutes is
  (a) $\frac{4}{1}$ **(b)** $\frac{30}{120}$ **(c)** $\frac{1}{4}$ **(d)** $\frac{120}{30}$ **(e)** none of these

21. The unit rate of 200 miles on 12 gallons is

  (a) $\frac{50}{3}$ **(b)** $\frac{50}{3}\dfrac{\text{miles}}{\text{gallon}}$ **(c)** $16\frac{2}{3}$ **(d)** $16\frac{2}{3}\dfrac{\text{miles}}{\text{gallon}}$ **(e)** none of these

22. Which of the following fraction pairs are directly proportional?
  (a) $\frac{1}{2}, \frac{50}{100}$ **(b)** $\frac{30}{3}, \frac{1}{90}$ **(c)** $\frac{12}{18}, \frac{20}{24}$ **(d)** all of these   **(e)** none of these

23. Which of the following fraction pairs are indirectly proportional?
  (a) $\frac{1}{2}, \frac{50}{100}$ **(b)** $\frac{30}{3}, \frac{1}{10}$ **(c)** $\frac{12}{18}, \frac{20}{24}$ **(d)** all of these   **(e)** none of these

24. The solution of $\dfrac{n}{12} = \dfrac{5}{8}$ is

  (a) $7\frac{1}{2}$ **(b)** $3\frac{1}{3}$ **(c)** 19.2   **(d)** all of these   **(e)** none of these

25. $\dfrac{n}{12} = \dfrac{5}{8}$ is equivalent to

  (a) $\dfrac{12}{n} = \dfrac{8}{5}$ **(b)** $\dfrac{5}{8} = \dfrac{n}{12}$ **(c)** $\dfrac{8}{5} = \dfrac{12}{n}$ **(d)** all of these   **(e)** none of these

26. $\frac{2}{3} \times \frac{2}{5} \times \frac{2}{5}$ in power notation is
  (a) $\frac{2}{5}^3$ **(b)** $\frac{8}{5}$ **(c)** $\frac{8}{125}$ **(d)** all of these   **(e)** none of these

27. $\frac{2}{5}^3$ evaluated is
  (a) $\frac{8}{125}$ **(b)** $\frac{8}{5}$ **(c)** $\frac{6}{5}$ **(d)** all of these   **(e)** none of these

28. 2^0 is just another way of writing
  (a) 0   **(b)** 2   **(c)** 1   **(d)** all of these   **(e)** none of these

29. The square root of $\frac{49}{100}$ is
  (a) $\frac{7}{10}$ **(b)** 0.7   **(c)** 70%   **(d)** all of these   **(e)** none of these

30. The perimeter of a rectangle with length 4 m and width 3 m is
  (a) 12 m   **(b)** 14 m   **(c)** 14 m^2 **(d)** all of these   **(e)** none of these

31. The approximate circumference of a circle with diameter 14 in. is
  (a) 44 in.   **(b)** 88 in.   **(c)** 154 in.   **(d)** all of these   **(e)** none of these

32. The area of a triangle with base 9 in. and height 6 in. is
  (a) 15 in.^{2} **(b)** 27 in.   **(c)** 27 in.^{2} **(d)** all of these   **(e)** none of these

33. The approximate area of a circle with radius 2 m is
  (a) 6.28 m^2 **(b)** 12.56 m   **(c)** 12.56 m^2 **(d)** all of these   **(e)** none of these

34. The volume of a rectangular prism with length 3 ft, width 4 ft, and height 6 in. is
    (a) 72 ft^3 (b) 13 ft^3 (c) 6 ft^3 (d) all of these   (e) none of these

35. The volume of a cylinder with radius 7 in. and height 5 in. is
    (a) 154 in.^{3} (b) 770 in.^{3} (c) 110 in.^{3} (d) all of these
    (e) none of these

36. The opposite of $\frac{3}{4}$ is
    (a) $\frac{3}{4}$ (b) $-\frac{3}{4}$ (c) $\frac{4}{3}$ (d) all of these   (e) none of these

37. The reciprocal of $\frac{3}{4}$ is
    (a) $\frac{3}{4}$ (b) $-\frac{3}{4}$ (c) $\frac{4}{3}$ (d) all of these   (e) none of these

38. The absolute value of $\frac{3}{4}$ is
    (a) $\frac{3}{4}$ (b) $-\frac{3}{4}$ (c) $\frac{4}{3}$ (d) all of these   (e) none of these

39. Which of the following are rational numbers?
    (a) integers   (b) whole numbers   (c) natural numbers   (d) all of these
    (e) none of these

40. Which of the following are not rational numbers?
    (a) $\frac{2}{3}$ (b) $\sqrt{4}$ (c) $-0.\overline{3}$ (d) all of these   (e) none of these

41. The correct symbol for ? in -5 ? -4 is
    (a) <   (b) >   (c) =   (d) all of these   (e) none of these

42. The sum of $-5 + 4$ is
    (a) 1   (b) -1 (c) 9   (d) -9 (e) none of these

43. The difference of $2 - (-7)$ is
    (a) 5   (b) -5 (c) 9   (d) -9 (e) none of these

44. The product of $-3(-2)$ is
    (a) -6 (b) -5 (c) 5   (d) 6   (e) none of these

45. The quotient of $\frac{-12}{3}$ is
    (a) -4 (b) 4   (c) $\frac{1}{4}$ (d) $-\frac{1}{4}$ (e) none of these

46. $\frac{3}{4}$ is equal to
    (a) $\frac{-3}{4}$ (b) $\frac{-3}{-4}$ (c) $-\frac{-3}{-4}$ (d) all of these   (e) none of these

47. $-\frac{3}{4}$ is equal to
    (a) $\frac{-3}{4}$ (b) $\frac{3}{-4}$ (c) $-\frac{-3}{-4}$ (d) all of these   (e) none of these

48. $-\frac{-12}{-18}$ simplified is
    (a) $-\frac{-2}{3}$ (b) $-\frac{2}{3}$ (c) $\frac{2}{3}$ (d) all of these   (e) none of these

49. The sum of $\frac{1}{2} + 0.75$ is
    (a) 1.25   (b) $1\frac{1}{4}$ (c) $\frac{5}{4}$ (d) all of these   (e) none of these

50. The difference of $2.5 - 3\frac{3}{4}$ is
    (a) 1.25   (b) 6.25   (c) -6.25 (d) all of these   (e) none of these

51. The product of $0.5(-\frac{1}{2})(-4)(0)(-\frac{3}{4})$ is
    (a) 0   (b) $-\frac{3}{4}$ (c) $\frac{3}{4}$ (d) all of these   (e) none of these

52. The quotient of $-\frac{1}{2} \div (-0.125)$ is
(a) -4 (b) 4 (c) 0.0625 (d) all of these   (e) none of these

53. $\dfrac{-1 - \sqrt{1^2 - 4(6)(-12)}}{2(6)}$ evaluated is

(a) $\frac{4}{3}$ (b) $-\frac{3}{2}$ (c) $-\frac{17}{12}$ (d) all of these   (e) none of these

54. "2 times x" can be written as
(a) $2 \cdot x$ (b) $2(x)$ (c) $2x$ (d) all of these   (e) none of these

55. "2 divide by x" can be written as

(a) $x \div 2$ (b) $x \cdot \frac{1}{2}$ (c) $\dfrac{x}{2}$ (d) all of these   (e) none of these

56. $2(3 \cdot 4) = (3 \cdot 4)2$ is an example of
(a) a commutative property   (b) an associative property
(c) a distributive property   (d) all of these   (e) none of these

57. $2(3 \cdot 4) = (2 \cdot 3)4$ is an example of
(a) a commutative property   (b) an associative property
(c) a distributive property   (d) all of these   (e) none of these

58. $2(3 - 4) = 2 \cdot 3 - 2 \cdot 4$ is an example of
(a) a commutative property   (b) an associative property
(c) a distributive property   (d) all of these   (e) none of these

59. $2(3 - 4) = 2(3 + (-4))$ is an example of
(a) a commutative property   (b) an associative property
(c) a distributive property   (d) all of these   (e) none of these

60. $-2(x - 3)$ equals
(a) $-2x + 6$ (b) $6 - 2x$ (c) $6 + (-2x)$ (d) all of these   (e) none of these

61. $2x + 3y + 5 - 3x - 2y - 4$ equals
(a) $-x + y + 1$ (b) $-x + y - 1$ (c) $x + y + 1$ (d) all of these
(e) none of these

62. $4m - (5m - 10n - m)$ equals
(a) $-2m - 10n$ (b) $-10n$ (c) $10n$ (d) all of these   (e) none of these

63. For $x = -2$; $3x^2 + 5x - 6$ equals
(a) 16 (b) -4 (c) -28 (d) -22 (e) none of these

64. For $a = 3$, $b = 5$, and $c = -2$; $\dfrac{-b + \sqrt{b^2 - 4ac}}{2a}$ equals

(a) $\frac{1}{3}$ (b) 2 (c) 1 (d) $-\frac{2}{3}$ (e) none of these

65. Which of the following are linear equations in one variable?
(a) $3x - 5 = 2x + 4$ (b) $2x + 3 = 2x - 1$ (c) $2x + 3y = 5$ (d) all of these
(e) none of these

66. -3 is a solution of

(a) $2x = 6$ (b) $\dfrac{x}{-3} = -1$ (c) $2x + 6 = 12$ (d) all of these

(e) none of these

67. The solution of $x + 5 = -2$ is
    (a) 3   (b) -7 (c) -3 (d) 7   (e) none of these

68. The solution of $y - 3 = 8$ is
    (a) 11   (b) 5   (c) -5 (d) -11 (e) none of these

69. The solution of $2w = -8$ is
    (a) 4   (b) -16 (c) -4 (d) -10 (e) none of these

70. The solution of $\dfrac{z}{-2} = -4$ is

    (a) 8   (b) 2   (c) -2 (d) 8   (e) none of these

71. The solution of $2m + 3 = 1$ is
    (a) -1 (b) 2   (c) -4 (d) 8   (e) none of these

72. The solution of $\dfrac{n}{5} - 3 = -8$ is

    (a) -1 (b) -25 (c) -55 (d) $-\frac{11}{5}$ (e) none of these

73. The solution of $3x - 5 = 8x + 5$ is
    (a) 0   (b) 2   (c) $\frac{11}{10}$ (d) -2 (e) none of these

74. The solution of $-2(x - 3) = -4$ is
    (a) 5   (b) -1 (c) 1   (d) -5 (e) none of these

75. The solution of $2(x - 3) = 5x - (3 + x)$ is
    (a) $-\frac{4}{3}$ (b) -3 (c) -1 (d) $-\frac{3}{4}$ (e) none of these

76. The solution of $\frac{1}{2}y + \frac{3}{4} = \frac{1}{8}$ is
    (a) $\frac{5}{4}$ (b) $-\frac{5}{4}$ (c) $\frac{5}{2}$ (d) $-\frac{5}{2}$ (e) none of these

77. The solution of $0.2w - 0.3 = 0.1$ is
    (a) 0.2   (b) -1 (c) 20   (d) 2   (e) none of these

78. The solution of $x + 33\frac{1}{3}\%x = 16$ is
    (a) 12   (b) 0   (c) 1   (d) 24   (e) none of these

79. The solution of $2x + 3y = 5$ for y is

    (a) $\dfrac{5 - 3y}{2}$ (b) $2x - 5$ (c) $\dfrac{5 - 2x}{3}$ (d) $\dfrac{2x - 5}{3}$ (e) none of these

80. The solution of $A = 2(l + w)$ for w is

    (a) $\dfrac{A}{2l}$ (b) $\dfrac{A}{2} + l$ (c) $l - \dfrac{A}{2}$ (d) $\frac{1}{2}A - l$ (e) none of these

Part 2: Problem Solving (20 questions)

81. The purchase price of a radio is $84. The sales tax is $4. What is the total price?

82. The purchase price of a radio is $84. The sales-tax rate is 4%. What is the sales tax?

83. The total price of a radio is $84. The sales tax is $4. What is the sales-tax rate?

84. The regular price of a suit is $120. After a discount of $40, what is the sale price?

85. The regular price of a suit is $120. If the suit is marked down 40%, how much is the cash discount?

86. The sale price of a suit is $120 after being discounted $40. What is the discount rate?

87. If a person earns $500 salary and $300 commission, how much is the total pay?

88. If a person has sales of $500 based on a commission rate of 30%, how much is the commission?

89. If a person gets $30 commission based on $500 in sales, what is the commission rate?

90. How much interest is paid on $100 at 5% per year simple interest for 120 days?

91. What is the total amount to be repaid for a loan of $1000 at 1% per month simple interest for 90 days?

92. What is the total amount in savings at the end of 1 year if you start with $100 and earn 6% per year compounded semi-annually?

93. A 120-mile trip took $7\frac{1}{2}$ gallons of gas. At that rate, how much gas will be needed for a 300-mile trip?

94. At the rate given in problem 93, how many miles can be traveled on a full 25-gallon gas tank?

95. At $4\frac{1}{2}$ mph, it takes 5 minutes to walk around a certain track. How long will it take to walk around the track at 6 mph?

96. At the rate given in problem 95, how fast will you need to run to get around the track in 2 minutes?

97. The length (l) of a certain man's foot is 12 inches when standing normally. Find the man's shoe-size number (n) by evaluating the formula: $n = 3l - 25$.

98. The height (h) of a certain woman is 64 inches (5 ft 4 in.). Find the woman's *ideal weight* (w) by evaluating the formula: $w = 5\frac{1}{4}h - 216$.

99. What distance (d) can a car travel in a time (t) of 15 minutes at a constant rate (r) of 50 mph using the distance formula: $d = rt$.

100. How much time (t) will it take a car to travel a distance (d) of 200 miles at a constant rate (r) of 60 mph using the distance formula: $d = rt$.

Final Examination Answers

Part 1

1. c	**2.** a	**3.** b	**4.** e	**5.** d
6. a	**7.** d	**8.** b	**9.** c	**10.** a
11. b	**12.** b	**13.** d	**14.** b	**15.** a
16. b	**17.** d	**18.** c	**19.** a	**20.** c
21. d	**22.** a	**23.** b	**24.** a	**25.** d
26. e	**27.** b	**28.** c	**29.** d	**30.** b
31. a	**32.** c	**33.** c	**34.** c	**35.** b
36. b	**37.** c	**38.** a	**39.** d	**40.** e
41. a	**42.** b	**43.** c	**44.** d	**45.** a
46. b	**47.** d	**48.** b	**49.** d	**50.** e
51. a	**52.** b	**53.** b	**54.** d	**55.** e
56. a	**57.** b	**58.** c	**59.** e	**60.** d
61. a	**62.** c	**63.** b	**64.** a	**65.** a
66. e	**67.** b	**68.** a	**69.** c	**70.** d
71. a	**72.** b	**73.** d	**74.** a	**75.** e
76. b	**77.** d	**78.** a	**79.** c	**80.** d

Part 2

81. $88	**82.** $3.36	**83.** 5%	**84.** $80
85. $48	**86.** 25%	**87.** $800	**88.** $150
89. 6%	**90.** $1.67	**91.** $30	**92.** $112.36
93. $18\frac{3}{4}$ gal	**94.** 400 mi	**95.** $3\frac{3}{4}$ min (or 3 min 45 s)	**96.** $11\frac{1}{4}$ mph
97. size 11	**98.** 120 lb	**99.** $12\frac{1}{2}$ mi	**100.** $3\frac{1}{3}$ hr (or 3 hr 20 min)

TABLE 1 Conversion Factors

U.S. Customary/Metric	From	To	Paper-and-Pencil Conversion Factors Multiply By	Calculator Conversion Factors Multiply By
Length	inches (in.)	millimeters (mm)	25	**25.4**
	inches	centimeters (cm)	2.5	**2.54**
	feet (ft)	meters (m)	0.3	**0.3048**
	yards (yd)	meters	0.9	**0.9144**
	miles (mi)	kilometers (km)	1.6	1.609
Capacity	drops (gtt)	milliliters (mL)	16	16.23
	teaspoons (tsp)	milliliters	5	4.929
	tablespoons (tbsp)	milliliters	15	14.79
	fluid ounces (fl oz)	milliliters	30	29.57
	cups (c)	liters (L)	0.24	0.2366
	pints (pt)	liters	0.47	0.4732
	quarts (qt)	liters	0.95	0.9464
	gallons (gal)	liters	3.8	3.785
Weight (Mass)	ounces (oz)	grams (g)	28	28.35
	pounds (lb)	kilograms (kg)	0.45	0.4536
	tons (T)	tonnes (t)	0.9	0.9072
Area	square inches (in.^{2})	square centimeters (cm^2)	6.5	6.452
	square feet (ft^2)	square meters (m^2)	0.09	0.09290
	square yards (yd^2)	square meters	0.8	0.8361
	square miles (mi^2)	square kilometers (km^2)	2.6	2.590
	acres (A)	hectares (ha)	0.4	0.4047
Volume	cubic inches (in.^{3})	cubic centimeters (cm^3 or cc)	16	16.39
	cubic feet (ft^3)	cubic meters (m^3)	0.03	0.02832
	cubic yards (yd^3)	cubic meters	0.8	0.7646
Temperature	degrees Fahrenheit (°F)	degrees Celsius (°C)	$\frac{5}{9}$ (after subtracting 32)	0.5556 (after subtracting 32)

Note: All conversion factors in bold type are exact: All others are rounded.

TABLE 1 (*Continued*)

Metric/U.S. Customary		Paper-and-Pencil Conversion Factors *Multiply By*	Calculator Conversion Factors *Multiply By*
From	*To*		
Length			
millimeters (mm)	inches (in.)	0.04	0.03937
centimeters (cm)	inches	0.4	0.3937
meters (m)	feet (ft)	3.3	3.280
meters	yards (yd)	1.1	1.094
kilometers (km)	miles (mi)	0.6	0.6214
Capacity			
milliliters (mL)	drops (gtt)	0.06	0.06161
milliliters	teaspoons (tsp)	0.2	0.2029
milliliters	tablespoons (tbsp)	0.07	0.06763
milliliters	fluid ounces (fl oz)	0.03	0.03381
liters (L)	cups (c)	4.2	4.227
liters	pints (pt)	2.1	2.113
liters	quarts (qt)	1.1	1.057
liters	gallons (gal)	0.26	0.2642
Mass (Weight)			
grams (g)	ounces (oz)	0.035	0.03527
kilograms (kg)	pounds (lb)	2.2	2.205
tonnes (t)	tons (T)	1.1	1.102
Area			
square centimeters (cm^2)	square inches (in.^{2})	0.16	0.1550
square meters (m^2)	square feet (ft^2)	11	10.76
square meters	square yards (yd^2)	1.2	1.196
square kilometers (km^2)	square miles (mi^2)	0.4	0.3861
hectares (ha)	acres (A)	2.5	2.471
Volume			
cubic centimeters (cm^3)	cubic inches (in.^{3})	0.06	0.06102
cubic meters (m^3)	cubic feet (ft^3)	35	35.31
cubic meters	cubic yards (yd^3)	1.3	1.308
Temperature			
degrees Celsius (°C)	degrees Fahrenheit (°F)	$\frac{9}{5}$ (then add 32)	**1.8** (then add 32)

Note: All conversion factors in bold type are exact. All others are rounded.

TABLE 2 Squares and Square Roots

Number N	Square N^2	Square Root $\sqrt{N}$	Number N	Square N^2	Square Root $\sqrt{N}$	Number N	Square N^2	Square Root $\sqrt{N}$
0	0	0	35	1225	5.916	70	4900	8.367
1	1	1	36	1296	6	71	5041	8.426
2	4	1.414	37	1369	6.083	72	5184	8.485
3	9	1.732	38	1444	6.164	73	5329	8.544
4	16	2	39	1521	6.245	74	5476	8.602
5	25	2.236	40	1600	6.325	75	5625	8.660
6	36	2.449	41	1681	6.403	76	5776	8.718
7	49	2.646	42	1764	6.481	77	5929	8.775
8	64	2.828	43	1849	6.557	78	6084	8.832
9	81	3	44	1936	6.633	79	6241	8.888
10	100	3.162	45	2025	6.708	80	6400	8.944
11	121	3.317	46	2116	6.782	81	6561	9
12	144	3.464	47	2209	6.856	82	6724	9.055
13	169	3.606	48	2304	6.928	83	6889	9.110
14	196	3.742	49	2401	7	84	7056	9.165
15	225	3.873	50	2500	7.071	85	7225	9.220
16	256	4	51	2601	7.141	86	7396	9.274
17	289	4.123	52	2704	7.211	87	7569	9.327
18	324	4.243	53	2809	7.280	88	7744	9.381
19	361	4.359	54	2916	7.348	89	7921	9.434
20	400	4.472	55	3025	7.416	90	8100	9.487
21	441	4.583	56	3136	7.483	91	8281	9.539
22	484	4.690	57	3249	7.550	92	8464	9.592
23	529	4.796	58	3364	7.616	93	8649	9.644
24	576	4.899	59	3481	7.681	94	8836	9.695
25	625	5	60	3600	7.746	95	9025	9.747
26	676	5.099	61	3721	7.810	96	9216	9.798
27	729	5.196	62	3844	7.874	97	9409	9.849
28	784	5.292	63	3969	7.937	98	9604	9.899
29	841	5.385	64	4096	8	99	9801	9.950
30	900	5.477	65	4225	8.062	100	10,000	10
31	961	5.568	66	4356	8.124			
32	1024	5.657	67	4489	8.185			
33	1089	5.745	68	4624	8.246			
34	1156	5.831	69	4761	8.307			

INDEX

Page numbers in bold type indicate solved problems.